Introduction
to
MATERIALS
SCIENCE

Bar code overlay

RUOFF, ARTHUR LOUIS
INTRODUCTION TO MATERIALS SCIE
000255729

620.19 R94

For

£6-75

# Introduction to MATERIALS SCIENCE

**Professor Arthur L. Ruoff**

Department of Materials Science
and Engineering
College of Engineering
Cornell University

Prentice-Hall, Inc., Englewood Cliffs, New Jersey

*Library of Congress Cataloging in Publication Data*

Ruoff, Arthur L.
  Introduction to materials science.

  Includes bibliographical references.
  1. Materials.    I. Title.
TA403.R86      620.1′12      72-2172
ISBN 0-13-487355-6

© 1972 by
Prentice-Hall, Inc.
Englewood Cliffs, N.J.

All rights reserved. No part
of this book may be reproduced
in any form or by any means
without permission in writing
from the publisher.

10 9 8 7 6 5 4 3 2 1

Printed in the United States of America

PRENTICE-HALL INTERNATIONAL, INC., *London*
PRENTICE-HALL OF AUSTRALIA, PTY., LTD., *Sydney*
PRENTICE-HALL OF CANADA, LTD., *Toronto*
PRENTICE-HALL OF INDIA PRIVATE LIMITED, *New Delhi*
PRENTICE-HALL OF JAPAN, INC., *Tokyo*

This book is dedicated to my parents.

# CONTENTS

**PREFACE**   xiii

**FOREWORD TO THE STUDENT**   xvii

**1 INTRODUCTION**   3

    1.1 What is Materials Science?   3
    1.2 What is Structure?   22
    1.3 What are Properties?   32
    1.4 How is Materials Science Relevant?   45
        REFERENCES   53
        PROBLEMS   54

**2 MECHANICAL PROPERTIES**   59

    2.1 Plastic Deformation of Solids   59
    2.2 Special Deformation Behavior of Solids   65
    2.3 The Flow of Fluids   74
    2.4 Internal Friction in Solids   77
    2.5 Ideal Plastic Behavior   85
        REFERENCES   86
        PROBLEMS   87

**3 ELECTRIC AND MAGNETIC PROPERTIES**   95

    3.1 Introduction   95
    3.2 Ohm's Law and the Hall Coefficient   99
    3.3 Metals and Semiconductors   104
    3.4 Insulators   109
    3.5 Special Dielectrics   117

3.6 Magnetic Properties  122
3.7 Soft Ferromagnetics  130
3.8 Hard Magnetic Materials  134
3.9 Special Magnetic Materials  136
3.10 Superconductors  137
REFERENCES  142
PROBLEMS  144

# 4 THERMAL AND CHEMICAL PROPERTIES   151

4.1 Space Rockets and Thermal Expansion Coefficient  151
4.2 Thermal Conductivity  153
4.3 Heat Capacity  159
4.4 Energy Changes During Phase Transformation  161
4.5 Temperature Effects on Properties  162
4.6 Electrical-Thermal Coupling  165
4.7 Heat Treatment  168
4.8 Chemical Properties  169
4.9 Summary of Properties  172
REFERENCES  174
PROBLEMS  175

# 5 BINDING IN ATOMS, MOLECULES AND CRYSTALS   179

5.1 Wave Mechanics  179
5.2 The Schrödinger Equation  187
5.3 The Hydrogen Atom  193
5.4 The Hydrogenlike Atom and the Periodic Table  196
5.5 Bonding of Atoms in Molecules and Condensed Phases  202
5.6 Covalent Bonding  208
5.7 Binding in Metals  213
5.8 Secondary Bonds  214
5.9 Polarization and Magnetization  220
REFERENCES  225
PROBLEMS  225

# 6 ATOMIC ARRANGEMENTS   233

6.1 Crystals and Lattices  233
6.2 Some Simple Crystals  238
6.3 Crystallographic Directions and Planes  242
6.4 Packing of Atoms in Crystals  247

- 6.5 Symmetry and its Relationship to Properties 261
- 6.6 Imperfections in Crystals 267
- 6.7 Glasses 283
- 6.8 Diffraction by Crystals 288
  - REFERENCES 293
  - PROBLEMS 295

# 7 POLYMERS 303

- 7.1 Introduction 303
- 7.2 An Idealized Random Chain 306
- 7.3 Degree of Polymerization 307
- 7.4 The Topology of Vinyl Polymers 310
- 7.5 Other Addition Polymers 314
- 7.6 Copolymers 318
- 7.7 Condensation Polymers 319
- 7.8 Network Polymers 324
- 7.9 Thermoplastics and Thermosetting Resins 326
- 7.10 Crystallinity in Polymers 326
- 7.11 Macromolecules in Living Matter 328
  - REFERENCES 332
  - PROBLEMS 333

# 8 MICRO- AND MACRO-STRUCTURE 337

- 8.1 Single Crystals 337
- 8.2 The Reflection Microscope 339
- 8.3 Polycrystalline Materials 347
- 8.4 Polyphase Materials 353
- 8.5 Composite Materials 355
- 8.6 Quantitative Microscopy 363
  - REFERENCES 365
  - PROBLEMS 366

# 9 EQUILIBRIUM AND KINETICS 369

- 9.1 Atom Motion and Temperature 369
- 9.2 Kinetics in an Ideal Gas 376
- 9.3 Internal Energy 380
- 9.4 Randomness and Entropy 382
- 9.5 Equilibrium in Chemical Systems 386
- 9.6 The Barometric Formula 389

9.7   Atom Vibrations   392
9.8   Kinetics of Reactions   396
9.9   Introduction to Diffusion   399
9.10  Special Cases of Diffusion   406
9.11  Applications of Diffusion Theory   415
9.12  Nucleation   419
      REFERENCES   424
      PROBLEMS   424

## 10  PHASE DIAGRAMS   431

10.1  Introduction   431
10.2  Binary Systems   435
10.3  Nonequilibrium Transformations   447
10.4  Age Precipitation Hardening   453
10.5  The Fe-C System   458
10.6  Segregation in Binary Alloys During Solidification   468
      REFERENCES   473
      PROBLEMS   474

## 11  ELECTROCHEMICAL PROPERTIES   479

11.1  Introduction   479
11.2  Half-Cell Potentials   485
11.3  Polarization and Overvoltage   491
11.4  Corrosion   496
11.5  Protecting Against Corrosion   501
      REFERENCES   505
      PROBLEMS   506

## 12  STRENGTHENING MECHANISMS   511

12.1  How Strong Can Materials Be?   511
12.2  Why are Bulk Materials so Weak?   518
12.3  General Strengthening Concepts   527
12.4  Solute Strengthening   529
12.5  Strain Hardening   530
12.6  Strengthening by Grain Boundaries   538
12.7  Second Phase Strengthening   540
12.8  Strengthening by Martensitic Transformation   547
12.9  Strengthening at High Temperatures   553
12.10 Strengthening Mechanisms in Polymers   559

12.11 Strengthening of Viscous Matrices  560
REFERENCES  561
PROBLEMS  562

## 13  ELECTRONS IN CONDENSED PHASES                                    567

13.1  The Electron Gas  567
13.2  The Quantized Electron Gas  571
13.3  Electrons in a Periodic Potential  579
13.4  Brillouin Zones  582
13.5  Conductivity  585
13.6  Intrinsic Semiconductors  593
13.7  Extrinsic Semiconductors  596
13.8  The p-n Junction  602
13.9  The Junction Transistor  608
13.10 Lasers  611
REFERENCES  615
PROBLEMS  615

## 14  MAGNETISM                                                         621

14.1  Diamagnetism  621
14.2  Paramagnetism  624
14.3  Ferromagnetism  632
14.4  Antiferromagnetism and Ferrimagnetism  636
14.5  Domains  639
14.6  Magnetization Processes According to Domain Theory  645
14.7  Magnetic Bubbles  650
REFERENCES  653
PROBLEMS  654

## 15  SUPERCONDUCTIVITY                                                 657

15.1  The Superconducting State  657
15.2  Fundamental Concepts  659
15.3  Collective de Broglie Wave  663
15.4  The Penetration Depth  665
15.5  Magnetic Flux Quantization  666
15.6  Type I vs Type II Superconductors  667
15.7  Fluxoid Pinning  672
REFERENCES  674
PROBLEMS  675

## EPILOGUE                                                              677

## INDEX                                                                 679

# PREFACE

### PURPOSE OF THE BOOK

The purpose of this text is to help the reader learn what properties materials have and why they have these properties. The student will quickly find that atomic, molecular, crystalline and other structures within solids determine properties and that properties can be altered, often drastically, by adjusting the internal structure of the material.

The prerequisites for this course are:
Chemistry, Differential Calculus with Integral Calculus being taken either previously or concurrently, and Physics.*

### CONTENT AND ARRANGEMENT

The book is divided into four parts. Chapter 1 is an introduction which illustrates the scope and purpose of Materials Science. The reader readily learns that metals, polymers, ceramics, and glasses can be treated quite generally.

Chapters 2–4 describe how materials behave and illustrate the magnitude of the properties exhibited by typical materials. The author feels that it is important that the student see the connection between properties of materials and the application of these materials in technology (prior to studying the details of structure), since this provides motivation to the engineering oriented student to learn why materials behave as they do.

"Let us first understand the facts and
then we may seek the causes."
—Aristotle

Chapters 5–10 are concerned with various aspects of structure including the kinetic and equilibrium phenomena which determine how

---

*At Cornell University the text is used in a second-semester, Freshman course. About one third of the students enter the course without any University Physics but with a background in physics obtained in a good College Physics course in high school (which covers the PSSC Physics book or its equivalent). Half of the students enter the course with both Differential and Integral Calculus and half enter the course with a background in Differential Calculus and are enrolled simultaneously in Integral Calculus.

that structure is obtained. The study of structure is built upward from electron and atoms to molecules, crystals, domains, polycrystalline structures, polyphase structures and composites.

Chapters 11–15 are studies of specific properties in more depth. This last part of the book builds on the earlier parts.

Historical notes, references in the text and at the end of the chapters and more than 350 Figures and 120 Examples with solutions are included. A unique feature of the book is the inclusion of 252 *Questions*.


## QUESTIONS

After a page or two of new material is presented, a question containing several parts is asked. These help the learner focus on important concepts and reinforce his learning. Answers to these are found at the end of the section.


## EXAMPLES

The text contains 121 examples with solutions. The student should always make a conscientious effort to solve these examples and should write down his answer before reading the given answer. These examples have several purposes:
- To illustrate theory with something more relevant.
- To clarify a concept.
- To reinforce learning.
- To help develop a feeling for the order of magnitude involved.
- To help develop the skill of making quick estimates.
- To relate to socioeconomic factors.
- To make the student use his thinking capabilities and his creative talents.


## PROBLEMS

There are 484 problems in this book. They are divided into three classes:
- Problems.
- More Involved Problems.
- Sophisticated Problems.

The Problems involve *informational knowledge* such as terminology, facts, classifications, sequences, methodology, etc. and *intellectual skills* such as comprehension, translation, interpretation and extrapolation.

The More Involved Problems involve *intellectual skills* at a deep level, *application abilities*, and at a modest level the abilities of *analysis*, *synthesis* and *evaluation*.

The Sophisticated Problems involve *analysis*, *synthesis* or *evaluation* at a deep level.

**REFERENCES**

Lists of books and articles for further study are given at the end of each chapter.

## SUGGESTIONS FOR COURSES

The author uses the text in an audiovisual-tutorial presentation with colored slides and sound tapes, and the course is available as a total package. This frees the author for individual work with the students in the learning center.

Different length courses with different objectives can be taught from the text (or the total package). Mechanical properties could be emphasized by covering Chapters 1, 2, 5–10 and 11–12. Electrical properties could be emphasized by covering Chapters 1, 3, 4, 5–10 and 13–15.

Using the packaged course in a learning center, it is possible to run three courses simultaneously.

## SELECTION OF TOPICS

The subject of materials science is so diverse that it is impossible to give due justice to all its many facets. Choices must be made. The topics are selected by the author with one major thought in mind: Eventually these materials are going to perform a function in a device. Therefore the student must develop a feeling for how materials behave. That is why the book starts with *how* materials behave. In the general area of phenomenology there are a number of laws of behavior (such as Hooke's law for elasticity or Ohm's law for conductivity) which should be part of an engineer's knowledge. The voltage-current characteristics of a p-n junction rectifier also falls into this class, but this is a relatively new (post World War II) behavior. It should be stated here that there are a very large number of effects (responses of materials to stimuli). Those which are introduced in this book are selected on the basis of the author's research experience, teaching experience, consulting experience, etc.

## FEEDBACK

A strong effort was made to produce a clear and well-illustrated text directed to the students' needs. Criticism by colleagues, research associates, graduate students and most importantly, feedback from the undergraduates who used the text, were solicited, evaluated and used, hopefully, to achieve these ends.

## ACKNOWLEDGEMENTS

The very able typing of Andrea Lucente and Nancy Brucker is much appreciated. I would like to express my gratitude to Ernest King and Benjamin Hawkins who spent many long hours after their freshman and senior years, respectively, reading and rereading the text and making detailed criticisms. I would also like to thank Jamie Chua for reading the final manuscript in detail and Dr. Prakash Rao (who was the principal tutor in our Materials Science learning center) for making many pertinent comments. Warren Yohe and Keh-Jim Dunn also deserve my acknowledgement for checking and criticizing the examples and the problems. Special thanks are due William Van Duzer for his help with the figures.

It is a pleasure to acknowledge the many individuals and publishers who made photographs and figures available. I also owe a debt to the many people who contributed to my more comprehensive book *Materials Science*.

Lastly, I am indebted to my colleagues and the many former students for stimulating ideas and discussion which provided the incentive for writing this book.

# FOREWORD TO THE STUDENT

Benjamin Bloom divides cognitive processes into six categories:

Knowledge (or information)
Skills
Application
Analysis
Evaluation
Synthesis (or creativity).

In simple language, it is necessary to learn some of the lore, some of the fundamental concepts, some of the skills of a field and to be able to apply these skills to new problems before it is possible to carry out analysis, evaluation or creative work in that field.

In studying a new field, it is always necessary to become familiar with the informational aspects of the field. One of the most important features here is to learn the vocabulary. In order to facilitate this, basic words in the present book are placed in bold face print. Some students may wish to make a table of vocabulary in a notebook with the word followed by either its definition or the page in the text where it is defined.

After a page or two of new material is presented in the text, a *Question* containing several parts is asked. Such questions can help the learner focus on the most important concepts covered in the material. Hence historical and other side issues can be discussed in the text without distracting from the main issues. The student should answer all these questions and understand and check the answers to each of these questions. Such study helps reinforce learning before a new body of material is approached.

Numerous *Examples* are given throughout the text. The examples are usually concerned with skills and applications of these skills although they are sometimes concerned with analysis, evaluation or synthesis. The student should always make a determined attempt to work these examples before looking at the answer. The answer should then be carefully studied because it often contains information not otherwise covered in the text.

The author recommends keeping a notebook with two types of questions:

(1) "Why" questions
(2) "I don't understand" questions.

The "I don't understand" question is required for clarification. For example, in Section 1.3, normal stress is defined. It is obvious to a specific student that he should understand this, but it is also clear to him that he does not (perhaps due to the choice of words used by the author and the manner in which the student happens to interpret these). So he puts the question in his notebook and after accumulating a reasonable number of these he sees his tutor, teaching assistant, professor or a fellow student to have these questions answered.

The "Why" question is a deeper question than the "I don't understand" question. Thus after studying stress and strain in Section 1.3, the student will learn that stress is often directly proportional to strain. Experimentally there is no doubt about it. But he may ask: Why is stress directly proportional to strain for small strains in the elastic region? Why isn't it proportional, say, to the square root of strain? Is the reason, perhaps, connected with the manner in which the atoms are bonded together? It should be emphasized that asking "Why" is not an idle exercise. All scientific and social progress owes its origin to someone who asks a probing question.

> Language was invented to ask questions. Answers may be grunts and gestures, but questions must be spoken. Humanness came of age when man asked the first question. Social stagnation results not from lack of answers but from absence of the impulse to ask questions.
>
> —Eric Hoffer

The student should discuss his "Why" questions with his peers, instructors, etc. Sometimes the instructor may say, as with the example of the "Why" questions used here, "that will be discussed in Chapter 12, and you may want to wait until then, or if you wish we can briefly discuss it now." This means that the student has anticipated the development of the subject; he is gaining perspective in the field.

For the convenience of the student and the instructor there are a number of problems at the end of the chapters, divided on the basis of difficulty into three groups. Every student should work a substantial majority of the category labelled *Problems*. Depending upon the purpose of the course and the student's aims he may wish to work some of the *More Involved Problems*. The *Sophisticated Problems* are for the highly motivated, gifted student.

The mks system is used throughout the discussions of the electrical and magnetic properties of materials. Moreover, there is a table in the frontispiece of the book which lists various fundamental constants in the mks system and the symbols for these constants (these constants and the symbols used there are used throughout the text). There is also a second table at the front of the book concerned with conversion to other units.

The author would prefer to have used the mks system of units throughout. However, inasmuch as this text is concerned with both science and technology, there are in some cases reasons for not doing this. For example, stress would be in units of Newtons per square meter, and there is probably not a single handbook in the United States concerned with yield stress, fatigue stress, etc., of engineering materials which uses such units. The author feels that in ten years the situation will be changed.

The author wishes the student success in his studies of materials, commensurate with the effort put forth. While all engineers should have an introduction to Materials Science, so that they can communicate in an intelligent fashion with the materials specialist, the author hopes that some of the readers of this book will devote themselves to solving materials problems of the future, such as inventing a substitute for the use of silver in photography.

Introduction
to
MATERIALS
SCIENCE

# Prologue

This chapter begins with a discussion of: What is materials science? To help answer this question and to help develop perspective, we shall take a quick view of the different areas of materials science which were involved in the development of large-scale integrated circuits. It should be emphasized that Section 1.1 is only a preview! (It is like a drive through a forest containing several different species of trees: We stop very briefly for a look at an oak, a maple, etc., to observe a few of their general features. Having done so, we can then proceed to make thorough studies of each later.) There will be many things in Section 1.1 which the reader will only vaguely understand; this is expected. If the reader can answer Questions 1.1.2 and 1.1.13 a week later, that would be adequate. The purpose of Section 1.2 is to help the reader develop a feeling for what the materials scientist means by the word *structure*. (This feeling could be properly demonstrated by being able to answer Questions 1.2.1–1.2.3 a week later.) Section 1.3 serves a similar purpose with respect to the word *property*. However, in Section 1.3 we are beginning to examine certain details. There are a number of specific definitions in boldface type. It is expected that these terms will become part of your knowledge, testable by Questions 1.3.1–1.3.8 or their equivalent. In particular, *Young's modulus* and *shear modulus* and the *tensile yield stress* are discussed. It is pointed out that the first two are *structure-insensitive* properties while the last is a *structure-sensitive* property. It is important that the student develop a feeling for the order of magnitude (factor of 10) of these properties. He should also ask *why* materials exhibit specific properties. Section 1.4 illustrates a few of the ways in which materials science is relevant to the needs of society. Students usually find this section quite interesting, and the ability to answer Question 1.4.1 is adequate.

# 1

# INTRODUCTION

## 1.1 WHAT IS MATERIALS SCIENCE?

Historically, separate introductory courses were taught in the areas of metals, ceramics, glasses, and polymers. There were two reasons for this: First, vastly different techniques were used in *producing* these materials. Second, their atomic structures were not understood so that the common origin of their behavior was unclear. In time different materials were *mechanically* combined permanently into integral structures (such as concrete or glass fiber reinforced plastics) and these were called composites.

### Q. 1.1.1

From a historical viewpoint, the five technological materials areas were (a)_____ (b)_____ (c)_____ (d)_____ (e)_____.

\* \* \*

*Note to the learner:* From a learning viewpoint it is best to write down your answer either here or in a notebook now. If it is incorrect, mark an × through it. The answers to questions are given at the end of each section.

In recent years other classes of materials such as semiconductors have become technologically important.

From a technological viewpoint there are millions of specific metals, ceramics, glasses, polymers, composites, semiconductors, etc. There has been a population explosion in the materials area. In fact the available number of commercial types of metals has increased by a factor of $10^3$ since 1900. The same can be said of ceramics and glasses. The growth of polymers and composites (and newer classifications, such as superconductors) has been even greater. Concurrent with these developments of new materials came advances in our knowledge of atomic structure and new developments in theory and experimental techniques which have made it possible to recognize similarities in seemingly diverse materials. The body of knowledge which makes it possible to study the behavior of all materials in a systematic manner is known as **materials science.**

In materials science we study the structure of matter, how a given structure is obtained and why that structure is obtained, and how materials behave and why they behave in this manner.

## Q. 1.1.2

What are the five areas of study of materials science?

_____
_____
_____
_____
_____

* * *

We shall further discuss what we mean by the word *structure* in the materials context in Section 1.2. An outline of topics covered in materials science is shown in Table 1.1.1. For the present purposes a property of a material is the way a material behaves under certain specified conditions. The idea of properties will be discussed further in Section 1.3.

**Table 1.1.1.** TOPICS STUDIED IN MATERIALS SCIENCE

| Fundamental Aspects | Types of Properties |
|---|---|
| Atomic binding (and electrons) | Thermal |
| Structure | Mechanical |
| Atomic vibrations | Chemical |
| Thermodynamics (and statistical mechanics) | Electrical |
| Phase equilibria | Optical |
| Kinetics | Magnetic |
| Molecular architecture | Radiation bombardment |

We shall now illustrate the interplay between these various topics by looking at a specific material, a silicon crystal with tiny amounts of boron or phosphorus added. (The reader should realize that this example is for the purpose of developing perspective; it is not expected that the reader should fully understand everything at this time.) We shall first consider pure silicon.

The silicon atom has 14 electrons surrounding a nucleus with a charge of 14$e$. As we shall see in the chapter on bonding, it is convenient to consider the silicon atom as an ion core (the nucleus plus a spherical charge distribution of ten electrons) and four valence electrons which take part in bonding to other atoms. These four electrons tend to be localized in $sp^3$ orbitals as shown in Figure 1.1.1. The contours in Figure 1.1.1 represent a

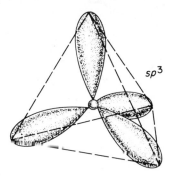

FIGURE 1.1.1. Equal electron density surfaces for $sp^3$ orbitals.

certain density of electrons (an electron in an atom behaves as a smeared-out charge distribution).

## Q. 1.1.3

The silicon atom can be considered to be (a)_____ ____ consisting of a nucleus and ten electrons having (b)_____ distribution and four (c)_____ electrons which take part in bonding.

\* \* \*

Solid silicon has the crystal structure shown in Figure 1.1.2. The length of the edge of the diamond cubic cell shown in Figure 1.1.2 is 5.43 Å. Recall that 1 Å = $10^{-8}$ cm = $10^{-10}$ m (the materials scientist uses X-ray diffraction to establish both the crystal structure and the length of the edge of the cubic cell). The bonds shown between a pair of silicons are overlapping $sp^3$ orbitals with the two electrons (one from each silicon) forming an electron pair bond or a covalent bond. In any case they tend to be localized in the region between the silicon atoms (the materials scientist uses X-ray diffraction to prove this). Each silicon is tetrahedrally bonded to four other silicons by such a bond (each silicon is at the centroid of a regular tetrahedron and there is a silicon at each vertex of the tetrahedron). Thus in the pure silicon crystal at low temperatures the valence electrons are localized in covalent bonds and are not free to move through the crystal. Since neither the valence electrons nor the ion cores are free

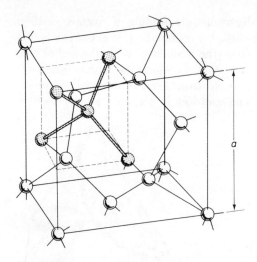

FIGURE 1.1.2. Silicon crystal. Note the tetrahedral bonding.

to move through the crystal, the material, at absolute zero temperature, is an electrical insulator (note how structure is related to a property, namely, electrical conductivity).

## Q. 1.1.4

In silicon each atom is (a)_____ bonded to (b)_____ other silicons by (c)_____ bond or (d)_____ bond. This results in (e)_____ crystal structure. Roughly the distance between atoms is (f)_____. The electrical conductivity at absolute zero temperature is related to the structural fact that both the (g)_____ and the (h)_____ are localized, i.e., are not free to move through the crystal.

\* \* \*

The picture of the static crystal just developed is not quite correct. Even at absolute zero temperature the atoms would be vibrating. There are $3N$ modes of vibration for $N$ atoms since the atoms vibrate in three directions. Associated with each of these modes is a characteristic frequency $\nu$. The highest frequency is about $10^{13}$ Hz ($10^{13}$ cycles/sec) and most of the vibrational frequencies are nearly this high. The energy associated with each mode (such as the $i$th mode) is

$$u_i = (n + \tfrac{1}{2})h\nu_i, \qquad n = 0, 1, 2, \ldots . \tag{1.1.1}$$

Here $h$ is a universal constant called **Planck's constant.** At absolute zero temperature, each mode has $n = 0$; i.e., it is in its lowest energy state (ground state). At higher temperatures, higher values of $n$ are involved, although some of the oscillators may still be in the ground state. When the temperature increases 1°, the average energy of the oscillators increases. The change in their average energy per degree Kelvin is the specific heat of the solid. This specific heat (a thermal property) can be calculated using the methods of statistical mechanics. The quantum of vibrational or elastic energy is called a phonon. Heat is transported in solids by phonons; hence the thermal conductivity (a thermal property) is determined by phonon behavior.

### Q. 1.1.5

Even at absolute zero temperature the atoms vibrate at a frequency of the order of (a)_____. The energy associated with each vibrational mode at absolute zero is (b)_____. At elevated temperature the energy of a given mode is (c)_____. The specific heat of a crystal is due to the increase in the (d)_____ of the oscillators as the temperature is raised 1°. The specific heat can be calculated using (e)_____. The tiny quanta of energy of the oscillators are (f)_____.

\* \* \*

Temperature also affects the valence electrons. It can be shown using concepts from quantum mechanics and statistical mechanics that at high temperatures a certain fraction of the valence electrons of silicon are freed from their localized positions and can move through the crystal. In fact the picture which emerges from quantum mechanics (which we shall study later and which you do not have to understand now) is that there are bands of energy in which there are very closely spaced allowable energy levels (these bands are called allowed energy bands) and bands of energy in which there are no allowable energy levels (forbidden bands). This is shown in Figure 1.1.3. The width of these energy bands is of the order of 1 eV (1 electron volt); however, the spacing between the energy levels within a band is a tiny fraction of this. One cubic centimeter of silicon contains about $10^{23}$ energy levels in each allowed band (a maximum of two electrons can occupy each level). At absolute zero temperature each energy level in the lowest allowable band (the valence band) of silicon is occupied by two electrons; the valence band of silicon is full; the conduction band is empty. This is illustrated in Figure 1.1.4(a). However, at room temperature (say), a certain small fraction of the electrons are thermally excited from near the top of the valence band to near the bottom of the conduction band. The empty state left in the valence band is called a hole. The number of thermally produced electrons in the conduction band equals the number

8  *Introduction*

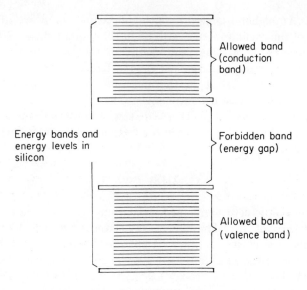

FIGURE 1.1.3. Allowed bands of energy levels and forbidden bands of energy levels in silicon.

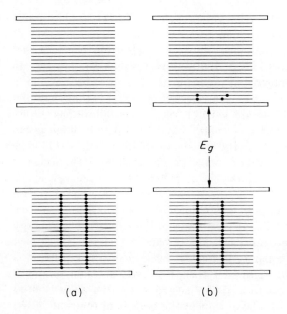

FIGURE 1.1.4. Electrons in silicon. (a) At absolute zero temperature. (b) At room temperature.

of thermally produced holes in the valence band. The number of conduction electrons per unit volume, $N_n$, is given by

$$N_n \propto e^{-E_g/2k_BT}. \qquad (1.1.2)$$

(The symbol $\propto$ means "is proportional to.") Here $E_g$, called the band gap energy, is the width of the forbidden energy band, $T$ is the absolute temperature, and $k_B$ is **Boltzmann's constant** (a universal constant); $k_B = 8.62 \times 10^{-5}$ eV/°K. The energy gap is a characteristic of the material; for silicon $E_g = 1.1$ eV. [It is not our purpose in this section to derive (1.1.2). Suffice it to say that (1.1.2) follows readily from statistical mechanics. Our purpose here is to show that the different topics in Table 1.1.1 are important in understanding the behavior of materials.]

The presence of these electrons and holes makes electrical conductivity possible (electrons in *completely* filled bands do not conduct electricity). It is pointed out here that holes behave as positive charge carriers (the materials scientist measures the Hall coefficient of the material to prove this). The conductivity is given by

$$\sigma = N_n e \mu_n + N_p e \mu_p. \qquad (1.1.3)$$

Here $N_n$ and $N_p$ are, respectively, the number per unit volume of negative and positive charge carriers, $e$ is the magnitude of the electronic charge, and $\mu_n$ and $\mu_p$ are the mobilities of the respective charge carriers. The mobility (or drift velocity per unit electrical field) is largely determined by interactions with phonons. These mobilities depend only slightly on temperature at room temperature.

Only a very tiny fraction of the electrons (one for each 10 billion silicons) is free, i.e., excited from the valence band to the conduction band. Because $N_n$ is still very small ($N_n \approx 10^{13}/\text{cm}^3$ while for a metal $N_n \approx 10^{23}/\text{cm}^3$) the silicon at room temperature is called a semiconductor. It is found experimentally that $\mu_n \approx 3\mu_p$, so the conductivity of pure silicon is dominated by the negative charge carriers (free electrons).

## Q. 1.1.6

At absolute zero temperature all the (many) energy levels of pure silicon in the (a)_____ band are filled while all of the energy levels in the (b)_____ band are empty. At absolute zero temperature pure silicon is an electrical (c)_____. At finite temperature (say room temperature), a small fraction of electrons are excited from the top of the (d)_____ band to the bottom of the (e)_____ band, leaving an equal number of (f)_____ in the former. The number of free electrons is related to the temperature according to (g)_____. The presence of charge carriers (charges which are free to move) leads to electrical (h)_____. The two factors which determine the electrical conductivity of a material are the (i)_____ of charge carriers and their (j)_____. Because the number of charge carriers per unit volume is small (at room temperature) compared to a metal, silicon is known as (k)_____.

* * *

Our discussion thus far has been concerned with pure silicon, but this has provided the background needed to discuss silicon with tiny quantities of phosphorus dissolved in it.

Phosphorus atoms are substituted for some silicon atoms (the crystal structure is still of the diamond cubic type). A phosphorus atom may be considered to be an ion core having ten electrons with a net spherical charge distribution plus five valence electrons. Four of these valence electrons are involved in making electron pair bonds with the four neighboring silicon atoms. The remaining electron is very easily thermally ionized (recall that ionization is the process of removing an electron from an atom) and is hence converted into a free electron (a charge carrier). The addition of an atom fraction of $10^{-5}$ of phosphorus (one phosphorus atom per $10^5$ silicon atoms) means that there will be a concentration of $10^{-5}$ free electrons per silicon atom (one free electron per $10^5$ silicons). Note how this completely overwhelms the thermal free electron concentration of $10^{-10}$ electrons per silicon atom. The addition of a small quantity of impurity (doping) of pentavalent materials produces an excess of electrons and an $n$-type material (conduction by negative charge carriers). If instead the silicon were doped with a trivalent element such as boron, there would be an excess of positive charge carriers (holes) and a $p$-type material would be produced.

Q. 1.1.7

The addition of a pentavalent element to silicon increases the concentration of (a)_____ while the addition of boron increases the concentration of (b)_____. The boron-doped silicon would be (c)_____ semiconductor. The conductivity at room temperature of silicon doped with $10^{-5}$ atom fraction of phosphorus would be about (d)_____ times as large as the conductivity of pure silicon.

* * *

Thus far we have referred to pure silicon as if it really existed. Other than in our minds there is probably no such thing; there are good reasons there never will be. These reasons involve thermodynamics and/or statistical mechanics. A very simple picture suffices for the present. Let us suppose (for pedagogical reasons) that pure silicon and pure boron are one-dimensional crystals as shown in Figure 1.1.5(a). Then when we substitute a boron for a silicon as in Figure 1.1.5(b) we break Si—Si bonds, we break B—B bonds, and we form Si—B bonds. If the latter are stronger and hence energetically stable, we would expect boron to dissolve. (If such stronger bonds are formed, the system would have a lower potential energy; the

(a)  —Si—Si—Si—Si—Si—Si—Si—Si—Si—Si—Si—

  —B—B—B—B—B—B—B—B—B—B—B—

(b)  —Si—Si—Si—Si—B—Si—Si—Si—B—Si—Si—

(c)  —Si—B—Si—Si—B—Si—Si—Si—Si—Si—Si—

FIGURE 1.1.5. One-dimensional crystals. (a) Pure crystals. (b) Impure one-dimensional crystal with 2B and 9Si. (c) Another arrangement of 2B and 9Si.

extra energy would be given off as heat.) Solubility would be favored by such a decrease in energy. But the opposite is actually true in the present case and so on the basis of energy alone we would not expect boron to dissolve (we find that we have to add an energy $E_B$ for each boron atom put in the silicon crystal). However, we know from experience that boron does dissolve and that the solubility increases with temperature. In fact we find that at low concentrations the atomic fraction of boron is

$$X_B \propto e^{-E_B/k_B T}. \tag{1.1.4}$$

The origin of this solubility behavior involves one of the most profound effects in all of science, namely, the concept of entropy. Entropy is simply a measure of the disorder of a system [the more configurations as in Figure 1.1.5(b) and (c), the more entropy]. Solubility is favored by an increase in entropy (as well as by a decrease in energy). If the latter increases (as in the present case), there still can be some solubility because of the increase in entropy. (The actual magnitude of the solubility is determined by an interplay between the energy and the product temperature times entropy.) Hence high temperature favors disorder and solubility. These considerations can be made quantitative and (1.1.4) can be derived theoretically.

### Q. 1.1.8

Solubility of atoms in silicon is favored both by (a)_____ in the energy of the system and (b)_____ of the product of temperature times entropy. Entropy is a measure of the (c)_____ of a system.

The more configurations which are possible, the larger is the (d)_____.
For low concentrations, the atom fraction of boron dissolved varies with
temperature according to (e)_____. The solubility (f)_____
rapidly as the temperature increases.

* * *

Let us now tackle the problem of putting the boron in the silicon.
Our first thought is to melt the silicon (in a container and surrounded by an
atmosphere which does not react chemically with the silicon), to add the

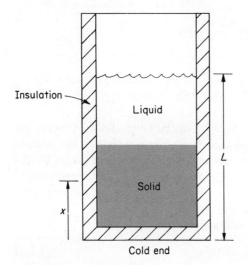

FIGURE 1.1.6. Unidirectional solidification.

required quantity of boron, and to cool. We would not get a uniform distribution of boron this way. In fact if the silicon were cooled entirely from one end as in Figure 1.1.6, we would find that the boron concentration would vary with distance according to

$$C(x) = K_S C_0 \left(1 - \frac{x}{L}\right)^{K_S - 1}, \tag{1.1.5}$$

where $K_S = 0.80$ in the present case and $C_0$ is the initial concentration of boron in the melt. We would find that the first solid to solidify ($x = 0$) has only 80% as much boron in it as the liquid from which it solidified.

The quantity $K_s$ is called the segregation coefficient. We could readily find the segregation coefficient from a phase diagram as shown in Figure 1.1.7; $K_s$ is the slope of the liquidus divided by the slope of the

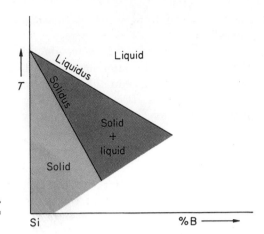

FIGURE 1.1.7. Sketch of a portion of a phase diagram of the silicon-boron system.

solidus. We shall discuss phase diagrams further in Chapter 10. The process of zone leveling which can be used to obtain a uniform distribution of the boron in silicon will be discussed then.

### Q. 1.1.9

Phosphorus has a segregation coefficient in silicon of $4 \times 10^{-1}$. When silicon containing only phosphorus of atom fraction $2.5 \times 10^{-4}$ begins to freeze the atom fraction of the first bit to solidify is (a)_____. This provides a technique for (b)_____ the silicon. If a phase diagram is available, it is easy to obtain the (c)_____ coefficient.

\* \* \*

The next fundamental aspect of materials science (see Table 1.1.1) is kinetics. A careful study of silicon reveals that sometimes a position in a crystal which should be occupied by a silicon is vacant (referred to as a vacancy). If the silicon is held at a high temperature for a sufficient time, the atom fraction of vacancies will be given by

$$X_v = A e^{-E_f/k_B T}, \qquad (1.1.6)$$

where $A \approx 1$ and $E_f \approx 2\text{--}3$ eV. $E_f$ is called the formation energy of a vacancy. Near the melting point of solids $X_v \approx 10^{-3}\text{--}10^{-4}$. Equation (1.1.6)

## 14 Introduction

can be derived from statistical mechanics; it is also found to be experimentally true.

These vacancies can move about in the crystal at high temperatures. A vacancy moves by interchanging with an atom. Thus an impurity atom (such as phosphorus) can diffuse into a pure silicon crystal by interchanging with the vacancies. It can be shown that the approximate distance a phosphorus atom moves in time $t$ is given by

$$x = 2\sqrt{Dt}, \qquad (1.1.7)$$

where $D$ is called the diffusion coefficient (for phosphorus in silicon). We shall show in Chapter 9 that the diffusion coefficient $D$ is proportional to the concentration of vacancies times the number of jumps per second (vacancy-atom interchanges). This will lead to the result that

$$D = D_0 e^{-Q_D/k_B T}. \qquad (1.1.8)$$

For phosphorus in silicon $D_0 = 10.5$ cm²/sec and $Q_D = 3.69$ eV. These values are obtained experimentally for each material as we shall see in Chapter 9.

### Q. 1.1.10

At high temperatures, crystals of pure silicon are not perfect but contain defects called (a)_____. The equilibrium vacancy concentration depends on temperature according to (b)_____. Impurity atoms can (c)_____ through a crystal by interchanging with vacancies. A good estimate of how far they move in the time $t$ is given by the expression (d)_____. The diffusion coefficient depends on temperature as follows: (e)_____.

\* \* \*

**EXAMPLE 1.1.1**

How far will phosphorus diffuse into silicon at 1200°C in 1 day? (The answers to examples are given immediately below the question but the student should make a reasonable effort to obtain the solution before looking at the answer.)

*Answer.* We first note that $T$ in Equation (1.1.8) is absolute temperature, e.g., °K. Hence in the present case $T = 273.16° + 1200° \doteq 1473°$K. Next we use Equation (1.1.8) and the sentence following it to find the diffusion coefficient.

Hence

$$\frac{Q_D}{k_B T} = \frac{3.69}{8.62 \times 10^{-5} \times 1473} = 29.0.$$

(The value of $k_B$ is found at the front of the book, as are other universal constants, conversion factors, etc.)

Since

$$e^{-x} = 10^{-x/2.3}$$

we have

$$D = 10.5 \times 10^{-29.0/2.3} \text{ cm}^2/\text{sec} = 10.5 \times 10^{-12.6}$$
$$= 10.5 \times 10^{0.4} \times 10^{-13} = 10.5 \times 2.5 \times 10^{-13}$$
$$= 2.6 \times 10^{-12} \text{ cm}^2/\text{sec}.$$

We can now use (1.1.7) to estimate the penetration depth. We note that 1 day = 86,400 sec. Hence $x = 2\sqrt{2.6 \times 10^{-12} \times 86,400} = 0.9 \times 10^{-3}$ cm.

Now let us suppose that we have a slice of a silicon crystal with a uniform distribution of phosphorus. This is an $n$-type semiconductor. We then pass $BCl_3$ gas over the hot surface of this semiconductor (actually a dilute concentration of $BCl_3$ in inert gas is used). The $BCl_3$ decomposes at the hot surface, with the $Cl_2$ being carried downstream by the inert gas, and the boron dissolving in the silicon. The concentration of boron in solution at the surface can be varied by changing the $BCl_3$ concentration. The boron diffuses inward from the surface. Let us say that the boron concentration (atom fraction) at the surface is $X_B = 10^{-4}$ and that the phosphorus throughout is $X_P = 10^{-6}$. Near the surface the semiconductor is now $p$-type, but farther from the surface it is $n$-type. We have therefore created a $p$-$n$ junction as shown in Figure 1.1.8. It passes current in one

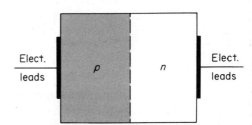

FIGURE 1.1.8. $p$-$n$ junction.

direction but not the other; this is an important electronic device called a diode. (Why it passes current in one direction but not the other can be readily explained and we shall do so in Chapter 13.)

## Q. 1.1.11

A *p-n* junction diode can be made by passing a gas of (a)_____ over phosphorus-doped silicon which is (b)_____ semiconductor. In this way we have a layer of *p*-type material and a layer of *n*-type material. The resultant device has what special characteristics? (c)_____.

* * *

If in the previous example, we had diffused boron in from both sides of the slice, we could have created a *p-n-p* junction transistor, a device capable of amplifying a signal. The *n-p-n* junction can do likewise.

Electronic circuits usually contain *p-n* junction diodes, *n-p-n* junction transistors, conductors, insulators, resistors, capacitors, etc. We have seen how it is possible to make one *p-n* diode from a slice of silicon or one *p-n-p* transistor. Might it be possible to carry the process of molecular architecture much further by building thousands of these various elements per square centimeter of a device? The answer is yes, and to see why a few more aspects of phase equilibria and kinetics need to be discussed. First, it is fairly easy to form an oxide layer of $SiO_2$ on silicon by passing water vapor over silicon at a high temperature. This oxide layer is an electrical insulator. Second, the rate at which boron or phosphorus (or other trivalent or pentavalent dopants such as aluminum or arsenic) diffuse in the oxide layer is negligible compared to the rate at which these elements diffuse into silicon. The oxide coating has masked the silicon. Third, this oxide layer can be dissolved by a hydrogen fluoride solution, a process called etching. Fourth, the oxide layer can be coated in some places with beeswax and these coated regions will not be dissolved by the hydrogen fluoride solution; this is also called masking. The beeswax can be readily removed when desired. The molecular architect now has all the requisite science to proceed. You should realize that his initial idea of placing thousands or tens of thousands of interconnected circuit elements (transistors, etc.) per square centimeter is itself very clever.

## Q. 1.1.12

Silicon is an ideal semiconductor for making integrated electronic circuits because in addition to making diodes and transistors in it, we can also treat it chemically to produce a layer which is an electrical (a)_____ _____. The diffusion of phosphorus or boron through this oxide layer is relatively (b)_____ compared to their diffusion rate in silicon. The oxide layer can be removed by a process called (c)_____. The chemical used to dissolve the silicon oxide is (d)_____. By (e)_____ the oxide layer (coating it with beeswax in certain areas) it is possible to remove the oxide layer in selected areas only.

\* \* \*

The molecular architect begins with a single crystal of silicon. He slices this into thin wafers and carefully polishes them. He then proceeds through the sequence of steps shown in Figure 1.1.9. This produces a

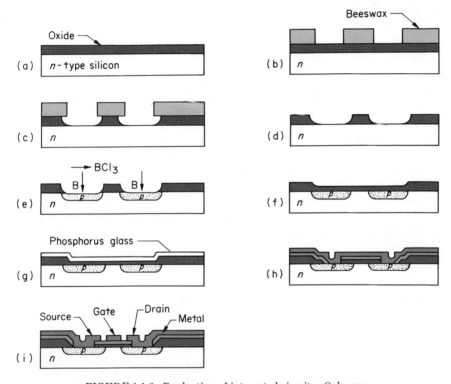

FIGURE 1.1.9. Production of integrated circuits. Only one tiny cross-sectional area of the wafer is shown. There may be $10^3$ or $10^4$ similar units per square centimeter. (a) Oxide layer formation. (b) Masking with beeswax. (c) Etching of oxide layer with HF solution. (d) Removal of beeswax mask. (e) Diffusion of boron into silicon at high temperature creating a $p$-type region. (f) Oxide layer formation. (g) Formation of a thin phosphorus glass (some phosphorus diffused into the $SiO_2$ layer). There is a reason for doing this which we shall not go into now. (h) Masking with beeswax, followed by etching, followed by removal of the beeswax mask, followed by metallization (a hot metal is evaporated in a vacuum and deposited on the cold surface). (i) Masking by beeswax, followed by selectively etching away certain areas of the metal layer. The result is a field effect transistor.

different type of device called a field effect transistor (which also can amplify a signal) rather than an $n$-$p$-$n$ transistor. Although we illustrated the molecular architecture of only one tiny area, similar processes can be carried on over the entire wafer simultaneously. The result is an integrated circuit as shown in Figure 1.1.10. Note that the little resultant device shown in Figure 1.1.9(h) on a wafer less than 1 mm thick and on an area of $10^{-4}$ cm$^2$ performs the same function as did the vacuum tube (the size of your thumb or bigger) in 1950. Solid state circuits can also be made incredibly reliable. They are responsible for the fantastic advances being made in computer technology.

The reader should reexamine Table 1.1.1. Note that we have covered all the fundamental aspects listed there and in fact it was necessary to do so to provide those fundamental concepts of science which have made large-scale integrated circuits possible.

> Those who rely on practice without science are like sailors without rudder or compasses.—Leonardo da Vinci

We have also shown how several of the types of properties listed in Table 1.1.1 are related to the fundamental aspects of materials science. In particular we have touched upon some of the thermal, chemical, and electrical properties. Some of the other types of properties will now be briefly discussed.

Because of the electron configuration of silicon we would not expect it to be strongly magnetic. With an elementary understanding of the energy band structure of silicon and of the nature of light we can say with some assurance that silicon will not be transparent at optical wavelengths but will be transparent at longer wavelengths (in the infrared). In fact the wavelength at which it becomes transparent provides one way of measuring the energy gap, $E_g$. There are other optical properties, whose basis we shall discuss later, such as the fact that a $p$-$n$ junction in silicon can operate as a solar battery. Since silicon is quite strong mechanically, its mechanical behavior is not of direct importance in most semiconductor applications; we should, however, mention that a quantitative description of the bonding in silicon would enable us to describe its elastic behavior. An understanding of the defects in the crystal would help us to understand why it fractures and why at higher temperatures it undergoes permanent deformation. Finally, this understanding helps us to see why high-energy radiation (neutrons, $\gamma$-rays, etc.) can generate various point defects (vacancies and interstitial atoms) in the atomic structure or can excite electrons from the valence band (leaving holes behind) to the conduction band. Large radiation doses can in fact destroy the crystal structure completely, cause void formation, cause diffusion of the dopant at room temperature, etc. One

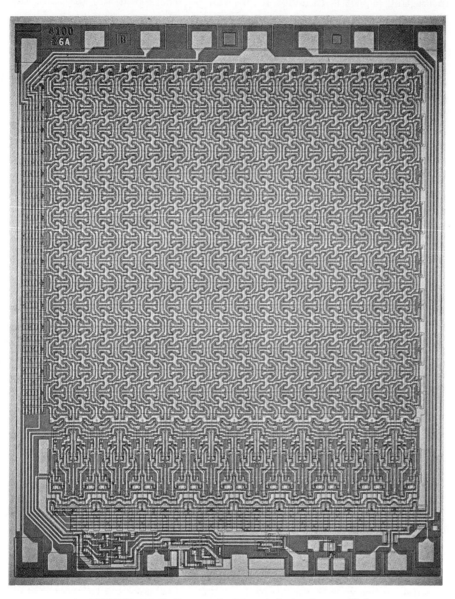

FIGURE 1.1.10. Large-scale integrated circuit magnified about 66 times. (Courtesy of Fairchild Semiconductor.)

interesting aspect is that boron ions can be accelerated and shot into the silicon; this is one way of doping. There is one more aspect of the microelectronic circuit or the *p-n* junction which deserves consideration. The boron and phosphorus atoms do not ordinarily diffuse much at room temperature. However, if the atoms could jump about, they would diffuse so as to uniformly distribute the dopants throughout the silicon, which would of course destroy the *p-n* junction. We say the *p-n* junction is in a metastable state (much like a book standing on its end with no side support) rather than in a stable state (like a book lying on its side) as in the sample with the uniformly distributed impurities. Most of the structures created by molecular architects are like that; i.e., they are metastable rather than stable. In the present example heating the *p-n* junction will allow the atoms to jump and the *p-n* junction will be destroyed. Metastability will be discussed further in Chapters 9 and 10.

## Q. 1.1.13

List the areas in which the materials scientist will be interested:

| Fundamental Aspects | Types of Properties |
|---|---|
| _____ | _____ |
| _____ | _____ |
| _____ | _____ |
| _____ | _____ |
| _____ | _____ |
| _____ | _____ |
| _____ | _____ |

\* \* \*

The example given here to illustrate the interplay between the fundamental aspects of materials science and the types of materials properties is but one of many which could have been used. Others will be discussed later.

To summarize, three special points deserve emphasis. First, that by slight modifications of structure it is (sometimes) possible to achieve enormous changes in behavior (1 part of phosphorus impurity per 100,000 silicons causes the room-temperature conductivity to increase by a factor of $10^5$). Second, many properties depend on temperature according to $\exp(-E/k_BT)$. This represents a *very* rapid variation with temperature. Third, the energies with which we deal are of the order of 1 eV.

## EXAMPLE 1.1.2

The diffusion coefficient of a metal just below the melting point $T_m$ is $10^{-8}$ cm²/sec. If $D_0 = 1$ cm²/sec, what is the value of the diffusion coefficient at $T_m/2$.

*Answer.* The diffusion coefficient varies with temperature according to (1.1.8). This could be written

$$D = 10^{-Q_D/2.3 k_B T} \text{ cm}^2/\text{sec},$$

with, in the present case,

$$\frac{Q_D}{2.3 k_B T_m} = 8.$$

Hence

$$\frac{Q_D}{2.3 k_B T_m/2} = 16$$

and

$$D\left(\frac{T_m}{2}\right) = 10^{-16} \text{ cm}^2/\text{sec}.$$

## ANSWERS TO QUESTIONS

**1.1.1** Metals, ceramics, polymers, glasses, composites.

**1.1.2** The structure of matter, how a structure is obtained, why that structure is obtained, how materials behave, why they behave in the manner they do.

**1.1.3** (a) An ion core, (b) a spherical charge, (c) valence.

**1.1.4** (a) Tetrahedrally, (b) four, (c) an electron pair, (d) a covalent, (e) a diamond cubic, (f) 1 Å = $10^{-10}$ m, (g) electrons, (h) ion cores.

**1.1.5** (a) $10^{13}$ Hz; (b) $\frac{1}{2}h\nu$; (c) $(n + \frac{1}{2})h\nu$, where $n$ is 0, 1, 2 ...; (d) average energy; (e) statistical mechanics; (f) phonons.

**1.1.6** (a) Valence, (b) conduction, (c) insulator, (d) valence, (e) conduction, (f) holes, (g) $\exp(-E_g/2k_B T)$, (h) conduction, (i) concentration, (j) mobility, (k) a semiconductor.

**1.1.7** (a) Electrons, (b) holes, (c) a $p$-type, (d) $10^5$.

**1.1.8** (a) A decrease, (b) an increase, (c) disorder (or the number of configurations), (d) entropy, (e) $\exp(-E_B/k_B T)$, (f) increases.

**1.1.9** (a) $10^{-4}$, (b) purifying, (c) segregation.

**1.1.10** (a) Vacancies, (b) $\exp(-E_f/k_B T)$, (c) diffuse (or move), (d) $2\sqrt{Dt}$, (e) $D = D_0 \exp(-Q_D/k_B T)$.

**1.1.11** (a) $BCl_3$; (b) an $n$-type; (c) it rectifies an ac current, i.e., passes an electrical current in one direction only.

**1.1.12** (a) Insulator, (b) slow, (c) etching, (d) hydrofluoric acid, (e) masking.

**1.1.13** See Table 1.1.1.

## 1.2 WHAT IS STRUCTURE?

> Now the smallest Particles of Matter may cohere by the strongest Attractions, and compose bigger Particles of weaker Virtue; and many of these may cohere and compose bigger Particles on which the Operations in Chymistry, and the Colours of natural Bodies depend, and which by cohering compose Bodies of a sensible Magnitude.—Sir Isaac Newton, *Opticks*

In the study of **structure** we are interested in

1. Specifying the positions of electrons and ions in atoms, molecules, and crystals. This is illustrated in Figures 1.2.1–1.2.5.

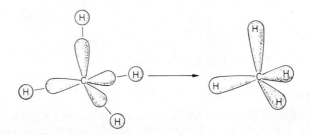

FIGURE 1.2.1. $CH_4$ molecules. The distance from the center of the carbon atom to the center of the hydrogen atom is 1.10 Å.

The C—H bond length (Figure 1.2.1) is 1.10 Å; this is the distance between the center of the carbon nucleus and the hydrogen nucleus (the proton). This is typical of atomic separations in molecules, liquids, and solids. The molecular configurations in Figure 1.2.2 illustrate different ways of attaching a $CH_3$ group to extremely long carbon backbones. These different arrangements *in* the molecule profoundly affect the efficiency with which these long chains can pack. The atactic form [see Figure 1.2.2(d)], which does not pack well, is a waxy material at room temperature; the isotactic form [see Figure 1.2.2(c)], which packs well, is a hard plastic at room temperature.

To specify atom positions, it is necessary to use the theory of binding along with certain experimental techniques such as field ion microscopy, X-ray (or neutron and electron) diffraction, etc. Thus the use of X-ray diffraction enables us to show that in NaCl, sodium is a positive ion and

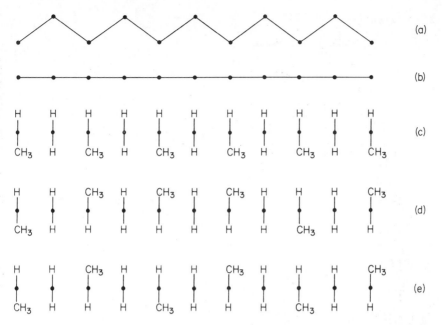

FIGURE 1.2.2. Polypropylene. (a) Planar view of the carbon backbone. (b) The same backbone viewed from above. (c) Isotactic arrangement of $CH_3$ groups (viewed from above). Note the regularity. (d) Atactic (random) arrangement of $CH_3$ groups. (e) Syndiotactic arrangement of $CH_3$ groups.

FIGURE 1.2.3. Prof. Müeller (inventor of field ion microscopy) with a field ion micrograph showing individual atoms. Magnification at the center is on the order of $10^9$ (magnification is not uniform). (Courtesy of Pennsylvania State University, College Station, Pa.)

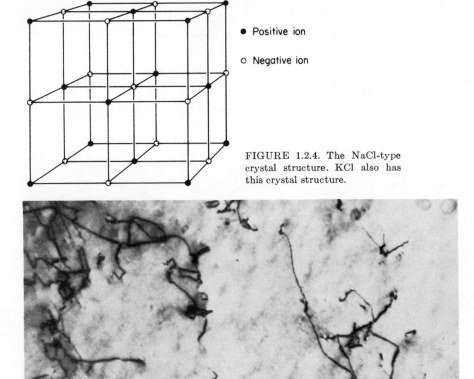

FIGURE 1.2.4. The NaCl-type crystal structure. KCl also has this crystal structure.

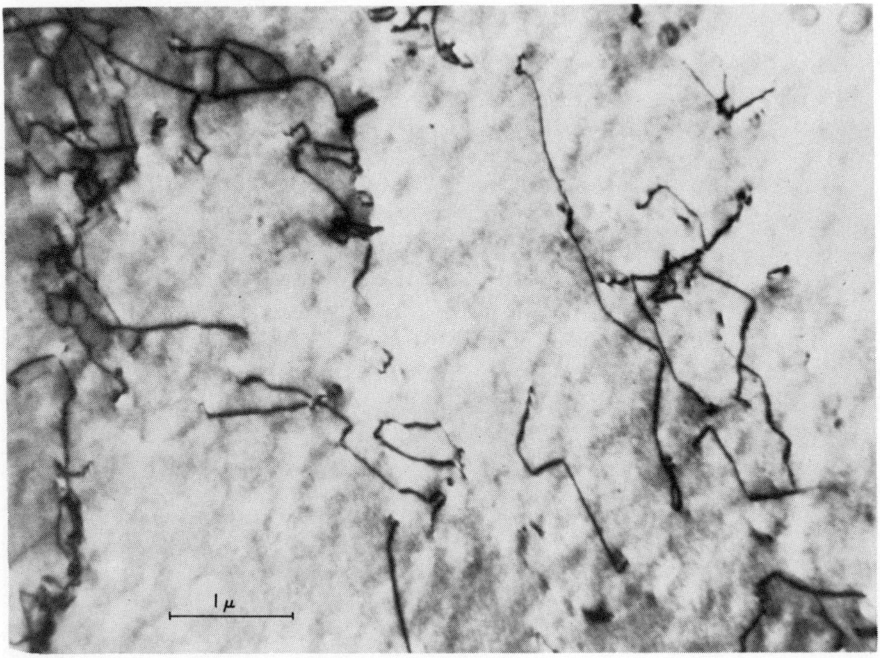

FIGURE 1.2.5. Electron micrograph of nickel showing defects within the crystal.

chlorine is a negative ion; i.e., NaCl is an alternating array of positive and negative ions.

2. Specifying the configurations of molecules. This is illustrated in Figures 1.2.6–1.2.8 for very large molecules (polymers).
3. Specifying the arrangement of domains in a crystal. This is illustrated in Figure 1.2.9, which shows magnetic domains (regions of a crystal magnetized in a certain

FIGURE 1.2.6. Aligned polymer chains.

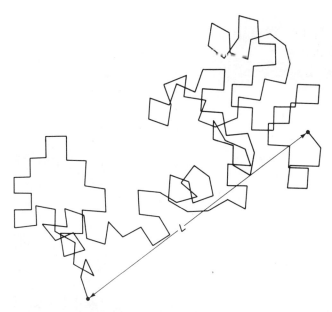

FIGURE 1.2.7. Random polymer chains.

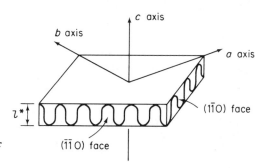

FIGURE 1.2.8. Folded polymer chains.

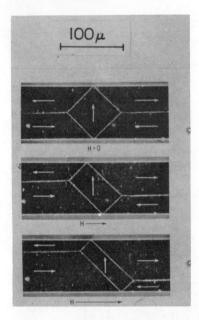

FIGURE 1.2.9. Domains in iron. Different regions are magnetized in different directions as shown. The crystal is 100 μ thick.

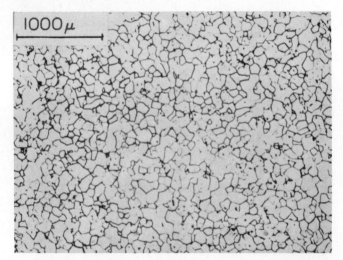

FIGURE 1.2.10. Polycrystalline iron. The specimen is carefully polished with a dispersion of fine alumina powder in water and is then etched.

direction). Configurations associated with cooperative phenomena involve ferromagnetic and ferroelectric domains, order domains in alloys, and fluxoids in superconductors.

4. Specifying an assemblage of crystals of the same materials. This is illustrated in Figures 1.2.10–1.2.12.

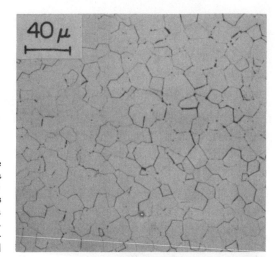

FIGURE 1.2.11. Polycrystalline magnesium oxide. This is a refractory (high-temperature) ceramic. The grain boundaries were revealed by etching with $H_3PO_4$. [R. E. Gardner and G. W. Robinson, *Journal of the American Ceramic Society* **45**, 46 (1962).]

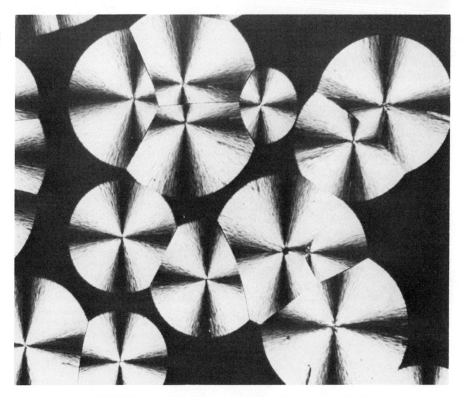

FIGURE 1.2.12. Spherulites (radially oriented crystals) in isotactic polystyrene (about halfway through the crystallization process) viewed under polarized light. (×900.) (Courtesy of H. D. Keith, Bell Telephone Laboratories, Murray Hill, N.J.)

If, for example, we allow a ladle of iron to freeze, we do not obtain a single crystal, but rather many tiny crystals which meet at grain boundaries. The size and shape of these grains can have a profound effect on material properties.

5. Specifying the arrangement of a mixture of different materials. Different examples of such mixtures are shown in Figures 1.2.13–1.2.17.

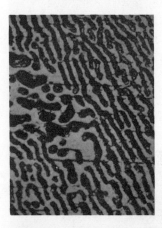

FIGURE 1.2.13. Pb–Sn lamellar structure, obtained by cooling an alloy of lead and tin from the melt. The dark regions are nearly pure Pb, the light regions nearly pure Sn. ($\times 200$.)

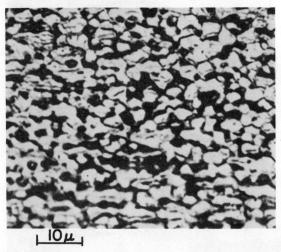

FIGURE 1.2.14. Pb–Sn microduplex structure, obtained by heavy plastic deformation (permanent deformation) of the Pb–Sn lamellar structure. (Courtesy of W. B. Morrison, U.S. Steel Corp., Monroeville, Pa.)

Some of these can readily be made by routinely cooling a liquid such as ordinary solder (Figure 1.2.13). The special microduplex structure of Figure 1.2.14 is necessary for the existence of superplasticity (ability to undergo very large deformation at low stresses), establishing the fact that structure is directly related to properties. The foam of Figure 1.2.15 can be made with 98% free volume. It has been used as a filler in automobile and airplane gasoline tanks to prevent splashing in crashes. Figure 1.2.16 shows a composite of oriented strong boron fibers (see Table 1.3.5) in an aluminum matrix. The resultant composite material has a strength and stiffness resembling that of the boron fibers (bulk boron is not nearly as strong).

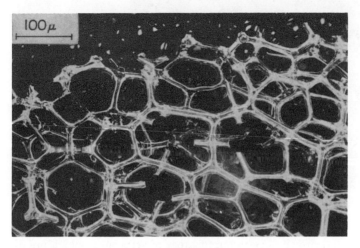

FIGURE 1.2.15. Polyurethane foam. (Courtesy of Scott Paper Co., Chester, Pa.)

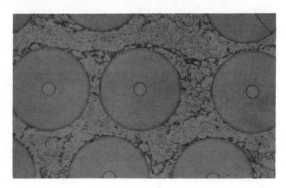

FIGURE 1.2.16. Boron fiber composite. Boron fibers (diameter, ~0.025 cm) in aluminum matrix. (Courtesy of Hamilton Standard Division of United Aircraft, Windsor Locks, Conn.)

Figure 1.2.17 shows a photograph of an ancient composite material, concrete. Here, a microscope is not needed to show the larger details of the structure.

The various mixtures may involve all sorts of configurations, e.g., spherical droplets (solid) in a solid matrix, rod-shaped precipitates in a solid matrix, or lamellar plates as in Figure 1.2.13. The shape, size, distribution, and distance between these particles can strongly affect properties.

30  *Introduction*

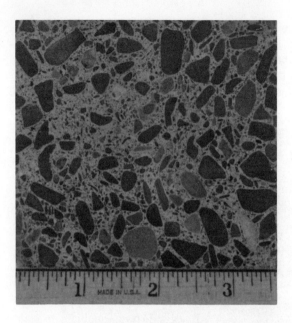

FIGURE 1.2.17. Concrete cross section. (Courtesy of F. Slate, Cornell University, Ithaca, N.Y.)

### Q. 1.2.1

What are the five different areas of structures which are discussed in materials science?

_____
_____
_____
_____
_____

\* \* \*

Structure is examined at various scales ranging from about 1 Å as in Figures 1.2.3 and 1.2.4 to 1 cm as in Figure 1.2.17 to 1 m as with reinforcement rods in concrete.

Table 1.2.1 shows the size resolution of different techniques.

From a scale viewpoint, structure can be considered on the atomic, microscopic, and macroscopic scale.

### Q. 1.2.2

Since optical light has a wavelength of the order of 5000 Å we might expect a good optical microscope to have a resolution of (a)_____

_____. To study crystal structure using X-ray diffraction, it is necessary to use X-rays having a wavelength of (b)_____. Many of the features of real crystals such as precipitates of 500-Å diameter or dislocations can be studied using (c)_____. The grain size of a crystal can usually be studied by polishing and (d)_____ the crystal and then examining it with the (e)_____. Structure can be discussed on the three size scales (f)_____.

\* \* \*

**Table 1.2.1.** SIZE RESOLUTION

| Structural Level | Size | Technique |
|---|---|---|
| Atomic | 1 Å | X-ray, neutron, or electron diffraction; field ion microscopy |
| Tiny precipitates | 10 Å | Small-angle X-ray scattering |
| Dislocations in crystals | 100–1000 Å | Transmission electron microscopy |
| Grains | $>1\ \mu$ | Optical microscopy |
| Large grains or particles | $>1000\ \mu$ | Direct visual observation |

STATES OF MATTER. Matter is often described in terms of the states of matter and in elementary discussions these are often the solid, liquid, and gaseous states. In a more general sense we might say that a transformation from one state of matter to another state of matter involves a finite change in volume or energy, or a radical change in the temperature derivatives of them (thermal expansion or specific heat). We might then also have the following states of matter: crystalline, glass, liquid crystal, monolayer films, rubber-like, ferromagnetic, ferroelectric, and superconductive. All these states of matter are associated with specific atomic or electronic arrangements, i.e., structure. They are likewise associated with specific behavior.

## Q. 1.2.3

In the simplest sense the three states of matter are determined by rigidity and the volume occupied by the matter. These states are the (a)_____, _____, and _____ states. In the (b)_____ state the atoms (or molecules) are close together and localized. In the (c)_____ state the atoms (or molecules) are close together and unlocalized. In the (d)_____ state the atoms (or molecules) are neither close together nor localized.

Thus these three states could be defined either on the basis of (e)_____ or (f)_____.

\* \* \*

## ANSWERS TO QUESTIONS

**1.2.1** Atomic arrangements, molecular configurations, domains (configurations associated with cooperative phenomena), polycrystalline assemblies, mixtures.

**1.2.2** (a) $\approx 1\ \mu$ parallel to the surface; (b) $\approx 1$ Å; (c) transmission electron microscopy; (d) etching; (e) optical microscope; (f) atomic, microscopic, and macroscopic.

**1.2.3** (a) Solid, liquid, gas; (b) solid; (c) liquid; (d) gas; (e) structure (as we just did); (f) properties.

## 1.3 WHAT ARE PROPERTIES?

A given crystal of silicon weighs 571 g (grams). Its mass $m$ is 571 g. This is the property of an object, not at all a unique feature of silicon. The same silicon crystal has a density, $\rho_m$, of 2.33 g/cm³. This is a property of the material. A copper wire has a resistance $R = 0.12\ \Omega$ (ohm). This is the property of an object. The same copper wire has a resistivity, $\rho_\xi = 1.67 \times 10^{-8}$ $\Omega$-m. This is a property of the material. Properties of materials and the corresponding properties of objects are related. Thus in the first case considered here,

$$m = \rho_m V, \qquad (1.3.1)$$

where $V$ is the volume of the specific silicon crystal. In the second case considered here,

$$R = \rho_\xi \frac{l}{A}, \qquad (1.3.2)$$

with $l$ and $A$ being the length and cross-sectional area, respectively, of the wire.

The free end of a diving board deflects 0.2 cm for each kilogram of mass it supports at its free end. We would say the compliance constant of the board is 0.2 cm/kg. The stiffness constant $k$ (which is the reciprocal of the compliance constant) is given by

$$k = \left(\frac{1}{4} \frac{BH^3}{L^3}\right) E \qquad (1.3.3)$$

Sec. 1.3    *What Are Properties?*    33

for the case where the base of the board is in a rigid foundation. Here $L$ is the length of the board, $H$ is the thickness of the board, $B$ is the width, and $E$ is the Young's modulus (a material property) of the glass-reinforced material of which the board is made. Equation (1.3.3) is derived in strength of materials books.

## Q. 1.3.1

The mass of a silicon crystal is a property of (a)_____ (an object or a material) while the density of a silicon crystal is a property of (b)_____. The length of a copper wire is the property of (c)_____. The resistance of a copper wire is the property of (d)_____. The resistivity of a copper wire is a property of (e)_____. The Young's modulus of a glass-reinforced plastic diving board is the property of (f)_____ while the stiffness constant is a property of (g)_____.

\* \* \*

MECHANICAL BEHAVIOR.    Consider an experiment in which the length of a piano wire is measured when a specific tensile force is applied. Table 1.3.1 shows the results for different initial lengths of the same wire.

**Table 1.3.1\***

| $l_0$ (initial length, cm) | $l$ (length, cm, with same force $F$) |
|---|---|
| 5.0000 | 5.0051 |
| 10.0000 | 10.0100 |
| 15.0000 | 15.0150 |
| 20.0000 | 20.0201 |
| 30.0000 | 30.0302 |
| 50.0000 | 50.0498 |
| 75.0000 | 75.0751 |
| 100.0000 | 100.0998 |

\* The values given may contain experimental errors. As a rule the lengths themselves would not be measured as accurately as shown.

### EXAMPLE 1.3.1

Is there a way of describing the state of deformation of these wires which does not depend on the length of the wire?

*Answer.* There are several ways. First, except for small experimental errors, $l/l_0$ (called the **extension ratio**) equals 1.001. Second, $(l - l_0)/l_0 =$

0.001. The change in length divided by the original length, $\Delta l/l_0$, is called the **nominal strain.** (Definitions which the student must learn are designated by boldface type.)

Inasmuch as the wire is subject to the same force in each case, we would expect the change in length per unit length of the wire to be the same in each case, as is found. It should be emphasized that the person who first had the idea illustrated by Example 1.3.1 (inventing strain to describe the state of deformation) was very creative. We usually designate the nominal tensile strain by the symbol $\epsilon$:

$$\epsilon = \Delta l/l_0 = (l - l_0)/l_0. \tag{1.3.4}$$

Let us now consider rods of steel (of exactly the same material) to which a tensile force $F$ is applied along the axis of a rod which causes a strain $\epsilon = 0.001$. If the rods have different cross-sectional areas, then different forces are needed to cause this strain as shown in Table 1.3.2.

Table 1.3.2

| $A_0$ (initial cross-sectional area of rod, cm²) | $F$ (force required to cause a strain $\epsilon = 0.001$, dynes) |
|---|---|
| 0.100 | $2.00 \times 10^8$ |
| 0.200 | $3.99 \times 10^8$ |
| 0.300 | $6.01 \times 10^8$ |
| 0.400 | $8.02 \times 10^8$ |
| 0.500 | $10.05 \times 10^8$ |
| 1.000 | $19.96 \times 10^8$ |
| 2.000 | $40.10 \times 10^8$ |

**EXAMPLE 1.3.2**

Find a quantity which causes a known tensile strain, but which, unlike force, is independent of the cross-sectional area.

*Answer.* As is clear from Table 1.3.2, the quantity $F/A_0$ has this property. The tensile force divided by the cross-sectional area (normal to the force) is called the **tensile stress.**

The **normal stress** is defined as

$$\sigma = \frac{F}{A_0}, \tag{1.3.5}$$

where $F$ is perpendicular to $A_0$ being either tensile or compressive. We shall consider $\sigma$ as positive if the normal stress is tensile and negative if compressive.

### Q. 1.3.2

Suppose that a cube of *unit* edge is deformed into a rectangular parallelepiped with edges (parallel to the original cube edges) of $\lambda_1$, $\lambda_2$, and $\lambda_3$. The quantities $\lambda_1$, $\lambda_2$, and $\lambda_3$ are called (a)_____. A wire is stretched in a tension machine from length $l_0$ to length $l$. The tensile strain is (b)_____. A rod of steel of 0.2-cm² cross-sectional area is pulled in tension by a force of $4 \times 10^8$ dynes. The normal stress is (c)_____. (Do not forget the units.) A cube of 1-cm edge is placed under a compressive load (normal to one pair of faces) of $8 \times 10^8$ dynes. The normal stress is (d)_____.

\* \* \*

There are many units of stress. The dyne per square centimeter is the cgs unit (the **bar** which equals $10^6$ dynes/cm² is often used). The newton per square meter (nt/m²) is the mks unit. The British unit, which is still used by engineers in the United States, is pounds per square inch or psi. The atmosphere is also a unit of stress. City water pressure is about 5 atm.

The presence of a tensile stress causes a tensile strain. If the stress is not too large, the deformation will decrease to zero when the stress is removed; the deformation under such conditions is called **elastic deformation**. If the stress is too large, some permanent deformation will remain when the stress is removed; this permanent deformation is called **plastic deformation**. The stress which causes a finite permanent strain, say $\epsilon = 0.001$ or 0.1%, is called the **tensile yield stress for 0.1% offset, $\sigma_0$**. Sometimes a different offset is used, such as 1% or 0.2%. Sometimes people get lazy and just refer to yield stress or yield strength. Piano wire (a steel wire which has undergone special processing) has a yield stress (0.1% offset) of 450,000 psi or $3 \times 10^{10}$ dynes/cm². A so-called mild steel may have a yield stress (0.1% offset) of $3 \times 10^9$ dynes/cm².

### Q. 1.3.3

If a wire is deformed reversibly, the deformation is (a)_____. Permanent deformation is called (b)_____. The tensile yield stress for 0.1% offset is defined as (c)_____. A mild steel has a tensile stress for 0.1% offset of (d)_____ psi or (e)_____ kbars.

\* \* \*

When a tensile stress is applied to a rod and this rod is elastically strained the stress (cause) leads to a strain (effect). We would find for most materials that the strain is directly proportional to the stress,

$$\epsilon = S\sigma. \qquad (1.3.6)$$

Here $S$ is called the tensile elastic compliance. It is more common in elementary courses to write

$$\sigma = E\epsilon, \qquad (1.3.7)$$

where $E = S^{-1}$. The proportionality constant $E$ is called the **Young's modulus,** after the English scientist Thomas Young (1773–1829). It is a material property.

The statement that stress is directly proportional to strain is **Hooke's law,** after the English scientist Robert Hooke (1660). Hooke's original statement, in fact, stated that the total deflection was proportional to the force. Elastic strains in metals, ceramics, and glasses are usually less than 0.01, i.e., 1%, because yielding or fracture occurs. Plastics can often be strained elastically more than this and many rubber-like materials can be strained elastically by 100–1000%. Hooke's law usually holds fairly well only for small strains, i.e., strains less than 1%. Like *all* the laws of science Hooke's law is only an approximation.

Young's modulus for steel is $2 \times 10^{12}$ dynes/cm$^2$; the value for gum rubber is only $2 \times 10^7$ dynes/cm$^2$.

### Q. 1.3.4

In a tensile test, in the regions of (a)_____ behavior the (b)_____ is usually found to be proportional to the (c)_____ with a constant of proportionality having units of stress called (d)_____. This behavior is an example of (e)_____ law. While most metals yield or fracture at strains of (f)____% or less, rubber-like materials often behave elastically for strains as large as (g)____%.

\* \* \*

The direct proportionality between stress and strain is known as a **linear relationship** and is an example of a **linear behavior.** This is illustrated in Figure 1.3.1(a). Let us suppose that the rod being stretched has an initial cross-sectional area $A_0$ and length $l_0$. Then $F = \sigma A_0$ and $\Delta l = \epsilon l_0$; let $\Delta l = x$. Then the $F$ vs. $x$ curve is as shown in Figure 1.3.1(b).

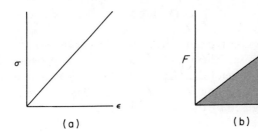

FIGURE 1.3.1. (a) Elastic stress-strain curve. (b) Force-extension curve for an elastic rod in tension. The shaded area represents the work done in stretching the rod.

**EXAMPLE 1.3.3**

If $F = kx$ for the elastic stretching of a rod, what is $k$?

*Answer.* Since $\epsilon = x/l_0$ and $\sigma = F/A_0$ and $\sigma = E\epsilon$ we have

$$k = \frac{EA_0}{l_0}. \tag{1.3.8}$$

The quantity $k$ is the **tensile stiffness constant** of the rod. It is not a material property as is $E$.

The work done when a force $F$ causes a parallel displacement $dx$ is

$$dW = F\,dx. \tag{1.3.9}$$

In the present case $F = kx$, so that

$$W = \int_0^x kx\,dx = \frac{kx^2}{2} = \frac{Fx}{2}. \tag{1.3.10}$$

It is the shaded area of Figure 1.3.1(b). This is potential energy which is stored in the stretched rod (as in a stretched rubber band). The student should know that the stored energy per unit volume, where the volume $V_0 = A_0 l_0$, is

$$u = \frac{1}{V_0}\int_0^x F\,dx = \int_0^x \frac{F}{A_0}\frac{dx}{l_0} = \int_0^\epsilon \sigma\,d\epsilon = \frac{E\epsilon^2}{2} = \frac{\sigma^2}{2E} = \frac{\sigma\epsilon}{2}. \tag{1.3.11}$$

This is called the **elastic energy density.**

## Q. 1.3.5

The direct proportionality between stress and strain is known as (a)_____ response. The tensile elastic stiffness constant of a rod is defined by the expression (b)_____. The energy stored in a stretched rod is given by the expression (c)_____. The elastic energy density in a stretched rod is given by the expression (d)_____. Two mechanical properties of materials which have been discussed thus far in this section are (e)_____ and (f)_____.

\* \* \*

In addition to **normal stress** as defined by

$$\sigma = \lim_{\delta A \to 0} \frac{\delta F_n}{\delta A} \qquad (1.3.12)$$

in Figure 1.3.2(a) there is also a **shear stress** defined by

$$\tau = \lim_{\delta A \to 0} \frac{\delta F_t}{\delta A} \qquad (1.3.13)$$

according to Figure 1.3.2(b). Here $\delta F_t$ is a tangential force. Suppose that the force $\delta F$ acting across the penny-shaped area of Figure 1.3.2 is at an angle $\theta$ from the normal to the area. Then $\delta F_n = \delta F \cos \theta$ and $\delta F_t = \delta F \sin \theta$; hence there is both a normal stress (component) and a shear stress (component).

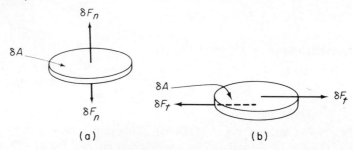

FIGURE 1.3.2. (a) Tensile stress (component). (b) Shear stress (component).

### EXAMPLE 1.3.4

A long circular cylindrical rod of cross-sectional area $A_0$ is pulled in tension by a force $F$. Are there any shear stresses present in the specimen?

*Answer.* Clearly, along the axis of the rod there is a tensile stress $\sigma = F/A_0$ and there are no shear stresses on a circular cross section normal to an axis. However, suppose we consider the stresses acting on a plane whose normal is at an angle $\theta$ to the axis of the rod as in Figure 1.3.3. The area $A$ of

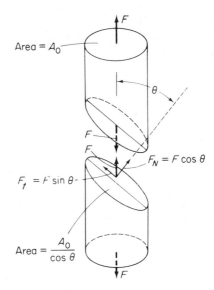

FIGURE 1.3.3. State of stress in a tensile specimen.

this elliptical section is $A_0/\cos\theta$. The force acting on this plane is $F$, with $F_n = F \cos\theta$ and $F_t = F \sin\theta$. Thus

$$\sigma = \frac{F_n}{A} = \frac{F}{A_0} \cos^2\theta \qquad (1.3.14)$$

and

$$\tau = \frac{F_t}{A} = \frac{F}{A_0} \cos\theta \sin\theta = \frac{F}{2A_0} \sin 2\theta. \qquad (1.3.15)$$

Recall that the maximum value of $\sin x$ is 1 when $x = \pi/2$. Hence $\tau$ has a maximum when $\theta = 45$ deg.

Thus shear stresses are indeed present in a tensile test! In fact shear stresses are present on all planes except those for which $\theta = 0$ or $\theta = 90$ deg. As we shall see later, it is the shear stresses which, if sufficiently large, actually cause plastic deformation.

In addition to the longitudinal strain $\epsilon$ associated with changes in length there is another strain $\gamma$ associated with changes in angle between two initially perpendicular lines in the specimen. This is illustrated in Figure 1.3.4. The **shear strain** is defined by

$$\gamma = \tan\psi. \qquad (1.3.16)$$

We shall usually be concerned with small strains for which it is justifiable to put

$$\gamma \doteq \psi. \tag{1.3.17}$$

## Q. 1.3.6

There are two kinds of stress components: (a)_____ and (b)_____.

A (c)____ stress component is caused by a force component normal to the area while a (d)____ stress component is caused by a force component parallel to the area. Although we usually say that in an ordinary tension test there is a normal stress $\sigma = F/A_0$, this is in fact dependent on which plane we choose to study. The maximum shear stress in a tension test (or compression test) lies on a plane whose normal is inclined at an angle of (e)____ to the axis of the rod. There are two kinds of strain components, the (f)_____ $\epsilon$, which is a measure of change of length, and the (g)_____ $\gamma$, which is a measure of the change of angle between two initially perpendicular lines.

\* \* \*

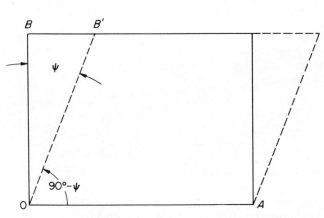

FIGURE 1.3.4. Shear strain.

It is found experimentally that there is a linear relation between shear stress and shear strain of the form

$$\tau = G\gamma, \tag{1.3.18}$$

where $G$ is the **shear modulus,** i.e., the proportionality constant.

It can be shown that a coil spring has a linear force versus deflection relation ($F = kx$) where the spring stiffness constant is given by

$$k = \left(\frac{d^4}{8D^3 n}\right) G. \tag{1.3.19}$$

Here $d$ is the wire diameter, $D$ is the coil diameter, and $n$ is the number of turns. [It must be emphasized that Equation (1.3.19) is given here only as an illustration that $k$ is proportional to $G$; the student is not expected to memorize (1.3.19).] It is also shown in strength of materials courses that the maximum force which can be applied to the spring without causing yielding is

$$F_{max} = \left(\frac{\pi d^3}{8D}\right) \tau_0, \tag{1.3.20}$$

where $\tau_0$ is the **shear yield stress** (which can be replaced by the **tensile fracture stress** $\sigma_f$ if the material fractures without yielding). It can be shown from the theory of plasticity that

$$\tau_0 = \frac{\sigma_0}{\sqrt{3}}, \tag{1.3.21}$$

where $\sigma_0$ is the tensile yield stress. Equations (1.3.19)–(1.3.21) can be used as design equations for a coil spring (tension or compression) such as the coil spring of an automobile.

The *two* elastic constants $E$ and $G$ suffice to describe the elastic behavior of an **elastically isotropic** material. A material will be elastically isotropic if the $E$'s measured in all directions in the material are the same. Single crystals are **anisotropic** (not isotropic) and have at least 3 elastic constants (cubic crystals) and as many as 21 elastic constants (triclinic crystals) depending on the symmetry nature of the crystals; however, polycrystalline aggregates of these crystals often behave in an elastically isotropic manner. This is another example of how structure affects the behavior of a material.

### Q. 1.3.7

For elastic behavior, there is (a)_____ relation between shear stress and shear strain. The proportionality constant is known as the (b)_____. The spring constant of a coil spring is directly proportional to (c)_____ (a material property). The maximum force which a coil spring can support without yielding is directly proportional to (d)_____ (a material property). The shear yield stress can be shown to be directly proportional to the (e)_____. Only (f)\_\_\_\_ elastic stiffness constants are needed to completely describe the behavior of an elastically isotropic material. Single crystals always require (g)\_\_\_\_ or more elastic stiffness constants.

Materials which have a property which varies with direction are said to be (h)_____ with respect to that property.

\* \* \*

Tables 1.3.3 and 1.3.4 list some values of $E$ and $G$ while Table 1.3.5 lists some values of $\sigma_0$ for different materials. Note the wide variation of properties among different materials. In particular note that rubber has values of $E$ and $G$ distinctly less than other materials. The reader should also examine the values of $E$ and $\sigma_0$ for three different aluminum materials;

**Table 1.3.3.** YOUNG'S MODULUS, $E$ (TENSILE STIFFNESS MODULUS), FOR VARIOUS BULK MATERIALS

| Material | $10^6$ psi | $10^{12}$ dynes/cm$^2$ |
|---|---|---|
| Gas | 0 | 0 |
| Liquid | 0 | 0 |
| Rubber (gum) | 0.00015 | 0.00001 |
| Rubber (hard) | 0.3 | 0.02 |
| Nylon | 0.3 | 0.02 |
| Concrete | 2.0 | 0.14 |
| Lead | 2.6 | 0.18 |
| Magnesium | 6.3 | 0.43 |
| Limestone | 8.3 | 0.57 |
| Glass | 10 | 0.69 |
| Aluminum (99.45% annealed) | 10 | 0.69 |
| Aluminum (99.45% cold rolled) | 10 | 0.69 |
| Aluminum (2024 heat-treated alloy) | 10.6 | 0.72 |
| Copper | 16 | 1.10 |
| Composite boron fibers 50% volume in a matrix of 6061 aluminum (along the direction of the fibers; the modulus perpendicular to the fibers is considerably less) | 30 | 2.06 |
| Steel | 30 | 2.06 |
| Beryllium | 37 | 2.5 |
| Boron fibers | 55–60 | 3.8–4.1 |
| Tungsten | 60 | 4.1 |
| Sintered carbide | 100 | 6.9 |
| Diamond | 162 | 11.0 |

it should be noted that the addition of about 6% of impurities to aluminum (the 2024 alloy) changes $E$ by about 6% while it changes $\sigma_0$ by 1150%. Since $E$ is relatively insensitive to changes in the molecular structure, $E$ is called a **structure-insensitive property**. However, $\sigma_0$, is highly sensitive to changes in molecular structure and is called a **structure-sensitive property**. The main objective of the molecular architect is to modify the structure of materials to optimize their properties (or even to obtain new properties). All the general classes of material behavior (Table 1.1.1) have some specific properties which are structure-insensitive and some which are structure-sensitive. Thus the thermal property, specific heat, is a structure-insensitive property at high temperature while thermal conductivity is a structure-sensitive property at low temperature.

Table 1.3.4. SHEAR STIFFNESS MODULUS, $G$, FOR ELASTICALLY ISOTROPIC MATERIALS

| Material | $10^6$ psi | $10^{12}$ dynes/cm$^2$ |
|---|---|---|
| Gas | 0 | 0 |
| Liquid | 0 | 0 |
| Gum rubber | 0.00005 | 0.000003 |
| Nylon | 0.1 | 0.007 |
| Aluminum | 4 | 0.28 |
| Copper | 6 | 0.41 |
| Steel | 12 | 0.83 |
| Hematite ($\alpha$-Fe$_2$O$_3$) | 13.7 | 0.94 |

### Q. 1.3.8

Elastic constants can vary by a factor of (a)\_\_\_\_\_ when stiff solids such as diamonds are compared to flexible materials such as gum rubber. As a rule the variation of Young's modulus among different aluminum alloys would be (b)\_\_\_\_\_ while the variation of tensile yield strength would be (c)\_\_\_\_\_. The ratio of yield strengths of the strongest copper to the weakest copper is about (d)\_\_\_\_\_. In general, Young's modulus is a structure- (e)\_\_\_\_\_ property while yield strength is a structure- (f)\_\_\_\_\_ property.

\* \* \*

We have thus far considered only isotropic elastic behavior (described in terms of $E$ and $G$) and isotropic plastic yielding (described by $\sigma_0$). There are many other interesting mechanical properties and some of these

**Table 1.3.5.** TENSILE YIELD STRESS, $\sigma_0$*

| Material | $10^3$ psi | | $10^9$ dynes/cm² (kbars) |
|---|---|---|---|
| Gas | 0 | | 0 |
| Liquid | 0 | | 0 |
| Nylon-66 bulk | 10 | (tensile strength) | 0.7 |
| Nylon-66 fiber | 100 | (tensile strength at 65% relative humidity) | 7.0 |
| 99.9999% single copper crystal (special care) | 0.001 | | 0.00007 |
| OFHC 99.95% annealed copper | 10 | | 0.7 |
| OFHC 99.95% cold-drawn copper | 40 | | 2.8 |
| Excellent copper whisker | 1,000 | | 70 |
| 99.45% annealed aluminum | 4 | | 0.28 |
| 99.45% cold-drawn aluminum | 24 | | 1.7 |
| 2024 (3.8–4.9% Cu, 0.3–0.9% Mn, 1.2–1.8% Mg) heat-treated aluminum alloy | 50 | | 3.4 |
| Molded phenolic (similar materials called Bakelite) | 5 | (tensile strength) | 0.34 |
| Bulk soda lime glass | 10 | (this is a typical tensile strength or fracture strength) | 0.7 |
| Freshly drawn (in vacuum) glass fiber | 2,000 | | 140 |
| Silicon coated commercial glass fiber for use in glass-reinforced plastics | 300 | (fracture strength) | 20.6 |
| Boron filament | 400 | (fracture strength) | 28 |
| Graphite filament | 300 | (fracture strength) | 20.6 |
| Sapphire whisker (strongest ones) | 10,000 | (fracture strength) | 700 |
| Gray cast iron (#20) | 20 | (fracture strength) | 1.4 |
| Pearlitic malleable cast iron | 45 | | 3.1 |
| AISI 1020 steel | 35–40 | | 2.4–2.8 |
| AISI 1095 steel (hardened) | 100–188 | | 7–13 |
| AISI 4340 annealed alloy steel | 65–70 | | 4.5–4.8 |
| AISI 4340 fully hardened alloy steel | 130–228 | | 9–16 |
| Maraging (300) steel | 290 | | 20 |
| Piano wire | 350–500 | | 24–34 |
| Iron whisker | 1,900 | | 131 |
| Concrete | 0.5 | (tensile strength varies considerably with nature of concrete and in fact we are usually more concerned with compressive strength) | 0.03 |

* At room temperature and at rates of loading typical of the usual static loading test (perhaps an hour to apply the load would be typical; more about this later). $\sigma_0$ is defined, except where noted, by 0.5% permanent extension. Certain of the quantities are tensile fracture strengths as noted rather than tensile yield stress.

will be discussed in Chapter 2; various electric, magnetic, and thermal properties will be discussed in later chapters. Many of these properties are highly structure-sensitive and hence the **molecular architect** can, by varying the structure, design materials with desired characteristics.

## ANSWERS TO QUESTIONS

**1.3.1** (a) An object, (b) a material, (c) an object, (d) an object, (e) a material, (f) a material, (g) an object.

**1.3.2** (a) Extension ratios, (b) $(l - l_0)/l_0$, (c) $2 \times 10^9$ dynes/cm$^2$, (d) $-8 \times 10^8$ dynes/cm$^2$.

**1.3.3** (a) Elastic deformation, (b) plastic deformation, (c) the stress which causes a permanent strain $\epsilon = 0.001$ or $0.1\%$, (d) 45,000, (e) 3.

**1.3.4** (a) Elastic, (b) stress, (c) strain, (d) Young's modulus, (e) Hooke's, (f) 1, (g) 1000.

**1.3.5** (a) A linear, (b) $F = kx$, (c) $W = kx^2/2$, (d) $u = E\epsilon^2/2$, (e) Young's modulus, (f) tensile yield stress.

**1.3.6** (a) Normal stress, (b) shear stress, (c) normal, (d) shear, (e) 45 deg, (f) tensile strain, (g) shear strain.

**1.3.7** (a) A linear, (b) shear modulus, (c) the shear modulus, (d) the yield stress, (e) tensile yield stress, (f) two, (g) three, (h) anisotropic.

**1.3.8** (a) $10^6$, (b) small (a few percent), (c) large ($1000\%$ or so), (d) $10^6$ (see Table 1.3.5), (e) insensitive, (f) sensitive.

**1.4 HOW IS MATERIALS SCIENCE RELEVANT?**

One maxim well known by scientists and engineers states that *the technological development of any age is limited by the available materials.* This will be illustrated with six relevant examples.

MICROELECTRONICS AND COMPUTER-BASED INSTRUCTION. The size and speed of computers is controlled by developments in materials science as was discussed in Section 1.1. Since the invention of the *n-p-n* transistor in 1948, the size of electronic devices has been reduced by a factor 100,000. A tiny chip of silicon, perhaps $0.1 \times 0.1$ in., now contains over 1000 active elements (transistors) and 1000 passive elements (resistors, etc.). Another major breakthrough involving three-dimensional microelectronic circuits might increase the density by another 1000 times. The use of large computers with visual display for educational purposes has been discussed by D. Alpert and D. L. Bitzer, "Advances in Computer-based Education," *Science* **167**, 1582 (1970). Computers will profoundly improve education as well as extend its capability. There is little doubt in the author's mind that the effect of computers on education in this age will be as great as the effect of Gutenberg's printing press was in bringing an end to the dark ages.

STRENGTHENING MECHANISMS AND AEROSPACE APPLICATIONS. Consider an airplane weighing 500 tons with 50 tons of cargo, 150 tons of structure, remainder fuel, and an 8000-mile range. Suppose we could triple the strength of all materials in the structure; then we would need only 50 tons of structure and hence could have a cargo of 150 tons. Each pound of cargo is worth $10 during the life of the plane. This means $2 \times 10^6$ per plane. If 1000 planes are produced, this means $2 \times 10^9$. Assuming that the new material costs only three times as much as the old and that fabricating expenses are the same, the $2 billion is a gain. What is the probability of this happening? We have already noted in Table 1.3.5 the wide variation of yield stress in aluminum. The process whereby the aluminum could be strengthened by about a factor of 10 was discovered by Dr. Alfred Wilm in 1906. Yet an aluminum alloy with a yield stress of 60,000 psi (4 kbars or $4 \times 10^9$ dynes/cm$^2$) is still far below the potential of aluminum. We shall see in Chapter 12 that *perfect* crystals of aluminum would have a yield stress of about 1,200,000 psi (80 kbars). Thus a gain of a factor of 20 still may be possible.

FAST BREEDER REACTOR.

Energy has always been the key to man's greatest goals and to his dreams of a better world.—Glenn T. Seaborg

In 1970, half of the new power stations on order in the United States were nuclear-powered. By the end of this century, more than half of the electricity which we produce will likely come from nuclear plants. In the nations with emerging technologies, nuclear energy will have an even more profound effect on the economy and standard of living. A superb discussion of the overall energy problem can be found in the article by A. M. Weinberg and R. P. Hammond, "Limits to the Use of Energy," *American Scientist* **58**, 412 (1970). Perhaps the most important feature of nuclear energy lies in the fact that it can be produced without pollution (except for thermal pollution).

The commercial development of the breeder reactor awaits the solution of materials problems. By the year 2000 the supply of $U^{235}$ will be gone. However, utilization of the much vaster supply of $U^{238}$ will be possible when commercial breeder reactors are developed which convert this material to fissionable plutonium.

Figure 1.4.1 shows the core assembly of such a reactor. The high-flux, high-energy (1 MeV) neutrons in the breeder reactor cause considerable

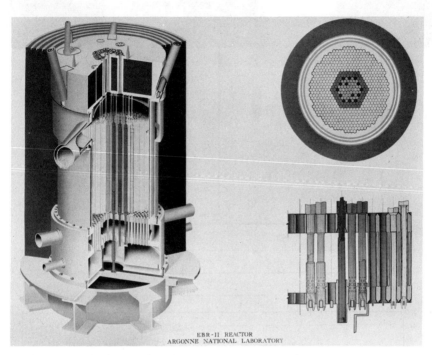

FIGURE 1.4.1. Cutaway view of the reactor core. The reactor core consists of an assembly of hexagonal-shaped ducts (top view shown in upper right-hand side of the picture). Inside each duct there is a bundle of fuel pins. The lower portion of the duct (shown in the lower right-hand side of the picture) is cylindrically shaped. Provisions are made so that they can be situated firmly on the grid plate. Sodium coolant is forced into the duct through the holes shown. The hexagonal shape is necessary to maintain the rigidity of the core assembly. (Courtesy of C. Y. Li, Argonne National Laboratory, Argonne, Ill.)

radiation damage as shown in Figure 1.4.2. The stainless steel cladding material undergoes volume increases of up to 20% due to the formation of voids and pores (see Figure 1.4.3). Because the radiation density and the temperature are not uniform in the reactor (these are the factors which determine void formation), the volume changes vary with position in the reactor. The fuel elements in the reactor will therefore become highly distorted and inoperative. At present this problem is unsolved but many

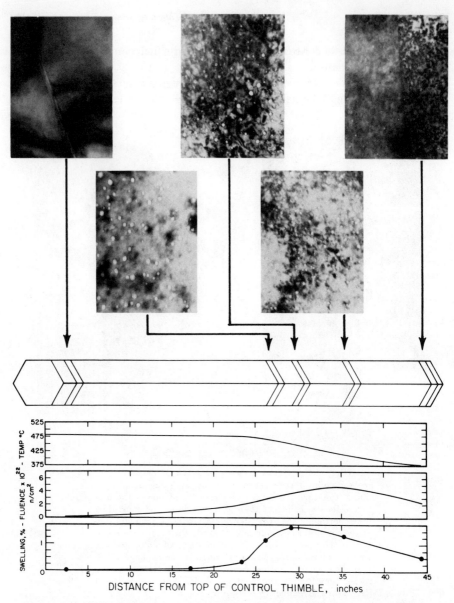

FIGURE 1.4.2. Void formation in stainless steels due to fast neutron irradiation. Electron microscope pictures of stainless steel specimens obtained from various positions of the hexagonal duct (control thimble) are shown. It is shown clearly that void formation depends on temperature, flux level of the neutrons, and the total exposure (fluence). The large voids shown in the picture are approximately 200 Å in diameter. Because the temperature and neutron flux distributions are not uniform inside the core, the extent of void formation (swelling) will therefore be nonuniform, causing the distortion of the core assembly. This type of dimensional instability is not acceptable. Similar void formation phenomena will occur in the material (stainless steel) of the metal can for fuel. There its effect on mechanical strength is also of concern. (Courtesy of C. Y. Li, Argonne National Laboratory, Argonne, Ill.)

research workers are involved in programs to find an answer to this swelling problem.

Eventually it also will be necessary to produce the fusion reactor (although the fusion reaction has not yet been controlled). This will also require a breeder reactor in which lithium is converted to tritium fuel. The fusion reactor will involve 14-MeV neutrons and will create an abundance of new challenging materials problems.

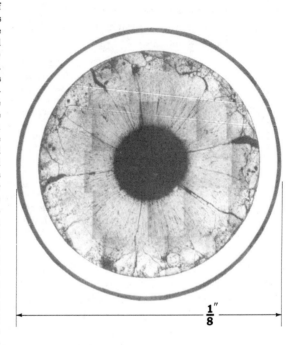

FIGURE 1.4.3. Cross section of a fuel pin (top view). The cross sections of the fuel pin show the restructuring of the oxide fuel pellet ($UO_2$ and $PuO_2$ mixture) and the reason for fuel swelling. The temperature in the fuel is very high with center temperature ($\sim 2800°C$) close to the melting point of the fuel. The gaseous fission products will form bubbles in the fuel. This is the reason for fuel swelling which will create a load on the metal can and cause failure. Some of the gas bubbles will migrate to the center of the fuel to form a void there because of the existence of a high-temperature gradient. After a short time in the reactor, solid oxide fuel pellets will become a continuous hollow cylinder. The interfaces between pellets are removed by high-temperature sintering processes. The cracks shown in the picture are caused by the thermal stress produced during the cooling down. (Courtesy of C. Y. Li, Argonne National Laboratory, Argonne, Ill.)

FUEL CELLS AND ELECTRIC BATTERIES. The development of cheap, rugged, commercial fuel cells to power automobiles, buses, and trucks also depends on developments in the materials area. Clearly, such a breakthrough would eliminate a major source of air pollution and noise pollution. An alternative solution would be the development of lightweight, powerful, rechargeable batteries (charged by electrical energy from clean nuclear reactors). Such a development is also a materials problem. Some basic problems with existing materials for batteries are that they are too costly, are too bulky, and have critical pressure and temperature ranges at which they must operate. For example, Ford has constructed a sodium-sulfur battery that delivers 150 W-hr/lb. The electrolyte in the battery is a sodium aluminate which is solid and is permeable only to sodium ions. Unfortunately, it does not start to operate until 245°C. Thus it must be kept hot at all times.

There is an excellent discussion of fuel cells by W. T. Grubb and L. W. Niedrach in the book *Direct Energy Conversion*, edited by George W. Sutton, McGraw-Hill Book Company, Inc., New York (1966).

BIOMEDICAL MATERIALS. There is presently a materials barrier which prevents the production of satisfactory artificial hip joints for arthritic patients and of artificial hearts. Figure 1.4.4 illustrates the problem

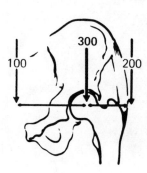

FIGURE 1.4.4. Upper part of the human femur, or thigh bone, which fits into the pelvic structure, and some of the loading stresses it undergoes. In older people, the neck of the bone (between the ball and shank) decreases markedly in radius, making that part of the bone very weak and susceptible to fracture under the loads. The ball which fits into the pelvic socket also develops a rough edge, resistant to motion, which puts even more pressure on the neck.

which occurs in the arthritic condition, and Figure 1.4.5 illustrates an artificial joint. One of the problems here is with stress-corrosion of stainless steel since the joints are often made of stainless steel. Also the Teflon used as joint lubricant is attacked by the body fluids and is unsatisfactory. The last fact is very interesting as it is possible to boil sulfuric acid in a Teflon beaker without any reaction between the Teflon and the sulfuric acid. Another important development on which some progress is being made is permanently implantable false teeth. It is in principle possible to

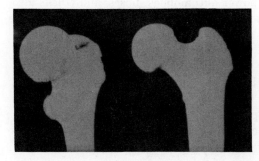

FIGURE 1.4.5. Artificial hip joint. One solution to the problem is an artificial bone. The two shown are made of polythene plastic. Others can be made of a core of stainless steel surrounded with a coating of Teflon to protect the core and reduce friction in the joint.

Sec. 1.4  How Is Materials Science Relevant?  51

make a ceramic which duplicates the hydroxyapatite structure of teeth (and bone). With an appropriate pore structure, tissue will attach itself.

A major problem in the development of a permanent artificial heart is that most materials tend to cause blood coagulation which eventually leads to thrombosis and death. The challenge here is to find out what causes blood coagulation and to develop materials which do not suffer from this drawback. Another problem is to develop materials which will not undergo fatigue (and break) after $10^9$ loading cycles (heart beats) in an unfriendly chemical.

Biomaterials engineering is also needed in the mental health field. One of the main causes of senility is the shortening of blood flow to the brain. Perhaps this could be overcome through the use of artificial arteries, but if so, new materials must be developed for them. The challenges to the biomaterials scientist are endless.

LOW-COST HOUSING. The heyday of the custom-built house is past. Housing costs can be halved by factory mass production. New materials and new ways of handling of both old and new materials play a vital role in this process. Included are the use of vinyl plastic plumbing; fiberglass walls, doors, and panels; foam-filled fiber-reinforced beams; sprayed-on urethane foams; sheet metal glued on to rigid foam panels; copper-laminated plywood for roofing; etc. Note the many *composite* materials.

## Q. 1.4.1

List six areas in which materials science is playing a vital role in technology.

_____
_____
_____
_____
_____
_____

\* \* \*

A summary of materials science and engineering and its role in society is shown in Table 1.4.1. In the present course we shall emphasize the materials science portion of Table 1.4.1. However, from time to time, applications of the concepts will be illustrated.

**Table 1.4.1.** MATERIALS SCIENCE AND ENGINEERING: METALS, SEMICONDUCTORS, POLYMERS, CERAMICS, GLASSES, COMPOSITES, ETC.

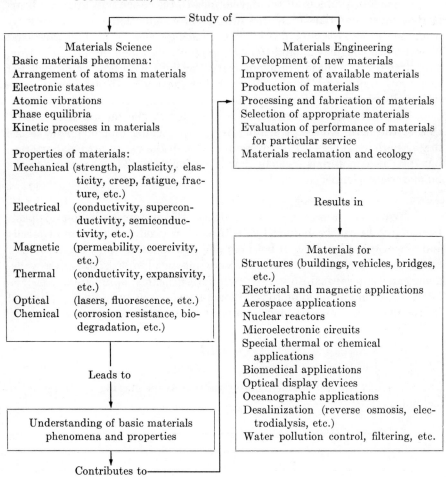

Q. 1.4.2

What is molecular architecture?

\* \* \*

# ANSWERS TO QUESTIONS

**1.4.1** Making computer memories more dense and faster. Making materials stronger. Making materials which will make the fast breeder reactor possible. Making new fuel cell electrodes, lighter batteries, body-compatible materials, materials for low-cost housing. There are many other answers, e.g., making good semipermeable membranes for use in desalinization of sea water by reverse osmosis.

**1.4.2** Designing and producing atomic arrangements which give a material with a desired property.

# REFERENCES

### General

*Materials*, A Scientific American Book, W. H. Freeman and Company, Publishers, San Francisco (1967). This is a monograph form of the September, 1967 issue of Scientific American which was completely devoted to materials. Photographs and drawings in the original articles were in color.

Moore, W. J., *Seven Solid States*, W. A. Benjamin, Inc., Reading, Mass. (1967). The author takes seven specific materials (salt, gold, silicon, steel, nickel oxide, ruby, and anthracene) and shows how they exemplify the different characteristics of the solid state.

### Microelectronics

Heath, F. G., "Large Scale Integration in Electronics," *Scientific American* (February, 1970) p. 22.

### Strengthening

Felbeck, D. K., *Introduction to Strengthening Mechanisms*, Prentice-Hall, Inc., Englewood Cliffs, N.J. (1968).

### Energy

Weinberg, A. M., and Hammond, R. P., "Limits to the Use of Energy," *American Scientist* **58**, 412 (1970).

Sporn, P., *Research in Electrical Power*, Pergamon Press, Inc., Elmsford, N.Y. (1966).

Seaborg, G. T., and Bloom, J. L., "Fast Breeder Reactors," *Scientific American* (Nov., 1970) p. 13.

### Fuel Cells

Grubb, W. T., and Niedrach, L. W., in the book *Direct Energy Conversion*, edited by G. W. Sutton, McGraw-Hill Book Company, Inc., New York (1966).

### Biomaterials

Stark, L., and Agarwal, G., Eds., *Biomaterials*, Plenum Publishing Corporation, New York (1969).

### Housing

Plant, A. F., "Scientists Build a House," *Industrial Research* (November, 1970) p. 48.

## PROBLEMS

**1.1** Suppose Avogadro's number $N_0$ of grains of sand were evenly distributed over the United States. If the sand grains have a diameter of about $5 \times 10^{-2}$ cm, how deep would be the sand?

**1.2** Give order of magnitude answers to the following:
  (a) Size of atoms in meters.
  (b) Energies associated with valence electrons in electron volts.
  (c) Vibrational frequency of atoms in hertz.

**1.3** Name four material properties which depend exponentially on the negative reciprocal of absolute temperature.

**1.4** Describe the process of molecular architecture as applied to the production of a microelectronic circuit.

**1.5** What experimental techniques are used to study
  (a) Arrangements of atoms in crystals.
  (b) Arrangement of crystals in polycrystalline aggregates.

**1.6** List the five different classifications of structure and give an example of each.

**1.7** Define
  (a) Normal stress.
  (b) Shear stress.
  (c) Normal strain.
  (d) Shear strain.
  (e) Hooke's law.
  (f) Linear response.

**1.8** (a) Explain why $E$ (and not $G$ or both $E$ and $G$) is involved in the stiffness constant of a beam.

(b) Explain why $G$ (and not $E$ or both $E$ and $G$) are involved in the stiffness constant of a coiled spring.

**1.9** Which material property is more important for the cutting tool of a lathe, $E$ or $\sigma_0$? Explain.

**1.10** (a) Would wood show isotropic elastic behavior? Explain.

(b) Isotropic tensile strength behavior? Explain.

**1.11** (a) How many elastic constants are needed to describe the elastic behavior (for small strains) of elastically homogeneous and isotropic bodies?

(b) Crystals?

**1.12** List one mechanical property which is structure-sensitive and one which is structure-insensitive.

**1.13** (a) Approximately what are the Young's moduli of the stiffest and the most flexible elastic materials?

(b) Express the above as a ratio.

**1.14** What is the range of tensile yield strength for the "same" material, e.g., copper?

**1.15** Make a table with three columns: material, $\sigma_0$, and $E$. Let the materials be 99.45% annealed aluminum, 99.45% cold-drawn aluminum, and 2024 heat-treated aluminum alloy. Discuss the significance of this table.

# MORE INVOLVED PROBLEMS

**1.16** Suppose the storage capacity of the brain involves $10^{11}$ diodes. Compare this density to the density of a computer memory based on the circuits of Section 1.1.

**1.17** Students, after reading a textbook, often are so impressed by what has been done that they are not aware of what still has to be done. By checking some journals, they can find that new things are being learned every day about materials. List several journals which are concerned with various aspects of materials science. Include at least one for each of the following specialized areas: polymers, ceramics, metals, glasses, and semiconductors.

**1.18** Design a diving board out of a glass-reinforced epoxy composite. The glass fibers run parallel to the length of the board. Assume $E = 300$ kbars and $\sigma_0 = 10$ kbars. Assume the base of the board is placed in a rigid foundation (in real life, a different scheme is often used). The maximum force allowed is $F_{max} = (BH^2/6L)\sigma_0$.

**1.19** A microduplex aluminum-zinc alloy has roughly spherical grains with a diameter of 5 $\mu$. About how many grains are there per cubic centimeter?

**1.20** A student is asked to design a coil spring which has a stiffness constant of $\sim 2$ ppi and which will stretch 4 in. without yielding. The spring also has to pass through a plastic tubing with a $\frac{1}{4}$-in. I.D. The spring is to be used to hold skate guards on hockey skates (to protect the blade while the wearer walks on concrete, etc.). He uses 130 turns of 0.031-in. diameter music wire wound into a tight spring with 0.21-in. O.D. He uses a double looped end.
(a) Does the spring meet the design requirements?
(b) Describe how the design criteria may have been chosen.

## SOPHISTICATED PROBLEMS

**1.21** Water can be extracted from salt water by applying a pressure to the salt water in a cylinder in which one end is made of a semipermeable membrane (water molecules go through but salt ions do not). Discuss the kinds of properties which this material must have. (This process is called desalinization by reverse osmosis.)

**1.22** Tiny crystals, called whiskers, have been grown with tremendous strengths, often approaching that expected of a truly perfect crystal. Write an essay on the current status of whisker growth.

**1.23** Write an essay on the projected energy needs in the United States by the year 2000.

**1.24** Measure the diameter of a steel coil spring in the front of an automobile. Obtain the static load on this spring. Assume that under dynamic conditions the load can be twice as large.
(a) How much would the spring deflect?
(b) What would be the minimum yield strength required?

# Prologue

Materials respond to mechanical stresses by undergoing displacements which are conveniently measured in terms of *strain*. The difference between *elastic* and *plastic deformation* is discussed, as is also the difference between *brittle* and *ductile* materials. *Strain hardening, ductility, toughness,* and *hardness* are aspects of plastic deformation which will be studied. *Annealing,* which can negate the effects of strain hardening, is described. Plastic instabilities are introduced. There are brief discussions of *stress concentrations, impact toughness, creep, fatigue,* and the *strength to weight* ratio.

Temperature is shown to be a vital parameter in mechanical behavior, especially in yielding, creep, and ductility and in brittle fracture phenomena.

In addition to the deformation of solids, the flow of liquids (and gases) is studied and the concept of a *viscosity coefficient* is introduced.

Finally, it is noted that materials are not perfectly elastic but that there is always a viscous component in the motion; this is called internal friction or mechanical damping and is measured by the *loss angle*.

The student should not only know which types of properties exist but should begin to develop a feeling for the *order of magnitude* of the quantities involved. He should also question why certain behavior occurs and he might also think about how his knowledge may be applied to everyday materials problems.

# 2

# MECHANICAL PROPERTIES

## 2.1 PLASTIC DEFORMATION OF SOLIDS

Certain aspects of mechanical behavior were introduced in Section 1.3. The quantities defined and discussed there were nominal normal stress, $\sigma$; shear stress, $\tau$; nominal longitudinal strain, $\epsilon$; shear strain, $\gamma$; Young's modulus, $E$; shear modulus, $G$; elastic strain energy density, $u$; and the tensile yield stress $\sigma_0$. In general mechanical behavior is the response to stress. The response to stress always involves a relative displacement of material. Such relative displacements may involve small elastic strains or they may involve plastic deformation or even fracture. In this chapter several different types of mechanical behavior, in addition to those already discussed in Section 1.3, will be described.

PLASTIC FLOW. A few materials behave elastically (or very nearly so) to the breaking point. Examples of such **brittle materials** are silica glass and white cast iron. However, most materials undergo **plastic deformation** or permanent deformation before breaking. Schematics of typical tensile stress-strain curves are shown in Figure 2.1.1. Copper, for example, shows the behavior of Figure 2.1.1(b) while steel and nylon exhibit the **yield point** behavior shown in Figure 2.1.1(c). We note that once yielding begins in the latter a lower stress is required to produce further deformation.

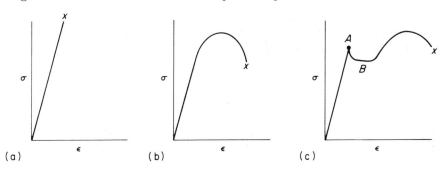

FIGURE 2.1.1. Various stress-strain curves. (a) Brittle. (b) Ductile. (c) Ductile with yield point. The upper yield point is at $A$ and the lower yield point which follows is at $B$.

Eventually, as deformation continues, a larger stress is required to produce further deformation, as is also true in the case of those ductile materials which do not exhibit yield points; this is called **work hardening** or **strain hardening**. Figure 2.1.2 shows an actual stress-strain curve for a copper rod which had been held at 800°C for 1 hr. Here the yield strength is 10,000 psi (0.6 kbars). Note the considerable work hardening which occurs. The nominal tensile stress (tensile force divided by the original area) at the maximum is about 35,000 psi.

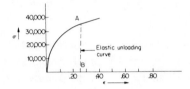

FIGURE 2.1.2. Tensile stress-strain curve for a copper rod held at 800°C for 1 hr prior to testing at room temperature.

Because of the large change in cross-sectional area and because of the large strains, it is useful to reconsider our original definitions of stress and strain. We define **real** (normal) **stress** by

$$\sigma_r = F/A; \qquad (2.1.1)$$

i.e., we divide the force by the actual area, $A$, not the original area, $A_0$. Similarly we define an increment of **real strain** by

$$d\epsilon_r = \frac{dl}{l} \qquad (2.1.2)$$

so that

$$\epsilon_r = \ln \frac{l}{l_0}, \qquad (2.1.3)$$

where $l_0$ is the original length.

During the initial elastic deformation, as the specimen stretches, its diameter shrinks. The **Poisson ratio** $\nu$ is

$$\nu = \frac{-\text{lateral strain}}{\text{longitudinal strain}}. \qquad (2.1.4)$$

For elastically isotropic media it can be shown [see H. D. Conway, *Mechanics of Materials*, Prentice-Hall, Inc., Englewood Cliffs, N.J. (1950), p. 24] that

$$\nu = \frac{E}{2G} - 1, \qquad (2.1.5)$$

where $E$ is Young's modulus and $G$ is the shear modulus, so that $\nu$ is not an independent elastic constant. $\nu$ has values between 0 and $\frac{1}{2}$ as shown in Table 2.1.1. However, it is found to a good approximation that the volume

Table 2.1.1. SOME VALUES OF POISSON'S RATIO FOR ELASTICALLY ISOTROPIC SOLIDS

| Material | $\nu$ |
|---|---|
| Beryllium | 0.05 |
| Glass | 0.20–0.25 |
| Rock | 0.25 |
| Steel | 0.28 |
| Copper | 0.33 |
| Nylon | 0.48 |
| Rubber | ~0.50 |

of a specimen remains constant during plastic deformation; i.e.,

$$Al = A_0 l_0. \tag{2.1.6}$$

The reader can show that this is equivalent to having $\nu = \frac{1}{2}$.

### Q. 2.1.1

Materials which exhibit only elastic deformation prior to fracture are called (a)_____. Materials which undergo more or less plastic deformation are known as (b)_____. As a result of plastic deformation, the yield strength of materials often increases appreciably; this is called (c)_____. Because of the large changes in shape, (d)____'____ stress and (e)_____ strain are used to describe plastic deformation. Poisson's ratio for steel during elastic deformation is 0.28 but during plastic deformation $\nu$ equals (f)____.

\* \* \*

During the early stages of plastic deformation (prior to the maximum in the nominal stress-strain curve of Figure 2.1.2) the cross-sectional area of the specimen decreases uniformly along the entire length (and so it can carry less load) but simultaneously the strength of the material is increasing (and so it can carry a greater load). At first the work-hardening effect prevails and the load-carrying ability increases. However, at the maximum of the stress-strain curve the two effects just cancel, and there is no increase in load-carrying ability. Beyond the maximum we would expect a slow decrease in load-carrying capability if the cross-sectional area continued to deform uniformly along the entire length; this is not the case. Rather, we find that at the maximum, **necking** begins as shown in Figure 2.1.3.

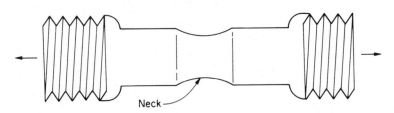

FIGURE 2.1.3. Necked tensile specimen.

It is an example of a **plastic instability.** In reality, we find that one part of the rod is slightly weaker than the rest. Since the maximum has been passed, the load that this weaker section can now carry will be even less than before, and it, therefore, deforms further with a *smaller* load than previously required. The stress at the maximum of the nominal tensile stress-strain curve is called the **tensile strength;** it is the maximum load which the tensile specimen can bear, divided by the *original* area. (The state of stress at the neck is not simple tension! See A. L. Ruoff, *Materials Science*, Prentice-Hall, Inc., Englewood Cliffs, N.J. (1973), Chap. 24.)

It is often found that plastic deformation can be represented by an empirical relation

$$\sigma_r = K_\epsilon \epsilon_r^n \tag{2.1.7}$$

where $K$ and $n$ are constants characteristic of specific materials and where $n$, called the **strain-hardening exponent,** is a fraction of 1. It then can be shown that the maximum in the *nominal* stress-strain curve occurs at a strain $\epsilon_r = n$.

## Q. 2.1.2

Up to the maximum of the nominal stress-strain curve the deformation is (a)_____ along the rod. Beyond the maximum (b)_____ occurs. This is an example of (c)_____. The maximum nominal tensile stress is known as the (d)_____.

\* \* \*

The extent of plastic deformation which a specimen undergoes is known as the **ductility.** A quantity often used to measure ductility is the percent reduction in cross-sectional area at fracture:

$$\text{PRA} = \frac{A_0 - A_f}{A_0} \times 100\%. \tag{2.1.8}$$

Here $A_f$ is the cross-sectional area of the neck at fracture. The real strain at fracture,

$$\epsilon_r = \ln \frac{A_0}{A_f}, \tag{2.1.9}$$

is also used as a measure of ductility.

Suppose that in Figure 2.1.2 the loading is stopped at the point $A$ where the cross-sectional area is $A$, and the specimen is then unloaded. This is an elastic process and occurs along the line $AB$ (with possible small deviation near the point $A$). Upon reloading the stress-strain curve would retrace the path $BA$ and at the point $A$ yielding would commence.

**EXAMPLE 2.1.1**

If the specimen at $B$ in Figure 2.1.2 is considered to be a new sample and $B$ is at $\epsilon = 0.25$, and the point $A$ is at a nominal stress of 32,000 psi, what is the yield strength of the new specimen?

*Answer.* Since $\epsilon = 0.25 = l/l_0 - 1$, we have $l/l_0 = 1.25 = A_0/A$. Thus the area of the original sample is 1.25 times the area of the new sample. Hence the stress in the new sample at yielding is $32,000 \times 1.25 = 40,000$ psi. The new sample would be less ductile than the original.

The shape of the stress-strain curve is affected by strain rate and temperature. In general as the strain rate is increased, larger stresses will be needed to reach the same strain. As the temperature is increased, smaller stresses will be needed to reach the same strain. The yield strength decreases with increasing temperature as shown in Figure 2.1.4. Note the precipitous drop around $0.5T_m$, where $T_m$ is the melting temperature.

The shape of the stress-strain curve is also greatly affected by pressure. For example, limestone, which is brittle in an ordinary tension test, behaves as a ductile material when tested in tension in a pressure vessel containing a fluid at a pressure of 150,000 psi.

The process of plastic deformation is important in geology (as in the formation of mountains, continental drifts, etc.) and in materials engineering (as in the forming of objects by materials movement, e.g., wire drawing or the forming of objects by material removal, e.g., milling, etc.).

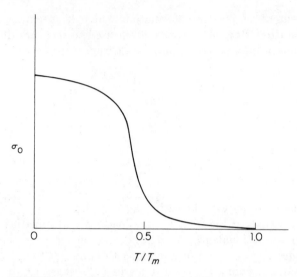

FIGURE 2.1.4. Typical variation of yield strength versus fraction of the melting point.

TOUGHNESS. The area under the nominal stress-strain curve up to the maximum is called the **toughness of infinite rods**. It is the work which must be done to deform a unit volume of the material up to the point that necking begins.

HARDNESS. **Hardness** is a measure of the resistance of a material to permanent indentation by an indenter such as a diamond (the latter behaves elastically during the test). The width of indentation left in the test specimen by a given indenter under a given force is often used as a measure of the hardness. Various devices such as the Rockwell, Vickers, and Brinell testers are used. The Rockwell tester has several scales, some of which are appropriate to testing soft materials such as plastics, and others which are applicable to testing hard materials such as carburized steel. The hardness of a *specific type of material* is related to the tensile strength. For example, the tensile strength of carbon steels (in pounds per square inch) is about 500 times the Brinell hardness number. Because hardness is much easier to measure than tensile strength (as well as being nondestructive), it is widely used for checking the properties of materials.

## Q. 2.1.3

As the temperature is raised, the stress needed to achieve a given strain (a)_____. As the velocity of straining increases, the stress needed to achieve a given strain (b)_____. At about $0.5T_m$, there is a (c)_____ in the yield stress of most materials. Hydrostatic pressure often greatly increases the (d)_____ of a material. The integral $\int_0^{\epsilon_{\sigma max}} \sigma \, d\epsilon$ defines (e)_____; the upper limit is the value of the strain $\epsilon$ at which the stress $\sigma$ is a maximum, i.e., $\sigma_{max}$. Resistance to indentation is called (f)_____. There is often a simple relation between resistance to indentation and (g)_____.

\* \* \*

ANNEALING. When the copper rod whose properties are shown in Figure 2.1.2 is cold-worked to $\epsilon = 0.50$, we find that it has (considered as a new specimen) a yield strength of $33,300 \times 1.5$ or 50,000 psi. If this cold-worked copper bar is heated to a high temperature (say 800°C for 1 hr), it will lose the strength gained during cold working, and the yield strength at room temperature will drop to 10,000 psi. The annealed material will behave in the same fashion as the annealed starting material described by Figure 2.1.2.

# ANSWERS TO QUESTIONS

**2.1.1** (a) Brittle materials, (b) ductile materials, (c) work hardening or strain hardening, (d) real, (e) real, (f) 0.50.

**2.1.2** (a) Uniform (or homogeneous), (b) necking, (c) a plastic instability, (d) tensile strength.

**2.2.3** (a) Decreases, (b) increases, (c) rapid drop, (d) ductility, (e) toughness, (f) hardness, (g) tensile strength.

## 2.2 SPECIAL DEFORMATION OF SOLIDS

STRESS CONCENTRATIONS. Consider a plate with a circular hole as shown in Figure 2.2.1. The tensile stress is $\sigma$. However, adjacent to the hole in the tensile direction there is a greater stress than throughout the rest of the specimen. We write

$$\sigma' = K\sigma, \qquad (2.2.1)$$

where $\sigma'$ is the stress adjacent to the hole as shown and $K$ is known as the **stress concentration factor**. In the present case, $K = 3$, as can be shown theoretically by elasticians or experimentally by experimental stress analysts

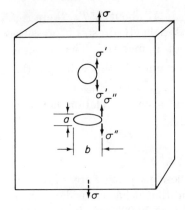

FIGURE 2.2.1. Plate loaded in tension. $\sigma'' > \sigma' > \sigma$.

using either photoelasticity techniques which measure the stresses directly or electrical strain gauges which measure strain which then must be converted to stress.

The stress concentration at the tip of the elliptical hole is much larger, and, in fact, for very sharp cracks (large $b/a$ ratio of the ellipse) this can take on values of the order of 1000 ($10^3$). Such sharp cracks are to be avoided by designers, and even the stress concentrations of a circular hole can be problems to the designer. There are some situations where very sharp cracks occur quite naturally, for example, on the surface of glass (glass here means simply a material, such as window glass, although later on the word *glasses* will refer to a group of materials with special atomic structure). The presence of these cracks makes window glass fracture at a stress of only 10,000 psi. In the absence of such cracks, glass has a strength in excess of 1,000,000 psi. The presence of a small crack in a highly ductile material might not cause too much concern, even though the tensile stress across the crack has reached 50,000 psi. The reason for this is shown in Figure 2.2.2.

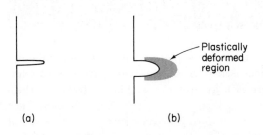

FIGURE 2.2.2. (a) Sharp crack with large stress concentration factor. (b) Less sharp crack due to plastic deformation around the tip of the crack. The higher the ductility, the more plastic deformation which can occur and the more rounded the tip of the crack becomes; hence the stress concentration factor decreases.

## Q. 2.2.1

A copper rod is formed by cold working at room temperature. This results in considerable (a)_____. When this rod is then held at a high temperature, a process called (b)_____, its yield strength (c)_____. When a crack is present in a material, the stress at the tip of the crack is much higher than in the bulk material. The ratio of these two stresses is known as the (d)_____. Sharp cracks are much more critical in (e)_____ materials than in (f)_____ materials.

* * *

IMPACT TOUGHNESS. Earl R. Parker begins his book *Brittle Fracture of Engineering Structures*, John Wiley & Sons, Inc., New York, (1957), as follows:

> On January 16, 1943, a T-2 tanker lying quietly at her fitting-out pier at Portland, Oregon, suddenly cracked in a brittle manner ... "without warning and with a report that was heard for at least a mile ...." The sea was calm, the weather mild, her computed deck stress was only 9,900 psi. There seemed to be no reason why she should have broken in two, but she did.

See Figure 2.2.3. We have already noted the role which ductility plays in affecting the propagation of a crack. Many steels (and also many other

FIGURE 2.2.3. Brittle fracture of a T-2 tanker.

materials) show a special behavior when notched specimens are impacted at high velocity and the energy absorbed is plotted versus temperature as in Figure 2.2.4.

The **Charpy V-notch impact test** is a specific type of test which refers to the sample size, notch size, and testing procedures. A schematic of the specimen and the loading is shown in Figure 2.2.5. The anvil is the head of a heavy pendulum which swings from its raised rest position when released.

At high temperatures, the specimen absorbs lots of energy because of the plastic deformation which occurs in the region surrounding the tip

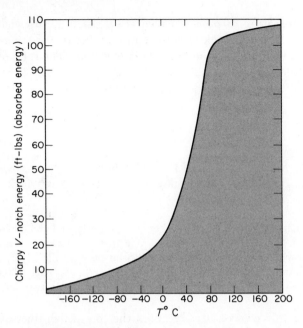

FIGURE 2.2.4. Impact test. Charpy V-notch energy versus temperature for a steel with $\sigma_0 = 60{,}000$ psi.

of the crack (see Figure 2.2.2), but as the temperature drops to around room temperature, there is a sudden sharp drop in the energy absorbed, i.e., in the toughness. Thus, at the higher temperatures, the steel is very ductile, while at the lower temperatures, it is brittle (or only slightly ductile). This peculiar behavior was responsible for the demise of several welded ships and of several welded bridges in Canada and Belgium during the winter of 1943 (which was especially cold). The temperature at which the drop in absorbed energy (inpact toughness) occurs is called the **ductile to brittle transition temperature.**

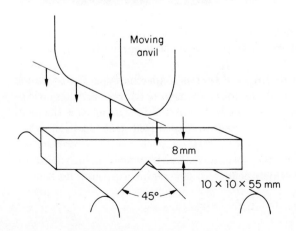

FIGURE 2.2.5. Schematic of Charpy V-notch specimen.

Several materials, such as austenitic 18-8 stainless steel, do not show such behavior and remain ductile to absolute zero. However, various materials in addition to ordinary carbon steel exhibit a ductile to brittle transition which the materials engineer must take into account.

**EXAMPLE 2.2.1**

A tank car is being built to transport liquid nitrogen (boiling point 78°K). Would the steel whose characteristics are shown in Figure 2.2.4 be used as the inner dewar?

*Answer.* This temperature corresponds to $-195$°C. The steel has a Charpy V-notch impact energy of only 2-3% of its high-temperature value. A small crack or notch in the vessel would be critical (because $K$ would effectively be very high). The answer is: No.

### Q. 2.2.2

Carbon steels exhibit (a)_____ transition temperature while (b)_____ steels do not. One measure of the toughness of a notched specimen is the (c)_____.

\* \* \*

CREEP. An engineer tries to run a heat engine at the highest possible temperature in order to maximize its efficiency. The maximum possible efficiency of an engine (as any student of thermodynamics can show) is

$$\text{max. eff.} = \frac{T - T_0}{T}.$$

Here $T$ is the internal temperature and $T_0$ is the exhaust temperature (which tends to be fixed by the environment, say room temperature or river temperature, etc.). The components of an engine, e.g., the turbine blades, are also under stress at these high temperatures. Under such conditions, the material continually deforms under fixed stress; this is called **creep.**

Typical behavior at constant stress is shown in Figure 2.2.6. Creep depends very strongly on temperature. A rough rule is that it becomes quite important above $T_m/3$ and critical above $T_m/2$, where $T_m$ is the melting temperature. Creep failure can occur in two ways: Either the magnitude of deformation becomes intolerable or the component ruptures. An empirical expression for the rate of **steady-state creep** (the region where $d\epsilon/dt$ is a constant in Figure 2.2.6) is

$$\dot{\epsilon} = B\sigma^n e^{-Q_c/RT}. \tag{2.2.2}$$

Here $B$ is a constant, $\sigma$ is the tensile stress, $n$ is a constant usually equal to about 5, $R$ is the gas constant, $T$ is the absolute temperature, and $Q_c$ is called the **activation energy for creep** (which has a characteristic value for each material).

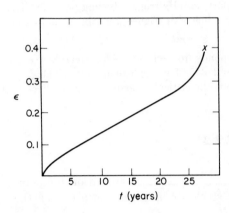

FIGURE 2.2.6. Creep curve. Strain, $\epsilon$, versus time, $t$.

**EXAMPLE 2.2.2**

A nickel sample shows a steady-state creep rate of

$$\dot{\epsilon} = 10^{-12} \text{ sec}^{-1}$$

at

$$T = 862°\text{K}$$

(which is half the melting temperature) at a tensile stress of only 1000 psi. If $Q_c = 69,000$ cal/mole, estimate the creep rate at $1293°\text{K}$ ($\frac{3}{4}T_m$) if the stress is the same.

*Answer.*

$$\frac{\dot{\epsilon}(1293°\text{K})}{\dot{\epsilon}(862°\text{K})} = \frac{B\sigma^n}{B\sigma^n} \frac{e^{-Q_c/R1293}}{e^{-Q_c/R862}}.$$

Hence, since $R = 1.987$ cal/mole-°K,

$$\dot{\epsilon}(1293°\text{K}) = 10^{-12} e^{-69,000/1.987[(1/1293)-(1/862)]}.$$

Remembering that $e^{-x} = 10^{-x/2.3}$ we have

$$\dot{\epsilon} = 10^{-12} \times 10^{5.8} \text{ sec}^{-1}.$$

The creep rate at the higher temperature is faster by nearly 1,000,000 times. If the creep rate at $0.5\,T_m$ is tolerable for only 30 years, at $0.75T_m$ the lifetime is only about 0.3 hr.

### Q. 2.2.3

If the low temperature of a process involving a thermal cycle is fixed, the efficiency of the process can be increased by (a)_____. At high temperatures materials (b)_____ under stress; this is known as creep. Creep usually becomes critical at what fraction of the melting temperature? (c)_____. Steady-state high-temperature creep depends on temperature according to (d)_____. This is a (e)_____ variation with temperature.

\* \* \*

CYCLIC STRESS FATIGUE. When a component is repeatedly stressed in tension and compression according to

$$\sigma = S \cos \omega t, \tag{2.2.3}$$

where $S$ is a stress amplitude, we find that it fractures at a tensile stress considerably below even the tensile yield strength as shown in Figure 2.2.7.

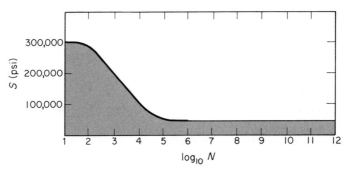

FIGURE 2.2.7. Plot of average fatigue strength versus the logarithm of the number of cycles of reversed loading, $\sigma = S \cos \omega t$, for an ultrastrength steel.

This is called **cyclic stress fatigue.** An example of such stressing would be the repeated bending back and forth of a ruler. (Any devoted paper clip bender could tell you about fatigue.) One very important attribute of fatigue behavior is the wide statistical spread of the results. Whereas the yield strength for each of 1000 specimens may fall within 2% of the mean, the fatigue strength of one of the specimens at a given number of cycles may be 50% below the average.

This is of considerable importance in machinery involving oscillations, e.g., airplane wings, which are buffeted by air turbulence, fuselage stresses due to air compression and decompression, landing struts which are flexed during landing, etc. Pressure cycles in the British Comet airplane caused catastrophic failures of the planes due to fatigue stressing at the corners of the rectangular window frames [see T. Bishop "Fatigue and the Comet Disasters," *Metal Progress* **67**, 77 (May, 1955)]. Fatigue causes the designer a number of problems, namely, that he can build a plane which is 99.99% safe, say, but not one which is 100% safe.

Some materials show a **fatigue limit;** i.e., after a certain large number of cycles, the stress amplitude $S$ stops decreasing and reaches a limiting value as in Figure 2.2.7. For many materials, however, $S$ simply continues to decrease. A rough rule is that for $10^8$ cycles, $\sigma_0/4 < S < \sigma_0/3$.

## Q. 2.2.4

When a material is loaded alternately in tension and compression, the material often fails at a stress considerably below $\sigma_0$. This is called (a)_____. A component on an automobile which is liable to fracture because of fatigue is (b)_____. When the fatigue stress $S$ reaches an asymptotic limit as the number of cycles $N$ increases to an arbitrarily large number, we say that the material exhibits (c)_____. A very rough rule for the magnitude of the fatigue stress for $10^8$ cycles is (d)_____.

\* \* \*

STIFFNESS TO WEIGHT AND STRENGTH TO WEIGHT RATIOS. The rocket or aircraft designer wants a structure which has the desired stiffness and the desired strength with the least weight (and cost). His desire to reduce the weight may be simply one of improving efficiency (so that cargo will replace the dead weight in a commercial aircraft) or one of feasibility (so that a certain space voyage can be made). Consequently, when designing columns, he is interested not only in Young's modulus but also in the **density of the material** $\rho$. He is actually interested in the ratio of stiffness of a component to its weight. In terms of materials properties, he is interested in the ratio $E/\rho$, i.e., tensile stiffness modulus/density; this is often called the **specific stiffness.** Similarly when designing a component on the basis of tensile fracturing, he is interested not only in knowing that the member is strong enough and hence in knowing tensile strength but also in the strength to weight ratio, which in terms of material properties means the ratio $\sigma_{T.S.}/\rho$; this is often called the **specific tensile strength**. In the earth's gravitational field this is simply the maximum length of a rod of material which can hang vertically from a support at one end without yielding. Values for three materials with high modulus/density and tensile strength/density ratios are shown in Table 2.2.1.

Table 2.2.1. DENSITY, STIFFNESS/DENSITY, AND TENSILE STRENGTH/DENSITY FOR SOME MATERIALS

| Material | $\rho$ | | $E/\rho$ | | $\sigma_{T.S.}/\rho$ | |
|---|---|---|---|---|---|---|
| | lb/in.$^3$ | g/cm$^3$ | $10^8$ in. | $10^8$ cm | $10^6$ in. | $10^6$ cm |
| Aluminum (2024-T3) | 0.100 | 2.76 | 1.06 | 2.7 | 1 | 2.5 |
| Steel (maraging) | 0.283 | 7.84 | 1 | 2.5 | 1.5 | 3.8 |
| Boron-reinforced epoxy (boron fibers parallel to rod) | 0.068 | 1.88 | 5 | 13 | 4 | 10 |

It should be obvious that the allowable fatigue stress/density ratio and the allowable creep strength/density ratio are also very important parameters in design.

### Q. 2.2.5

When designing components such as the landing gear for airplanes, it is important to have columns and beams with a given stiffness for the least weight. The property of interest is the (a)_____. It is also necessary that the landing gears have the desired strength for the least weight so we are interested in the (b)_____.

\* \* \*

## ANSWERS TO QUESTIONS

**2.2.1** (a) Work hardening, (b) annealing, (c) decreases, (d) stress concentration factor, (e) brittle, (f) ductile.

**2.2.2** (a) A ductile to brittle, (b) austenitic stainless, (c) Charpy V-notch impact energy.

**2.2.3** (a) Increasing the high temperature, (b) deform continuously, (c) $0.5T_M$, (d) $\dot{\epsilon} \propto \exp(-Q_c/RT)$, (e) very rapid.

**2.2.4** (a) Cyclic stress fatigue; (b) springs, front wheel spindle arrangement, piston rods, etc.; (c) a fatigue limit; (d) $\sigma_0/4 < S < \sigma_0/3$.

**2.2.5** (a) Specific stiffness, (b) specific tensile strength.

## 2.3 THE FLOW OF FLUIDS

We have noted in Chapter 1 that gases and liquids do not support shear stresses in the static case, i.e., the shear modulus $G = 0$. However, liquids and gases are more or less **viscous media** in that they do resist flow under dynamic conditions. For example, in a long transmission pipeline there are periodic pumping stations which build up the pressure and provide the driving force for moving the oil to the next pumping station. Pressure-measuring devices could be installed at a distance $L$ apart (with no pumping station in between); say the measured pressure drop is $\Delta P = P_2 - P_1$. Then we find that the flow rate $Q$ (volume/time) passing a particular cross section is given by

$$Q = \frac{dV}{dt} = \frac{1}{\eta}\left[\frac{\pi R^4}{8L}\right]\Delta P, \qquad (2.3.1)$$

where $R$ is the pipe radius. Thus $dV/dt$ is found to be directly proportional to the pressure drop, to a geometrical factor, and to the inverse of $\eta$, where $\eta$ is a *material property* known as the viscosity coefficient. Equation (2.3.1) holds for incompressible fluids and is known as Poiseuille's equation (1844). Newton envisaged the process of viscous flow as the movement of successive layers of fluid with different velocities $v$ over each other as shown in Figure 2.3.1. Newton proposed that there is a viscous drag proportional to the derivative of the velocity with respect to distance in a direction normal to the layers which are flowing over each other. Thus the shear stress $\tau$ is

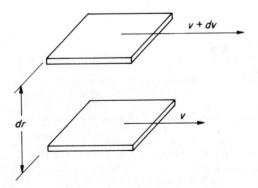

FIGURE 2.3.1. Adjacent layers of fluid flowing over each other.

related to $dv/dr$ according to

$$\tau = -\eta \frac{dv}{dr},  \qquad (2.3.2)$$

where $\eta$ is a constant characteristic of the material called the **viscosity coefficient**. Materials which behave in this fashion are called **Newtonian fluids**. It is an elementary problem in hydrodynamics to derive Poiseuille's equation starting with Newton's expression.

Table 2.3.1 shows the viscosity coefficient of some materials; $\eta$ has units of (force-time/area). A system commonly used by hydrodynamicists is the cgs system and the unit for $\eta$ is then the poise: $1\text{ P} = 1$ dyne-sec/cm². Note the wide range of values for $\eta$.

Table 2.3.1. VISCOSITY COEFFICIENTS AT 20°C

| Material | Millipoise |
|---|---|
| Hydrogen (1 atm) | 0.084 |
| Ethylene (1 atm) | 0.093 |
| Oxygen (1 atm) | 0.167 |
| Nitrogen (1 atm) | 0.192 |
| Hexane | 3.26 |
| Octane | 5.42 |
| Carbon tetrachloride | 9.09 |
| Water | 10.09 |
| Mercury | 15.47 |
| Light machine oil | $\approx 1,000$ |
| Heavy machine oil | $\approx 5,000$ |
| Glycerol (glycerine) | 10,690 |
| Shoe wax | $\approx 5 \times 10^9$ |
| Pitch (for roads) | $\approx 10^{13}$ |
| Soda-silica glass | $\approx 10^{25}$ |

Hydrodynamicists also use the **fluidity coefficient** $\Phi$ where

$$\Phi = \frac{1}{\eta}. \qquad (2.3.3)$$

The viscosity coefficient for ideal gases is independent of pressure and increases slowly with absolute temperature ($\propto T^{3/4}$). However, the viscosity coefficient of liquids varies strongly with the temperature and pressure. The viscosity of liquids decreases rapidly with increasing temperature often varying according to

$$\eta = \eta_0 e^{Q_{vis}/RT}, \qquad (2.3.4)$$

where $\eta_0$ and $Q_{vis}$ are characteristics of each material. In terms of fluidity this would be

$$\Phi = \Phi_0 e^{-Q_{vis}/RT}. \qquad (2.3.5)$$

Compare Equation (2.3.5) with the temperature dependence of steady-state creep, i.e., the viscous flow of solids, given by (2.2.2). For a number of applications it would be desirable to have liquids for which $\eta$ does not vary with $T$, e.g., lubricants for engines.

Viscosity is both foe and friend of the design engineer. Because of viscosity, work is required to send gas or oil through a pipeline. However, it is also true that because of viscosity, hydrodynamic bearings are possible so that a shaft rotates in a sleeve without rubbing against the sleeve. As the shaft rotates, a pressure is built up in the fluid and if the shaft tends not to be centered in the bearing (because of loading normal to its axis), the fluid pressure builds up even higher at the point of smallest clearance. The shaft therefore rides on this pressurized layer of fluid. The pressure buildup often exceeds 10 kbars. The viscosity of air makes possible the air bearing. In another case, pumps supply a continuous flow of oil to the region between the flat oil pads at the bottom of the carriage of the Hale telescope and the flat foundation surface on which they rest; this makes possible the easy rotation of this enormous structure which weighs hundreds of tons using only a small motor since the structure is actually riding on oil. Many adhesives are simply high-viscosity fluids which, because they do not flow readily, hold the members together. Duco cement is an example.

Note that the flow through a single tube given by Poiseuille's equation could be written as

$$\frac{1}{\pi R^2} \frac{dV}{dt} = \frac{1}{\eta} \frac{R^2}{8} \frac{\Delta P}{L}.$$

The term on the left is called the flux, $J$; **flux** is the flow rate across unit area. If we replace $L$ by $x$ and consider $P$ to be a function of $x$, then we have over a length of pipe $dx$,

$$\frac{\Delta P}{L} = \frac{P(x) - P(x + dx)}{dx} = -\frac{dP}{dx} \qquad \text{as } dx \to 0$$

and hence we have

$$J = -\frac{R^2}{8\eta} \frac{dP}{dx}.$$

Since a porous media such as soil consists of a lot of interconnected tubes, we might expect, for such a material, to have a relationship in which flux is directly proportional to the negative of the **pressure gradient,** i.e., to the maximum rate of change of pressure with direction, which in this case is given by

$$J = -k_P \frac{dP}{dx}. \qquad (2.3.6)$$

This is known as **Darcy's equation** and the $k_P$ is known as the **permeability coefficient.** The flow of fluids through porous foams, porous steels, porous glass, etc., obeys Darcy's equation and is important in filtering, pollution control, impregnation of self-lubricating bearings, production of synthetic hard superconductors, etc.

### Q. 2.3.1

The fundamental equation for viscous flow relates the viscous shear stress to the (a)_____
_____. Fluids for which the viscosity coefficient does not vary with velocity are known as (b)_____. The viscosity coefficients of gases vary (c) (slowly or rapidly) with temperature while the viscosity coefficients of liquids vary (d)_____. The fluidity coefficient is the (e)_____ of the viscosity coefficient. Darcy's equation states that the (f)_____ is proportional to the negative of the pressure gradient.

\* \* \*

## ANSWERS TO QUESTIONS

**2.3.1** (a) Velocity gradient normal to the layers on which the shear stress acts, (b) Newtonian fluids, (c) relatively slowly, (d) very rapidly, (e) reciprocal, (f) flux.

## 2.4 INTERNAL FRICTION IN SOLIDS

Suppose that a mass $M$ is hung in a vacuum chamber on a linear coiled steel spring with a stiffness constant $k$. The mass is then pulled downward from its normal rest position ($x = 0$) by a distance $x_0$, held at $x = x_0$ and then released. Its velocity at the instant of release is zero. The potential energy stored in the spring at the instant it is released is $kx_0^2/2$; the kinetic energy of the mass at that instant is zero. At some later time when the displacement of the mass from the rest position is $x$ and the velocity is $\dot{x}$, the potential energy of the spring is $kx^2/2$ and the kinetic energy of the mass is $M\dot{x}^2/2$. Therefore, the total energy of the system (the mass and the spring) is represented by the sum of these. Hence

$$\frac{kx^2}{2} + \frac{M\dot{x}^2}{2} = \frac{kx_0^2}{2}. \tag{2.4.1}$$

Equation (2.4.1) is an expression which states that the total energy of the system remains the same. A system which behaves in this manner is known as a **conservative system.** The system was placed in the vacuum chamber so that there were no external viscous forces tending to retard the motion of the mass (aerodynamicists refer to the retarding force of air on a moving body as drag). The displacement of the mass is given by

$$x = x_0 \cos \omega_0 t, \tag{2.4.2}$$

where $\omega_0$, the **natural frequency,** is given by

$$\omega_0 = \sqrt{\frac{k}{M}}. \tag{2.4.3}$$

Here $x_0$ is called the **amplitude** of the vibration. The motion described by (2.4.2) is called **simple harmonic motion.**

**EXAMPLE 2.4.1**

Show that the position of the mass is indeed given by $x = x_0 \cos \sqrt{k/M}\, t$, starting with Equation (2.4.1).

*Answer.* By differentiating the expression for $x$ with respect to $t$, we find that the velocity is $\dot{x} = dx/dt = -x_0 \sqrt{k/M} \sin \sqrt{k/M}\, t$. Substitution of the expressions for $x$ and $\dot{x}$ into (2.4.1) gives an identity as required.

The force exerted on the ideal spring obeys $F = kx$, so that a plot of $F$ vs. $x$ is a straight line. We might imagine an experimental arrangement in which the force on the spring is measured by one device and the position of the mass measured by another. Suppose the output of both devices is voltage and that these respective voltages are fed into an oscilloscope, with $F$ on the vertical axis and $x$ on the horizontal axis. The picture on the oscilloscope will be a spot representing the instantaneous value of $(F, x)$. The spot will move back and forth along the line shown in Figure 2.4.1. We examined this apparently trivial phenomenon with the purpose of keeping it in mind for later studies, particularly those of damped vibrations.

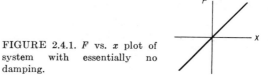

FIGURE 2.4.1. $F$ vs. $x$ plot of system with essentially no damping.

## Q. 2.4.1

The motion of a mass $M$ attached to a linear spring with stiffness constant $k$ and stretched initially by $x_0$ is described by the energy equation (a)_____. By differentiation of the terms in this energy equation with respect to time, the equation $M(d^2x/dt^2) = $ (b)_____ is obtained. This is simply an expression of the fact that the mass times its acceleration equals (c)_____.
An equation for the displacement versus time is (d)_____. This is an example of (e)_____ motion. The frequency of this motion (measured in radians) is given by (f)_____. There are no energy losses in this system; such a system is known as (g)_____ system.

\* \* \*

In an actual experiment the student probably would not notice any deviations from a straight line in the previously described oscilloscope experiment. However, the student would note that after a large number of cycles the amplitude (one half the total length of the line) would become less and less. This effect is much more dominant if the steel spring is replaced by a rubber band. Then we find $x$ vs. $t$ behavior as shown in Figure 2.4.2 and

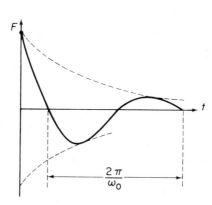

FIGURE 2.4.2. Free damped oscillations.

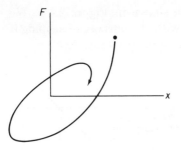

FIGURE 2.4.3. $F(x)$ for free damped oscillations.

$F(x)$ behavior as shown in Figure 2.4.3. The amplitude, rather than remaining fixed at $x_0$, decreases exponentially with time according to

$$\text{amplitude} = x_0 e^{-t/\tau}, \quad (2.4.4)$$

where $\tau$ is a constant, characteristic of the system, known as the **relaxation time**. The system is said to be damped. Moreover, as is clear from Figure 2.4.3, displacement is not zero when the force is zero as was the case for pure harmonic motion shown in Figure 2.4.1; rather the force reaches zero first. Experimentally it would be found that the displacement lags the force by an angle $\delta$; the displacement and the force are said to be out of phase; thus if force equals a damped amplitude times $\cos \omega t$, the displacement equals a damped amplitude times $\cos(\omega t - \delta)$. The angle $\delta$ is called the **loss angle**. This type of motion is called **damped motion**. The phenomenon is called **internal friction**, or internal damping, anelastic behavior or viscoelastic behavior. Examples of the loss angle for different materials are shown in Table 2.4.1. In general $\delta$ is a small fraction of 1 rad. It is often found experimentally that

$$\tau \omega_0 = \frac{2}{\delta} \quad (2.4.5)$$

so that a single damping parameter such as the loss angle suffices to describe the damping completely. The loss angle is a structure-sensitive property so that the values given in Table 2.4.1 must be considered as typical only.

### Q. 2.4.2

An actual spring-mass system is never truly conservative; there is always some (a)_____ present. Two convenient parameters for describing internal friction are (b)_____. In damped oscillatory motion the amplitude decreases (c)_____ with time. Moreover, there is a difference in (d)_____ between the force in the viscoelastic material and the displacement. We might expect the loss angle to be (e)_____ affected by small changes in composition, by mechanical deformation, and by heat treatment because it is a (f)_____ property.

\* \* \*

Table 2.4.1. ANGLE OF LOSS FOR DIFFERENT MATERIALS

| Material | Angle of Loss (rad) |
|---|---|
| Quartz | 0.0002 |
| Steel | 0.0004 |
| Brick | 0.003 |
| Concrete | 0.007 |
| Wood | 0.02 |
| Cork | 0.07 |
| Rubber | (hard) 0.01–0.1 (soft) |

It is easy to show for small damping ($\delta \ll 1$) that the fractional decrease in amplitude per cycle is

$$\frac{\Delta A}{A} = \pi \delta \qquad (2.4.6)$$

and that the fractional decrease in stored elastic energy per cycle

$$\frac{\Delta u}{u} = 2\pi \delta. \qquad (2.4.7)$$

This stored elastic energy is converted into heat. Thus the motion of a damped spring is a **dissipative process** while the motion of the ideal spring is a conservative process. Damped mechanical behavior often can be represented by a drag force proportional to the velocity [this is analogous to Poiseuille's equation for fluid flow where the force needed to push the fluid through the pipe ($\Delta P \pi R^2$) was proportional to the flow velocity ($Q/\pi R^2$)]. Thus the force on the mass in the present case is $-kx - \mu \dot{x}$, where $\mu$ is a damping coefficient. This is called a damped linear system. Analysis of this mathematical problem is given in the previously cited text by Ruoff, *Materials Science*. See the learning sequence there in Problems 2.39–2.42. A simple example of damped oscillatory motion is the bouncing of a rubber ball on a nearly rigid base (which approximates a perfectly rigid base; unless the base is rigid, energy will be dissipated in the base as well as in the rubber ball).

## Q. 2.4.3

The fractional amplitude loss per cycle, the fractional energy loss per cycle, and the reciprocal of the relaxation time are all directly proportional to the (a)_____. The internal friction can often be ascribed to a drag which is directly proportional to the (b)_____. In making bells, a material with a (c)_____ loss angle would be desirable. In making automotive tires, a material with a (d)_____ loss angle would be desirable.

\* \* \*

In many practical applications a periodic force such as $F = F_0 \cos \omega t$ is applied to the mass in a damped system. The amplitude of the displacement then varies as shown in Figure 2.4.4. The peak in the curve occurs at

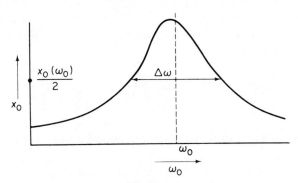

FIGURE 2.4.4. Amplitude-frequency curve for forced vibration showing the resonance peak.

a frequency called the **resonance frequency,** $\omega_r$. If the damping is small ($\delta \ll 1$), the resonance frequency is very close to the natural frequency. The ratio of the amplitude of the displacement at resonance when a force $F_0 \cos \omega_r t$ is applied to the displacement when a static force $F_0$ is applied is called the **resonance amplification factor.** For the case of a linear system with small damping, the resonant amplification factor is found to be

$$\frac{x_0(\omega_r)}{x_0(0)} = \frac{1}{\delta}. \tag{2.4.8}$$

## EXAMPLE 2.4.2

A bar of steel of 1-in.² cross-sectional area is stretched in tension as it supports a weight of 2000 lb. If a cyclic force of 200 lb were applied at the resonance frequency (which would be 70 Hz for a bar 100 ft long), what additional stress would be present?

*Answer.* In Table 2.4.1 we noted that the loss angle of steel was 0.0004. Hence, by (2.4.8) the resonance amplification is 2500 so that the additional force is $200 \times 2500 = 500{,}000$ lb and therefore the additional stress is 500,000 psi. Even a very strong steel would be near its breaking point. Steels can be made with a much higher loss angle and one of the problems of the materials scientist is to see how variations in various aspects of structure affect the loss angle.

The resonance amplification factor (and hence the loss angle) is a very important quantity in design, not only in design of dynamic equipment, such as lathes, airplane wings, engines, tires, etc., but for what might appear to be static structures, such as bridges. The student has probably already read of marching soldiers breaking cadence as they cross a bridge, of the story of a tiny dog causing near failure of a foot bridge as it trotted across, and of the infamous bridge "Galloping Gertie" (see "Bridges" in *Encyclopaedia Britannica* and note the discussion of the Tacoma Narrows bridge).

The **peak width** (width of the peak at half the resonance amplitude) for a linear system is found to be

$$\Delta\omega = \omega_0 \sqrt{3}\, \delta. \qquad (2.4.9)$$

The **Q-factor** or quality factor of the system is given by $Q = \omega_r/(\omega_1 - \omega_2)$, where $\omega_1$ and $\omega_2$ are the frequencies where the power dissipation in the system has fallen to one half its resonance value. Power is given by $F\dot{x}$; it is the rate at which work is done. Power dissipation is the product of the dissipative force and the velocity. For a linear system $Q$ is related to the loss angle $\delta$ by

$$Q = \frac{1}{\delta}. \qquad (2.4.10)$$

$Q$ is a fundamental quantity which measures the energy dissipation of an oscillatory system and has broad applications in various areas of science as shown in Figure 2.4.5.

### Q. 2.4.4

If a mass rests on a viscoelastic foundation and is driven by a force $F_0 \cos \omega_r t$, its maximum displacement will be (a)_____ times as large as the displacement as when a static force $F_0$ is applied. This is called (b)_____. The loss angle of the best synthetically grown quartz crystals is $10^{-6}$; the $Q$ is (c)_____.

\* \* \*

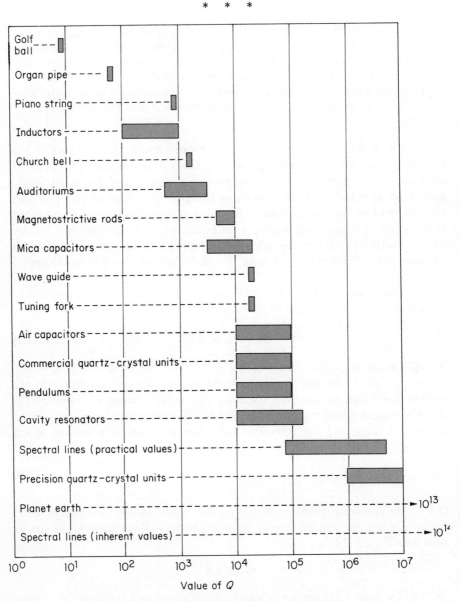

FIGURE 2.4.5. $Q$'s for various phenomena and devices. [From E. I. Green, *American Scientist* **43**, 584 (1955).]

# ANSWERS TO QUESTIONS

**2.4.1** (a) $M\dot{x}^2/2 + kx^2/2 = kx_0^2/2$, (b) $-kx$, (c) the force exerted by the spring on the mass, (d) $x = x_0 \cos \omega_0 t$, (e) simple harmonic motion, (f) $\sqrt{k/M}$, (g) a conservative.

**2.4.2** (a) Damping, (b) loss angle and relaxation time, (c) exponentially, (d) phase angle, (e) strongly, (f) structure-sensitive.

**2.4.3** (a) Loss angle, (b) velocity, (c) low, (d) high.

**2.4.4** (a) $1/\delta$, (b) resonant amplification, (c) 1,000,000.

## 2.5 IDEAL PLASTIC BEHAVIOR

In certain materials, such as putty, certain moulding clays, etc., the stress-strain curve appears as in Figure 2.5.1. These are called **ideal elastoplastic materials.** Note that plastic flow occurs with no strain hardening. It is also assumed to occur independent of the strain rate. Moreover, yielding in compression occurs at the same magnitude of the stress as in tension.

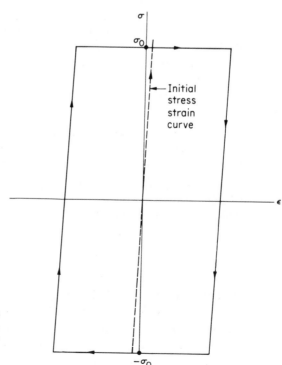

FIGURE 2.5.1. Stress-strain loop of ideal elastoplastic material. The area represents the energy dissipated as heat during one cycle.

## Mechanical Properties

The energy per unit volume dissipated as heat when this material is plastically deformed through one whole cycle is just the area within the stress-strain loop (shown shaded), i.e.,

$$\Delta u = \int_{\substack{\text{closed} \\ \text{loop}}} \sigma \, d\epsilon = \oint \sigma \, d\epsilon. \tag{2.5.1}$$

Although crystalline materials behave approximately in this way, it is usually true that after having yielded in, for instance, tension, these materials exhibit a somewhat smaller initial yield stress in compression. This effect is known as the **Bauschinger effect.** For purposes of mathematical convenience, real materials are often assumed to behave as ideal elastoplastic materials.

# REFERENCES

### General Reference

Dieter, G. E., Jr., *Mechanical Metallurgy*, McGraw-Hill Book Company, Inc., New York (1961). This book has an extensive discussion of stress and strain; plasticity; the tension, torsion, and hardness tests; fracture; fatigue; creep; and stress rupture as well as a discussion of the plastic forming of materials.

Jaeger, J. C., *Elasticity, Fracture and Flow*, John Wiley & Sons, Inc., New York (1956). This little monograph describes stress in simple but thorough fashion, pp. 1–20. Strain is similarly discussed, pp. 20–49.

Timoshenko, S., and Young, D. H., *Elements of Strength of Materials*, Van Nostrand Reinhold Company, Princeton, N.J., (1968). This book analyzes the stresses in and the deflections of simple mechanical components such as the tension rod, torsion bar, beam, column, pressure vessel, coil spring, etc.

Wulff, J., et al., *Structure and Properties of Materials*, Vol. III, *Mechanical Behavior*, John Wiley & Sons, Inc., New York (1965). Chapter 1 gives a summary of the mechanical tests which are performed to study the mechanical properties of solids.

### Handbooks

"Materials Selector Issue" of *Materials Engineering*. This annual issue contains extensive tables of properties, in particular, mechanical properties.

*Handbook of Chemistry and Physics*, ed. by R. C. Weast, Chemical Rubber Co., Cleveland (1969). Extensive lists of material properties, particularly those which are *not* structure-sensitive.

*Handbook of Tables for Applied Engineering Science*, ed. by R. E. Bolz and G. L. Tuve, Chemical Rubber Co., Cleveland, (1970).

Parker, E. R., *Materials Data Book for Engineers and Scientists*, McGraw-Hill Book Company, Inc., New York (1967).

**Special Topics**

Finnie, I., and Heller, W. R., *Creep of Engineering Materials*, McGraw-Hill Book Company, Inc., New York (1959).

Parker, E. R., *Brittle Fracture of Engineering Structures*, John Wiley & Sons, Inc., New York (1957).

Sines, G., and Waisman, J. L., *Metal Fatigue*, McGraw-Hill Book Company, Inc., New York (1959).

Tetelman, A. S., and McEvily, A. J., Jr., *Fracture of Structural Materials*, John Wiley & Sons, Inc., New York (1967).

# PROBLEMS

**2.1** Make a list of questions (at least 12) for mechanical properties of *why* materials behave as they do, e.g., why is stress directly proportional to strain for small displacements (e.g., why is stress not proportional to the square root of strain). This should be an extensive list and should include topics discussed in Section 1.3 as well as those discussed in this chapter.

> There is a reason for man's search for understanding which extends beyond his immediate needs for food and shelter. This reason is his inherent urge to know. We share the great adventure with one another. Let us together pursue science for the practical benefits it may yield. But let us pursue it also because we would rise above the sordid daily struggle with nature and with ourselves, look out upon the universe in which we find ourselves, and down into its intricacies. Let us pursue science because we respond to the urge which lies deep in all of us to press on toward greater understanding.—Vannevar Bush, *Today's Research and Tomorrow's World*. An address before guests of the Board of Directors of Stanford Research Institute, Los Angeles (1954).

**2.2** Why is flexural rigidity (or the stiffness constant of a beam) important in the design of a cutting tool for a lathe?

**2.3** Give some reasons both torsional and flexural rigidity are important in the design of an airplane wing.

**2.4** (a) If there is so much variation of $\sigma_0$ among different irons and steels, why is not the strongest steel always used in applications?
(b) Suggest some reasons there might be such variation in $\sigma_0$.

**2.5** Show that a material which is plastically isotropic has Poisson's ratio $\nu = \frac{1}{2}$ for plastic flow.

**2.6** What is work hardening? What practical value does it have?

**2.7** (a) Explain why necking occurs in a tensile specimen.
(b) Necking is an example of what phenomenon?

**2.8** Give two different measures of ductility in a tension test.

**2.9** When copper rod is drawn into wire, it repeatedly is passed through a die where it is cold drawn and a heating chamber where it is annealed. Explain why.

**2.10** A glass shaft is twisted and fractures when the maximum shear stress reaches 10,000 psi. A geometrically similar copper shaft loaded in the same manner yields when the maximum shear stress reaches 10,000 psi. Suppose the latter does not work harden but plastically deforms until the shear strain at the surface is $\gamma = 6$. How much tougher is the copper than the glass?

**2.11** (a) What is the stress concentration factor?
(b) How large is it at a rivet hole on an airplane skin?
(c) How large is it in minute cracks on the surface of glass?
(d) Why are cracks more dangerous in brittle materials than in ductile materials?

**2.12** Name four applications in which cyclic fatigue is likely to be very important.

**2.13** Name four applications in which creep is likely to be one of the main design-limiting properties.

**2.14** Why are the results of an impact energy test useful to the design engineer who is building bridges, ships, etc.?

**2.15** How long can a uniform steel rod be if it is supported at the top end only and is not to yield? Assume that $\sigma_0 = 300{,}000$ psi and that the specific gravity is 7.80.

**2.16** Viscosity data for $CCl_4$ is shown below:

| T°C | 0 | 20 | 40 | 60 | 80 |
|---|---|---|---|---|---|
| $\eta$ (mP) | 13.47 | 9.09 | 7.38 | 5.84 | 4.68 |

Show that these data obey Equation (2.3.4) where $T$ in (2.3.4) is the absolute temperature and evaluate $Q_{vis}$.

**2.17** The following table from Table 20 of M. D. Hersey, *Theory of Lubrication*, Chapman & Hall, Ltd., London (1936), shows the relation between the logarithm of the reduced viscosity coefficients and pressure for an oil.

| Pressure (atm) | ln $\eta/\eta_1$ |
|---|---|
| 1 | 0 |
| 500 | 0.32 |
| 1000 | 0.63 |
| 2000 | 1.24 |
| 3000 | 1.84 |
| 4000 | 2.45 |
| 5000 | 2.94 |
| 6000 | Solid |

What simple relationship exists between viscosity and pressure?

**2.18** Fatigue cracks appear on the portion of a shaft which rotates within a heavy-duty hydrodynamic bearing. Why?

**2.19** Describe an experience in your life where mechanical resonance in a forced vibration either caused failure or concern.

**2.20** In driving an automobile down a concrete thruway at 65 mph, a student notices a discomfiting subsonic vibration at about 3 Hz. Discuss the possible origin of this.

**2.21** Describe the meaning of the term loss angle. How is it related to relaxation time? Fractional energy loss per cycle? Q-factor? Fractional change in amplitude?

**2.22** Suppose all the components of an airplane could be readily and cheaply welded together. Why might you still want to use rivets or other fasteners?

**2.23** Suppose we ignored the mechanical difficulty associated with using a rigid rod of material to lift an elevator. Give two good reasons (from the materials properties viewpoint) a cable would be preferred to the rod.

# MORE INVOLVED PROBLEMS

**2.24** The **bulk modulus**, $K$, is defined by

$$dP = -K \frac{dV}{V},$$

where $dV/V$ is the fractional volume change due to an incremental pressure change.
(a) Show that the expression for the bulk modulus of an ideal gas is $K = P$.
(b) The bulk modulus of elastically isotropic solids is related to $E$ and $G$ by

$$K = \frac{E}{9 - 3E/G}.$$

Find $K$ for aluminum.

**2.25** The deepest point in the ocean is about 35,000 ft.
(a) Calculate the hydrostatic pressure there in pounds per square inch and kilobars.
(b) Would such a pressure change the volume of an aluminum bar relative to its volume at atmospheric pressure? An ideal gas? *Hint:* See Problem 2.24.

**2.26** It can be shown (see Jaeger in the References) that a longitudinal sound wave passes through an elastically isotropic material with a velocity $v_l$, where

$$K + \tfrac{4}{3}G = \rho v_l^2.$$

$K$ is defined in Prob. 2.24. A shear sound wave has a velocity $v_s$ given by

$$G = \rho v_s^2.$$

Describe briefly how the elastic constants of solids can be measured using sound waves.

**2.27** Describe seismology and its applicability in prospecting for oil and in studying the structure of the earth. See Problem 2.26. Also see A. F. Fox, *The World of Oil*, Pergamon Press, Inc., Elmsford, N.Y. (1964).

**2.28** Calculate the velocity of shear waves in polycrystalline quartz at 1 atm pressure if quartz has a density of 2.648 g/cm³ and a shear modulus of 377 kbars (1 kbar = $10^9$ dynes/cm²). See Problem 2.26.

**2.29** Show how each of the mechanical properties discussed in the present chapter is involved in the design of
(a) A motorcycle.
(b) An automobile.
(c) A motor boat.
(d) An airplane.

**2.30** Prove that the two stress states shown are equivalent if $\sigma = \tau$.

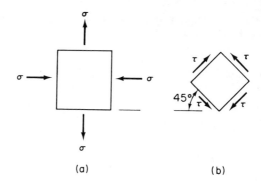

PROBLEM FIGURE 2.30.   (a)   (b)

**2.31** The stress at the surface of a shaft is as given in Problem 2.30. Suppose that a solid circular shaft is twisted carefully (no bending) to failure. Sketch the form of the break if
(a) The material is ductile, e.g., copper.
(b) The material is brittle, e.g., chalk.

**2.32** The columns in an ordinary airplane wing are made of aluminum.
(a) If their buckling load is given by

$$F_{max} = \frac{\pi^2 EI}{4L^2},$$

should they be hollow or solid from an equal weight basis to maximize $F_{max}$ if

$$I = \frac{\pi(R^4 - R_i^4)}{4}?$$

Here, $R$ is the outer radius and $R_i$ is the inner radius of a hollow pipe.
(b) In addition to configurational changes (hollow versus solid), what other possibilities does the design engineer have for increasing the buckling load without increasing the weight of a column?

**2.33** Assume the wooden beams under the floor of a house are uniformly loaded, with a weight $w$ per unit length. Assume also that their ends simply rest on the foundation. The maximum deflection is given by

$$\delta = \frac{5wL^4}{384EI},$$

where $I$ is the cross-sectional moment of inertia of the beam, $E$ is the stiffness modulus, and $L$ is the length of the beam. Assume $E = 1.2 \times 10^6$ psi. For a rectangular beam whose width is $B$ and whose height is $H$ (in the direction of loading),

$$I = \frac{BH^3}{12}.$$

How many times stiffer is a $2 \times 12$ in. wooden beam with the 12-in. dimension vertical than one with the 2-in. dimension vertical?

**2.34** Many adhesives are simply high-viscosity liquids which wet the solid. A butt joint with liquid adhesive is shown:

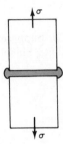

PROBLEM FIGURE 2.34.

It can be shown, starting with Newton's law of viscosity, that the time versus adhesive thickness, $h$, for a given tensile stress $\sigma$ is

$$t = \frac{3HR^2}{4\sigma}\left(\frac{1}{h_0^2} - \frac{1}{h^2}\right),$$

where $h_0$ is the initial thickness and $R$ is the radius of the rod. The assumption is made that there is sufficient excess adhesive to flow into the joint to prevent necking.

(a) Calculate the lifetime of a butt joint between two rods whose radii are 2 cm if $h_0 = 10^{-5}$ cm and glycerine is used at 20°C, and if $\sigma = 1000$ bars.

(b) Same as (a) but with pitch.

(c) Using pitch, how large a stress could be supported for 5 min? Is this reasonable?

## SOPHISTICATED PROBLEMS

**2.35** Use the results of Problem 2.30 to show that $G = E/2(1 + \nu)$. *Hint:* Since the stresses are equivalent, they must cause the same strain.

**2.36** Derive the equation for $K$ given in Problem 2.24(b).

**2.37** In visualizing the lamellar flow through a pipe one has a tube of liquid of radius $r$ being pushed past the adjacent layer by the force $\pi r^2 P$.
  (a) According to Newton's viscosity equation the resistive force (the friction force) is given by what?
  (b) Use the balance between these two forces to show that $v = v_{max}[1 - (r^2/R^2)]$ and find $v_{max}$.
  (c) Use the results of (b) to derive Poiseuille's equation.

**2.38** Derive the equation given in Problem 2.34. Assume that $h \ll R$.

# Prologue

The most important electrical property is the *electrical resistivity*. Based on resistivity, materials are classified as *metals, semiconductors, insulators*, or *superconductors*. The concepts of *charge carriers* and *Hall coefficient* are introduced. Several important properties of insulators or dielectrics in addition to resistivity are described: *permittivity, dielectric constant, dielectric strength*, and *dielectric loss angle*. In addition some dielectric materials which exhibit special properties, namely, *ferroelectricity* and *piezoelectricity*, are studied. Ferroelectric materials always lose their ferroelectric property at high temperatures.

Magnetic materials are classified according to whether they attract or expel magnetic flux. Ordinary *diamagnetic materials* expel magnetic flux slightly, *paramagnetic materials* attract flux somewhat, while *ferromagnetic materials* attract flux strongly. From an engineering viewpoint the last are the most important materials. The behavior of magnetic materials is described in terms of their *magnetic induction* versus *magnetic field* curves, i.e., their *B-H curves*. In general the *B-H* curve for a material will be a *hysteresis loop*. Important properties of magnetic materials are the *permeability coefficient*, the *saturation induction*, the *remanent induction*, and the *coercive force*. Ferromagnetic materials are classified as *soft* and *hard*. Magnetic materials become strained in the presence of magnetic fields, an effect called *magnetostriction*. Increased temperature causes ferromagnetism to disappear in a characteristic way.

The reader should ask why these properties exist and he should develop a feeling for the order of magnitude of these properties for different materials. Moreover, it is helpful to think of how these properties can be effectively utilized in constructing apparatus, machinery, devices, etc.

# 3

# ELECTRIC AND MAGNETIC PROPERTIES

## 3.1 INTRODUCTION

In studying electric and magnetic phenomena, it is necessary to use several vector quantities. Two of these, electric intensity **E** and magnetic induction **B**, are defined now. The boldface font, as used for **E** and **B**, is used to designate a vector quantity. The magnitude of their vector quantities is designated by the ordinary font as with $E$ and $B$, respectively. The **electric intensity E** is defined by the force it exerts on a charge $q$, namely,

$$\mathbf{F} = q\mathbf{E}. \tag{3.1.1}$$

(To measure **E** at a point we need only measure the force on a tiny static charge $q$ at that point.) If the vector **E** is defined at every point in a region of space, a **vector field,** called the electric (intensity) field, exists.

The most widely used system of units in discussing electric and magnetic behavior is the mks system. The unit of force is the newton (nt). This is the force which causes a mass of 1 kg to be accelerated 1 m/sec.² The unit of charge is the coulomb (C). Thus **E** has units of newtons per coulomb; it also has units of volts (V) per meter. Hence force also has the units of volt-coulombs per meter.

What is the origin of **E**? An electrical field **E** arises because of the presence of a static electric charge. Thus a charge $q_2$ in a vacuum generates an electric intensity

$$\mathbf{E}_2 = \frac{q_2}{4\pi\epsilon_0 r^2}\left(\frac{\mathbf{r}}{r}\right), \tag{3.1.2}$$

where $r$ is the distance from the point at which the charge is located and $\epsilon_0$ (called the permittivity constant of vacuum) has a value of $8.854 \times 10^{-12}$ C/V-m (the coulomb per volt is also called a farad). The vector $\mathbf{r}/r$ is a unit vector directed radially outward from the charge. If $q_2$ is a positive charge, the electric intensity **E** is directed away from $q_2$. If there is a charge

$q_1$ located a distance $r$ from the charge $q_2$, then the force exerted by $q_2$ on $q_1$ is

$$\mathbf{F}_1 = q_1 \mathbf{E}_2 = \frac{q_1 q_2}{4\pi\epsilon_0 r^2} \mathbf{n}_r, \qquad (3.1.3)$$

where $\mathbf{n}_r$ is a unit vector pointing from $q_2$ to $q_1$. This is **Coulomb's law**. It is a fundamental law of physics which is proved by experiment. If both $q_1$ and $q_2$ are positive, the force is repulsive.

The **magnetic induction B** (often called just the magnetic field in secondary school books) is defined by the force it exerts on a charge $q$ moving with a velocity **v**. This force is

$$\mathbf{F} = q(\mathbf{v} \times \mathbf{B}). \qquad (3.1.4)$$

(For students not familiar with vector products this means that the magnitude of the force equals $q$ times the magnitude of **v** times the magnitude of **B** times the sine of the angle between **v** and **B**; and the vector **F** is perpendicular to the plane of the two vectors **v** and **B**. The direction of the force is, for the case of a positive charge, in the direction in which a right-hand screw lying along **F** would advance if it were rotated in the sense of **v** rotating into **B**.)

What is the origin of **B**? A moving charge generates a magnetic induction **B** (in addition to the electric intensity **E**). Thus there is a magnetic induction field associated with a current carrying wire. The units of $B$, as is clear from the definition, are newton-seconds per coulomb-meter, which equals volt-seconds per square meter. It is common to define the unit called the weber as 1 Wb = 1 V-sec so that the unit of $B$ is webers per square meter.

If both $E$ and $B$ are present, then the force on a charge $q$ is

$$\mathbf{F} = q(\mathbf{E} + \mathbf{v} \times \mathbf{B}). \qquad (3.1.5)$$

This is known as the **Lorentz force equation**.

## Q. 3.1.1

A static positive point charge $q$ gives rise to (a)_____ field whose magnitude is (b)_____. The force which an electric intensity **E** exerts on a charge $q$ is (c)_____. Coulomb's law states that the force between the two charges $q_1$ and $q_2$ is (d)_____. The force which a magnetic induction **B** exerts on a charge $q$ moving with a velocity **v** is (e)_____.

\* \* \*

We also shall on occasion use the term *electric potential*. It is helpful to first review potential energy. As an example, when you studied the linear spring in Section 1.3 you noted that the potential energy stored in the spring was

$$U = \frac{kx^2}{2}.$$

The force which the spring exerts on the mass in a conservative (nondissipative) spring-mass system is

$$F = -\frac{dU}{dx} = -kx.$$

In a more general case when the potential energy may vary with $x$, $y$, or $z$ we say that the force in a conservative system is the negative gradient of the potential energy, which we write as

$$\mathbf{F} = -\text{grad } U. \tag{3.1.6}$$

The gradient of $U$ is a vector whose magnitude equals the maximum rate of change of $U$ with distance and whose direction is the direction of the maximum rate of change of $U$.

Considering the fact that electrical intensity is force per unit charge it appears that it may be useful to define an electric potential $V$ which is potential energy per unit charge such that

$$\mathbf{E} = -\text{grad } V. \tag{3.1.7}$$

(This definition is possible only when magnetic fields, if present, do not vary with time.) If $V$ varies with $x$ only, then

$$E = -\frac{dV}{dx}.$$

If $V$ varies linearly with $x$ and if $V_{ab}$ is the net drop in $V$, then across a distance $l$,

$$E = \frac{V_{ab}}{l}. \tag{3.1.8}$$

The mks unit for electrical potential is the volt.

## EXAMPLE 3.1.1

Prove that the electric potential associated with a charge $q$ in a vacuum is

$$V = \frac{q}{4\pi\epsilon_0 r}. \tag{3.1.9}$$

*Answer.* Since $V$ varies only in the radial direction the gradient of $V$ is $dV/dr$. Since differentiation of (3.1.9) gives $dV/dr = -q/4\pi\epsilon_0 r^2$ and $E = -dV/dr$, we have

$$E = \frac{q}{4\pi\epsilon_0 r^2}$$

as required by (3.1.2).

Because the electric potential due to a point charge $q_2$ in a vacuum is $V_2 = q_2/4\pi\epsilon_0 r$, the potential energy of the charges $q_1$ and $q_2$ where $q_1$ is located a distance $r$ from the charge $q_2$ is $q_1 V_2$ or

$$U = \frac{q_1 q_2}{4\pi\epsilon_0 r}. \tag{3.1.10}$$

We could simply say that this is the potential energy of a pair of charges separated by $r$.

In general the change in potential energy of a system when a charge $dQ$ is taken from zero potential to a potential $V$ is

$$dU = V\, dQ$$

so the potential energy is

$$U = \int V\, dQ. \tag{3.1.11}$$

The mks unit for energy is the joule (J).

## Q. 3.1.2

The potential energy $U$ of a conservative system is related to the force **F** by (a)_____. The electrical potential $V$ is related to the electrical intensity **E** by (b)_____ if the magnetic induction does not vary with time. If $V$ varies linearly along one straight path only, the relation among the potential drop $V_{ab}$, the length $l$, and the magnitude of the electric intensity $E$ is (c)_____. The electric potential due to a point charge $q$ is (d)_____. The potential energy of a pair of charges separated by a distance $r$ is (e)_____.

\* \* \*

Although the basic units in the mks system are kilogram, meter, second, and coulomb, it is convenient to define certain other units because of the quantities which they measure, which are shown in parentheses in Table 3.1.1.

Table 3.1.1. DEFINITIONS OF UNITS

| | |
|---|---|
| Newton ≡ kilogram-meter per square second | (force) |
| Joule ≡ newton-meter | (energy) |
| Volt ≡ joule per coulomb | (electric potential) |
| Ampere ≡ coulomb per second | (current) |
| Ohm ≡ volt per ampere | (resistance) |
| Farad ≡ coulomb per volt | (capacitance) |
| Henry ≡ volt-second per ampere | (inductance) |
| Weber = joule per ampere | (magnetic flux) |

## ANSWERS TO QUESTIONS

**3.1.1** (a) An electric, (b) $q/4\pi\epsilon_0 r^2$, (c) $q\mathbf{E}$, (d) $q_1 q_2/4\pi\epsilon_0 r^2$ and is repulsive if positive, (e) $q(\mathbf{v} \times \mathbf{B})$.

**3.1.2** (a) $\mathbf{F} = -\operatorname{grad} U$, (b) $\mathbf{E} = -\operatorname{grad} V$, (c) $E = V_{ab}/l$, (d) $q/4\pi\epsilon_0 r$, (e) $q_1 q_2/4\pi\epsilon_0 r$.

## 3.2 OHM'S LAW AND THE HALL COEFFICIENT

OHM'S LAW. It is found by experiment that an applied electric potential difference $V_{ab}$ (a voltage drop) across the length of a wire causes a steady current $i$ in the wire such that

$$V_{ab} = Ri, \tag{3.2.1}$$

where the proportionality constant $R$, called the resistance, is a property of a specific object (and is not a material property). The quantity $R$ depends on the dimensions of the object. [Equation (3.2.1) is sometimes called Ohm's law in elementary books, although it is not the fundamental form of Ohm's law. The situation is analogous to calling the force-displacement relation for a rod, $F = kx$, Hooke's law, rather than calling the more fundamental relation, $\sigma = E\epsilon$, Hooke's law.]

If we now study how the resistance varies with the length and cross-sectional area of the wires, we find that $R$ is directly proportional to the length and inversely proportional to the cross-sectional area. We write this result in the form

$$R = \frac{\rho l}{A}, \qquad (3.2.2)$$

where $\rho$ is a proportionality constant which does not depend on the size of the object but does depend on the material used. Therefore $\rho$ must be a material property; it is called the **electrical resistivity.** We can now rewrite

$$V_{ab} = Ri = \frac{\rho l}{A} i$$

in the form

$$\frac{V_{ab}}{l} = \rho \frac{i}{A}. \qquad (3.2.3)$$

The **electric current density J** is a vector whose magnitude is the current (charge per unit time) flowing across unit area and whose direction is the normal to the area in the direction of current flow. Hence

$$J = \frac{i}{A}. \qquad (3.2.4)$$

**J** is also called the **charge flux.** $J$ has units of coulombs per square meter-second or amperes per square meter. We note that (3.2.3) can be combined with (3.2.4) and (3.1.8) to give

$$\mathbf{E} = \rho \mathbf{J}. \qquad (3.2.5)$$

This is the fundamental form of **Ohm's law.** $\rho$ has units of volt-meters per ampere or ohm-meters.

It is also common to write Ohm's law in the form

$$\mathbf{J} = \sigma \mathbf{E}, \qquad (3.2.6)$$

where the proportionality constant $\sigma$, called the **electrical conductivity,** is the reciprocal of $\rho$.

### Q. 3.2.1

The drop in electric potential when current passes through a resistor is (a)_____. The fundamental form of Ohm's law is (b)_____. The resistance of a wire of length $l$ and cross-sectional area $A$ is (c)_____.

\* \* \*

Electrical resistance (and hence resistivity) is important because it determines the rate at which energy is dissipated as heat (the power dissipation). The power is given by $P = V_{ab}i$ and the power dissipated as heat is therefore

$$P = Ri^2. \qquad (3.2.7)$$

The power dissipation per unit volume is simply

$$\frac{P}{lA} = \frac{V_{ab}i}{lA} = EJ = \rho J^2. \qquad (3.2.8)$$

Table 3.2.1 shows some values of resistivity for common electrical materials. The first column shows values for **metals** which are very good electrical conductors and the second column shows two different groups:

Table 3.2.1. ELECTRICAL RESISTIVITIES IN mks UNITS OF METALS AND NONMETALS AT 20°C*

| Metals | Resistivity ($10^{-8}$ Ω-m) | Nonmetals | Resistivity (Ω-m)† |
|---|---|---|---|
| Silver | 1.6 | Semiconductors | |
| Copper | 1.67 | Silicon | 0.00085 |
| Gold | 2.3 | Germanium | 0.009 |
| Aluminum | 2.69 | Insulators | |
| Magnesium | 4.4 | Diamond | $10^{10}$–$10^{11}$ |
| Sodium | 4.61 | Quartz | $1.2 \times 10^{12}$ |
| Tungsten | 5.5 | Ebonite | $2 \times 10^{13}$ |
| Zinc | 5.92 | Sulfur | $4 \times 10^{13}$ |
| Cobalt | 6.24 | Mica | $9 \times 10^{13}$ |
| Nickel | 6.84 | Selenium | $2 \times 10^{14}$ |
| Cadmium | 7.4 | Paraffin wax | $3 \times 10^{16}$ |
| Iron | 9.71 | | |
| Tin | 12.8 | | |
| Lead | 20.6 | | |
| Uranium | 29 | | |
| Zirconium | 41 | | |
| Manganin | 44 | | |
| Titanium | 55 | | |
| Lanthanum | 59 | | |
| 96% iron—4% Si | 62 | | |
| Cerium | 78 | | |
| Nichrome | 100 | | |

* Values selected from the *American Institute of Physics Handbook*, pp. 4-90, 9-38 (1963).
† Note the different units in the two columns.

**semiconductors,** which have intermediate values of $\rho$, and **insulators,** which have very high values of $\rho$ or are very poor electrical conductors (note the different scales on the two columns). Observe the enormous range of properties; the data of Table 3.2.1 suggest that

$$\frac{\rho_{\text{insulator}}}{\rho_{\text{metals}}} \approx 10^{24}.$$

In addition, there are materials called **superconductors** which carry dc electrical current without any detectable loss. Not all materials have superconducting properties, but those which do are superconductors only at very low temperatures (**cryogenic temperatures**), below about 20°K. Superconductors are discussed further in Section 3.10.

### Q. 3.2.2

Electrical power is given by (a)_____. The power dissipation in a resistor is therefore (b)_____. The power dissipation per unit volume is (c)_____. Based on their electrical resistivity, there are (d)_____ classes of materials. The room-temperature resistivity of a metal is of the order of (e)_____ and of a semiconductor (f)_____. Insulators may have resistivities of (g)_____ times that of good metals at room temperature. Some materials called (h)_____ carry dc current at (i)_____ temperatures without any power losses at all.

\* \* \*

HALL EFFECT. There is another type of behavior known as the Hall effect which is important in the study of various conductors (metals, semiconductors, etc.) because it enables us to determine the charge and concentration of the charge carrier. The Hall effect is also the basis of a number of electronic devices. We consider a bar of material as shown in Figure 3.2.1. A current flows through the material in the $x$ direction; this current equals the charge flux times the area; i.e., $i_x = J_x tl$. The flux $J_x$ is given by the

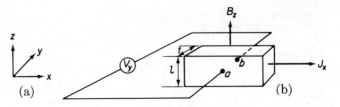

FIGURE 3.2.1. Hall effect experiment.

product of the number of charge carriers per unit volume, $N$, their charge $q$, and their net velocity $v_x$:

$$J_x = Nqv_x.$$

In the experiment, the current flow in the $x$ direction takes place in the presence of a magnetic induction in the $z$ direction, $B_z$. In the presence of the magnetic induction $B_z$, a charge moving with velocity $v_x$ is subject to a force in the $y$ direction of

$$F_y = -qv_xB_z;$$

therefore positive charges are forced toward face $a$ and negative charges to face $b$. This charge buildup produces an electric potential difference $V_{ab}$.

This can be measured by the voltmeter as shown in Figure 3.2.1 across faces $a$ and $b$. This phenomenon is the **Hall effect,** discovered by E. H. Hall in 1888. The electric field produced by these charges is

$$E_y = \frac{V_{ab}}{t}.$$

This Hall field $E_y$ produces a force on the charge carries in the conductor given by

$$F_y = qE_y.$$

Since there is no current flow through the voltage-measuring device, i.e., in the $y$ direction, there must be no net force on the charges in the $y$ direction. Hence

$$qE_y - qv_xB_z = 0,$$

from which it follows that

$$v_x = \frac{E_y}{B_z}.$$

The **Hall coefficient** is defined by

$$R_H = \frac{E_y}{J_xB_z}. \qquad (3.2.9)$$

Combining this definition with the expressions for $J_x$ and $v_x$ gives

$$R_H = \frac{1}{Nq}. \qquad (3.2.10)$$

The Hall effect therefore provides a means of measuring the sign of the charge and the concentration of the charge carrier. (The analysis given here applies to the case in which only a single charge carrier is present; the more general case is considered in Chapter 13.)

In the mks system the units for $R_H$ are volt-cubic meters per ampere-weber or cubic meters per coulomb. (This makes use of the following identity: 1 C = 1 A-Wb/V.) In the mks system we have $E_y$ (volts per meter), $J_x$ (amperes per square meter), and $B_z$ (webers per square meter). The Hall coefficient is positive for some metals (calcium, beryllium) and negative for others (copper, sodium). It is negative for an $n$-type semiconductor such as phosphorus-doped silicon and positive for a $p$-type semiconductor such as boron-doped silicon.

### Q. 3.2.3

The definition of charge flux is (a)_____. If the velocity of all the charged particles is $v_x$, then the charge flux is given by (b)_____ if all charges have the same sign. Assuming all the charge carriers are of the same sign, it can be shown that $R_H$ equals (c)_____. In such a case the Hall effect measurement enables us to distinguish between an $n$-type and (d)_____ semiconductor.

\* \* \*

## ANSWERS TO QUESTIONS

**3.2.1** (a) $V_{ab} = Ri$, (b) $\mathbf{E} = \rho \mathbf{J}$ (or $\mathbf{J} = \sigma \mathbf{E}$), (c) $\rho l/A$, where $\rho$ is the electrical resistivity.

**3.2.2** (a) $P = V_{ab}i$, (b) $Ri^2$, (c) $\rho J^2$, (d) four, (e) $10^{-8}$ $\Omega$-m, (f) $10^{-2}$ $\Omega$-m, (g) $10^{24}$, (h) superconductors, (i) cryogenic or very low.

**3.2.3** (a) The charge flow across (normal to) a unit area in unit time, (b) $Nqv_x$, (c) $1/Nq$, (d) a $p$-type.

## 3.3 METALS AND SEMICONDUCTORS

METALS. Good conductors play an important role in electricity generation and transmission equipment. Their primary role in electrical generators is the creation of a magnetic induction due to current flow in a wire. As an example it can be shown that the generation of a magnetic induction of 10 Wb/m² (or 100 kG) in a solenoid of copper wire with an internal diameter of 0.1 m and outer diameter of 0.3 m and a length of 1 m requires a power of 6300 kW. (See the previously cited Ruoff, *Materials Science*, Example 3.2.1.) (The standard furnace for heating a three-bedroom house uses about 10 kW at maximum capacity.) The heat dissipation in the magnet is enormous.

To solve these problems the designer might

1. Cool by passing through large amounts of water. This is a major thermal engineering problem. This is precisely the procedure used at the National Magnet Laboratories. See Figure 3.3.1.

FIGURE 3.3.1. 250-kG (25-Wb/m²) magnet at National Magnet Laboratories. Inner bore is 2 in. in diameter. Magnet contains *3 tons of copper* (hollow tubing), has a power consumption up to 16 MW, and requires 2000 gal of cooling water per minute. (Courtesy of Francis Bitter National Magnet Laboratory, Cambridge, Mass.)

2. Use a superconductor (more about that later).
3. Use a material with a smaller $\rho$. A rapid check of Table 3.2.1 shows that silver is only slightly better than copper. However, the resistivity of metals drops considerably as temperature is decreased as illustrated in the log-log plot in Figure 3.3.2. Potassium has a melting temperature of 333°K; note that the resistivity varies slowly with temperatures above $T_m/5$. This temperature variation is a characteristic of metals.

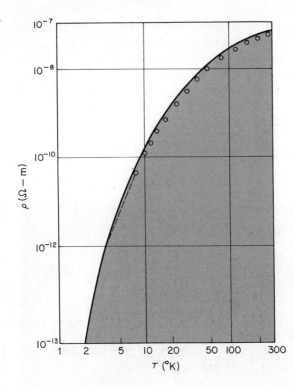

FIGURE 3.3.2. Log-log plot of bulk resistivity versus absolute temperature for 99.99% potassium. Circles and dashed line are experimental. Solid line is computed. [From G. T. Meaden, *Electrical Resistance of Metals*, Plenum Publishing Corporation, New York (1965).]

Very high-purity aluminum has been made for which the resistivity at 4.2°K (the boiling point of liquid He) is only $2 \times 10^{-5}$ times the room-temperature (20°C) value. The **resistivity ratio** is $\rho_{293.2°K}/\rho_{4.2°K}$; for high-purity aluminum this is 50,000. The higher the purity, the higher the resistivity ratio. The room-temperature resistivity is only slightly affected by impurities but the low-temperature resistivity is greatly affected. Thus operation of magnets, large generators, and motors (using high-purity wire) at cryogenic temperatures can greatly reduce the ohmic heating. However, it is costly to maintain a temperature of, say, 4.2°K. To do so requires liquid helium which costs about $5/liter. One watt-hour of energy will boil off 1.3 liters of helium.

## Q. 3.3.1

The resistivity of metals varies (a) (rapidly or slowly) with temperature at 20°C. The room-temperature resistivity is a structure- (b)_____ property. The resistivity at 4.2°K is a structure- (c)_____ property. The resistivity ratio for very pure aluminum is of the order of (d)_____. One measure of the purity of metals is their (e)_____

_____. The resistivity of metals as a rule (f)_____ as the temperature increases.

\* \* \*

SEMICONDUCTORS. Pure metals are very good conductors at very low temperature but their resistivity *increases* as temperature increases. *Pure* semiconductors (such as pure germanium) are insulators at very low temperatures but their resistivity *decreases very rapidly* as temperature increases, although they are still poor conductors at high temperatures relative to metals.

This behavior is shown in Figure 3.3.3 for a gallium-doped germanium crystal. Note that the behavior from 100°C to 900°C is of the form

$$\rho = \rho_0 e^{E_g/2k_B T}. \tag{3.3.1}$$

Here $\rho_0$ and $E_g$ are parameters characteristic of given materials, $k_B$ is **Boltzmann's constant,** and $T$ is the absolute temperature. For gallium-doped germanium $E_g = 0.72$ eV (electron volts) and $\rho_0 = 2.63 \times 10^{-7}$ Ω-m. The quantity $E_g$ is called the band gap energy. It is determined by the electronic properties of the material as discussed in a later chapter. (Note that the plot for semiconductors in Figure 3.3.3 involves $\log_{10} \rho$ vs. $1/T$ while the plot for metals in Figure 3.3.2 involves $\log_{10} \rho$ vs. $\log_{10} T$. The variation with temperature is *very* different in the two cases!)

Note that while the temperature increases from 400°K to 900°K for the semiconductor the resistivity changes by a factor of $10^{-3}$. Because of this very strong change with temperature, semiconductors are useful for making temperature-measuring devices called **thermistors.** The point where deviations from the straight line occur (at low temperatures) depends strongly on the amount of gallium added and small fractions of 1% cause profound changes. Hence the resistivity of semiconductors is a structure-sensitive property. At high temperature this semiconductor is called an **intrinsic** semiconductor because the conduction is intrinsic to the germanium (and not to the impurities present); at low temperatures the semiconductor is said to be **extrinsic.**

### Q. 3.3.2

In general the resistivity of an extrinsic semiconductor is (a)_____ _____ of temperature at low temperature. A semiconductor is intrinsic at 500°K; at lower temperatures it becomes (b)_____. How does the resistivity of an intrinsic semiconductor depend on temperature? (c)_____. The quantity which defines the rate of change of resistivity with temperature in the intrinsic range is called the (d)_____.

* * *

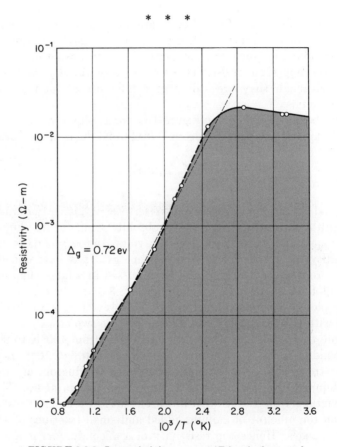

FIGURE 3.3.3. Log resistivity versus $1/T$ for single crystals of gallium-doped germanium in the temperature range 25–900°C. [W. C. Dunlap, *An Introduction to Semiconductors*, John Wiley & Sons, Inc., New York (1957).]

## ANSWERS TO QUESTIONS

**3.3.1** (a) Slowly, (b) insensitive, (c) sensitive, (d) 50,000, (e) resistivity ratio, (f) increases.

**3.3.2** (a) Nearly independent, (b) extrinsic, (c) $\rho \propto \exp(E_g/2k_BT)$, (d) band gap energy.

## 3.4 INSULATORS

In addition to the fact that insulators have a resistivity which differs by a factor of about $10^{24}$ from metals and $10^{10}$ from semiconductors they have several other properties which are important. These include the dielectric constant, the loss angle, and the dielectric breakdown strength which are common to all insulators (or **dielectrics** as they are sometimes called) plus some properties unique to special dielectrics (such as ferroelectricity and piezoelectricity). Consider a simple parallel plate capacitor with plates of area $A$ separated by a distance $d$ as in Figure 3.4.1. When a voltage is

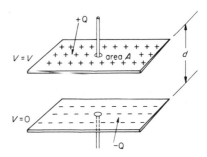

FIGURE 3.4.1. Parallel plate capacitor.

applied across the plates one of the plates attains a net charge of $+Q$ while the other attains a net charge of $-Q$; i.e., there is a *displacement* of charge from one plate to the other. The charge $Q$ is found to be directly proportional to the applied potential $V$,

$$Q = C_0 V. \qquad (3.4.1)$$

Here $C_0$ is a proportionality constant called the capacitance. The capacitance is the charge which can be held separated on the capacitor plates by unit electric potential difference. We use the subscript zero to designate that the space between the plates is a vacuum. Note that Equation (3.4.1) describes linear behavior. The quantity $C_0$ depends on the area $A$ and separation $d$ of the plates according to

$$C_0 = \frac{\epsilon_0 A}{d}$$

when the space is vacuum and the lateral plate dimensions are very large with respect to $d$. Here $\epsilon_0$ is called the **permittivity of vacuum** and, in mks units, has a value of $8.854 \times 10^{-12}$ F/m. The unit of capacitance is the farad. If a homogeneous isotropic electrically insulating material fills the

110    *Electric and Magnetic Properties*

space between the plates, the capacitance will increase (as first shown by Henry Cavendish in about 1770) and will have a value

$$C = \frac{\epsilon A}{d}, \qquad (3.4.2)$$

where $\epsilon > \epsilon_0$.

We next define a quantity $\kappa$, called the **dielectric constant,** by

$$\kappa = \frac{\epsilon}{\epsilon_0}. \qquad (3.4.3)$$

Typical values are given in Table 3.4.1; note the necessity of specifying

**Table 3.4.1.** VALUES OF DIELECTRIC CONSTANTS AT ROOM TEMPERATURE

| Material | Dielectric Constant | Frequency (Hz) |
|---|---|---|
| Air (1 atm, 0°C) | 1.00059 | 0 |
| Mineral oil | 2.2 | 0 |
| Polytetrafluoroethylene (PTFE) | 2.1 | $0$–$10^{10}$ |
| NaCl | 2.25 | $10^{10}$ |
| NaCl | 5.62 | 0 |
| Poly(vinyl chloride) (PVC) | 2.6 | $10^{10}$ |
| Poly(vinyl chloride) (PVC) | 6.5 | $10^2$ |
| Porcelain | 8 | 0 |
| Mica | 65–85 | 0 |
| Water | 80 | 0 |
| Titanates | $\sim 1000$ | 0 |

frequency. Note that the electrical energy stored in a capacitor is given, after (3.1.11), by

$$U = \int V \, dQ = \frac{Q^2}{2C} = \frac{CV^2}{2} = \frac{\kappa C_0 V^2}{2}. \qquad (3.4.4)$$

Thus the energy stored in a capacitor of a given volume at a given voltage is increased by the factor $\kappa$ by the presence of the dielectric material. Hence an extremely small capacitor can be built using materials such as the titanates.

From (3.4.2) and (3.4.4) it follows that the stored energy per unit volume is

$$u = \frac{U}{Ad} = \frac{\epsilon V^2}{2d^2},$$

and because $E = V/d$, we have

$$u = \frac{\epsilon E^2}{2}. \tag{3.4.5}$$

Recall that we are dealing with a linear phenomenon. It is useful to review a previous example of linear behavior, namely, the elastic stretching of a rod. There we found that the potential energy of a stretched rod was given by the expression $Fx/2$, the product of the applied force $F$ times the displacement $x$ divided by 2 [see Equation (1.3.10)]. We also found that the stored energy density in that case was given by the expression $\sigma\epsilon/2$, i.e., the product of the stress times the strain divided by 2 [see Equation (1.3.11)]. It is common to speak of stress as being a generalized force and strain a generalized displacement. Might we do the same thing with electric behavior? Then we might call $E$ a generalized force (it is, after all, the force per unit charge) and let the variable $D$ represent an electric displacement such that

$$D = \epsilon E. \tag{3.4.6}$$

Then it is clear from (3.4.5) and (3.4.6) that

$$u = \frac{ED}{2}. \tag{3.4.7}$$

The quantity $D$ is actually the magnitude of a vector, called the **electric displacement vector,** and is related to the electric intensity by

$$\mathbf{D} = \epsilon \mathbf{E}. \tag{3.4.8}$$

More generally, as shown in courses in electromagnetism, the electric energy density is given by

$$u = \int E \, dD \tag{3.4.9}$$

even when $\epsilon$ varies with $E$, i.e., when the behavior is nonlinear. This should be compared with the integral for stored elastic energy [see Equation (1.3.11)]. The detailed physical significance of the vector $\mathbf{D}$ is associated with Gauss' law and is described in texts in electromagnetism. It is a fundamental quantity in Maxwell's equations.

The origin of the dielectric constant will be studied later. Measurements of dielectric constants provide valuable data to the scientist attempting to understand molecular structure.

## Q. 3.4.1

The relationship between charge and voltage on a parallel plate capacitor in a vacuum is an example of (a)_____ relationship. The capacitance of a capacitor is a measure of its ability to store (b)_____. When a dielectric slab is inserted between the plates of the capacitor, the capacity to store charge is increased by the factor (c)____. Many common insulators have a dielectric constant equal to about (d)____.

\* \* \*

The variables in Equation (3.4.4) can be considered to be analogs of the variables in Equation (1.3.9) with $V$ and $F$ as analogs and $Q$ and $x$ as analogs. Just as a spring is not a purely elastic member but contains a viscous element so a capacitor (containing a dielectric) is not a pure capacitance but contains a resistive element; i.e., the resistance is finite and not infinite as it would be if the dielectric material were a perfect insulator. If the electric potential difference across a capacitor is proportional to $\cos \omega t$, and if the capacitor contains a *perfect* dielectric, the charge is also proportional to $\cos \omega t$; i.e., $Q$ and $V$ are in phase. However, when a real (or imperfect) dielectric is used the charge is then found to be proportional to $\cos(\omega t - \delta_{el})$; i.e., the charge lags the electric potential by a small angle $\delta_{el}$, called the **dielectric loss angle.**

The current flow through the resistance results in energy dissipation. It is found that the energy loss per cycle with the dielectric present divided by the maximum energy stored in the capacitor with vacuum with the same applied voltage is

$$\frac{\Delta u(\text{dielectric})}{u_{\max}(\text{vacuum})} = 2\pi \kappa \delta_{el}. \qquad (3.4.10)$$

The product $\kappa \delta_{el}$ is called the **dielectric loss factor.** The electric loss angle, like its mechanical counterpart, is a structure-sensitive property. At $10^{10}$ Hz pure $SiO_2$ has $\delta_{el} = 5 \times 10^{-4}$ while $SiO_2$ with 2% $K_2O$ has $\delta_{el} = 7.5 \times 10^{-3}$, a change of a factor of 15 for a 0.02 fractional composition change. As in the mechanical case the energy dissipated is converted into heat. Note that heat can cause degradation of the dielectric or insulating material which leads to eventual breakdown.

The dielectric loss angles are *approximately* frequency-independent. Some values are shown in Table 3.4.2.

Table 3.4.2. ELECTRIC LOSS ANGLE OF DIELECTRIC MATERIALS

| Material | Loss Angles (60 Hz) |
|---|---|
| Pure silica | 0.0001 |
| Polystyrene | 0.0002 |
| Paraffin | 0.002 |
| Porcelain | 0.01–0.02 |
| Hard rubber | 0.01 |
| Dry paper | 0.01 |
| Lucite | 0.03 |
| Bakelite | 0.05 |

One interesting aspect of the electric loss angle and the mechanical loss angle is shown in Figure 3.4.2. The nearly one-to-one correspondence

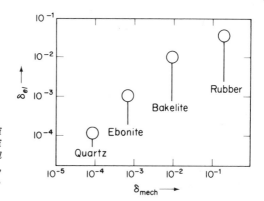

FIGURE 3.4.2. Comparison of mechanical and electrical angle of loss. [From C. Zwikker, *Physical Properties of Solid Materials*, Pergamon Press, Inc., Elmsford, N.Y. (1955).]

between these two loss angles suggests the same molecular origin of the loss angle for the two processes.

### Q. 3.4.2

The dielectric loss angle is an analog of the (a)_____. The dielectric loss angle is a structure- (b)_____ property.

The energy dissipated per cycle in a dielectric is (c)————————————.
Because $\delta_{el} \simeq \delta_{mech}$ for several materials, (d)————————————
————————————.

\* \* \*

There is another dielectric property which is an analog of mechanical behavior. The electric analog of the tensile fracture strength in plastic flow is the **dielectric strength.** It is defined as the voltage per unit length (voltage gradient or electric field) at which failure occurs. Failure or breakdown is characterized by a rapid increase in current and may be associated

Table 3.4.3. DIELECTRIC STRENGTH

| Material | kV/cm |
|---|---|
| Air (1 atm) | 30 |
| Dry paper | 45 |
| $SF_6$ (1 atm) | 80 |
| Porcelain | 15–120 |
| Mineral oil | 150 |
| Varnish | 80–350 |
| $SF_6$ (7 atm) | 400 |
| Air (7 atm) | 480 |
| Mica | 1200–2000 |

with melting, burning, and vaporization of the dielectric. It is irreversible for solid dielectrics. Some values of dielectric strength are shown in Table 3.4.3.

Pressurized gases are often used as insulating dielectrics in very high-voltage X-ray machines used to make radiographic studies of thick structural sections. They are also used in 1-MeV Van de Graaff generators whose electron or ion output may be utilized in such phases of materials processing as paint drying on automobiles, polymerization of polymers, and ion implantation to make *p-n* junctions. Liquids such as mineral oil may be used in heavy transformers and regulators where heavy heat transfer losses necessitate the excellent heat transfer brought about by convection. In such cases the presence of a small concentration of water is highly detrimental since it greatly reduces the dielectric strength. Examples of solid insulators in use are shown in Figures 3.4.3 and 3.4.4.

Sec. 3.4  Insulators  115

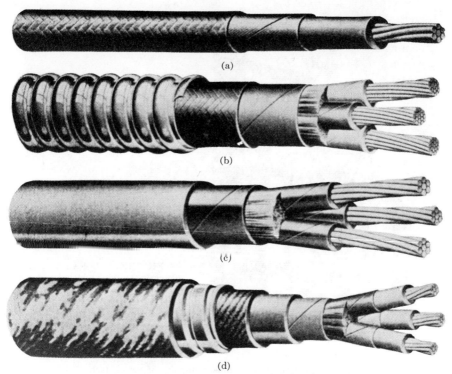

FIGURE 3.4.3. Cable construction (*polymeric insulators*). (a) Rubber-insulated cable with weatherproof braided finish (single conductor stranded). (b) Varnished-cambric insulated cable with filler, belt, braided finish, and interlocked spiral steel armor (three-conductor stranded). (c) Varnished cambric insulated with filler, belt, and lead sheath. (d) Parkway cable with rubber-insulated conductors belted, lead covered, projected by jute, interlocked armor, and jute overall. [From J. F. Young, *Materials and Processes*, John Wiley & Sons, Inc., New York (1954).]

### Q. 3.4.3

The electric field at which breakdown occurs is known as the (a)_____. About the highest electric field which materials will tolerate is (b)_____. The three important properties of ordinary dielectrics are (c)_____
_____.

\* \* \*

**EXAMPLE 3.4.1**

Compare the maximum elastic energy density in piano wire with the maximum electric energy density in mica.

FIGURE 3.4.4. Five-stage Cockcroft-Walton multiplier circuit 1200-kV dc generator for testing high-voltage cables. Note *ceramic* insulator. (Courtesy of Ferranti Ltd., Chadderton, Lancs., England.)

*Answer.* The maximum elastic energy density is given [see (1.3.11)] by $u = \sigma_0^2/2E$, where $\sigma_0$ is the tensile yield stress [see Table 1.3.5] and $E$ is Young's modulus [see Table 1.3.3]. Using $\sigma_0 = 3.4 \times 10^{10}$ dynes/cm² and $E = 2.1 \times 10^{12}$ dynes/cm², we get $u = 2.8 \times 10^8$ dynes/cm² = $2.8 \times 10^8$ ergs/cm³ = $2.8 \times 10^7$ J/m³.

The maximum electric energy density is given [see (3.4.3) and (3.4.5)] by $u = \kappa\epsilon_0 E_0^2/2$, where $\kappa$ is the dielectric constant [see Table 3.4.1] and $E_0$ is the dielectric strength [see Table 3.4.3]. Using $E_0 = 2 \times 10^8$ V/m and $\kappa = 85$, we get $u = 1.5 \times 10^7$ J/m³.

These energy densities should be compared to the energy density of a petroleum fuel of approximately $2 \times 10^{10}$ J/m³. It is unlikely that there will be clock-spring-powered automobiles or capacitor-bank-powered automobiles.

## ANSWERS TO QUESTIONS

**3.4.1** (a) A linear, (b) charge, (c) $\kappa$, (d) 5.

**3.4.2** (a) Mechanical loss angle, (b) sensitive, (c) converted into heat, (d) we suspect that they have the same molecular origin.

**3.4.3** (a) Dielectric strength, (b) $10^6$ V/cm, (c) dielectric constant, loss angle, and dielectric strength.

## 3.5 SPECIAL DIELECTRICS

To discuss special dielectrics it is useful to have a variable which compares $D$ for the case in which a dielectric is present to the case in which it is absent; in the latter case $\mathbf{D}_0 = \epsilon_0 \mathbf{E}$. This variable is defined as

$$\mathbf{P} = \mathbf{D} - \mathbf{D}_0 = \mathbf{D} - \epsilon_0 \mathbf{E} = (\epsilon - \epsilon_0)\mathbf{E} \qquad (3.5.1)$$

and is called the **polarization.** We can also write

$$\mathbf{P} = \epsilon_0 \chi \mathbf{E}, \qquad (3.5.2)$$

where $\chi$, called the **dielectric susceptibility,** is given by

$$\chi = \kappa - 1. \qquad (3.5.3)$$

Polarization has a simple physical significance. Consider an electrically neutral solid. It consists of an equal number of positive and negative charges, more or less evenly distributed so there is no net charge of one kind in any large region of the specimen. If the solid is placed in an electric field $\mathbf{E}$, positive charges of the material are displaced slightly in the direction of $\mathbf{E}$ and negative charges of the material are displaced slightly in the opposite direction. We say that the solid has become polarized. The product of the sum of the charges times the magnitude of their displacement equals the magnitude of the polarization. The polarization is directed from the negative charge to the positive charge, i.e., in the direction of $\mathbf{E}$. From Equations (3.4.9) and (3.5.1) it follows that the electric energy density in the dielectric can be written in the form

$$u = \int E \, dP. \qquad (3.5.4)$$

For the simple dielectric materials which we have discussed thus far $\chi$ does not vary with the magnitude or sign of $\mathbf{E}$; as the electric field in such a solid increases the material becomes more polarized in direct proportion to the field and as the field decreases it retraces the original path. This is another example of a *linear response*.

There is an important class of materials for which $\chi$ not only varies with $\mathbf{E}$ but is not even a single-valued function of $\mathbf{E}$. These are the **ferroelectrics** and their behavior is illustrated by the $P$-$E$ curve of Figure 3.5.1. The path followed as $E$ is increased is $OA$. At the point $A$ the polarization closely approaches its maximum value, $P_s$, called the **saturation polarization.** Upon cycling $E$ the path subsequently traced is $[A \text{ to } P_R \text{ to} -E_C \text{ to } B \text{ to } (-P_R)]$ $[A \text{ to } P_R \text{ to} -E_C \text{ to } B \text{ to } (-P_R)] \cdots$. $P_R$ is called the **remanent polarization** and $E_C$ the **coercive field.** Note that when $E$ goes to zero, the polarization is either $P_R$ or $-P_R$; i.e., the solid remains polarized even in the absence of a field. There is actually a decay of polarization with time after the electric field is turned off so that

$$P_R = P_R(0)e^{-t/\tau}. \tag{3.5.5}$$

Here $\tau$ is the **relaxation time** for the process. $\tau$ depends strongly on temperature; as temperature increases, $\tau$ decreases. An increase in temperature tends to destroy ferroelectric behavior. One of the most common ferro-

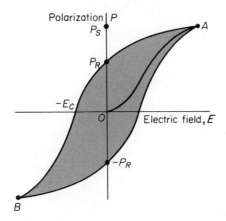

FIGURE 3.5.1. *P-E* curve for ferroelectric behavior.

electrics is $BaTiO_3$ (barium titanate). Above 120°C it loses its ferroelectric characteristics; this temperature is called the **ferroelectric Curie point,** $T_C$. Some values of $P_S$ and $T_C$ are shown in Table 3.5.1.

**Table 3.5.1.*** FERROELECTRIC CRYSTAL DATA

| Material | Formula | $T_C$ (°K) | $P_S$ (C/m²) | at $T$°K |
|---|---|---|---|---|
| Rochelle Salt | $NaK(C_4H_4O_6)\cdot 4H_2O$ | 297 | 0.0024 | 278 |
| KDP | $KH_2PO_4$ | 123 | 0.053 | 96 |
| Perovskite | $BaTiO_3$ | 393 | 0.26 | 296 |
| TGS | Triglycine sulphate | 322 | 0.028 | 293 |

* See F. Jona and G. Shirane, *Ferroelectric Crystals*, Pergamon Press, Inc., Elmsford, N.Y. (1962).

Figure 3.5.2 illustrates how the saturation polarization varies as $T$ increases toward $T_C$. Note that high temperature destroys ferroelectricity.

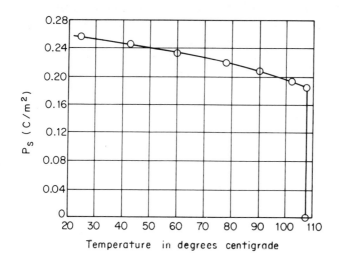

FIGURE 3.5.2. Spontaneous polarization referred to the cube edge of cubic $BaTiO_3$. [After J. W. Merz, *The Physical Review* **91**, 513 (1953).]

## Q. 3.5.1

In an ordinary dielectric, the polarization is (a)_____ proportional to the electric field. For ferroelectrics, however, the polarization is not linearly related to the electric field, nor is it even (b)_____ function of the field. The polarization which remains in a ferroelectric when the field is removed is called the (c)_____; this polarization decreases (d)_____ with time. The rate of this decrease (e)_____ as the temperature increases. One of the most common ferroelectrics is (f)_____.

\* \* \*

PIEZOELECTRIC CRYSTALS. Piezoelectric materials are another special class of dielectric materials. Certain crystals when stressed mechanically in the elastic range develop a polarization. Quartz and Rochelle salt are examples of this (also all ferroelectric crystals show this effect). This is known as the **piezoelectric effect.** Alternatively, if the same crystals are placed in an electrical field, they will be strained (and if constrained, will develop stresses). This is known as the **inverse piezoelectric effect.**

These effects have numerous applications in industry, e.g., as electromechanical transducers in sonar (to send and receive sound pulses, which are really stress waves, through the water). Also, one way to measure the elastic constants of crystals is to send a stress pulse into the crystal and to measure velocity of sound in the crystal. To do this a small quartz crystal is glued to the sample and is pulsed electrically. The inverse effect sends a stress pulse into the sample, which reflects off the other parallel end of the sample back to the transducer where it causes a direct effect and thus a second voltage pulse. Since the path traveled is known and the time between the two electrical pulses is measured, the velocity of sound is obtained. Now see Problem 2.26.

Hundreds of millions of tiny quartz crystals are used as frequency references. A quartz crystal will have a resonance frequency for mechanical oscillations determined by its elastic constants, density, and geometry. Quartz crystals have been grown synthetically with quality factor $Q > 10^6$ so the mechanical resonance frequency will be very sharp since the peak width [see Equations (2.4.9) and (2.4.10)] is given by

$$\frac{\Delta\omega_0}{\omega_0} = \frac{\sqrt{3}}{Q}.$$

Because the mechanical oscillations are coupled to the electrical oscillations, this provides a fixed electrical frequency reference; see Figure 3.5.3. Piezoelectric behavior is usually described by a linear relationship

FIGURE 3.5.3. Quartz crystal resonator.

between polarization and strains and between strains and electric field. The coefficients which describe this relationship are called the piezoelectric coefficients. A mathematical framework using Cartesian tensors is used to describe this behavior.

A field of $10^5$ V/m (100 V across 1 mm of sample) along a certain crystallographic direction of quartz called the diad axis would produce a strain of $-2.3 \times 10^{-7}$. A compressive mechanical stress of $1.8 \times 10^5$ dynes/cm² (2.6 psi) would yield the same strain.

Figure 3.5.4 shows a small piezoelectric motor capable of delivering 15 hp (to and fro motion, not rotary) at its tip.

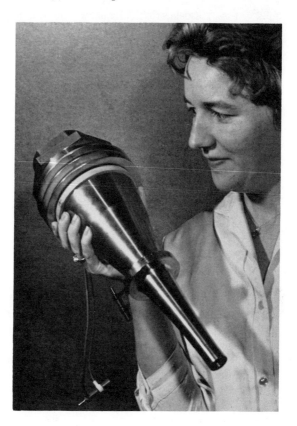

FIGURE 3.5.4. 15-hp piezoelectric motor weighing 22 lb. (Courtesy of C. C. Libby, Department of Welding Engineering, Ohio State University, Columbus.)

## Q. 3.5.2

A material which becomes polarized when stressed is called (a)_____ material; an example is (b)_____. Such materials have various applications as signal senders and receivers, etc. One of their important applications is as frequency references. The natural frequency of the mechanical vibration of a plate is determined by the density, (c)_____, and the geometry; the width of the resonance line is determined by the (d)_____ or the (e)_____.

\* \* \*

## ANSWERS TO QUESTIONS

**3.5.1** (a) Directly, (b) a single-valued, (c) remanent polarization, (d) exponentially, (e) increases, (f) $BaTiO_3$.

**3.5.2** (a) A piezoelectric, (b) quartz, (c) the elastic coefficients, (d) $Q$-factor, (e) mechanical loss angle.

### 3.6 MAGNETIC PROPERTIES

The electrical potential drop across a long solenoidal coil in which the current is changing is

$$V = L \frac{di}{dt}, \tag{3.6.1}$$

if we assume the $i^2R$ losses are negligible because the winding is large-diameter wire of high conductivity. The quantity $L$ is called the **self-inductance** of the coil, $i$ is the current, and $t$ is the time. If the coil is in a vacuum, we use a subscript zero, $L_0$. It can be shown from electromagnetic theory or experimentally that

$$L_0 = \mu_0 n^2 (Al). \tag{3.6.2}$$

Here $\mu_0$ is the **permeability coefficient of a vacuum** [$4\pi \times 10^{-7}$ H/m (henry per meter) in rationalized mks units], $n$ is the number of turns per unit length, $A$ is the cross-sectional area of the coil, and $l$ is the length of the coil (not to be confused with the length of wire used in forming the coil). We assume $l$ is very large relative to the coil diameter.

**EXAMPLE 3.6.1**

Find the energy stored in the above coil.

*Answer.* The power is $Vi$ so that the stored magnetic energy is

$$U_{\text{mag}} = \int Vi\, dt = \int L_0 i\, di = \tfrac{1}{2} L_0 i^2. \tag{3.6.3}$$

If a rod of homogeneous isotropic material (from the magnetic viewpoint) fills the space within the coil, then the self-inductance is

$$L = \mu n^2 (Al), \tag{3.6.4}$$

where $\mu$ is the **permeability coefficient** of the substance. It is useful to introduce the term **relative permeability** coefficient for the ratio $\mu/\mu_0$.

Note that this gives the ratio of the self-inductance of the coil with magnetic material in the coil and without, i.e., $L/L_0$; it also gives the ratio of the energy in the coil in these two cases. For the present, we have assumed that $\mu$ does not vary with $di/dt$, so that $L$ does not vary with $di/dt$. Therefore, the behavior described by (3.6.1) is linear behavior.

The magnitude of the magnetic induction within the long solenoidal coil is

$$B = \mu n i \qquad (3.6.5)$$

as can readily be proved by experiment. The magnetic induction **B** is parallel to the axis of the solenoid, as shown in Figure 3.6.1. Note that

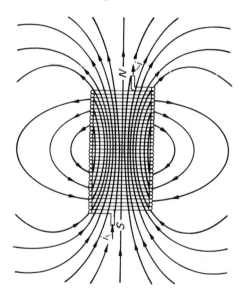

FIGURE 3.6.1. Solenoid with the corresponding closed lines of **B**. The end from which **B** lines emerge behaves like the north pole of a bar magnet. If the length of the solenoid is very much greater than its diameter, then **B** is uniform inside the coil except near to its ends.

each line of **B** is a closed loop. This is a fundamental property of any **B** field.

The stored magnetic energy per unit volume, or the magnetic energy density, is

$$u = \frac{U}{Al} = \frac{\tfrac{1}{2}Li^2}{Al} = \frac{\mu n^2 i^2}{2}. \qquad (3.6.6)$$

From Equation (3.6.5) this can be written in the form

$$u = \frac{(ni)B}{2} = \frac{B^2}{2\mu}.$$

As with electric energy, we would expect to have the energy density of a linear system described by the product of a generalized force times a generalized displacement divided by 2. We therefore write the linear relation

$$B = \mu H. \tag{3.6.7}$$

Hence

$$u = \frac{HB}{2}. \tag{3.6.8}$$

The quantity $H$ is actually a vector, called the **magnetic intensity,** and is given by

$$\mathbf{H} = \frac{\mathbf{B}}{\mu}. \tag{3.6.9}$$

More generally, as shown in courses in electromagnetism, the magnetic energy density is

$$u = \int H \, dB, \tag{3.6.10}$$

a relation which holds not only for linear behavior, but also for nonlinear behavior in which $\mu$ varies with $H$. Strong magnetic materials often behave in a highly nonlinear fashion.

The detailed physical significance of the magnetic intensity $\mathbf{H}$ is associated with Maxwell's extension of Ampere's law and is described in texts on electromagnetism.

Materials can be classified from the magnetic viewpoint according to the magnitude of $\mu/\mu_0$ as follows:

$\mu \lesssim \mu_0,$     **diamagnetic;**

$\mu \gtrsim \mu_0,$     **paramagnetic;**

$\mu \gg \mu_0,$     **ferromagnetic.**

For diamagnetic materials the magnitude of the fractional variation of $\mu$ from $\mu_0$ [i.e., $(\mu - \mu_0)/\mu_0$, which is also called the **magnetic susceptibility**] is about $-10^{-5}$ while for paramagnetic materials it is about $10^{-4}$. Thus diamagnetic materials tend to weakly repel the lines of $\mathbf{B}$ whereas paramagnetic materials attract them. Hence liquid oxygen, which is paramagnetic, clings to a magnet. For such materials $\mu$ does not vary with $H$ (except for unusually intense magnetic fields). For many applications such materials can be considered as nonmagnetic. However, the actual study of their magnetism provides in many cases detailed knowledge of their electronic structure. Figure 3.6.2 shows the lines of $\mathbf{B}$ for the cross section of a rod placed in a uniform magnetic field for a nonmagnetic material (a) and a ferromagnetic material (b). Note how the lines of $\mathbf{B}$ are attracted by the ferromagnetic material. A practical application of this is shown in Figure 3.6.3. Here a piece of soft iron is present within a solenoid. In this case,

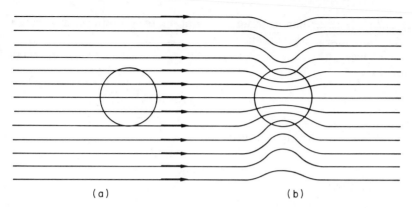

FIGURE 3.6.2. Lines of **B** in a rod normal to a longitudinal magnetic field. (a) Nonmagnetic rod. (b) Ferromagnetic rod.

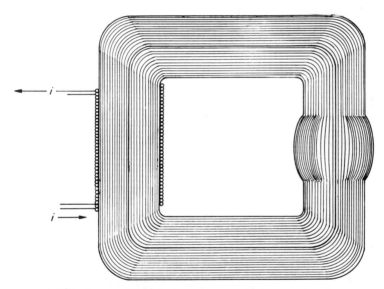

FIGURE 3.6.3. Solenoid with a soft iron core (not a permanent magnetic material) showing the lines of **B** being attracted by the iron and hence being led around the path shown.

when the lines of **B** leave the ends of the solenoid, rather than spreading out as in Figure 3.6.1, they are led around the path as shown in Figure 3.6.3.

### Q. 3.6.1

The ratio of the self-inductance of a coil with magnetic material filling the coil and without is a measure of the (a)_____

_____. The fractional variation of $\mu$ from $\mu_0$ is written as (b)_____ and is called the (c)_____. The latter is very (d)_____ and (e)_____ for ordinary diamagnetic materials.

\* \* \*

From a technological viewpoint, the ferromagnetic materials are extremely important; note that $\mu$ is not generally a constant for these materials. In fact, the $B$-$H$ relationship is not even reversible; its behavior is thus not at all like the $F$-$x$ relation for an ideal spring but is more like that of the ideal plastic discussed in Section 2.5. A typical $B$-$H$ curve is shown in Figure 3.6.4. The first $B$-$H$ curves were measured by Ewing in 1885 and

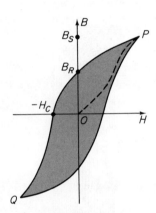

FIGURE 3.6.4. Typical $B$-$H$ curve for ferromagnetic material.

he called them **hysteresis curves.** The path followed is $O$ to $P$ to ($B_R$ to $-H_C$ to $Q$ to $P$) to ($B_R$ to $-H_C$ to $Q$ to $P$). We have the following important definitions:

$B_S$                                **saturation induction**
                                   (the limiting value of $B$-$\mu_0 H$ for large $H$).

$B_R$                                **remanent induction**
                                   (magnitude of $B$ when $H = 0$).

$H_C$                                **coercive force**
                                   (magnitude of $H$ when $B = 0$).

$\mu_i = \lim\limits_{H \to 0} dB/dH$ along line $\overline{OP}$      **initial permeability.**

maximum of $dB/dH$ along line $\overline{OP}$      **maximum permeability.**

**Magnetization** $M$ can be defined by

$$M = \frac{B}{\mu_0} - H \qquad (3.6.11)$$

It is the analog of polarization in dielectric materials.

Magnetization has a simple physical meaning in atomic terms and is discussed in Chapter 5.

### Q. 3.6.2

For (a)_____ materials, the permeability coefficient usually does not vary with $H$. For (b)_____ materials the permeability coefficient not only depends on $H$ but also depends on (c)_____. Although $\mu$ can be very large for ferromagnetic materials, it is always true that for sufficiently large fields $dB/dH$ approaches (d)_____. The value of $B-\mu_0 H$ for which this occurs is known as (e)_____. When the magnetic field is reduced from a very large value to zero, the material remains magnetized; the value of $B$ is then called the (f)_____. If the magnetic field is then made negative, the value of $B$ will reach zero when $H$ equals (g)_____. The complete path of $B$ vs. $H$ when $H$ is cycled is called the (h)_____.

\* \* \*

An important feature of ferromagnetism is the saturation induction, $B_S$; beyond the point $P$, $dB/dH \doteq \mu_0$; i.e., the ferromagnetic material

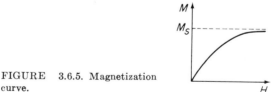

FIGURE 3.6.5. Magnetization curve.

makes zero contribution to the cause of increasing $B$ as $H$ increases. $B_S$ has a characteristic value for a given material. $B_S$ is relatively independent of impurities, mechanical deformation history, thermal history, etc., and is hence called a structure-insensitive property. The **saturation magnetization** $M_S$ is the maximum value of the magnetization. The $M$-$H$ curve hence has the shape shown in Figure 3.6.5. The magnetic susceptibility, $\chi_m =$

$(\mu - \mu_0)/\mu_0$, relates **M** and **H** according to

$$\mathbf{M} = \chi_m \mathbf{H}. \tag{3.6.12}$$

Temperature has a strong effect on ferromagnetism; it tends to destroy it. See Figure 3.6.6. All ferromagnetism vanishes for $T > T_C$. $T_C$ is the **Curie**

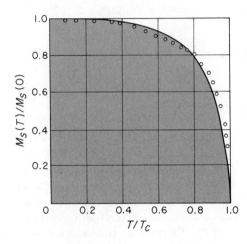

FIGURE 3.6.6. Saturation magnetization versus temperature for nickel. [Experimental values are from P. Weiss and R. Forrer, *Annales de Physique* **5**, 153 (1926).] $M_S$ at $T = 0$ is $4.8 \times 10^5$ A/m and $T_C = 631°$K.

**temperature.** It could also be called the **ferromagnetic transition temperature.**

Table 3.6.1 shows some values of $T_C$ and $M_S$ for strong magnetic materials.

**Table 3.6.1.** CURIE POINTS AND SATURATION CONSTANTS

| Material | $T_C$ (°K) | $M_S$ ($10^5$ A/m) |
|---|---|---|
| Fe | 1043 | 17.1 |
| Co | 1400 | 14 |
| Ni | 631 | 4.85 |
| Gd | 289 | 19.8 |
| MnBi | 630 | 6.75 |
| $3Y_2O_3$–$5Fe_2O_3$* | 560 | 2.0 |

* Also called YIG for yttrium iron garnet.

$B_R$ and $H_C$ (and hence the shape of the hysteresis loop) are structure-sensitive properties. We note that there is a dissipation of magnetic energy per unit volume per one hysteresis loop given by

$$\Delta u = \int_{\substack{\text{closed} \\ \text{loop}}} H \, dB = \oint H \, dB \qquad (3.6.13)$$

which is converted into heat. This is called the **hysteresis loss**. It equals the area within the $B$-$H$ loop. Ferromagnetic materials can be divided into two classes: soft magnetic materials and hard magnetic materials. We shall discuss each class in more detail.

### Q. 3.6.3

The quantities $B_S$, $M_S$, and $T_C$ are all structure- (a)_____ properties. The general shape of the $M$-$H$ curve is (b)_____ _____. In general, $M_S$ varies how with $T$? (c)_____ _____. The effect of increasing temperature on ferromagnetism is (d)_____. All ferromagnetism vanishes above a temperature called the (e)_____. The mks units for $H$ and $M$ are (f)_____. The saturation induction for iron at $T = 0°$K is about (g)____Wb/m². The area inside the hysteresis loop represents (h)_____.

\* \* \*

**EXAMPLE 3.6.2**

An ideal capacitor of capacitance $C$ is charged so that there are charges $Q_0$ and $-Q_0$ on the plates. An ideal inductor of inductance $L$ is attached across the plates. Write the expression for the energy of the system.

*Answer.* Initially the energy of the system was $Q_0^2/2C$. At some later instant, the charge is $Q$ (rather than $Q_0$) and a current $i$ flows in the circuit. Then the energy is

$$\frac{Li^2}{2} + \frac{Q^2}{2C} = \frac{Q_0^2}{2C}.$$

Now

$$i = \frac{dQ}{dt} = \dot{Q}$$

so that this equation becomes

$$\frac{L\dot{Q}^2}{2} + \frac{Q^2}{2C} = \frac{Q_0^2}{2C}. \qquad (3.6.14)$$

Note the analogy with the linear spring mass-system described by Equation (2.4.1). The inductor is analogous to the mass and the capacitor to the spring. Note that $Q = Q_0 \cos \omega_0 t$, where $\omega_0 = (LC)^{-1/2}$.

## ANSWERS TO QUESTIONS

**3.6.1** (a) Relative permeability of the material, (b) $(\mu - \mu_0)/\mu_0$, (c) magnetic susceptibility, (d) small, (e) negative.

**3.6.2** (a) Diamagnetic and paramagnetic; (b) ferromagnetic; (c) the previous values of $H$; (d) $\mu_0$; (e) the saturation induction, $B_S$; (f) remanent induction, $B_R$; (g) the coercive force, $H_C$; (h) hysteresis loop.

**3.6.3** (a) Insensitive; (b) $M$ increases as $H$ increases and for large $H$, $dM/dH$ approaches zero; (c) it decreases slowly from the value $M_S(0)$ as $T$ increases from absolute zero, but decreases rapidly toward zero as $T$ approaches $T_C$; (d) to destroy it; (e) Curie temperature; (f) amperes per meter; (g) 2; (h) energy dissipated as heat per cycle.


### 3.7 SOFT FERROMAGNETICS

A material for which $B$ is nearly a single-valued function of $H$ is a **soft magnetic material**. (Such materials are usually mechanically soft also;

Table 3.7.1.* IMPORTANT PROPERTIES OF SOME SOFT FERROMAGNETICS

| Material | $\mu_i/\mu_0$ | $\oint H\, dB$ (J/m³) | $B_S$ (Wb/m²) |
|---|---|---|---|
| Commercial iron ingot | 250 | 500 | 2.16 |
| 4% Si–96% Fe (random texture) | 500 | 50–150 | 1.95 |
| 3.25% Si–96.75% Fe (oriented) | 15,000 | 35–140 | 2.0 |
| 45 Permalloy† | 2,700 | 120 | 1.6 |
| Mumetal‡ | 30,000 | 20 | 0.8 |
| Supermalloy§ | 100,000 | 2 | 0.79 |

\* From J. Wulff et al., *The Structure and Properties of Materials*, John Wiley & Sons, Inc., New York (1966).

† 45 % Ni–55 % Fe.

‡ 75 % Ni–5 % Cu–2 % Cr–18 % Fe.

§ 80 % Ni–15 % Fe–5 % Mo.

i.e., they have a low yield strength.) They are characterized by a high initial permeability, $\mu_i$, and a small hysteresis loop, i.e., small coercive force $H_C$. See Table 3.7.1.

Silicon-iron has important applications in power equipment involving ac currents (such as the 60 Hz of our utilities). Such equipment includes

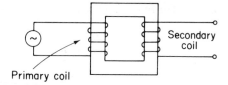

FIGURE 3.7.1. Schematic of iron core transformer. Actually each coil would have many layers of windings.

transformers, generators, motors, controllers, meters, etc. A schematic of a transformer is shown in Figure 3.7.1.

**EXAMPLE 3.7.1**

What is the reason for using the iron core in the transformer shown in Figure 3.7.1?

*Answer.* The emf (voltage) produced in the secondary coil is found by experiment to be the product of a quantity known as the mutual inductance $M$ and the rate of change of current in the primary coil. The mutual inductance is given by the expression

$$M = k\sqrt{L_1 L_2},$$

where $k$ is called the coupling constant (if the entire magnetic flux of one coil passes through the other, $k = 1$; in the absence of the iron core $k$ would be small in the above case). The iron core serves the dual purpose of enhancing the magnetic flux (i.e., increasing both $L_1$ and $L_2$ by the factor $\mu/\mu_0$) and of leading the flux from the primary to the secondary coil (i.e., increasing $k$). We note that $\mu/\mu_0$ can have values of several hundred or more. If the magnetic core is not present, the number of windings would have to be greatly increased. Hence compare the size of a transformer with and without the use of a ferromagnetic core. The size would be increased by a factor of $10^3$ to $10^6$ if the magnetic core were not used.

Figure 3.7.2 shows a transformer station. Imagine the cost of the copper and the size of the station if magnetic core material were not available! In 1968 about 670,000 tons of such materials were produced.

Philip Sporn in a fascinating little book (64 small pages) [*Research in Electric Power*, Pergamon Press, Inc., Elmsford, N.Y. (1966)] predicts an increase by a factor of 6.6 in power generation by the year 2000. The critical parameter to the design engineer is the energy loss per cycle. The two important contributions to this are (1) the hysteresis loss given by Equation (3.6.13) and (2) the eddy current losses. Eddy current losses arise as follows. The changing magnetic field induces a flow of electrical current (eddy current) in the magnetic material. This leads to ordinary $i^2R$ drops, i.e., ordinary ohmic heating. The annual loss due to these hysteresis losses in the United States approaches $10^8$. For use at higher frequencies (e.g.,

FIGURE 3.7.2. Power transformer station. (Courtesy of Pacific Gas and Electric Company, San Francisco, Cal.)

in communication equipment) nickel-iron alloys have found widespread use. They have smaller hysteresis loops and hence less power loss. (Recall that power loss is the product of the energy loss per cycle times the frequency). These nickel-iron alloys are not used in the lower-frequency utility applications because of the high initial cost.

Impurities can have an enormous effect on the relative permeability ratio as shown in Table 3.7.2. The relative permeability ratio is a highly structure-sensitive property which is strongly affected by impurities, heat treatment, and plastic deformation.

Magnetically soft ceramics have the same macroscopic $B$-$H$ characteristic as the soft ferromagnetic materials, although they are called ferrimagnetic. The most common of these materials are ferrites and garnets. These are made by mixing powders of metallic oxides which are then formed into a solid block of material by a process known as sintering (which we shall study later after we have studied the surfaces of solids and the motion of atoms in solids). A typical composition is Ferroxcube A (48% MnO–$Fe_2O_3$, 52% ZnO–$Fe_2O_3$). Its electrical resistivity is large (0.5 × $10^6$ Ω-m) compared to metals such as iron-silicon (see Table 3.2.1 for typical values for metals) and hence its eddy current losses are small compared to the silicon-irons. For this reason it is used in transformers in high-frequency (∼1 MHz, i.e., $10^6$ cycles/sec) communication equipment.

Table 3.7.2. RELATIVE PERMEABILITY AND CARBON CONTENT OF IRON

| Material | Relative Permeability |
|---|---|
| Ultra pure iron | 1,500,000 |
| Fe with 0.004% C | 16,000 |
| Fe with 0.01% C | 8,000 |

An important garnet, yttrium iron garnet (YIG), $3Y_2O_3$–$5Fe_2O_3$, has hysteresis losses sufficiently small that it can be used at microwave frequencies.

### Q. 3.7.1

Soft magnetic materials are characterized by high (a)_____ _____ and a small (b)_____. Iron cores are placed in transformers because they increase the (c)_____ _____ by a very large factor. The relative permeability is a structure-(d)_____ property. Two important causes of losses in transformer cores are (e)_____. Ceramic magnetic materials are electrical (f)_____ and hence exhibit (g)_____ eddy current losses.

\* \* \*

MAGNETOSTRICTION. The magnetization of a ferromagnetic material is accompanied by a small change in length; this effect is called **magnetostriction**. This is the origin of another energy loss phenomenon during cyclic magnetization. Magnetostrictive effects set up strong unwanted mechanical oscillations in many cases and so represent another factor of concern to the design engineer. The sample may either shrink in the direction of **H** (cobalt) or stretch in the direction of **H** (as iron does for small fields, although it shrinks for large fields).

Magnetostriction provides one mechanism of converting electrical energy into mechanical energy (via an intermediate, magnetic energy). The resultant device is called an electromechanical transducer. For a discussion see W. P. Mason, *Electromechanical Transducers and Wave Filters*, Van Nostrand Reinhold Company, New York (1948).

## ANSWERS TO QUESTIONS

**3.7.1** (a) Initial permeability, (b) hysteresis loop (or small coercive force), (c) mutual inductance, (d) sensitive, (e) hysteresis losses and eddy current losses, (f) insulators, (g) tiny.

### 3.8 HARD MAGNETIC MATERIALS

Materials characterized by huge hysteresis losses, large $B_R$, and large $H_C$ are called **hard magnetic materials.** These are permanent magnet materials. The designer of a permanent magnet wants $B_R$ to be large so that the greater part of the magnetization will remain when the magnetizing field is removed and he wants $H_C$ to be large in order that the magnet will not be easily demagnetized.

The coercive force is perhaps the ferromagnetic property most sensitive to control through changes in composition, heat treatment, and mechanical deformation; further, it is a most important property to consider in selection of ferromagnetic materials for practical applications. Coercive force values in commercial materials range from 50,000 A/m in a loudspeaker magnet (Alnico V) and 200,000 A/m in a special high-stability magnet (Fe–Pt) to 20 A/m in an iron-silicon transformer core and 0.3 A/m in a Supermalloy pulse transformer [for students familiar with other units, 1 A/m = $4\pi \times 10^{-3}$ Oe (oersteds)]. Thus, the coercive force may vary by a factor of approximately $10^6$. Coercive force values are closely related to the saturation hysteresis loss at low frequencies, since the area of the hysteresis loop is approximately the product of the coercive force and the saturation magnetic induction.

A close correlation exists between the initial relative permeability and the coercive force as shown in Figure 3.8.1 which spans the scale from soft to hard magnetic materials.

The permanent magnet design engineer is particularly interested in the maximum value of the product $BH$ along the demagnetization curve (along line $\overline{B_R H_C Q}$ in Figure 3.6.4). This is called the **maximum energy product** and some typical values are shown in Table 3.8.1. The maximum energy product is represented by the largest rectangle which can be in-

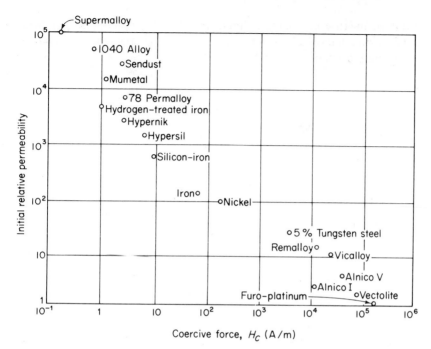

FIGURE 3.8.1. Correlation between initial permeability and coercive force. Note the log-log plot. [After C. Kittel, *Introduction to Solid State Physics*, John Wiley & Sons, Inc., New York (1956).]

scribed in the second quadrant of the *B-H* curve. Research in the last two decades into the nature of the rare earths and into their chemistry (they essentially all have the same valence and hence are very difficult to separate

Table 3.8.1. MAXIMUM ENERGY PRODUCT FOR HARD MAGNETIC MATERIALS

| Material | Max. Energy Product $(J/m^3)$ |
|---|---|
| Samarium-cobalt | 120,000 |
| Platinum-cobalt | 70,000 |
| Alnico V* | 36,000 |
| Ferroxdur† | 12,000 |
| Fe–Co sintered powder | 7,000 |
| Carbon steel | 1,450 |

* Alnico V has a composition 51 % Fe, 24 % Co, 14 % Ni, 8 % Al, 3 % Cu.
† Ferroxdur is a hard ferrite, $BaFe_{12}O_{19}$.

by standard means) has paid off handsomely in the case of the samarium-cobalt alloy which costs about one-eightieth of its nearest competitor Pt–Co and has a higher energy product.

## Q. 3.8.1

Because the designer of a permanent magnet wants a large (a)_____ and a large (b)_____, the quantity of most interest to him is the (c)_____. This can be defined as (d)_____. The hardest magnetic material is (e)_____; this is characterized by (f)_____. This should be compared to some of the softest ferromagnetic materials which have hysteresis losses of only (g)_____. Roughly, for all soft and hard magnetic materials, the initial relative permeability is (h)_____ proportional to the coercive force.

\* \* \*

## ANSWERS TO QUESTIONS

**3.8.1** (a) $B_R$, (b) $H_C$, (c) maximum energy product, (d) the largest rectangle which can be fitted in the second quadrant of the $B$-$H$ curve, (e) samarium-cobalt, (f) a maximum energy product of 120,000 $J/m^3$, (g) 2 $J/m^3$, (h) inversely.

### 3.9 SPECIAL MAGNETIC MATERIALS

Special methods of fabrication make possible a number of different kinds of magnetic materials. **Thin magnetic films** (usually of Fe–Ni which can be formed in a thickness of $\approx 3000$ Å by vacuum deposition of the vapor on a nonmagnetic substrate) can have switching times of $10^{-8}$ sec or better and hence have applications in computers.

Sufficiently small particles of ferromagnetic materials always exist as magnetized particles when the material is in the ferromagnetic state. Tiny elongated particles of iron having a diameter of only a few hundred Ångstroms (**elongated single domains**) are placed in a nonmagnetic molten binder such as lead (or epoxy resin) which is then hardened in the presence of a magnetic field. Such a *composite* material can have a very high maximum energy product ($10^4$ $J/m^3$) compared to the value for bulk carbon steel.

Magnetic tape of our common tape recorders is also a composite consisting of oriented elongated fine particles of $\gamma\text{-Fe}_2\text{O}_3$ on a plastic film.

## 3.10 SUPERCONDUCTORS

In 1911 Kamerlingh Onnes discovered that the dc resistivity of mercury *vanished* at 4.15°K. This temperature is called the **superconducting critical temperature,** $T_c$. Since that time $T_c$ has been measured for a number of elements and the results for a few of these are shown in Table 3.10.1. It has also been noted that above a certain magnetic field the superconducting property is destroyed. We call this the **superconducting critical field,** $H_c$. It varies with temperature; the value at absolute zero is $H_c(0)$. Some values of $H_c(0)$ are given in Table 3.10.1.

Table 3.10.1. $T_c$ AND $H_c(0)$ VALUES FOR LOW FIELD SUPERCONDUCTORS

| Type | Material | $T_c$ (°K) | $H_c(0)$ (A/m) |
|---|---|---|---|
| I | Tin | 3.72 | $2 \times 10^4$ |
| I | Lead | 7.19 | $7 \times 10^4$ |
| II | Vanadium | 5.03 | $9 \times 10^4$ |
| II | Niobium | 9.1 | $20 \times 10^4$ |

Figure 3.10.1 shows the temperature dependence of the critical field. We note that both magnetic and temperature fields have a strong effect

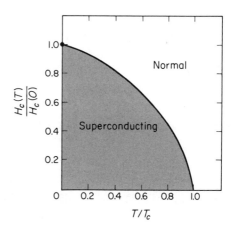

FIGURE 3.10.1. Reduced critical field versus reduced temperature.

on superconductivity; they tend to destroy it. Note the similarity with the effect of temperature on saturation magnetization. Roughly

$$H_c(T) = H_c(0)\left(1 - \frac{T^2}{T_c^2}\right). \tag{3.10.1}$$

The materials shown in Table 3.10.1 would not be of much use in the bulk form in making high field magnets ($H \sim 10^7$ A/m) since $H_c(0)$ is much less than this. However, they could be used as memory storage devices for computers.

TYPE I SUPERCONDUCTORS. Sn and Pb are examples of Type I superconductors. A long cylinder of a **Type I superconductor** parallel to a magnetic field will completely exclude magnetic induction ($B = 0$)

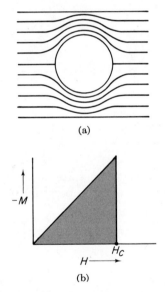

FIGURE 3.10.2. Perfect diamagnetism. (a) Lines of **B** for a perfectly diamagnetic rod normal to a longitudinal magnetic field. (b) $-M$ vs. $H$ curve for perfectly diamagnetic material with $H$ parallel to rod.

if the field is below $H_c$ in which case $M = -H$; the material is said to be perfectly diamagnetic; above $H_c$ it is completely penetrated. This behavior, known as the **Meissner effect,** is illustrated in Figure 3.10.2. The path is retraced as $H$ decreases; the process is reversible.

TYPE II SUPERCONDUCTORS. **Type II superconductors** show a Meissner effect only for small $H$ and then deviations occur. See Figure 3.10.3.

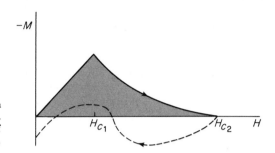

FIGURE 3.10.3. Magnetization of a Type II superconductor. Dashed line shows behavior of nonideal Type II as $H$ decreases.

For such materials the quantity $H_c$ is defined by

$$\frac{H_c^2}{2} = -\int_0^{H_{c_2}} M \, dH.$$

Below $H_{c_1}$ the material is in the superconducting state; above $H_{c_2}$ it is a normal conductor; and in between it is mixed.

The ideal Type II superconductors will retrace the $M$-$H$ path when $H$ is decreased. Most real Type II superconductors, however, are irreversible and follow a path such as shown by the dashed line in Figure 3.10.3. Values of $H_{c_2}$ as large as $3 \times 10^7$ A/m have been found.

HIGH FIELD SUPERCONDUCTORS. In 1961 Kunzler showed that $Nb_3Sn$ had an $H_c(0)$ value in excess of $10^7$ A/m. The critical current density versus field intensity for a specific $Nb_3Sn$ wire at 4.2°K is shown in Figure 3.10.4.

Superconductors which can carry large currents at large fields are known as **high field superconductors.** $Nb_3Sn$ has a critical temperature

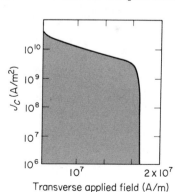

FIGURE 3.10.4. Critical current density versus transverse magnetic field for a heat-treated $Nb_3Sn$ superconductor at 4.2°K.

of 18.05°K. The highest critical temperature known is slightly above 20°K. (A high field superconductor with a transition temperature above room temperature is still a magnet-builder's dream.) Typical high critical temperature high field superconductors are either primarily Nb- or V-based compounds or alloys. They are Type II superconductors. See Table 3.10.2.

Table 3.10.2. $T_c$ FOR HIGH FIELD SUPERCONDUCTORS

| Compound | $T_c$ (°K) | Compound | $T_c$ (°K) |
|---|---|---|---|
| $Nb_3Al_{0.8}Ge_{0.2}$* | 20.98 | $V_3Si$ | 17.1 |
| $Nb_3Al_{0.8}Ge_{0.2}$ | 20.05 | $V_3Ga$ | 16.5 |
| $Nb_3Sn$ | 18.05 | | |
| $Nb_3Al$ | 17.5 | | |

* After special annealing treatments.

Two of the technologically important high field materials are $Nb_3Sn$ and an alloy of Nb–Zr.

The $J_c$-$H$ curves (critical current density versus magnetic intensity) of superconductors (like the $B$-$H$ curves of magnetic materials or the $\sigma$-$\epsilon$ curves of mechanical behavior) are *highly dependent* on impurity content, thermal history, and mechanical working. This is illustrated in Figure 3.10.5.

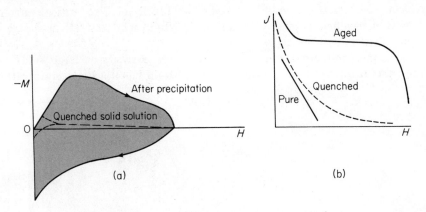

FIGURE 3.10.5. (a) Magnetization curves at 4.2°K of a Pb–8.3 at. % Na alloy as quenched from solution temperature and after aging 4 hr at room temperature to allow precipitation. (b) Sketch of corresponding critical current curves.

Recall that pure lead is a Type I superconductor. The addition of 8.3 at. % Na followed by quenching gives a nearly ideal Type II superconductor while the age precipitation heat treatment drastically alters this behavior. Cold working would introduce similar nonideal Type II behavior, i.e., make the superconductor harder. Cold working also makes the material mechanically harder (stronger) as does the age precipitation heat treatment.

The high dc current carrying capacity with no resistance heating makes possible the construction of high field superconducting magnets which are midgets compared to normal conducting magnets. Compare the magnet of Figure 3.10.6 with that of Figure 3.3.1.

FIGURE 3.10.6. 15 $Wb/m^2$ superconducting magnet. Compare with normal coil magnet of Figure 3.3.1. (Courtesy of RCA Laboratories, Princeton, New Jersey.)

**EXAMPLE 3.10.1**

How large is the allowable current density, $J$, in a #18 copper wire exposed to room-temperature air? Such wire has a diameter of $10^{-3}$ m and can carry currents up to 5 A (handbook value).

*Answer.* Hence $J = i/A \sim 6 \times 10^6$ $A/m^2$. Compare with $J_c \sim 10^{10}$ $A/m^2$ values for superconducting $Nb_3Sn$ at 4.2°K.

Hard superconductors will also be used in the field coils of the next generation of electric generators; this will eliminate the need for heavy iron cores in the armature. In addition superconductors may serve for long-distance power transmission lines, not withstanding the significant costs involved in cooling the conductor.

### Q. 3.10.1

Superconductivity is destroyed by high (a)_____ and high (b)_____. In Type I superconductors, the magnetic induction is (c)_____; i.e., the material is perfectly (d)_____; this is known as the (e)_____. In Type II superconductors, the magnetic induction lines begin to penetrate slowly above (f)_____ and do not penetrate completely until (g)_____ is reached. Certain (h)_____ superconductors can carry very high current densities in excess of (i)_____. Their $J_c$-$H$ curves are structure- (j)_____.

\* \* \*

## ANSWERS TO QUESTIONS

**3.10.1** (a) Temperatures, (b) magnetic fields, (c) zero, (d) diamagnetic, (e) Meissner effect, (f) $H_{c_1}$, (g) $H_{c_2}$, (h) high field, (i) $10^{10}$ A/m², (j) sensitive.

## REFERENCES

**Electrical Materials**

Duckworth, H. E., *Electricity and Magnetism*, Holt, Rinehart and Winston, Inc., New York (1961). Systematically develops and applies theory to a number of simple electronic and magnetic components and devices.

Young, J. F., *Materials and Processes*, John Wiley & Sons, Inc., New York (1954). Pages 179–200 contain a description of conductor contacts, resistor materials, while pp. 439–485 are concerned with dielectric materials.

Von Hippel, A. R., Ed., *Dielectric Materials and Applications*, John Wiley & Sons, Inc., New York (1954). Von Hippel states in the preface "...we hope to establish alliances between research worker, development engineer, manufacturer, field engineer and actual user of 'non-metals.' " He succeeds! The student might well read this book over lightly just to develop a feeling for the many applications of dielectrics.

Jona, F., and Shirane, G., *Ferroelectric Crystals*, Pergamon Press, Inc., Elmsford, N.Y. (1962). Has excellent crystal data on ferroelectric and piezoelectric behavior.

*Handbook of Chemistry and Physics*, ed. by R. C. Weast, The Chemical Rubber Co., Cleveland (1969). Has tables of properties.

"Materials Selector Issue" (annual issue of *Materials Engineering*). Has tables of properties.

## Magnetic Materials

Young, J. F., *Materials and Processes*, John Wiley & Sons, Inc., New York (1954), pp. 200–240. This contains a good discussion of the phenomenological aspects of magnetic behavior.

Bardell, P., *Magnetic Materials in the Electrical Industry*, Macdonald & Co. (Publishers) Limited, London (1960). The student should look over this book (or the next one) just to become more familiar with the many applications of magnetic materials.

Say, M. G., Ed., *Magnetic Alloys and Ferrites*, George Newnes Limited, Tower House, South Hampton St. London (1954). Printed by Whitefriars Press Limited, London.

Jacobs, I. S., "The Role of Magnetism in Technology," *Journal of Applied Physics* **40**, 917 (1969). This paper nicely describes the developments taking place in magnetic technology.

Becker, J. O., "Permanent Magnets," *Scientific American* (December, 1970), p. 92. This article discusses the breakthrough associated with the new samarium-cobalt magnets.

*Metals Handbook*, "Magnetic Materials," American Society for Metals, Cleveland (1948), pp. 587–600. Contains a good description of the structure-sensitive nature of many magnetic properties.

*Handbook of Chemistry and Physics*, ed. by R. C. Weast, The Chemical Rubber Co., Cleveland (1969). Has tables of properties.

## Superconductors

Fishlock, D., *A Guide to Superconductivity*, American Elsevier Publishing Company, Inc., New York (1969). Contains an excellent discussion of the technology of superconductors including their use in motors and generators, for power line transmission, and for memories in computers.

Newhouse, V. L., *Applied Superconductivity*, John Wiley & Sons, Inc., New York (1964). The student should at first use this book to become familiar with the many applications of superconductors.

*Handbook of Chemistry and Physics*, ed. by R. C. Weast, The Chemical Rubber Co., Cleveland (1969). Has tables of properties.

Roberts, B. W., "Superconducting Materials," *Progress in Cryogenics* **4**, 161 (1964). Has tables of properties.

## PROBLEMS

**3.1** List at least ten questions concerned with the reason *why* certain electrical properties exist. For example, why is there a linear relation between charge flux and electric field? Why is the resistivity of a metal relatively insensitive to structure at room temperature?

**3.2** The copper wire used for overhead electrical transmission is usually heavily cold worked. Does this affect the economics of electrical transmission greatly? Would it, if the transmission were at 4.2°K rather than around room temperature?

**3.3** Sodium metal is a very good conductor. Moreover, NaCl is much more available than CuS. Is it ever likely that sodium will be used extensively to transmit power?

**3.4** (a) Define Ohm's law in fundamental terms.
(b) When does this lead to

$$R = \rho \frac{l}{A}?$$

**3.5** What are typical room-temperature resistivity values for
(a) Metals?
(b) Semiconductors?
(c) Insulators?

**3.6** (a) What is the typical behavior of $\rho(T)$ for metals?
(b) What is meant by the resistivity ratio?
(c) How is $\rho$ (4.2°K) affected by impurities? Cold working?

**3.7** Assuming that there is only one type of charge carrier present, what quantities does the Hall coefficient measurement yield?

**3.8** (a) What is the typical behavior of $\rho(T)$ for semiconductors?
(b) How is this function affected by impurities?

**3.9** Give the analog for dielectric behavior of the following aspects of mechanical behavior:
(a) Hooke's law.
(b) Tensile fracture strength.
(c) Internal damping.

**3.10** Sketch the *P-E* curve (given that $E = E_0 \cos \omega t$) for
(a) Ideal dielectrics.
(b) Real dielectrics.
(c) Ferroelectrics.

**3.11** List at least ten questions concerned with the reason *why* certain magnetic properties exist. For example, why do magnetic materials reach a maximum magnetization as the field increases?

**3.12** For a 60-Hz transformer which operates at high fields, which is the more important design parameter:
(a) The maximum relative permeability?
(b) The initial relative permeability?

**3.13** Why are soft ferromagnetic cores used in transformers?

**3.14** Which of the quantities are essentially the same for magnetically hard and soft iron: $\mu_i$, $B_S$, $B_R$, $H_C$?

**3.15** Discuss why Si–Fe is used for utility equipment, Permalloy at audio frequencies, ferrites at higher frequencies, and garnets at very high frequencies.

**3.16** (a) Describe the advantage of using a ferromagnetic core in an electromagnet operating to $B = 1$ Wb/m².
(b) Would the same be true if $B = 15$ Wb/m²?

**3.17** Let $H_x$ be the value of $H$ for which $B = B_S$. Then the **squareness ratio** is defined as the ratio of $B$ evaluated at $-H_x/2$ divided by $B_S$.
(a) Why is this an important property for a magnetic core in a digital computer?
(b) The material Ferroxcube A has a high squareness ratio. Sketch its $B$ vs. $H$ curve.

**3.18** You are given a small uniform magnetic field $\mathbf{H} = (H_x, 0, 0)$. A sphere of ferromagnetic material is placed in this field. Sketch the flux lines.

**3.19** In some scientific experiments, it is necessary to have a region that is free of magnetic field. How would you achieve this?

**3.20** List at least five questions concerned with the reason *why* certain superconductive properties exist. For example, why does superconductivity exist?

**3.21** You are given a small uniform magnetic field $\mathbf{H}$. A sphere of Type I superconductor is placed in this field. Sketch the flux lines.

**3.22** You believe from theoretical reasons that $B = 0$ in superconducting lead. How would you show this experimentally?

**3.23** (a) What is the highest value of $T_c$ currently available?
(b) Why would superconductors be more interesting if $T_c > 77°$K? If $T_c > 300°$K?

# MORE INVOLVED PROBLEMS

**3.24** A solenoid is tightly wound with repeated layers. Its internal bore diameter is $D$, its length is $L$, and the thickness of all the windings is $t$. Assuming that the insulation is of negligible thickness the power dissipated is

$$P = \frac{\rho}{4\pi^2 \times 10^{-14}} \frac{D+t}{2} \frac{B^2}{t} L.$$

146   Electric and Magnetic Properties

Calculate the power needed to generate a magnetic induction of 10 Wb/m² in a solenoid with inner diameter 0.1 m and a length of 1 m if the windings are copper at room temperature.

**3.25** Design a device which will continuously multiply two variables together. Consider that variable $A$ is in the form of a current output from one device and variable $B$ is in the form of a current output from a second device. *Hint:* Consider the Hall effect.

**3.26** Given in the year 2000 that $2.4 \times 10^6$ tons of silicon iron will be used showing a hysteresis loss per cycle of 35 W-sec/m³ in 60-Hz utility equipment, estimate the annual kilowatt-hour dissipation due to $\oint H\, dB$ losses. How much would be saved if the dissipation could be reduced by 50% (assume 1 kWh costs $0.02)?

**3.27** The figure below shows an electromagnet. The gap $l_g$ is very small, say 1/100th of the average magnetic flux path ($2\pi R = l_m$). $l_g$ is also small with respect to the pole face diameter, so that there is very little fringing of the flux. In this case the magnetic induction at any point in the magnetic circuit can be written

$$B \doteq \frac{Ni}{[(2\pi R - l_g)/\mu] + (l_g/\mu_0)}.$$

Here $N$ is the total number of turns.

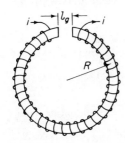

PROBLEM FIGURE 3.27.

(a) The effect of the presence of a core with $\mu/\mu_0 = 2000$ is to increase $B$ by what factor over the case of no core?
(b) What would the factor be if there were no gap with $\mu/\mu_0 = 2000$ (of course, this would be useless as an electromagnet)?

**3.28** The table below lists corresponding values of $H$ and $B$ for a specimen of commercial hot-rolled silicon steel, a material widely used in transformer cores
(a) Construct graphs of $B$ and $\mu$ as functions of $H$, in the range from $H = 0$ to $H = 1000$ A-turns/m.
(b) What is the maximum permeability?
(c) What is the initial permeability ($H = 0$)?
(d) What is the permeability when $H = 800{,}000$ A-turns/m?

| Magnetic Field Intensity, $H$ (A-turns/m) | Magnetic Induction, $B$ (Wb/m$^2$) | |
|---|---|---|
| 0 | 0 | |
| 10 | 0.050 | |
| 20 | 0.15 | |
| 40 | 0.43 | |
| 50 | 0.54 | |
| 60 | 0.62 | |
| 80 | 0.74 | |
| 100 | 0.83 | Magnetic |
| 150 | 0.98 | properties |
| 200 | 1.07 | of silicon |
| 500 | 1.27 | steel |
| 1,000 | 1.34 | |
| 10,000 | 1.65 | |
| 100,000 | 2.02 | |
| 800,000 | 2.92 | |

**3.29** Discuss the possible advantage of a composite coil magnet for reaching ultrahigh static magnetic fields (say 25 Wb/m$^2$): Make the inner core of ultrahigh-purity aluminum and the outer core of superconducting wire.

**3.30** Discuss the possibility of making a motor from hard superconductors—all magnetic fields are generated—no permanent ones.

# SOPHISTICATED PROBLEMS

**3.31** At very low temperatures, there is a size effect on resistivity.
  (a) Explain what is meant by this. This is discussed from a theoretical viewpoint by R. B. Dingle, *Proceedings of the Royal Society (London)* **A201**, 545 (1950), and E. H. Sondheimer, *Advances in Physics* **1**, 1 (1952); and from an experimental viewpoint by K. Forsvoll and I. Holweck, *Journal Applied Physics* **34**, 2230 (1963), and F. Dworschak, H. Schuster, H. Wollenberger, and J. Wurm, *Physica Status Solidi* **21**, 741 (1967).
  (b) Calculate the actual resistivity of a 0.003-cm-diameter wire of copper (whose bulk resistivity ratio is 50,000) at 4.2°K.
  (c) Calculate the apparent resistivity at 4.2°K of a thin foil of copper (1000 Å) whose bulk resistivity ratio is 50,000.

**3.32** At very high frequencies, electrical current moves to the surface of a conductor. The skin depth (layer in which most of the current passes) is readily derivable from Maxwell's equations [J. C. Slater and N. H. Frank, *Electromagnetism*, McGraw-Hill Book Company, Inc., New York (1947), p. 126].

What is the skin depth for copper at room temperature at 1 MHz? For copper at 4.2°K with a resistivity ratio of 100,000 at 10 MHz?

**3.33** Write an essay on the current state of development of magnetic tapes, including their production.

**3.34** What is meant by a synthetic hard superconductor?

**3.35** Superconducting solenoid magnets provide a convenient source of large magnetic fields. However, their inductance is large and they are subject to instabilities that limit the voltage that can be imposed across them. These limitations dictate some special characteristics for their power supplies. A simple series $RL$ circuit connected to a variable dc voltage source $V(t)$ gives a good representation of the problem in which the magnet provides the pure inductance $L$ and the lead resistance is $R$.

(a) Find the source $V(t)$ that will increase the current linearly with time at the rate of $dI/dt = \alpha$ until a current $I_m$ is reached and then hold the current steady at $I_m$. Plot it. *Hint:* Differential equation is

$$L\frac{dI}{dt} + IR = V(t) \quad \text{or} \quad L\frac{d^2q}{dt^2} + R\frac{dq}{dt} = V(t).$$

(b) If the maximum voltage that can be imposed across the magnet ($L$ alone) is 2 V, $L = 10$ H, and $R = 0.01$ Ω, find the maximum value of $\alpha$ and the time required to reach $I = 20$ A.

**3.36** Union Carbide has announced a development program for using very pure niobium wire in the superconducting state for transmission of ac current across a country. Describe the status of this project and the economic feasibility relative to presently used transmission methods.

**3.37** If all the electrical power consumed by New York City were transmitted by a single superconducting cable (dc), what diameter would be required?

# Prologue

The effects of temperature can be characterized as those due to a change in temperature such as *thermal expansion*, those due to a gradient of temperature such as *thermal conductivity*, and those in which the absolute temperature itself plays a major role in determining the existence or magnitude of a given property. Thermal expansion is shown to be an important parameter to the design engineer. *Thermal conductivity* and *thermal diffusivity* are important material properties in describing heat conduction in solids. Thermal conductivity is a structure-sensitive property of solids at low temperature. We have previously observed that ferroelectricity, magnetism, and superconductivity disappear at high temperatures. So does the solid state; the *melting temperature* of a solid is an important parameter. A number of material properties are described in terms of coefficients which depend exponentially on the negative reciprocal of temperature. Certain *thermoelectric properties* which form the basis for the thermoelectric refrigerator and the thermoelectric generator are introduced. Certain materials, when subjected to an appropriate temperature cycle (*heat treatment*), exhibit drastically altered properties. The concepts of *equilibrium* and *kinetics* in chemical processes such as transformation of diamond to graphite, oxidation of metals, etc., are described, as is the important concept of *metastability*. Chemical reaction rates usually depend strongly on temperature. Temperature fields clearly have a profound effect upon the behavior of materials. Finally, in the last section, there is a summary of the properties of materials which have been discussed thus far in the book.

# 4

# THERMAL AND CHEMICAL PROPERTIES

## 4.1 SPACE ROCKETS AND THERMAL EXPANSION COEFFICIENT

A space rocket with a pyroceram nose cone plummets into the atmosphere whereupon its surface temperature is quickly raised thousands of degrees. We say the material is subjected to **thermal shock.** To understand the significance of this to the design engineer we consider a more mundane problem, namely, rapidly placing a hot glass circular rod at temperature $T_2$ into water at temperature $T_1$. The outside surface readily approaches the temperature of the water (and attempts to shrink) while the rest of the glass remains hot (and hence does not shrink and tends to keep the outer layer from shrinking). The outer layer is therefore in tension and the maximum circumferential tensile stress (which occurs at the beginning of the ideal or instantaneous quench) is given by

$$\sigma = \frac{E\alpha}{(1-\nu)} \Delta T, \qquad (4.1.1)$$

where $E$ is Young's modulus, $\nu$ is Poisson's ratio, $\alpha$ is the thermal expansion coefficient, and $\Delta T = T_2 - T_1$.

**EXAMPLE 4.1.1**

The thermal expansion coefficient for a glass rod is $5 \times 10^{-6}/°K$. Assume $E = 10^7$ psi, $\nu = \frac{1}{4}$. What is the maximum tensile stress when a rod is quenched from 770°C into water at room temperature.

*Answer.*

$$\sigma = \frac{10^7}{0.75} \times 5 \times 10^{-6} \times 750$$
$$= 50{,}000 \text{ psi.}$$

Table 1.3.5 suggests a fracture stress of 10,000 psi for glass.

Since $\Delta T$ is fixed and $E$ does not vary much with temperature, we must choose $\alpha$ small to minimize the stresses due to thermal shock; in fact the ideal choice would be to choose a material for which $\alpha = 0$ over the temperature range of concern (which is why pyroceram was used for the nose cone).

The **linear thermal expansion coefficient** $\alpha$ is defined as the rate of fractional change of length with temperature

$$\alpha = \frac{1}{l}\frac{dl}{dT} = \frac{d\epsilon_r}{dT}. \tag{4.1.2}$$

Here $\epsilon_r$ is the real strain defined by Equation (2.1.3). For small temperature ranges $\alpha$ may be considered a constant and so there exists a *linear* strain response to the temperature change

$$\epsilon_r = \alpha \, \Delta T. \tag{4.1.3}$$

Some materials behave isotropically; i.e., the strain in all directions is the same when the temperature of the entire body is changed from $T_2$ to $T_1$, i.e., by $\Delta T$. However, many materials show anisotropic behavior, i.e., their linear thermal expansion coefficient varies with direction.

The **volume thermal expansion coefficient** $\beta$ is

$$\beta = \frac{1}{V}\left(\frac{dV}{dT}\right). \tag{4.1.4}$$

For isotropic materials $\beta = 3\alpha$.

There are some applications in which a value of $\alpha \approx 0$ is necessary: the thermal shock situation, tapes for surveying, and gyroscopes. There are

**Table 4.1.1.*** LINEAR THERMAL EXPANSION COEFFICIENT AT 20°C

| Material | $\alpha$ (°K$^{-1}$) |
|---|---|
| Pyroceram | $0.4 \times 10^{-6}$ |
| Invar | $0.9 \times 10^{-6}$ |
| Porcelain | $(1\text{–}4) \times 10^{-6}$ |
| Glass (crown) | $9 \times 10^{-6}$ |
| Steel | $11 \times 10^{-6}$ |
| Ice (at 0°C) | $50 \times 10^{-6}$ |
| Sodium | $60 \times 10^{-6}$ |

* From *Handbook of Chemistry and Physics*, Chemical Rubber Co., Cleveland (1969).

other applications in which $\alpha = $ constant $\neq 0$ is desired, e.g., liquid in thermometers, metals for bimetallic strip temperature indicators, etc. There are many applications in which a value of $\alpha = 0$ would be desirable but for which it is not possible. Examples are housing materials, bridge materials, highway materials, etc. Some values of thermal expansion coefficients are given in Table 4.1.1. Near room temperature, the variation of the thermal expansion coefficient of solids with temperature is small. At cryogenic temperatures (near absolute zero) the thermal expansion coefficient is quite small and is found to be roughly proportional to $T^3$; it goes to zero as $T$ goes to zero.

### Q. 4.1.1

When materials are quenched they are subjected to (a)_____ _____. The maximum stresses attained in a material in an ideal quench is proportional to the Young's modulus times the (b)_____ _____. The (c)_____ _____ is called the linear thermal expansion coefficient. It varies (d)_____ with temperature near room temperature and above but approaches (e)_____ as the temperature approaches absolute zero.

\* \* \*

## ANSWERS TO QUESTIONS

**4.1.1** (a) A thermal shock, (b) linear thermal expansion coefficient, (c) fractional change of length divided by the change in temperature, (d) slowly or not at all, (e) zero.

## 4.2 THERMAL CONDUCTIVITY

Just because the pyroceram nose cone of the previous section withstood the thermal shock does not mean the astronauts' problems are over. The surface of the nose cone is now at a very high temperature and this heat will be conducted into the interior. Assuming that the temperature on the outside is now fixed, the time $t_L$ in which the inside temperature changes by a fraction, approximately $1/e$, of the outside temperature is given by

$$t_L \approx \frac{L^2}{4D_T} \quad (4.2.1)$$

where $L$ is the thickness of the heat barrier and $D_T$ is the **thermal diffusivity**, which is given by

$$D_T = \frac{k_T}{\rho C_P}. \qquad (4.2.2)$$

The quantity $t_L$ is called the **thermal diffusion time.** Here $k_T$ is the thermal conductivity, $\rho$ is the density, and $C_P$ is the heat capacity per unit mass at constant pressure. It should be noted that $D_T$ is more nearly independent of temperature than are the other variables which will be studied later.

**EXAMPLE 4.2.1**

Assuming that $D_T = 8 \times 10^{-3}$ cm²/sec, that the initial temperature of the heat shield was 0°C, and that the outside temperature became 1000°C (and remained 1000°C), calculate the thermal diffusion time if $L = 2$ cm.

*Answer.* From (4.2.1) we have

$$t_L = 125 \text{ sec.}$$

This is an example of a **time-dependent heat flow problem.**

Equation (4.2.1) is derived in the previously cited book by Ruoff, *Materials Science*, Section 6.4. Actually a real nose cone would be subject to a much higher external temperature than 1000°C.

We are concerned here with the thermal conductivity. It was proposed by Fourier in 1824 that the fundamental law regarding the conduction of heat in one dimension, say the $x$ direction, is

$$J_x = -k_T \frac{dT}{dx}. \qquad (4.2.3)$$

Here $J_x$ is the quantity of heat passing a unit area whose normal is in the $x$ direction in unit time and $dT/dx$ is the *gradient* of the temperature for the one-dimensional case. Here $k_T$ is called the **thermal conductivity coefficient**; it is often found that over a moderate temperature range $k_T$ does not vary with temperature and can be considered a constant. The negative sign accounts for the fact that heat flows from hot to cold. A common unit for $k_T$ is Joules cm⁻¹ sec⁻¹ °K⁻¹ = watts cm⁻¹ °K⁻¹.

Table 4.2.1 shows some values of the coefficient $k_T$ for solids. Equation (4.2.3) is known as the **Fourier heat conduction law** for the one-dimensional case. The general form is

$$\mathbf{J} = -k_T \text{ grad } T. \qquad (4.2.4)$$

Sec. 4.2                                                          Thermal Conductivity    155

**Table 4.2.1.** THERMAL CONDUCTIVITY AT 20°C; THERMAL CONDUCTIVITIES OF METALS AND NONMETALS AND GASES AT 20°C

| Element | Thermal Conductivity (W/cm-°K) | Material | Thermal Conductivity (W/cm-°K) |
|---|---|---|---|
| Silver | 4.19 | Quartz | 0.05 |
| Copper | 3.94 | Gas carbon | 0.042 |
| Gold | 2.93 | Marble | 0.029 |
| Aluminum | 2.38 | Ice (at 0°C) | 0.021 |
| Magnesium | 1.68 | Mica | 0.008 |
| Tungsten | 1.63 | Glass (window) | 0.008 |
| Zinc | 1.13 | Sulfur | 0.0025 |
| Nickel | 0.88 | Paraffin wax | 0.0025 |
| Cadmium | 0.84 | Ebonite | 0.0017 |
| Iron | 0.71 | Sand | 0.0004 |
| Tin | 0.65 | Rockwool insulation | 0.0005 |
| Lead | 0.33 | Hydrogen (1 atm) | 0.0017 |
| Uranium | 0.25 | Oxygen (1 atm) | 0.00025 |
| Zirconium | 0.17 | Nitrogen (1 atm) | 0.00025 |
| Titanium | 0.15 | Argon (1 atm) | 0.00016 |
|  |  | Ethylene (1 atm) | 0.00017 |
|  |  | Vacuum (perfect) | (Zero) |

## EXAMPLE 4.2.2

Assume that the walls of a house have a 10-cm thick rockwool insulation and a total area of $2.5 \times 10^6$ cm². Calculate the rate at which heat is lost if the outside temperature is $-15°C$ and the inside temperature is $25°C$.

*Answer.*

$$Q_x = J_x A = k_T \times \tfrac{40}{10} A = 0.0005 \times 4 \times 2.5 \times 10^6$$
$$= 5000 \text{ J/sec} = 5 \text{ kW}.$$

In this particular problem we assume that the temperature of the wall does not change with time (this is called a **steady-state heat flow** problem). Hence the heat flux at each value of $x$ (normal to the wall) is the same.

## Q. 4.2.1

The quantity which determines the rate of temperature change inside a body when its surface temperature changes is called the (a)_____

_____. A heat conduction problem in which the temperature in the body changes with time is called (b)_____. The thermal diffusion time is given by (c)_____. The fundamental law of heat conduction is called (d)_____. It states that (e)_____. The proportionality coefficient is called the (f)_____. At room temperature this coefficient usually varies so slowly with temperature that it can be assumed not to vary with temperature; in that case the rate of heat flow is (g)_____ response.

* * *

The thermal conductivity of solids at low temperatures is a highly structure-sensitive property, as illustrated for copper of different purity in Figure 4.2.1. It should, however, also be noted that for all of the copper types $k_T$ will drop rapidly as $T \to 0$. Moreover, it should be noted that with increasing temperature $k_T$ for all types of copper approaches the same value (about 4 W cm$^{-1}$ °K$^{-1}$) near room temperature and is fairly constant thereafter.

Figure 4.2.2 shows the thermal conductivity versus temperature behavior of different $SiO_2$ compounds: quartz, quartz which has displaced atoms, and fused quartz (quartz glass). Although the composition is the same in all cases, the thermal conductivity varies by a factor of $10^4$.

Because of its high thermal conductivity and its low viscosity which aids heat transport by convection, hydrogen gas is often used as a coolant fluid in electrical generators. It is also a reducing agent so it helps prevent the degradation by oxidation of insulating plastics, etc., in the generator.

The most effective solid thermal insulators involve porous solids (preferably without interconnecting pores) since these can be filled with a gas which has a low conductivity (except for He and $H_2$ gas) relative to solids. A dilemma which aquanauts face is the need for good thermal insulating suits since ordinary insulating materials are crushed by the pressure.

## Q. 4.2.2

The thermal conductivity of materials at 1 atm varies by a factor of about (a)_____. The thermal conductivity of materials at high temperatures is a structure- (b)_____ property and at low temperatures a structure- (c)_____. The thermal conductivity of quartz at 4.2°K (liquid helium temperature) can be increased by a factor of about (d)_____ by heavy irradiation which causes (e)_____.

* * *

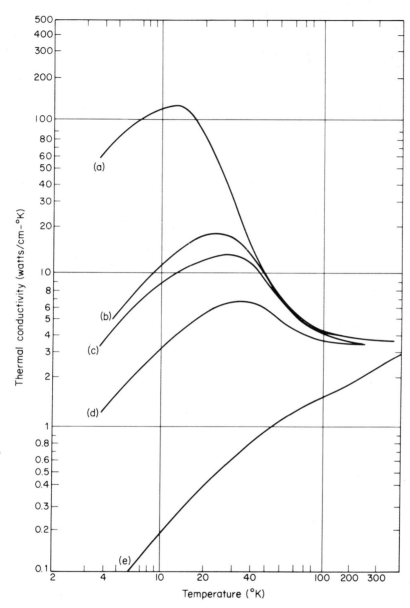

FIGURE 4.2.1. Low-temperature thermal conductivities of different copper specimens. (a) High-purity copper. (b) Coalesced copper. (c) Copper, electrolytic tough pitch. (d) Free-machining tellurium copper. (e) Copper, phosphorous deoxidized. [From R. B. Scott, *Cryogenic Engineering*, Van Nostrand Reinhold Company, New York (1959).]

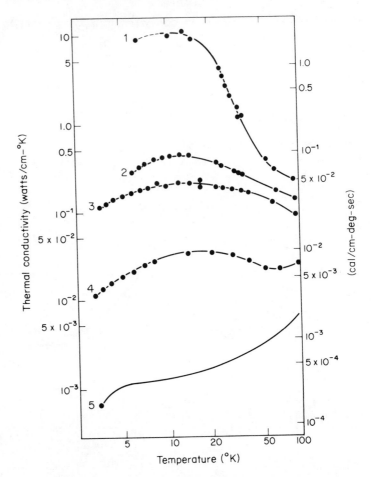

FIGURE 4.2.2. Thermal conductivity of quartz crystal, irradiated quartz crystal, and quartz glass. (1) Quartz crystal perpendicular to axis of specimen with 5 mm-square cross section. (2) Concentration of displaced atoms $1 \times 10^{-4}$. (3) $2 \times 10^{-4}$. (4) $2 \times 10^{-3}$. (5) Quartz glass. [From C. Kittel, *Introduction to Solid State Physics*, John Wiley & Sons, Inc., New York (1956).]

# ANSWERS TO QUESTIONS

**4.2.1** (a) Thermal diffusivity, (b) a time-dependent problem, (c) $t = L^2/4D_T$, (d) Fourier's law, (e) heat flux is proportional to the negative gradient of temperature, (f) thermal conductivity coefficient, (g) a linear.

**4.2.2** (a) $10^4$, (b) insensitive, (c) sensitive property, (d) $10^3$, (e) displaced atoms.

## 4.3 HEAT CAPACITY

The total energy of a block of material may be divided into the external energy and the internal energy. The external energy is associated with the rigid body motion of the block: There may be kinetic energy associated with the translation of the entire mass, kinetic energy associated with the rotation of the mass about the center of gravity, and potential energy associated with the height of the block. The internal energy is the energy present within the solid due to the presence of various fields, such as stress, electric, magnetic, and temperature fields.

As an example, the electric internal energy density is $u_{\text{elect}} = \int E\, dP$. $u_{\text{elect}}$ may vary from point to point in the solid. The total electric internal energy in the block of material is $U_{\text{elect}} = \int u_{\text{elect}}\, d\tau$, where $d\tau$ is an element of volume.

The thermal energy associated with a block of material is another example of internal energy. Whereas the change in internal energy associated with stress, electric, and magnetic fields is due to the work done on the material, the change in internal energy density due to temperature is a result of heat added to the system. If this heat is added under conditions in which the volume of the material is held constant, then the change in internal energy density is

$$\Delta u = \int C_v\, dT. \qquad (4.3.1)$$

To be consistent with our previous notation $\Delta u$ would have units of joules per cubic meter and $C_v$ would have units of joules per cubic meter-degrees Kelvin. However, it is also common to measure $\Delta u$ in units of energy per mass, or in units of energy per mole, where a mole refers to Avogadro's number of molecules. It is the latter definition which is used here. $\Delta u$ is then called the change in **molar internal energy** and the quantity $C_v$ is called the **molar heat capacity at constant volume.** It is defined by the rate of change of internal energy per mole with respect to temperature when volume is kept constant. It is useful to define $C_v$ because there are fairly wide temperature ranges over which it is often very nearly a constant and hence we can write

$$\Delta u = C_v\, \Delta T. \qquad (4.3.2)$$

In experiments and in engineering processes it is common to add heat while the pressure remains constant. We call the heat added to a mole of material at constant pressure, $\Delta h$, the change in **enthalpy** and we write

$$\Delta h = \int C_p\, dT, \qquad (4.3.3)$$

where $C_p$ is called the **molar heat capacity at constant pressure**. It is the amount of heat which must be added to a mole of material to increase the temperature by 1° at constant pressure.

IDEAL GASES. An ideal gas obeys the law $PV = nRT$. Here $P$ is pressure, $V$ is volume, $n$ is the number of moles, $R$ is the universal gas constant, and $T$ is the absolute temperature. Values for the universal gas constant are given in the table in the frontispiece. For an ideal monatomic gas such as argon,

$$C_v = \frac{3R}{2}. \tag{4.3.4}$$

For diatomic and other polyatomic gases, $C_v$ is larger than this. We shall not discuss this further here (the student can look in a good book on statistical mechanics). For an ideal gas

$$C_p = C_v + R. \tag{4.3.5}$$

SOLIDS. For solids the heat capacity at constant volume per mole of atoms at high temperature is

$$C_v \doteqdot 3R. \tag{4.3.6}$$

This is the **"law" of Dulong and Petit;** by high temperature we mean a temperature greater than the Debye temperature $\theta_D$ which is a characteristic temperature for every solid; it is roughly $\theta_D \simeq 0.2 T_M$, where $T_M$ is the melting temperature. Near the Debye temperature, $C_v$ begins to drop as the temperature is decreased and at very low temperatures $C_v$ is proportional to $T^3$. For solids $C_p \doteqdot C_v$.

## Q. 4.3.1

When heat is added to a material under conditions of constant volume there is a change of (a)_____ of the material; the change of this quantity per degree per mole of material is called the (b)_____. If heat is added at constant pressure there is a change of (c)_____ of the material. For an ideal monatomic gas $C_v =$ (d)_____ while for a solid at high temperatures $C_v =$ (e)_____. Below the (f)_____ temperature the molar heat capacity (or the specific heat per mole of atoms) of solids (g)_____

\* \* \*

## ANSWERS TO QUESTIONS

**4.3.1** (a) Internal energy, (b) molar heat capacity or the specific heat per mole, (c) enthalpy, (d) $\tfrac{3}{2}R$, (e) $3R$, (f) Debye, (g) decreases rapidly toward zero.

## 4.4 ENERGY CHANGES DURING PHASE TRANSFORMATION

When materials undergo simple phase transformations such as melting, vaporization, or sublimation the process is associated with a finite quantity of heat absorbed over an infinitesimal temperature range at constant pressure. We define

$\Delta h_m$ = heat absorbed at constant pressure per mole of solid when it melts; this is called the **heat of melting** (fusion).

$\Delta h_v$ = heat absorbed at constant pressure per mole of liquid when it vaporizes; this is called the **heat of vaporization.**

$\Delta h_s$ = heat absorbed at constant pressure per mole of solid when it vaporizes; this is called the **heat of sublimation.**

We saw in Example 4.2.1 that the inside wall of the astronauts' nose cone heated up very rapidly in a few seconds. One way to prevent this would be to coat the nose cone with an ablative compound which would absorb considerable heat in *decomposing* and *vaporizing;* this heat would simply be carried away with the air rushing past. We shall consider the energies associated with the decomposition of molecules in Chapter 5. In general $\Delta h_m$ is much less than $\Delta h_v$ and $\Delta h_s$; the latter are nearly equal. A few typical examples are shown in Table 4.4.1 in units of kilocalories per mole and in

Table 4.4.1. HEATS OF MELTING AND VAPORIZATION

| Material | $\Delta h_m$ | | $\Delta h_v$ | |
|---|---|---|---|---|
| | kcal/mole | $R°K$ | kcal/mole | $R°K$ |
| $H_2$ | 0.028 | 14 | 0.22 | 110 |
| $H_2O$ | 1.43 | 718 | 11.3 | 5,780 |
| Na | 0.63 | 304 | 24.6 | 12,400 |
| Al | 2.55 | 1280 | 67.6 | 34,000 |
| Fe | 3.56 | 1790 | 96.5 | 48,500 |
| NaCl | 7.22 | 3630 | 183 | 92,000 |
| KCl | 6.41 | 3220 | 165 | 83,000 |

**162** *Thermal and Chemical Properties*

units of $R°K$ where $R$ is the universal gas constant. The quantity $\Delta h_v$ and $\Delta h_s$ are good measures of the strength of bonding between the atoms in the liquid and solid, respectively.

**EXAMPLE 4.4.1**

Using the fact that $T(°K) = T(°C) + 273.16$, compute the melting temperature $T_m$ for H$_2$O, Na, Al, Fe, and NaCl using the data of Table 4.5.1. Then use this result to establish an approximate rule for the temperature change the solid would have to undergo without melting to absorb the heat which it absorbs at a constant temperature upon melting.

*Answer.* The heat capacity per mole of solid Na, Al, and Fe is $3R$, for NaCl it is $6R$, and for ice it is $9R$. Your studies should show that

$$\Delta h_m \simeq C_p \frac{T_m}{3}.$$

Thus the temperature change is about $T_m/3$.

## Q. 4.4.1

The energy which must be added to a mole of solid at the melting point to melt it is called the (a)_____; for ordinary solids this is about equal to the energy added when heating the solid so that its temperature increases by about (b)_____. The heat of sublimation or heat of vaporization is about (c)_____ times as large as the heat of fusion. The heat of vaporization (or sublimation) is a good measure of the (d)_____ _____.

\* \* \*

# ANSWERS TO QUESTIONS

**4.4.1** Heat of fusion, (b) one third of the absolute melting temperature, (c) 25 (see Table 4.4.1), (d) bonding energy of the atoms in the solid.

## 4.5 TEMPERATURE EFFECTS ON PROPERTIES

So far we have discussed the effects of change of temperature fields and of gradients in temperature fields. We now want to discuss some of the direct effects of the temperature field itself, i.e., $T$ (and not just $\Delta T$ or grad $T$).

EFFECTS DESTROYED BY TEMPERATURE. We have already noted that sufficiently high temperatures destroy

1. Ferromagnetism.
2. Superconductivity.
3. Ferroelectricity.
4. Shear rigidity (solids melt or vaporize).
5. Togetherness (liquids which occupy a fixed volume vaporize to gases which occupy as large a volume as they are given).

Temperature also has a profound effect on the mechanical yield strength, on the rate of creep [see Equation (2.2.2)], on the fluidity of liquids [see Equation (2.3.5)], on the electrical conductivity (or resistivity) of metals (see Figure 3.3.2) and semiconductors [see Equation (3.3.1)], and on the thermal conductivity of solids (see Figure 4.2.1).

As temperature increases, the rate of degradation of rubber increases, the rate of oxidation in various substances increases, and the rate of chemical reactions in general increases.

Because temperature so strongly affects most of the properties of materials, be they gases, liquids, or solids, it is very important to always specify the temperature field. Along with the temperature field other fields such as magnetic, electrical, stress, and gravitational fields may have more or less importance in determining the properties of materials. The detailed study of temperature and its significance is called thermodynamics and statistical mechanics.

One of the important general properties of a material is its **melting temperature.** As we shall see in Chapter 5, solids in which the atoms or molecules are weakly bonded to each other have low melting points, while the strongly bonded solids have high melting points (and high boiling points). Examples of melting temperatures are shown in Table 4.5.1. Materials which have very high melting points (above 1600°C) are called **refractory materials.**

Many properties are simple functions of $T/T_m$; i.e., the properties of different materials are primarily determined by the **fraction of the melting point** $T/T_m$ at which they are held. As an example the yield stress in a wide variety of materials drops off rapidly when $T/T_m \approx 0.5$.

EXPONENTIAL TEMPERATURE EFFECTS. A number of material properties are macroscopically defined in terms of material coefficients which depend exponentially on the negative reciprocal of *absolute* temperature:

$$\text{material coefficient} \propto e^{-Q/RT}.$$

Examples are creep rate [Equation (2.2.2)], fluidity of liquids [Equation (2.3.5)], conductivity of semiconductors [see Equation (3.3.1) for the resistivity], diffusion coefficients, chemical reaction rates, rate of electron emission from a filament (hot wire), vapor pressure above a liquid (where $Q$ means the heat of vaporization) or solid (where $Q$ means the heat of sublimation), etc. We particularly emphasize the exponential temperature

Table 4.5.1. MELTING TEMPERATURES

| Material | °C | Material | °C |
|---|---|---|---|
|  |  | Al | 660 |
| $H_2$ | −259.14 | NaCl | 801 |
|  |  | Cu | 1083 |
| Ar | −189.2 | Fe | 1535 |
| $H_2O$ | 0 | $\alpha$-$Al_2O_3$ | 2015 |
| Na | 97.8 | Mo | 2610 |
|  |  | MgO | 2800 |
| Polyethylene* | 140 | W | 3380 |
| Nylon 610 | 220 | Graphite† | ~3400 |
| Nylon 66 | 260 |  |  |
| Pb | 327.3 | HfC | 3890 |

\* Infinite chain extrapolation.
† Sublimes.

dependence because the properties vary so strongly with temperature in these cases. For example the self-diffusion coefficient for copper is $10^{-8}$ cm²/sec just below $T_M$, $10^{-16}$ cm²/sec at $T_M/2$, and $10^{-32}$ cm²/sec at $T_M/4$.

### Q. 4.5.1

Five states of matter or effects destroyed by high temperature are (a)_____. Name five properties which are strongly affected by temperature (b)_____. Materials which have very high melting temperatures are known as (c)_____. Two materials whose melting points are in excess of 2000°C are (d)_____. The temperature dependence of many properties is primarily determined by (e)_____. List at least five properties which depend exponentially on the negative reciprocal of absolute temperature: (f)_____.

\* \* \*

# ANSWERS TO QUESTIONS

**4.5.1** (a) Ferroelectric, ferromagnetic, superconductive, solid, liquid; (b) yield strength, creep rate, fluidity or viscosity, electrical conductivity, thermal conductivity; (c) refractories; (d) $Al_2O_3$ and $MgO$; $Mo$ and $W$; graphite; (e) $T/T_m$; (f) creep, fluidity, conductivity of semiconductors, diffusion in solids, thermionic emission, vapor pressure above a liquid or solid.

## 4.6 ELECTRICAL-THERMAL COUPLING

We have seen previously how a mechanical stress creates a voltage across certain materials (piezoelectric effect). A temperature change causes a strain (thermal expansion), and if the material is constrained, it causes a stress; this is the **thermoelastic effect**.

A tourmaline crystal (an electrical insulator) whose temperature is changed by 1°C becomes polarized; this is the **pyroelectric effect** (a voltage of 740 V across a 1 cm-thick crystal would produce the same polarization). Only certain materials show a pyroelectric effect.

Temperature and electrical behavior are also coupled in interesting and useful ways in conductors.

THERMOCOUPLES. Figure 4.6.1 shows two dissimilar electrical conductors with junctions at different temperatures $T_1$ and $T_2$. An electric

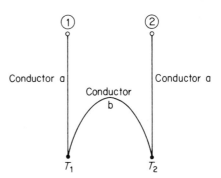

FIGURE 4.6.1. Thermocouple.

potential difference is generated between point one and point two. This electric potential difference depends on the materials of conductors $a$ and $b$ and on the difference between the temperatures $T_1$ and $T_2$. Assuming $T_1$ is fixed at 0°C, the ice point, the electric potential difference will be directly related to $T_2$. With suitable calibration this device can be used to measure temperature potentiometrically. It is known as a **thermocouple**. The

**Thermal and Chemical Properties**

thermal generation of electric potential differences is known as the **Seebeck effect.** Typical thermocouple materials are shown in Table 4.6.1. The

Table 4.6.1.* THERMAL EMF'S OF METALS RELATIVE TO PLATINUM (MILLIVOLTS; COLD JUNCTION AT 0°C)

| Temp. of Hot Junction (°C) | Copper | Iron | Constantan | 90% Pt, 10% Rh | Chromel | Alumel |
|---|---|---|---|---|---|---|
| 100 | 0.76 | 1.98 | −3.51 | 0.643 | 2.81 | −1.29 |
| 200 | 1.83 | 3.69 | −7.45 | 1.436 | 5.96 | −2.17 |
| 300 | 3.15 | 5.03 | −11.71 | 2.316 | 9.32 | −2.89 |
| 400 | 4.68 | 6.08 | −16.19 | 3.251 | 12.75 | −3.64 |
| 500 | 6.41 | 7.00 | −20.79 | 4.221 | 16.21 | −4.43 |
| 600 | 8.34 | 8.02 | −25.47 | 5.224 | 19.62 | −5.28 |
| 700 | 10.49 | 9.34 | −30.18 | 6.260 | 22.96 | −6.18 |
| 800 | 12.84 | 11.09 | −34.86 | 7.329 | 26.23 | −7.08 |
| 900 | 15.41 | 13.10 | −39.45 | 8.432 | 29.41 | −7.95 |
| 1000 | 18.20 | 14.64 | −43.92 | 9.570 | 32.52 | −8.79 |
| 1100 | — | — | — | 10.741 | 35.56 | −9.58 |
| 1200 | — | — | — | 11.935 | 38.51 | −10.34 |
| 1300 | — | — | — | 13.138 | 41.35 | −11.06 |
| 1400 | — | — | — | 14.337 | 44.04 | −11.77 |

* From *American Institute of Physics Handbook*, McGraw-Hill Book Company, Inc., New York (1957).

thermal emf for copper-constantan with the hot junction at 800°C and cold junction at 0°C would be $12.84 - (-34.86) = 47.70$ mV.

**EXAMPLE 4.6.1**

Calculate the average thermal emf per degree between 700 and 800°C for copper-constantan.

*Answer.* At 700°C the thermal emf is $10.49 - (-30.18) = 40.67$ mV and hence

$$\frac{47.70 - 40.67}{100} = \frac{0.0703 \text{ mV}}{\text{deg}}.$$

The emf/degree is called the **thermoelectric power.**

HEAT PUMP. Consider now the thermocouple of Figure 4.6.2 in which we pass a current $i$. In this case heat is generated at one of the junctions and absorbed at the other according to

$$\dot{Q}_1 = -\pi_{ab}J, \qquad (4.6.1)$$

$$\dot{Q}_2 = \pi_{ba}J. \qquad (4.6.2)$$

Here $\dot{Q}$ is the rate of heat production at a junction and $J$ is the electrical current density. $\pi_{ab}$ and $\pi_{ba}$ are known as the **Peltier coefficients.** The effect is known as the **Peltier effect.** The device is a thermoelectric heat pump, or a thermoelectric refrigerator, since one of the junctions will be

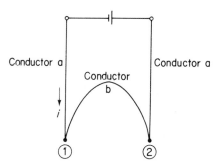

FIGURE 4.6.2. Thermoelectric heat pump.

cooled and the other will be heated. The efficiency of such a device depends on $i^2R$ losses in the circuit (and hence on electrical conductivities) and on losses due to heat flow down the temperature gradients (and hence on the thermal conductivities) and on radiation. A junction of $p$-type versus $n$-type $Bi_2Te_3$ (semiconductor) has a thermoelectric power of about 0.25 mV/°C.

Using a semiconducting alloy ($Bi_2Se_3$–$Bi_2Te_3$) for one conductor with platinum as the other conductor a maximum cooling to $-37°C$ can be obtained from $0°C$. Operated in the reverse fashion (one junction kept cold, the other heated) the thermocouple becomes a thermoelectric generator. Thermoelectric devices are described by S. I. Freedman in G. W. Sutton, Ed., *Direct Energy Conversion*, McGraw-Hill Book Company, Inc., New York (1966), p. 105.

## Q. 4.6.1

If a clamped solid object is heated, it develops (a)_____. This is known as the (b)_____. An example of a pyroelectric material is (c)_____. The (d)_____ effect is the basis of a temperature-measuring device known as the thermocouple. If a current is passed through a thermocouple, (e)_____ is produced at one junction and (f)_____ at the other, an effect known as the (g)_____ effect.

The efficiency of a thermoelectric refrigerator is increased if the (h) _____ conductivity is high but the (i) _____ conductivity is small.

\* \* \*

# ANSWERS TO QUESTIONS

**4.6.1** (a) Stresses, (b) thermoelastic effect, (c) tourmaline, (d) Seebeck, (e) heat, (f) absorbed, (g) Peltier, (h) electrical, (i) thermal.

## 4.7 HEAT TREATMENT

The metallurgist, the ceramicist, the glass maker, and the polymer chemist often subject a material to a certain temperature cycle to achieve desired properties. In the simplest case a solid is melted and the lower viscosity fluid can then be cast into a desired shape. We have already mentioned the annealing of cold-worked materials in which the result of the annealing process is to decrease the yield stress. There are two other very important heat treatment processes which are discussed briefly here.

HEAT TREATMENT OF STEEL. Steel is an alloy of iron and carbon (up to 2% by weight) and possibly other elements. When certain of these alloys are held near 850°C for, say, 1 hr (this is called **austenitizing**) and then **quenched** rapidly in water (or oil) at room temperature, there is a large increase in strength and hardness. The tensile yield stress may increase from 80,000 to 240,000 psi. Whereas the material could be readily machined before the heat treatment, it can only be machined by grinding after the heat treatment. Because it usually is too hard and brittle after this process, it is often **tempered** (held at an intermediate temperature for a while) to increase its toughness. In later chapters we shall study from the atomic viewpoint how and why this three-step heat treatment works.

AGE PRECIPITATION HARDENING. Certain alloys such as aluminum with 4% Cu (by weight) and perhaps some other alloying elements can have their yield stress increased in the following way. They are first held at 550°C for 1 hr (**solution heat treatment**) and then **quenched** in water to room temperature. Unlike the steels mentioned earlier they are now very soft. If the temperature is increased to say 100°C and held there for several hours (**aging**) and then cooled to room temperature, the alloy will be hard and have a high yield stress. The reason this three-step heat treatment works will also be studied later from the atomic viewpoint.

Sec. 4.8	Chemical Properties	169

**EXAMPLE 4.7.1**

You want to waste as little energy flattening rivet heads on an airplane construction job as possible. Suggest a way of using an appropriate age precipitation hardenable alloy for this job.

*Answer.* Choose an alloy which ages at room temperature. Solution heat-treat the rivets, quench the rivets, and store them cold until used. After installation they will age at room temperature and will reach high strength in several hours. *Note*: Certain aircraft have over 100,000 rivets.

OTHER PROPERTIES. Heat treatments not only have profound effects on various mechanical properties, but also on thermal, magnetic, electric, superconducting, optical properties, etc. We have already seen some examples of this and in later chapters we shall see more examples and the reasons these effects occur.

### Q. 4.7.1

In a heat treatment, a material is (a)_____ _____. Heat treatments can often profoundly affect the internal structure of a material and hence can profoundly affect the (b) _____ _____. The strengthening of steel involves a (c)_____ process. Certain aluminum alloys can be hardened by a heat treatment process known as (d)_____ _____.

\* \* \*

## ANSWERS TO QUESTIONS

**4.7.1** (a) Subjected to a specified temperature cycle; (b) properties, particularly the structure-sensitive properties; (c) three-step; (d) age precipitation hardening.

## 4.8 CHEMICAL PROPERTIES

Rapid chemical reactions are desirable for chemical etching (as in the making of microelectronic circuits in Section 1.1), chemical milling, or etching for microscopic examination of grain boundaries, for producing power in a battery or fuel cell, or for producing iron by the reduction of iron oxides. On the other hand a chemical reaction may be undesirable if it leads to excess oxidation of a material, damaging corrosion, photodecomposition, biodegradation, etc.

In studying chemical reactions, there are two factors to consider: equilibrium and kinetics. This will be illustrated with the reaction

$$4Cu + O_2(g) \rightarrow 2Cu_2O. \tag{4.8.1}$$

It is found for many metals in which the solubility of oxygen in the metal is negligible and in which the oxide is stoichiometric that the equilibrium oxygen pressure is given by

$$P_0 \text{ (atm)} = e^{-\Delta S^0/R} e^{\Delta H^0/RT}. \tag{4.8.2}$$

Here the quantities $\Delta S^0$ and $\Delta H^0$ are characteristic of the specific reaction and to a first approximation can be considered independent of temperature.

If at a specific temperature, the oxygen pressure is greater than the equilibrium pressure at that temperature, then the oxygen will *tend* to react with copper to produce the oxide until the pressure is reduced to the equilibrium pressure at that temperature. Thus if

$p > p_0$ oxide *tends* to form,

$p < p_0$ oxide *tends* to decompose.

For the above reaction $p_0 \approx 10^{-42}$ atm at 0°C and $p_0 \approx 10^{-8}$ atm at 1000°C. Clearly there is a strong *tendency* for copper to react with the oxygen in air (0.2 atm) to produce $Cu_2O$; however, from everyday experience we realize that it must react only very slowly at room temperature or else all copper pipe and copper containers would be virtually completely converted to the oxide. However, we find experimentally that at 1000°C, the copper reacts more rapidly (but not so rapidly that the rate cannot be easily measured). It is also true that if the oxygen pressure is *very* low at 1000°C, the $Cu_2O$ will decompose at a reasonably rapid rate until the oxygen pressure eventually increases to the equilibrium pressure. Thus for a noticeable reaction to actually take place two conditions must be satisfied; first, there must be a tendency for the reaction to occur; second, the rate must be sufficiently rapid. Another example which illustrates the importance of kinetics involves two crystalline forms of carbon, diamond and graphite. Experiments show that at pressures below

$$P_d = (0.027T(°K) + 7.1) \text{ kbars}$$

graphite is stable while at pressures above $P_d$ diamonds are stable. Thus at 300°K (room temperature), $P_d = 15.2$ kbars. Graphite is therefore the equilibrium structure of carbon at 300°K and 1 atm; graphite is called the

**equilibrium structure** or the **stable form**. However, diamonds also exist at room temperature. There is a tendency for these to convert to graphite but no perceptible conversion takes place because the rate of the reaction is so slow; diamond is called the **metastable structure** of carbon at 300°K and 1 atm. It is called metastable, instead of unstable, because while it has the tendency to change to graphite it does not do so. At 1700°C and 1 atm, the conversion takes place at a reasonable rate (a few hours). Usually the rate of chemical reaction varies extremely rapidly with temperature, often being of the form

$$\text{rate} \propto e^{-Q_{act}/RT}, \tag{4.8.3}$$

where $Q_{act}$ is the activation energy which is a characteristic of a specific reaction.

### Q. 4.8.1

There are two important factors in studying chemical reactions: (a)_____. For a noticeable reaction to actually take place there must (b)_____. If both Cu and $Cu_2O$ are present in a closed (inert) chamber with oxygen at $10^{-1}$ atm at 1000°C more (c)_____ will be produced. If both Cu and $Cu_2O$ are present at 1000°C in a closed (inert) chamber with oxygen at $10^{-14}$ atm more (d)_____ will be produced. Copper metal in the presence of air at 1 atm at 300°K could be called (e)_____. Diamonds at 1 atm and 300°K are (f)_____ while graphite is (g)_____. It is not possible to convert graphite to diamonds at 300°K at 20 kbars because (h)_____ even though there is (i) _____. The rate of a chemical reaction often varies with temperature according to (j)_____.

\* \* \*

Various aspects of chemical behavior will be discussed further in Chapters 8, 9, 10, and 11. Specific aspects of corrosion will be covered in Chapter 11. It is pointed out that mechanically stressing a specimen in a corrosive environment leads to even more deleterious effects than the chemical environment alone. An example of the results of stress corrosion is shown in Figure 4.8.1.

Although corrosion of metals accounts for a loss of perhaps $10 billion a year, degradation of polymers is also a costly problem. For example, in the development of electrical cable insulation, certain electrical properties must be present in required magnitudes (dielectric constant, loss angle,

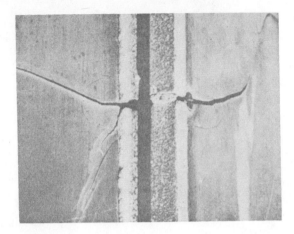

FIGURE 4.8.1. Example of stress corrosion.

dielectric strength) not only initially but for periods of many years, say half a century or more for buried cable. The development of dielectric materials whose electrical properties survive the test of time is a major task.

## ANSWERS TO QUESTIONS

**4.8.1** (a) Equilibrium and kinetics, (b) be a tendency for the reaction to occur and the rate must be sufficiently rapid, (c) $Cu_2O$, (d) $Cu$ and $O_2$, (e) a metastable material, (f) a metastable structure, (g) a stable structure, (h) the rate of conversion is too slow, (i) a tendency for this transformation to occur, (j) $\exp(-Q_{act}/RT)$.

## 4.9 SUMMARY OF PROPERTIES

LINEAR CONSERVATIVE PROPERTIES. Examples of linear conservative properties are

$$\left.\begin{array}{r}\sigma = E\epsilon,\\ \tau = G\gamma,\end{array}\right\} \text{Hooke's law} \qquad (1.3.7)$$
$$(1.3.18)$$

$$D = \epsilon E, \qquad (3.4.6)$$

$$B = \mu H. \qquad (3.6.7)$$

We recall, however, that (1.3.7) and (1.3.18) apply for small strains only, that (3.4.6) does not hold for ferroelectric materials (unless $\epsilon$ is allowed to vary with $E$), and that (3.6.7) holds only approximately for soft magnetic materials and not at all for hard magnetic materials (unless $\mu$ is allowed to vary with $H$).

LINEAR DISSIPATIVE PROPERTIES.  Examples are

$$J = \sigma E = -\sigma \operatorname{grad} V \quad \text{Ohms law} \quad (3.2.6)$$

$$J = -k_T \operatorname{grad} T \quad \text{Fourier's law} \quad (4.2.4)$$

$$J = -k_p \operatorname{grad} P \quad \text{Darcy's law} \quad (2.3.6)$$

$$\tau = -\eta \frac{dv}{dr} \quad \text{Newtonian flow} \quad (2.3.2)$$

LIMITS.  Many of the properties which we have studied are simple limiting values beyond which a material undergoes drastic changes in its behavior. Examples are

Yield stress.
Fracture strength.
Fatigue limit.
Dielectric strength.
Polarization saturation.
Saturation induction.
Critical field.
Transformation temperatures.
Transformation pressure.

Transformation temperatures include melting and boiling points; ferromagnetic, ferroelectric, and superconducting transition temperatures; ductile to brittle transition temperature in steel; temperature at which the yield strength drops rapidly; etc.

HYSTERESIS LOOPS.  There are hysteresis loops associated with

Ideal plastic behavior.
Internal friction.
Dielectric losses.
Ferromagnetism.
Ferroelectricity.
Nonideal superconductivity.

### Q. 4.9.1

Properties are often characterized by (a)_____ relations. This is true both for conservative behavior and (b)_____. Properties are also characterized by (c)_____ and (d)_____ _____.

\* \* \*

STRUCTURE SENSITIVITY. We have previously noted the distinction between structure-sensitive and structure-insensitive properties. Some examples are shown in Table 4.9.1.

**Table 4.9.1.** STRUCTURE-SENSITIVE AND STRUCTURE-INSENSITIVE PROPERTIES

| Type of Property | Structure-Insensitive | Structure-Sensitive |
|---|---|---|
| Mechanical | Density<br>Elastic moduli | Mechanical loss angle<br>Tensile yield stress<br>Fracture strength<br>Plasticity |
| Thermal | Thermal expansion (high temperature)<br>Melting point<br>Specific heat (high temperature)<br>Heat of fusion | Thermal conductivity (at low temperatures) |
| Electrical | Resistivity (metallic) at high temperatures<br>Dielectric constant<br>Saturation polarization of ferroelectric materials | Resistivity at low temperatures in semiconductors and metals<br>Electrical loss angle |
| Magnetic | Paramagnetic and diamagnetic properties<br>Saturation induction of ferromagnetic materials | Ferromagnetic properties (including magnetostriction) except saturation induction |
| Superconductive | Transition temperature | Current carrying capacity |

## ANSWERS TO QUESTIONS

**4.9.1** (a) Linear, (b) dissipative behavior, (c) limits, (d) hysteresis losses.

## REFERENCES

*Handbook of Chemistry and Physics*, edited by R. C. Weast, Chemical Rubber Co., Cleveland (1969). Contains extensive data on melting points, specific heats, heats of vaporization, thermal conductivities, etc.

Scott, R. B., *Cryogenic Engineering*, Van Nostrand Reinhold Company, New York (1959). Contains considerable data on thermal properties, particularly in the low-temperature range where these properties may vary rapidly with temperature. Also contains a convenient table for conversion of units.

Gebhart, B. I., *Heat Transfer*, McGraw-Hill Book Company, Inc., New York (1961).

Carslaw, H. S., and Jaeger, J. C., *Conduction of Heat in Solids*, Oxford University Press, Inc., New York (1959). A large number of problems involving heat conduction through differently shaped objects and methods for solving them are discussed.

# PROBLEMS

**4.1** List 12 "why" questions about thermal properties

**4.2** (a) Explain why many glasses fracture when taken from scalding water and placed in cold water.
(b) How would you solve this problem?

**4.3** Suppose that in addition to the insulated wall space, the house of Example 4.2.2 has an additional area of $1.25 \times 10^5$ cm² of windows which are 0.3 cm thick. Compute the maximum rate at which heat is lost through the glass.

**4.4** Design two devices which depend on difference in thermal expansion coefficient.

**4.5** Ice cubes at $-10°C$ fracture when dropped into toluene at $-35°C$. Estimate the maximum tensile strength of the ice. Assume $\nu = \frac{1}{3}$ and $E = 3 \times 10^5$ psi.

**4.6** Why is the thermal diffusivity and not just thermal conductivity the important parameter in time-dependent heat flow?

**4.7** Why do $\rho$ (mass density) and $C_P$ (heat capacity per unit mass) not enter into steady-state heat flow problems?

**4.8** There is an interesting relationship for metals, the Wiedemann-Franz relationship, which states that $k_T/\sigma$ (the ratio of thermal to electrical conductivity) or $k_T\rho$ = constant times $T(°K)$. Show for five metals that the product $k_T\rho$ is approximately equal to the same constant.

**4.9** To what do you attribute the different $k_T(T)$ curves for copper in Figure 4.2.1?

**4.10** Is there any difference in $k_T$ of crystalline quartz ($SiO_2$) and noncrystalline silica ($SiO_2$)?

**4.11** Clothing, the thermal insulation in the walls of houses, etc., have a $k_T$ approximately equal to what?

**4.12** What is the value of the molar heat capacity of crystals at high temperatures, say, $T_m/2$?

**4.13** (a) Illustrate the Seebeck and Peltier effects.
(b) What possible application does each have?

**4.14** What three factors determine the efficiency of a Peltier refrigeration system?

**4.15** Criticize the statement: Diamonds are unstable at room temperature.

**4.16** Suppose that the graphitization of diamond obeys (4.8.3) with $Q_{act} = 150$ kcal/mole. How many times faster is this reaction at 2000°K than at 300°K?

## MORE INVOLVED PROBLEMS

**4.17** Show for materials in which the linear thermal expansion coefficient is isotropic that $\beta = 3\alpha$.

**4.18** In Section 3.3 we considered a 10-Wb/m² solenoid magnet. This magnet generated 6300 kW of heat. Assuming that water is used to cool it and that the water temperature is changed by 20°C, how many gallons are needed per hour?

**4.19** Describe three temperature-measuring devices where the temperature is measured electrically.

## SOPHISTICATED PROBLEMS

**4.20** (a) Show that if heat flows in the $x$ direction only,

$$\frac{\partial J}{\partial x} = -C_P \rho \frac{\partial T}{\partial t}.$$

(b) Combine this with Fourier's law to show that

$$D_T \frac{\partial^2 T}{\partial x^2} = \frac{\partial T}{\partial t}$$

for the case where $k_T$, $C_P$, and $\rho$ do not vary with temperature.

**4.21** Show that $x^2/D_T t$ or a multiple thereof is a fundamental quantity in one-dimensional time-dependent heat flow. See Carslaw and Jaeger in the References.

# Prologue

In previous chapters, the macroscopic behavior of materials was discussed. To understand why these properties exist (and how to modify them in a systematic scientific fashion, as well as to make possible the discovery of new types of behavior), it is necessary to study materials at the atomic level. We begin with a study of the *Bohr model* of the hydrogen atom. We then introduce the concepts of modern quantum mechanics. Next the hydrogen atom is studied in this context. Then we proceed to consider more complicated atoms and the atomic table. We note that the outer electron is held to the atom by an energy of about 10 eV and that atoms have a radius of about 1 Å. The *particle in a box model* is studied.

The concepts of the nature of bonding between atoms which may result in the formation of molecules or condensed phases is discussed. *Ionic bonding* and the Born model are discussed. *Covalent bonding* and *metallic bonding* are described. *Secondary-type* bonds and their origin are studied. It is important to develop a feeling for the nature and the order of magnitude of these different types of bonds.

The origins on the atomic scale of *electric dipoles, polarization, magnetic dipoles,* and *magnetization* are discussed. Finally, the fact that the electron has an intrinsic magnetic dipole moment (just as it has an intrinsic charge or mass) is discussed.

# 5

# BINDING IN ATOMS, MOLECULES, AND CRYSTALS

## 5.1 WAVE MECHANICS

No great discovery was ever made without a bold guess.—Ascribed to Isaac Newton (1642–1727).

Classical Newtonian mechanics provides an adequate mathematical description of the behavior of large slow-moving objects. Thus the motion of objects with a fixed mass is described by

$$M\mathbf{a} = \mathbf{F}, \tag{5.1.1}$$

where $M$ is the rest mass and $\mathbf{a}$ is the acceleration caused by the force $\mathbf{F}$. We recall that force $\mathbf{F}$ has both direction and magnitude. When, however, the relative velocity of interacting objects approaches the speed of light, classical mechanics is no longer adequate and we must use relativistic mechanics. There are also difficulties with classical mechanics if the particles are tiny and close together, such as in the case of an electron and a proton in a hydrogen atom. The model proposed for this atom by Niels Bohr in 1913 involved the planetary motion of an electron about a fixed proton as shown in Figure 5.1.1. (The proton has a rest mass 1836 times that of the electron.) Bohr was led to consider this type of model as a

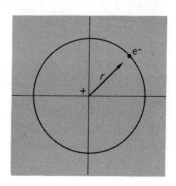

FIGURE 5.1.1. Circular planetary orbit of electron in the hydrogen atom.

result of the α-particle scattering experiment (performed by E. Marsden and H. Geiger and explained by Ernest Rutherford in 1910) which showed that essentially the entire mass of a gold atom was contained in a very tiny region of space relative to that occupied by a gold atom. This indicated that the majority of the volume must be occupied by the electrons surrounding the tiny nucleus. Bohr chose to consider only circular orbits in the hydrogen atom. An electron moving along a circular path with a constant speed $v$ along the path is actually being accelerated toward the center of the circle with an acceleration $v^2/r$ (see Problem 5.20). According to (5.1.1)

$$\frac{mv^2}{r} = F,$$

where $F$ is the magnitude of the force acting on the electron toward the center of the circle. In the present case the attraction of the electron toward the center is due to the coulombic attractive force $e^2/4\pi\epsilon_0 r^2$ (in mks units; $\epsilon_0$ is the permittivity constant; in cgs units the $4\pi\epsilon_0$ factor would be absent, i.e., equal to 1). Hence we have

$$\frac{mv^2}{r} = \frac{e^2}{4\pi\epsilon_0 r^2} \qquad (5.1.2)$$

or $mv^2 = e^2/4\pi\epsilon_0 r$. Consequently, for *any* orbital velocity there is a corresponding radius. Thus it appears that any radius is possible for an orbit; we shall shortly see that this is not so in the Bohr theory. There is an additional shortcoming of this classical theory; an accelerating charged particle should radiate energy according to electromagnetic theory; hence the electron should lose energy and spiral into the proton, which it clearly does not do.

### Q. 5.1.1

An example of a large slow-moving object whose rest mass changes with time or velocity is (a)_____. In such a case Newton's second law should be stated in the form $d\mathbf{p}/dt = \mathbf{F}$, where $\mathbf{p}$ is the momentum. If the mass of an object does not change with time, Newton's law can be written in the form (b)_____. A particle moving at constant speed $v$ along a circular path of radius $r$ undergoes (c)_____ directed toward the center of the path equal in magnitude to (d)_____. A charge $-e$ is attracted to a charge $+e$ a distance $r$ away with a force (e)_____. This is called (f)_____ law. In the Bohr atom $mv^2 =$ (g)_____.

\* \* \*

Bohr noted that the total energy of the atom (electron and proton) is the sum of the kinetic and the potential energies,

$$\text{K.E.} + U = E. \qquad (5.1.3)$$

The potential energy of an electron separated by $r$ from a proton is $U = -e^2/4\pi\epsilon_0 r$ (in mks units). [The potential energy is the work done (by an outside agent) in bringing a charge $q_2$ from $r = \infty$ to a distance $r$ from the charge $q_1$ located at $r = 0$. It equals in general $q_1 q_2/4\pi\epsilon_0 r$. If both $q_1$ and $q_2$ have the same sign, they repel each other, and the work will be positive. It is stored in the pair of charges. In the present case $q_1 = -e$ and $q_2 = e$.] Since $mv^2/2 = e^2/8\pi\epsilon_0 r$ [from (5.1.2)] we note that the kinetic energy is half the magnitude of the potential energy and equal in magnitude to the total energy. We have

$$E = \frac{-e^2}{8\pi\epsilon_0 r}. \qquad (5.1.4)$$

A widely separated electron and proton ($r \to \infty$) would have zero energy. A negative energy occurs if the atom is stable (relative to the widely separated particles); i.e., there is a decrease in the total energy of the system when the atom is formed.

There appears to be nothing in the theory so far to stop $r$ from approaching zero in which case $E \to -\infty$; in fact, as already pointed out, we should expect the electron to spiral into the proton; i.e., there would be no stable orbits. This clearly does not happen. In fact from the behavior of gases such as helium and argon it was known that atoms have radii of the order of an Ångstrom while from Rutherford-type experiments we know the nucleus of the atom has a radius several orders of magnitude smaller. (The volume of atoms such as helium and argon had been estimated, from the deviations of their gases from the ideal gas law and from the viscosity of the gases, prior to Bohr's model.)

### Q. 5.1.2

The total energy of a system equals its (a)_____ plus its (b)_____. The potential energy of the electron and proton separated by $r$ is (c)_____. The kinetic energy is (in terms of $e^2/r$) (d)_____. The total energy of the hydrogen atom is $E = $ (e)_____.

\* \* \*

When Bohr developed his model, there were data available on the possible frequencies of light emitted or absorbed by the hydrogen atom (the **emission spectra** and the **absorption spectra**). Experimentally obtained frequencies, $\nu$, obeyed the relation

$$\nu(\text{sec}^{-1}) = 3.287 \times 10^{15} \left(\frac{1}{n_a^2} - \frac{1}{n_b^2}\right), \quad n_b > n_a, \tag{5.1.5}$$

where $n_a$ and $n_b$ are integers.

Also it was known from the work of Planck and Einstein that although light could often be considered as waves there were situations in which it had to be considered as particles (which we call **photons**) with energy

$$E = h\nu, \tag{5.1.6}$$

where $h$ is **Planck's constant** (obtainable by experiment, $h = 6.62 \times 10^{-34}$ J-sec). This is called the **Einstein relation** and is one of the fundamental relations of science. The relationship between the frequency and wavelength $\lambda$ of light is

$$\lambda\nu = c, \tag{5.1.7}$$

where $c$ is the speed of light.

### Q. 5.1.3

The possible emission frequencies for hydrogen atoms equal a constant times (a)_____. Light sometimes behaves as a particle called (b)_____ with energy given by (c)_____. This is called the (d)_____. In general the wavelength, frequency, and velocity of a wave are related by (e)_____.

\* \* \*

**EXAMPLE 5.1.1**

Visible light has a wavelength of around 6500 Å. (a) What is its frequency? (b) Its energy?

*Answer.* (a) $\nu = 3 \times 10^8/(6500 \times 10^{-10}) = 4.62 \times 10^{14}$ sec$^{-1}$. (b) $E = 6.62 \times 10^{-34} \times 4.62 \times 10^{14} = 3.06 \times 10^{-19}$ J. This is 1.91 eV.

By *postulating* (making a bold guess) that stable orbits do exist if and only if the angular momentum of the orbiting electron $mvr$ equals an integer times $h/2\pi$, i.e.,

$$mvr = \frac{nh}{2\pi}, \quad n = 1, 2, 3, \ldots, \tag{5.1.8}$$

Bohr, in 1911, was able to derive (5.1.5). Equation (5.1.8) is called the **Bohr postulate**. Combining (5.1.2) and (5.1.8) gives the possible orbit radii

$$r = \frac{\epsilon_0 h^2 n^2}{\pi m e^2}, \quad n = 1, 2, 3, \ldots. \tag{5.1.9}$$

From (5.1.4) and (5.1.9) we obtain

$$E = \frac{-me^4}{8h^2\epsilon_0^2} \cdot \frac{1}{n^2} = \frac{-2.18 \times 10^{-18}}{n^2} \text{ (J)} = \frac{-13.6}{n^2} \text{ (eV)},$$
$$n = 1, 2, 3, \ldots. \tag{5.1.10}$$

Thus the lowest energy state (strongest binding) occurs for $n = 1$; this is called the **ground state**. For the case of no binding (completely separated electron and proton), $n = \infty$ and $E = 0$. Thus it should take 13.6 eV to ionize a hydrogen atom, i.e., to completely remove the electron from the atom in its ground state. This is called the **ionization potential**. The experimental value agrees with the theoretical value. The integer $n$ is called a **quantum number**.

### EXAMPLE 5.1.2

Suppose the hydrogen atom is in an excited state $n = 3$ and drops to the state $n = 2$.
(a) What happens to the energy of the atom?
(b) What else happens?

*Answer.* (a) The energy changes from $E_3 = -13.6/3^2$ to $E_2 = -13.6/2^2$. Hence

$$\Delta E = E_2 - E_3 = -13.6 \left( \frac{1}{2^2} - \frac{1}{3^2} \right) \text{eV} = -1.89 \text{ eV}.$$

The atom in the $n = 2$ state is more stable than in the $n = 3$ state. (b) A photon is emitted with energy equal to 1.89 eV. Its frequency is obtained from $E = h\nu$. Its wavelength is then obtained from (5.1.7). In the present case $\lambda = 6560$ Å. The wavelength and energy can be related by (5.1.6) and (5.1.7). This leads to the relation

$$\lambda \text{ (Å)} = \frac{12{,}400}{\Delta E \text{ (eV)}}. \tag{5.1.11}$$

In general when the energy state changes from $n_a$ to $n_b$ (or the reverse) the *magnitude* of the energy change is

$$\Delta E = \frac{me^4}{8h^2\epsilon_0^2}\left(\frac{1}{n_a^2}-\frac{1}{n_b^2}\right) = 13.6\left(\frac{1}{n_a^2}-\frac{1}{n_b^2}\right)(\text{eV}), \qquad n_b > n_a, \qquad (5.1.12)$$

and the frequency of the absorbed (or emitted) photon is

$$\nu = \frac{\Delta E}{h} = \frac{me^4}{8h^3\epsilon_0^2}\left(\frac{1}{n_a^2}-\frac{1}{n_b^2}\right), \qquad n_b > n_a. \qquad (5.1.13)$$

Table 5.1.1 gives calculated and experimental values for the case where electrons change from excited states, $n > 2$ to the $n = 2$ state and a photon of a given wavelength is emitted. The fine structure (note the two values of measured wavelength) is studied in courses in quantum mechanics and modern physics. Various possible emission lines are shown in Figure 5.1.2.

**Table 5.1.1.** BALMER SPECTRAL SERIES

| Name of Line | Value of $n$ | Calculated Wavelength (Å) | Measured Wavelength (Å) |
|---|---|---|---|
| $H_\alpha$ | 3 | 6562.793 | 6562.8473 |
|  |  |  | 6562.7110 |
| $H_\beta$ | 4 | 4861.327 | 4861.3578 |
|  |  |  | 4861.2800 |
| $H_\gamma$ | 5 | 4340.466 | 4340.497 |
|  |  |  | 4340.429 |
| $H_\delta$ | 6 | 4101.738 | 4101.7346 |
| $H_\epsilon$ | 7 | 3970.075 | 3970.0740 |
|  | 8 | 3889.052 | 3889.0575 |
|  | 9 | 3835.387 | 3835.397 |
|  | 10 | 3797.900 | 3797.910 |
|  | 11 | 3770.633 | 3770.634 |
|  | 12 | 3750.154 | 3750.152 |
|  | 13 | 3734.371 | 3734.372 |
|  | 14 | 3721.948 | 3721.948 |
|  | 15 | 3711.973 | 3711.980 |

## Q. 5.1.4

Bohr postulated that stable orbits exist if the (a)_____ _____ equals $n$ times (b)_____. This led to $E =$

$13.6/n^2$ (eV). For the ground state $n = $ (c)_____. The ionization potential of the hydrogen atom is (d)_____. If the hydrogen atom is in an excited state $n = 3$ and changes to $n = 2$, the difference in energy is $\Delta E = $ (e)_____. The photon emitted has energy $\Delta E = $ (f)_____.

\* \* \*

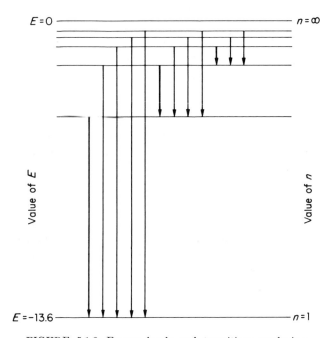

FIGURE 5.1.2. Energy levels and transitions producing photons in the hydrogen atom.

It can be shown that the radii of Bohr orbits is

$$r = \frac{\epsilon_0 h^2 n^2}{\pi m e^2}.$$

For $n = 1$, this is usually designated by $a_0$, where

$$a_0 = \frac{\epsilon_0 h^2}{\pi m e^2} = 0.529 \text{ Å}; \qquad (5.1.14)$$

$a_0$ is called the **Bohr radius.**

Electrons and nuclei have radii of $\sim 10^{-13}$ cm, atoms of $\sim 10^{-8}$ cm (1 Å), large molecules such as a nylon polymer chain of $\sim 10^{-4}$ cm, and crystals in steel have "diameters" of $\sim 10^{-3}$ cm.

The Bohr model was based on the rather arbitrary postulate of Equation (5.1.8). The physical significance of the Bohr postulate was finally pointed out by L. de Broglie in 1923. He recognized that matter had a dual nature, that it sometimes could be described by particles with mass, momentum, etc., and sometimes by waves with wavelength, frequency, etc. (just as Planck and Einstein had noted earlier for light). The recognition of this by de Broglie in 1923 led to the development of quantum mechanics by Schrödinger, Heisenberg, Dirac, and others. de Broglie notes in his Nobel Prize speech why he was led to the wave-particle duality of matter:

> When I began to consider these difficulties [of contemporary physics] I was chiefly struck by two facts. On the one hand the quantum theory of light cannot be considered satisfactory, since it defines the energy of a light corpuscle by the equation $E = h\nu$, containing the frequency $\nu$. Now a purely corpuscular theory contains nothing that enables us to define a frequency; for this reason alone, therefore, we are compelled, in the case of light, to introduce the idea of a corpuscle and that of periodicity simultaneously.
> 
> On the other hand, determination of the stable motion of electrons in the atom introduces integers; and up to this point the only phenomena involving integers in Physics were those of interference and of normal modes of vibration. This fact suggested to me the idea that electrons too could not be regarded simply as corpuscles, but that periodicity be must assigned to them also.

See L. de Broglie, *Matter or Light*, Dover Publications, Inc., New York (1946).

de Broglie suggested that the fundamental equation connecting these two types of behavior (particle and wave) was

$$p = \frac{h}{\lambda}, \qquad (5.1.15)$$

where $p$ is the momentum and $\lambda$ is the wavelength. This is now called the **de Broglie relation** and is one of the fundamental equations of science.

## Q. 5.1.5

The radius of the electron in the ground state of the hydrogen atom is about (a)_____ Å. This is to be compared with the radii of nuclei and individual electrons of (b)_____ Å. de Broglie's great accomplishment was his recognition of the (c)_____ of matter. de Broglie's relation states that the momentum of a particle equals Planck's constant times (d)_____. de Broglie was led to the conclusion that particles sometimes behave as waves, because wave motion is often associated with (e)_____.

\* \* \*

## ANSWERS TO QUESTIONS

**5.1.1** (a) A rocket, (b) $M\mathbf{a} = \mathbf{F}$, (c) an acceleration, (d) $v^2/r$, (e) $e^2/4\pi\epsilon_0 r^2$ in mks units, (f) coulomb's, (g) $e^2/4\pi\epsilon_0 r$.

**5.1.2** (a) Kinetic energy, (b) potential energy, (c) $-e^2/4\pi\epsilon_0 r$ in mks units, (d) $e^2/8\pi\epsilon_0 r$, (e) $-e^2/8\pi\epsilon_0 r$.

**5.1.3** (a) $[(1/n_a^2) - (1/n_b^2)]$, where $n_a$ and $n_b$ are integers with $n_b > n_a$; (b) a photon; (c) $E = h\nu$; (d) Einstein relation; (e) $\lambda\nu = c$; if you forget the exact relation, use dimensional analysis.

**5.1.4** (a) Angular momentum, (b) $h/2\pi$, (c) 1, (d) 13.6 eV, (e) 13.6 $[(1/2^2) - (1/3^2)]$, (f) $h\nu$.

**5.1.5** (a) 0.5, (b) $\sim 10^{-5}$, (c) dual nature, (d) $\lambda^{-1}$, (e) integers.

## 5.2 THE SCHRÖDINGER EQUATION

Schrödinger in 1924 took the work of de Broglie a step further and thereby formulated modern quantum mechanics. He knew that integers quite often arose naturally in the solution of differential equations and particularly those involving vibrations and waves. It is useful to digress at this point and to discuss differential equations briefly so that the reader can better appreciate Schrödinger's contribution.

A differential equation is a mathematical equation involving derivatives. Thus the equation

$$\frac{d^2y}{dx^2} + k^2 y = 0,$$

where $k$ is some constant, is a differential equation. In general this specific differential equation has (in addition to the trivial solution $y = 0$) a solution

$$y = A \sin kx + B \cos kx.$$

Here $A$ and $B$ are unknown constants. (To prove that this function is a solution, simply substitute it into the differential equation and show that it leads to an identity. Thus, although one may not know how to find the solution of a differential equation, one can check to see whether or not an alleged solution is a solution. This is the only kind of knowledge of differential equations which is required in this book.)

The reader should also become familiar with a type of problem involving differential equations known as an eigenvalue problem. Such a problem arises, for example, in the buckling of a column. The reader may on occasion have held a ruler or scale lengthwise between his two hands and pressed on the ends of the ruler in the lengthwise direction. He may have noted that the ruler always remained straight until a certain load was reached and then the deflections rapidly became very large (buckling). L. Euler had shown (in 1744) that when loads $P$ are applied to the ends of a column of length $L$ along the axis of the column the deflection $y$ is described by the differential equation

$$EI \frac{d^2y}{dx^2} + Py = 0.$$

Here $E$ is Young's modulus, $I$ is the cross-sectional moment of inertia (which depends on geometry only), and $x$ is the distance along the column. Suppose that the ends of the column are supported so that they do not deflect laterally; i.e.,

$$y = 0 \text{ at } x = 0 \quad \text{and} \quad y = 0 \text{ at } x = L.$$

Then the differential equation has a nontrivial solution

$$y = A \sin \sqrt{\frac{P}{EI}} x$$

if and only if $P$ takes on special values

$$P = n^2 \frac{EI\pi^2}{L^2}, \quad \text{where } n = 1, 2, 3, \ldots.$$

In other words the column shows no lateral deflection until the load reaches

$$P = \frac{EI\pi^2}{L^2}$$

and then it buckles.

Note that $P$ is a parameter in the original differential equation; what makes this an eigenvalue problem is the fact that a nontrivial solution

$y(x)$ can be found only if $P$ is chosen in a certain way. Note that all of the possible values of $P$ are characterized by an integer; i.e., they equal a constant times the square of any integer.

Often, in eigenvalue problems, we are more interested in finding the possible values of the parameter ($P$ in this case) for which a nontrivial solution exists [$y(x)$ in this case] than we are in finding the solution itself. Thus, in the present example we are not particularly interested in the shape of the buckled column but rather we are concerned with the load which causes buckling.

Schrödinger was, of course, familiar with the partial differential equations associated with wave phenomena. Such equations involve partial derivatives which we shall briefly discuss here. As an example, consider the function $\psi(x, y, z)$. It is a function of three variables. We may be interested in the rate of variation of this function in one specific direction only, say the $x$ direction at fixed values of $y$ and $z$. We call this the partial derivative of $\psi$ with respect to $x$ and write $\partial\psi/\partial x$. We read this as follows: the partial (derivative) of $\psi$ with respect to $x$. (The function $\partial\psi/\partial x$ itself is a function of the three variables $x$, $y$, and $z$.) The vibration of a violin string is approximately described by the partial differential equation

$$\frac{\partial^2 y}{\partial x^2} = \left(\frac{\rho}{T}\right) \frac{\partial^2 y}{\partial t^2}.$$

Here $\rho$ is the density of the wire, $T$ is the tensile force stretching the string, $x$ is the distance along the string, $t$ is time, and $y$ is the lateral displacement of the string. In this case $y$ is a function of the two variables $x$ and $t$. The student familiar with partial differential equations could readily show that the string of length $L$ may vibrate with one of the allowed frequencies

$$\nu = \frac{n}{2L}\sqrt{\frac{T}{\rho}}, \quad n = 1, 2, 3, \ldots,$$

or the vibration may be a combination of the harmonic ($n = 1$) and subharmonics ($n > 1$).

With this brief background, let us proceed. Schrödinger accepted the conservation of energy law: Kinetic energy plus potential energy equals total energy; i.e., K.E. $+ U = E$. The kinetic energy of a particle is given by K.E. $= \frac{1}{2}mv^2$. Since linear momentum $\mathbf{p} = m\mathbf{v}$ we have K.E. $= p^2/2m = (p_x^2 + p_y^2 + p_z^2)/2m$. Thus rearrangement of the energy equation gives

$$\frac{-(p_x^2 + p_y^2 + p_z^2)}{2m} + (E - U) = 0,$$

where the momentum is given by $(p_x, p_y, p_z)$.

It was with such knowledge of classical mechanics that Schrödinger postulated the wave mechanical form of the energy equation, namely,

$$\frac{1}{2m}\frac{h^2}{4\pi^2}\left[\frac{\partial^2\psi}{\partial x^2}+\frac{\partial^2\psi}{\partial y^2}+\frac{\partial^2\psi}{\partial z^2}\right]+(E-U)\psi=0. \qquad (5.2.1)$$

It must be emphasized that he did *not* derive this equation. He postulated it just as Newton postulated **F** = *m***a**, just as de Broglie postulated $p = h/\lambda$, etc. **Schrödinger's equation** is one of the fundamental laws of physics (which are stated, not derived). Such an equation is either right or wrong depending on whether or not it agrees with experiment.

There are two unknowns in Equation (5.2.1), $E$ and $\psi$. Equation (5.2.1) is an eigenvalue problem in which we are concerned primarily with finding the values of $E$ for which nontrivial solutions for $\psi$ exist, and not $\psi$ itself. For example, if we examine the hydrogen atom, the potential energy $U = -e^2/4\pi\epsilon_0 r$ is known. If Equation (5.2.1) is rewritten in spherical coordinates, then someone knowledgeable in the subject of differential equations could show that Equation (5.2.1) has a nontrivial solution (for $\psi$) if and only if the energy $E$ is given by Equation (5.1.10); i.e., only certain allowed energy levels are possible and these are characterized by integers.

What is the physical significance of $\psi$? In Equation (5.2.1) $\psi$ is a function of the spatial coordinates $x$, $y$, and $z$ and has the following meaning: $\psi^2\,d\tau$ is the probability of finding the electron in the volume element $d\tau$. In other words the electron in the hydrogen atom cannot be thought of in terms of specific orbits but rather we can only speak of the probability of finding the electron in some element of volume. The preciseness of classical mechanics is replaced by a type of uncertainty in quantum mechanics. The quantity $\psi^2$ is known as the **probability density.** It is a quantity which is analogous to a concentration. The quantity $\psi$ which is called the **wave function,** by itself, has no physical significance.

To illustrate the application of Schrödinger's equation we study the following example. Consider argon gas in a large cubic container of edge $a$ at a low pressure and at high temperature. Because the gas is dilute, the atoms are widely separated and their interactions are negligible except immediately before, during, and after collisions. Moreover, the time spent between collisions is very large compared to the time spent in collisions. Consequently, on the average the interaction between the particles, and hence the potential energy, is negligible. This point is discussed further in Example 5.8.2. If a particle is independent of the others, the behavior of the particle is described by Equation (5.2.1) with $U = 0$. It is instructive for pedagogical reasons to consider the case where the particle moves in one dimension only, the $x$ direction, and is restricted to the domain $0 \leq x \leq a$. Consideration of the one-dimensional case greatly reduces the mathematical manipulations but still enables one to see how the integers arise. Since the particle is never present at $x \leq 0$, $\psi^2$ and hence $\psi$ must be zero there; this is also true for $x \geq a$. For this one-dimensional problem Schrödinger's equation reduces to

$$\frac{d^2\psi}{dx^2} + \frac{8\pi^2 mE}{h^2}\psi = 0; \qquad (5.2.2)$$

this applies in the domain $0 \leq x \leq a$ and this is subject to

$$\psi = 0 \text{ at } x = 0, \qquad (5.2.3)$$
$$\psi = 0 \text{ at } x = a, \qquad (5.2.4)$$

and

$$\int_0^a \psi^2 \, dx = 1. \qquad (5.2.5)$$

The latter condition states that the probability of finding the particle somewhere in the domain $0 \leq x \leq a$ is 1. An expert in differential equations could readily show that nontrivial solutions exist if and only if

$$E = \frac{n^2 h^2}{8ma^2}, \quad n = 1, 2, 3, \ldots; \qquad (5.2.6)$$

the possible solutions are

$$\psi(x) = \left(\frac{2}{a}\right)^{1/2} \sin \frac{n\pi x}{a}. \qquad (5.2.7)$$

[The reader may directly verify for his own satisfaction that $\psi(x)$ as given by (5.2.7) and $E$ as given by (5.2.6) is indeed a solution by substituting these expressions into (5.2.2) and showing that it leads to an identity as required; moreover, the conditions of (5.2.3)–(5.2.5) are satisfied.] The problem just considered involves the **one-dimensional particle in a box.** This model is used to estimate the color of dye molecules (see Problem 5.31).

## Q. 5.2.1

To Schrödinger, the fact that the energy levels of the hydrogen atom were associated with integers suggested that (a)_____. The starting point for the development of Schrödinger's equation was (b)_____. The quantity $\psi^2 \, d\tau$ represents (c)_____

_____. The allowable energy levels for the one-dimensional particle in a box are (d)_____.

\* \* \*

If the particle had been enclosed in a rectangular parallelepiped of dimensions $a$, $b$, and $c$ with $U = 0$ in the box and it were impossible for the particle to leave the box, then it could be shown that a nontrivial solution to (5.2.1), with $U = 0$, occurs if and only if

$$E = \frac{h^2}{8m}\left(\frac{n_1^2}{a^2} + \frac{n_2^2}{b^2} + \frac{n_3^2}{c^2}\right), \tag{5.2.8}$$

where each of $n_1$, $n_2$, and $n_3$ are integers called quantum numbers. This is called the **particle in a box problem.** As we shall see later, the valence electrons in a metal are free to wander through the metal (the electron gas). Equation (5.2.8) provides a good first approximation for the possible energy states of these electrons. We note that there are three distinct quantum numbers in the three-dimensional case. Actually the electron has a fourth quantum number, the spin quantum number $m_s = +\frac{1}{2}$ or $m_s = -\frac{1}{2}$. Thus the state of an electron is specified by four quantum numbers. For a discussion of how (5.2.8) is obtained, see the learning sequence in Ruoff, *Materials Science*, Problems 8.15–8.18 (previously cited).

### Q. 5.2.2

The particle in a box expression for $E$ is a good approximation for the (a)_____. The expression for $E$ in this case involves (b)_____ quantum numbers. Actually, the state of the electron is specified by (c)_____ quantum numbers. The origin of the quantum number $m_s$ is discussed in Section 5.9.

\* \* \*

**EXAMPLE 5.2.1**

The kinetic energy of the electron in the hydrogen atom in the ground state is 13.6 eV. How large would a cubic box be which contained an electron in its ground state moving in a potential $U = 0$ with a kinetic energy of 13.6 eV?

*Answer.* We note that $1 \text{ eV} = 1.6 \times 10^{-19}$ C-V $= 1.6 \times 10^{-19}$ J. Hence in mks units

$$13.6 \times 1.6 \times 10^{-19} = \frac{(6.63 \times 10^{-34})^2(1^2 + 1^2 + 1^2)}{8 \times 9.11 \times 10^{-31} a^2}$$

where $a$ is measured in meters. We get $a = 2.9 \times 10^{-10}$ m $= 2.9$ Å. It would be impossible to put an electron with energy $E = 0$ in such a box. It would always have an energy $E = 13.6$ eV or greater. This is contrary to our experience with golf balls which, if placed in a box, remain at rest.

**EXAMPLE 5.2.2**

A golf ball has a mass of about 45.8 g. It is placed in a cubical box whose edge is 10 cm. The golf ball appears to be at rest. Is that consistent with the results mentioned in Example 5.2.1?

*Answer.* In this case

$$E = \text{K.E.} = \frac{h^2 3}{8ma^2} \approx 4 \times 10^{-64} \text{ J}.$$

According to classical mechanics

$$\text{K.E.} = \tfrac{1}{2}mv^2$$

so that in the present case $v \approx 10^{-31}$ m/sec. Is the golf ball moving or at rest? Are you uncertain? Note that it would take about $10^{13}$ years to move 1 Å.

## ANSWERS TO QUESTIONS

5.2.1 (a) A differential equation might be involved, especially one related to wave motion and vibrations; (b) $p^2/2m + U = E$; (c) the probability of finding the particles in the volume element $d\tau$; (d) $E = h^2n^2/8ma^2$.

5.2.2 (a) Allowed energy levels of valence electrons in a metal, (b) three, (c) four.

## 5.3 THE HYDROGEN ATOM

In this case we write Schrödinger's equation in spherical coordinates rather than Cartesian coordinates (this seems reasonable since in earlier concepts of the atom the electron traveled around the nucleus):

$$\frac{1}{r^2}\frac{\partial}{\partial r}\left(r^2 \frac{\partial \psi}{\partial r}\right) + \frac{1}{r^2 \sin^2 \theta}\frac{\partial^2 \psi}{\partial \phi^2} + \frac{1}{r^2 \sin \theta}\frac{\partial}{\partial \theta}\left(\sin \theta \frac{\partial \psi}{\partial \theta}\right)$$
$$+ \frac{8\pi^2 m}{h^2}\left(E + \frac{e^2}{4\pi\epsilon_0 r}\right)\psi = 0. \quad (5.3.1)$$

Here, $r$, $\theta$, and $\phi$ have their usual significance as shown in Figure 5.3.1.

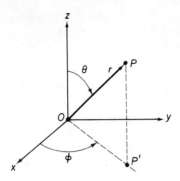

FIGURE 5.3.1. Spherical coordinate system.

Note that for $U$ in Schrödinger's equation we have used

$$U = -\frac{e^2}{4\pi\epsilon_0 r},$$

which is the ordinary potential energy in mks units of the negatively charged electron and the positively charged proton a distance $r$ apart.

The additional conditions on $\psi$ in the present case are

$$\psi(r, \theta, \phi) = \psi(r, \theta, \phi + 2\pi),$$
$$\psi(r, \theta, \phi) = \psi(r, \theta + 2\pi, \phi),$$
$$\lim_{r \to \infty} \psi(r, \theta, \phi) = 0, \qquad (5.3.2)$$
$$\int_{r=0}^{\infty} \int_{\theta=0}^{\pi} \int_{\phi=0}^{2\pi} \psi^2 \, d\tau = 1.$$

The first two equations express the fact that $\psi$ must be a periodic function of $\phi$ and $\theta$, respectively, while the last equation states that the probability of finding the electron somewhere in space is 1. We shall not go through the mathematical details of obtaining the solution here [see, e.g., L. Pauling and E. B. Wilson, *Introduction to Quantum Mechanics*, McGraw-Hill Book Company, Inc., New York (1935)].

Suffice it to say that a nontrivial solution exists if and only if the energy is quantized according to

$$E = -\frac{me^4}{8\epsilon_0^2 h^2} \frac{1}{n^2}, \qquad n = 1, 2, 3, \ldots \qquad (5.3.3)$$

(as with the Bohr model). The corresponding solutions for $\psi$ will have the form

$$\psi(r, \theta, \phi) = R(r)\Theta(\theta)\Phi(\phi). \qquad (5.3.4)$$

Sec. 5.3                          The Hydrogen Atom

From our previous experience with the particle in the box, we might expect that these wave functions would be characterized by three quantum numbers, i.e., one quantum number for each coordinate. In the present case these quantum numbers are called $n$, $l$, and $m_l$, where the form of the solution dictates that $n = 1, 2, 3, \ldots$, and $l = n - 1, \ldots, 0$, while $m_l = -l, \ldots, 0, \ldots, l$. Here $n$ is called the **principal quantum number**, $l$ is called the **angular momentum quantum number**, and $m_l$ is called the **magnetic quantum number**. In addition there is a spin quantum number with $m_s = \pm\tfrac{1}{2}$. There is a **spectroscopy notation** for describing the electrons in the hydrogen atom; the $l$ states are specified by either a number or a letter as follows:

$$
\begin{array}{ccccccccc}
l = 0 & 1 & 2 & 3 & 4 & 5 & 6 & 7 & 8 \\
s & p & d & f & g & h & i & k & l
\end{array}
$$

Thus an electron with $n = 1$, $l = 0$ is called a 1s electron. There are two possible 1s electrons; both have $m_l = 0$ and one has a spin quantum number $m_s = \tfrac{1}{2}$, and the other $m_s = -\tfrac{1}{2}$. In the ground state $n = 1$, $l = 0$, $m_l = 0$. The wave function for the ground state is

$$\psi_{1s} = \frac{1}{\sqrt{\pi}} \frac{1}{a_0^{3/2}} e^{-r/a_0}, \tag{5.3.5}$$

where $a_0$ is the Bohr radius given by (5.1.14).

### Q. 5.3.1

To find the possible wave functions and energy levels for the hydrogen atom, Schrödinger's equation is written in (a)_____ coordinates and the potential is (b)_____. It is found that solutions exist if and only if (c) _____. It is also found that the wave functions are characterized by three quantum numbers called (d)_____. These are related in the following way: (e)_____. In all, the state of the electron is described by (f)_____ quantum numbers. There are altogether (g)_____ 2p electrons.

\* \* \*

The ground state wave function of Equation (5.3.5) has spherical symmetry; i.e., it does not vary with $\theta$ and $\phi$.

## EXAMPLE 5.3.1

The volume $d\tau$ between two concentric shells of radius $r$ and $r + dr$ is $4\pi r^2\, dr = d\tau$. The probability of finding the ground state electron in this volume element is $\psi^2 4\pi r^2\, dr$, where $\psi$ is given by (5.3.5). The quantity $\psi^2 4\pi r^2$ is called the **radial probability density**. What is the most probable radial position of the electron?

*Answer.* In differential calculus you learned that one of the conditions for finding a maximum of a function of $r$ was that its first derivative with respect to $r$ must be zero. You can now proceed on your own to show that the radial probability density for the 1s state has a maximum when $r = a_0$.

Although the results from Schrödinger's equation bear some relation to the Bohr model, there are some significant differences. In the Bohr model the electron moved in a plane and its position was uniquely characterized by the radius of that orbit; thus in the ground state the electron was a distance $a_0$ from the proton. In the present theory, it is possible to speak only of the density distribution of the electron, i.e., of its probability of being in a certain volume element. Thus with the atom in the ground state there is a certain probability of finding the electron between spherical shells of radius $a_0/2$ and $a_0/2 + dr$, there is a higher probability of finding it between spherical shells of radius $a_0$ and $a_0 + dr$, and there is some probability of finding it between $3a_0$ and $3a_0 + dr$.

## ANSWERS TO QUESTIONS

**5.3.1** (a) Spherical; (b) $-e^2/4\pi\epsilon_0 r$ (in mks units); (c) the energy is quantized in the same manner as in the Bohr model; (d) $n$, $l$, and $m_l$; (e) $l = n - 1$, $n - 2, \ldots 0$, and $m_l = -l, \ldots, 0, \ldots, l$; (f) four; (g) six.

## 5.4 THE HYDROGEN-LIKE ATOM AND THE PERIODIC TABLE

A **hydrogen-like atom** has a nucleus of charge $Ze$ (where $Z$ is the atomic number) with a single electron, e.g., H, $He^+$, $Li^{2+}$, $Be^{3+}$, .... The behavior of this atom is described by Schrödinger's equation but with $U$ replaced by $-Ze^2/4\pi\epsilon_0 r$. Its possible energy states are given by

$$E = -\frac{mZ^2 e^4}{8\epsilon_0^2 h^2}\frac{1}{n^2}, \tag{5.4.1}$$

i.e., by

$$E = -13.6\frac{Z^2}{n^2} \text{ (eV).} \tag{5.4.2}$$

Table 5.4.1 gives a few of the hydrogen-like wave functions. The quantum number $n$ is described by either a number or a letter (spectroscopy notation) as follows:

$$n = 1 \quad 2 \quad 3$$
$$K \quad L \quad M.$$

The 1s, 2s, 3s, and all s-state electrons have spherical symmetry. As $n$ increases, the maximum of the radial distribution function would move to

**Table 5.4.1. THE HYDROGEN-LIKE WAVE FUNCTION**

*K Shell*

$n = 1, l = 0, m_l = 0$:

$$\psi_{1s} = \frac{1}{\sqrt{\pi}} \left(\frac{Z}{a_0}\right)^{3/2} e^{-Zr/a_0}$$

*L Shell*

$n = 2, l = 0, m_l = 0$:

$$\psi_{2s} = \frac{1}{4\sqrt{2\pi}} \left(\frac{Z}{a_0}\right)^{3/2} \left(2 - \frac{Zr}{a_0}\right) e^{-Zr/2a_0}$$

$n = 2, l = 1, m_l = 0$:

$$\psi_{2p_z} = \frac{1}{4\sqrt{2\pi}} \left(\frac{Z}{a_0}\right)^{3/2} \frac{Zr}{a_0} e^{-Zr/2a_0} \cos\theta$$

$n = 2, l = 1, m_l = \pm 1$:

$$\psi_{2p_x} = \frac{1}{4\sqrt{2\pi}} \left(\frac{Z}{a_0}\right)^{3/2} \frac{Zr}{a_0} e^{-Zr/2a_0} \sin\theta \cos\phi$$

$$\psi_{2p_y} = \frac{1}{4\sqrt{2\pi}} \left(\frac{Z}{a_0}\right)^{3/2} \frac{Zr}{a_0} e^{-Zr/2a_0} \sin\theta \sin\phi$$

larger $r$ values. The nature of the s-state electrons is illustrated in Figure 5.4.1.

Non-s states lack radial symmetry. The $p$ states each have a dumbbell shape; e.g., $\psi_{2p_z}^2 = 0$ at the origin and has a maximum along the $z$ axis at a distance of $\pm 2a_0/Z$. The main axis of the dumbbell is the $z$ axis for the $2p_z$ state, i.e., it has rotational symmetry around the $z$ axis. Likewise the $y$ axis is the main axis of the dumbbell for the $2p_y$ state. The general nature of the electron distribution for different orbitals is shown in Figure 5.4.2.

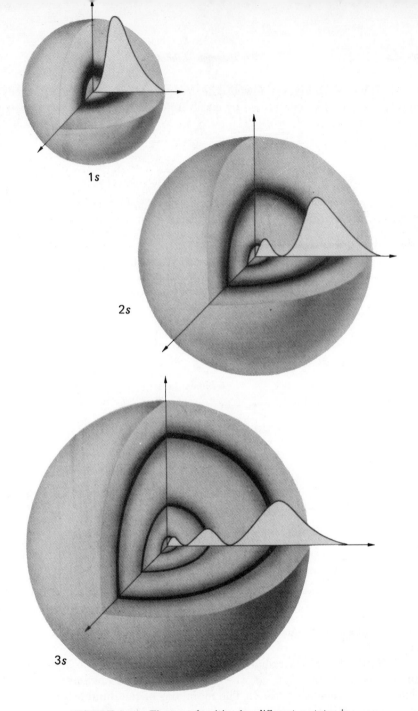

FIGURE 5.4.1. Electron densities for different $s$ states in hydrogen-like atoms. The heavy solid curves show the most probable radial positions; these are the positions where the radial distribution function has maxima. The radial probability densities $\psi^2 4\pi r^2$ are also plotted as a function of $r$ along one of the axes.

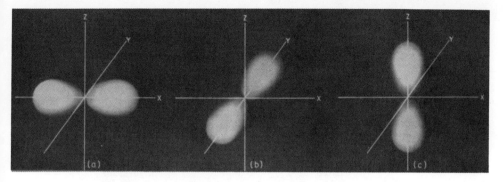

FIGURE 5.4.2. Polar representation of angular dependence of $\psi_{2p}^2$. The magnitude of a vector from the origin to a point on the surface represents the value of the angular dependence of $\psi_{2p}^2$. (a) $\psi_{2p_x}$. (b) $\psi_{2p_y}$. (c) $\psi_{2p_z}$.

**EXAMPLE 5.4.1**

How many $2p$ states are there?

*Answer.* We have $n = 2$, $l = 1$. Hence $m_l$ can take on the values $-1, 0, 1$ and the electron spin can be $m_s = -\frac{1}{2}$ or $\frac{1}{2}$. Hence there are six $2p$ electrons as follows:

| $n$ | $l$ | $m_l$ | $m_s$ |
|---|---|---|---|
| 2 | 1 | 0 | $\frac{1}{2}$ |
| 2 | 1 | 0 | $-\frac{1}{2}$ |
| 2 | 1 | 1 | $\frac{1}{2}$ |
| 2 | 1 | 1 | $-\frac{1}{2}$ |
| 2 | 1 | $-1$ | $\frac{1}{2}$ |
| 2 | 1 | $-1$ | $-\frac{1}{2}$ |

**Q. 5.4.1**

$s$-state electrons are represented by electron clouds for which constant density surfaces are (a)_____. $p$-state electrons are described by polar representations. These surfaces are shaped like (b) _____ whose axes are oriented at (c)_____ for $p_x$, $p_y$, and $p_z$ states.

\* \* \*

ATOMS ARE COMBINATIONS OF HYDROGEN-LIKE ELECTRONIC STATES. The **Pauli exclusion principle** states that each electron in an atom has a different wave function, i.e., a different set of *four* quantum numbers.

The first approximation of an atom is that all its electrons have the general form of hydrogen-like wave functions. Thus the **electron configuration** of the light elements is as shown in Table 5.4.2. At least for the lighter atoms, successive atoms are seen to have the electron configuration of the preceding atom plus one more electron.

Table 5.4.2. ELECTRON CONFIGURATIONS OF THE LIGHT ELEMENTS

| Atom | Atomic Number | Electron Configuration |
|---|---|---|
| Hydrogen | 1 | $1s^1$* |
| Helium | 2 | $1s^2$ |
| Lithium | 3 | $1s^2 2s^1$ |
| Beryllium | 4 | $1s^2 2s^2$ |
| Boron | 5 | $1s^2 2s^2 2p^1$ |
| Carbon | 6 | $1s^2 2s^2 2p^2$ |
| Nitrogen | 7 | $1s^2 2s^2 2p^3$ |
| Oxygen | 8 | $1s^2 2s^2 2p^4$ |
| Fluorine | 9 | $1s^2 2s^2 2p^5$ |
| Neon | 10 | $1s^2 2s^2 2p^6$ |
| Sodium | 11 | $1s^2 2s^2 2p^6 3s^1$ |
| Magnesium | 12 | $1s^2 2s^2 2p^6 3s^2$ |

\* The exponent equals the number of electrons in the atom having this kind of hydrogen-like wave function. If the exponent is one it is often not used.

The electronic configuration of the atoms are noted on periodic charts of the elements.

### Q. 5.4.2

The first approximation to the structure of an atom is that all its electrons have (a)_____. The (b)_____ states that every electron in an atom must have a unique set of quantum numbers. The electron configuration of carbon whose atomic number is 6 is (c)_____. The electron configuration of sodium (atomic number 11) is (d)_____.

\* \* \*

### EXAMPLE 5.4.2

Suppose the $2p$ electrons are all filled as in neon. Show that the sum of the $\psi^2$ for all these $p$ electrons is spherically symmetric.

*Answer.* We have

$$\psi^2 = \psi_{2p_x}^2 + \psi_{2p_y}^2 + \psi_{2p_z}^2$$

for each spin state. Each of these functions (see Table 5.4.1) has the same radial factor which we call $R_{2p}$. Hence

$$\psi^2 = R_{2p}^2(\cos^2 \theta + \sin^2 \theta \cos^2 \phi + \sin^2 \theta \sin^2 \phi)$$
$$= R_{2p}^2.$$

It is also true that a full complement of $d$ electrons (ten in all) or $f$ electrons gives a spherically symmetric distribution.

More generally it is true that a full "shell" of electrons (same $n$) has a spherically symmetric distribution. For example, He and Ne have this structure which is characteristic of chemical stability. Thus, for example, sodium tends to ionize to Na$^+$ which has the Ne configuration and fluorine readily becomes F$^-$ which has the Ne electron configuration when the ionic compound NaF is formed. Table 5.4.3 gives some values of the first ionization potential (energy needed to take one electron from the atom to infinity leaving a singly charged ion behind). The high values for He and Ne illustrate the stability of full shell configurations (which have spherically symmetric charge distributions). The second ionization potential is the energy needed to take one electron from the singly charged ion to produce a doubly charged ion. We would expect this to be very high for Li$^+$ and Na$^+$.

Note the increase in ionization potential as a given state $(n)$ is filled. Note how the tendency to form a positive ion increases as the ionization potential decreases.

**Table 5.4.3. IONIZATION POTENTIALS (eV)**

| Element | First Ionization Potential | Second Ionization Potential |
|---|---|---|
| H  | 13.60 |       |
| He | 24.58 | 54.41 |
| Li | 5.39  | 75.62 |
| Be | 9.32  | 18.21 |
| B  | 8.30  | 25.12 |
| C  | 11.27 | 24.38 |
| N  | 14.55 | 29.61 |
| O  | 13.62 | 35.08 |
| F  | 17.42 | 34.98 |
| Ne | 21.56 | 40.96 |
| Na | 5.14  | 47.29 |
| Mg | 7.64  | 15.03 |

It should be noted that our model of the atom is only approximate since it ignores the interactions of electrons with each other. This approximation could be expected to be good for inner electrons and very poor for the outer electron. Thus in an element with a medium atomic number such as copper ($Z = 29$) the present model would nearly correctly describe a transition from the $L$ shell ($n = 2$) to the $K$ shell ($n = 1$) which results in an X-ray of wavelength $\lambda = 1.54$ Å. The energy of the $K$ shell electron in such elements is of the order of 10 keV.

The student is now in a position to appreciate the experiment which provides the basis for the fourth quantum number, electron spin (see Section 5.9 on magnetization).

### Q. 5.4.3

A full shell (all possible electrons with a specific $n$) has (a)_____ _____. A full subshell such as the $2p$ subshell with its six electrons has (b)_____. Full shells correspond to (c)_____ configurations. This is why we would expect the first ionization potential of lithium (atomic number 3) to be (d)_____ and its second ionization potential to be (e)_____. The hydrogen-like model is a particularly good approximation for (f) (inner or outer) electrons of medium atomic number elements.

\* \* \*

## ANSWERS TO QUESTIONS

**5.4.1** (a) Spheres, (b) dumbbells, (c) 90 deg.

**5.4.2** (a) Hydrogen-like states; (b) Pauli exclusion principle; (c) $1s^2 2s^2 2p^2$; (d) $1s^2 2s^2 2p^2 3s$; this is sometimes written Ne$3s$, where Ne represents the neon configuration.

**5.4.3** (a) A spherical distribution of electronic charge, (b) a spherical electron charge distribution, (c) stable, (d) small, (e) very large, (f) inner.

### 5.5 BONDING OF ATOMS IN MOLECULES AND CONDENSED PHASES

I beg of you, therefore, to grant the request of Simplicio, which is also mine; for I am no less curious and desirous than he to learn what is the binding material which holds together the parts of solids so that they can scarcely be separated.—Galileo Galilei, *Dialogues Concerning the Two New Sciences.* Translated by Henry Crew and Alfonso de Salvio. Published in Great Books of the Western World, Vol. 28. Encyclopedia Britannica, Inc., Chicago (1952) p. 134.

Atoms are bonded together to form molecules or condensed phases by several different types of bonding. These are (1) ionic bonding, (2) covalent or homopolar bonding, (3) metallic bonding, and (4) secondary-type bonding.

It is only in a few special cases that the actual bonding is exactly of one type or another (more often it is a combination). In the sections which follow each of these bond types will be taken up individually. The student should attempt to develop a feeling for the origin and the strength of these different types of bonding and for how the behavior of the material depends on the nature of the bonds. The first three types of bonds have strengths of about 100 kcal/mole while the secondary-type bonds have strengths of only 1–10 kcal/mole. Note that

$$1 \text{ eV/atom} = 23.01 \text{ kcal/mole},$$

so the bond energy of primary bonds is several electron volts while for secondary bonds it is on the order of a tenth of an electron volt.

### Q. 5.5.1

The three kinds of primary bonding are (a)_____. Primary bonds have energies of the order of (b)_____ eV; while secondary bonds have energies of about (c)_____ eV.

\* \* \*

In the first approximation, the simplest type of bond is the ionic bond; this will be considered first.

THE IONIC BOND. Certain atoms readily lose or gain an electron to form an ion, with a resultant electron configuration of an inert gas atom but with a net negative or positive charge. Thus sodium readily loses an electron to form the sodium ion [with electron configuration $1s^2 2s^2 2p^6$] while chlorine readily gains an electron to form the chloride ion [with electron configuration $1s^2 2s^2 2p^6 3s^2 3p^6$]. The calculation of the energy change in each case is a detailed quantum mechanical problem; alternatively these quantities can be measured. We shall assume that we know these quantities and ask, Can these separate $Na^+$ and $Cl^-$ ions now be put together to form a stable molecule? We find that a true NaCl molecule exists in the vapor in which the bonding is due mainly to the electrostatic attraction between one $Na^+$ and one $Cl^-$ ion.

The force between two point charges with charges $Q_1$ and $Q_2$ can be represented at moderate distances of separation $r$ by the coulombic force $Q_1Q_2/4\pi\epsilon_0 r^2$ or by the coulombic potential energy $Q_1Q_2/4\pi\epsilon_0 r$. We can write for the net potential energy of the two ions

$$U = \frac{Q_1Q_2}{4\pi\epsilon_0 r} + \frac{b}{r^n}, \qquad (5.5.1)$$

where the second term is called the **Born repulsive potential** (energy); here $b$ and $n$ are positive constants which are characteristics of the particular ions. Usually $n \approx 10$. The ions repel each other at very close distances because the electron clouds of the ions overlap; this is the origin of the $b/r^n$ term. The function $U(r)$ is plotted in Figure 5.5.1 for NaCl; the

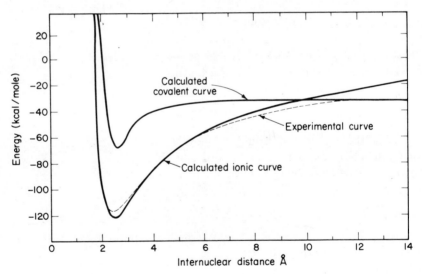

FIGURE 5.5.1. Potential energy of sodium chloride as a function of internuclear separation.

minimum in the curve represents the stable internuclear separation for the molecule held together by an **ionic bond**. Also shown is the result for a NaCl molecule calculated on the basis of covalent bonding; the latter is discussed in the next section. Note that at large separations separate Na and Cl atoms are a more stable system than separate $Na^+$ and $Cl^-$ ions, and thus we find that the molecule NaCl dissociates into atoms instead of ions.

The general form of these potential energy curves is of interest. They are *parabolic* near their minimum but rise much more rapidly for $r$ less than the equilibrium distance than for $r$ greater than the equilibrium distance.

## Q. 5.5.2

The ionic bond between a sodium ion and a chloride ion involves (a)_____ attraction between two ions which are considered as (b)_____ charges and a repulsive term due to the reluctance of the electronic clouds of the ions to (c)_____. The net potential interactions between two ions of charges $e$ and $-e$ is (d)_____, where the exponent $n$ is about (e)_____. At the minimum (i.e., at the *equilibrium* distance) there is no net (f)_____ acting on an ion. Near the minimum of the potential curve, it has a (g)_____ shape; this can be derived from (5.5.1).

* * *

**EXAMPLE 5.5.1**

Let $r_0$ be the equilibrium distance between the centers of the ions in an ionic bond. Calculate the potential energy $U(r_0)$.

*Answer.* The minimum of the potential energy curve corresponds to $dU/dr = 0$ for $r = r_0$.

By differentiation,

$$\frac{dU}{dr} = -\frac{1}{r}\left(\frac{-e^2}{4\pi\epsilon_0 r} + \frac{nb}{r^n}\right).$$

Hence

$$\frac{b}{r_0^n} = \frac{1}{n}\frac{e^2}{4\pi\epsilon_0 r_0}.$$

If this is substituted into the expression for $U$ evaluated at $r_0$, the result is

$$U = \frac{-e^2}{4\pi\epsilon_0 r_0}\left(1 - \frac{1}{n}\right).$$

The fact that the potential curve is parabolic near the minimum implies that the bond behaves like a linear spring and the molecule like a harmonic oscillator. The coefficient $n$ can be expressed in terms of the frequency of the harmonic oscillator; the latter can be measured spectroscopically as we shall see in Chapter 9. Also, an expert in quantum mechanics could calculate $n$, although it is not an easy problem and, because he would be forced to make approximations, the answer may be only fair.

We next ask, Is it possible to form a stable solid from the Na⁺ and Cl⁻ ions? A model of the solid which does form is shown in Figure 5.5.2. It consists of a periodic regularly alternating array of ions, of which a portion is shown in Figure 5.5.2, known as the **sodium chloride structure**. All

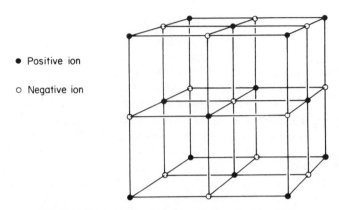

● Positive ion

○ Negative ion

FIGURE 5.5.2. Portion of the sodium chloride structure. The rows would have about $10^8$ atoms in them if the crystal was a cube with a 1-cm edge.

the lithium, sodium, potassium, and rubidium halides have this crystal structure at room temperature and pressure. So does the important refractory material MgO. The distance between the ions in this structure is of the order of a few Ångstroms ($10^{-10}$ m).

The ions are electrostatically attracted to each other. The lattice does not collapse, because when the electronic charge clouds of the ions begin to overlap, they repel each other. As a first approximation we can consider the ions as *rigid* spheres. We can define the radius of these spheres in such a way that the sum of the cation (positive ion) radius and anion (negative ion) radius equals the interionic distance. These radii are called **ionic radii**. We can consider that the electrostatic attraction brings the ions (spheres) together until they just touch; since the spheres are rigid they can get no closer. Such a solid would be incompressible (while real solids are compressible) and would have a binding energy which is about 10% too large. Thus we must remember that the idea of a rigid sphere ionic radius is a fiction but a useful fiction.

It would be more accurate to consider the ions to be compressible as is the case for the Born model. We therefore assume that a potential of the type

$$u_{ij} = \frac{Q_i Q_j}{4\pi\epsilon_0 r_{ij}} + \frac{b}{r_{ij}^n} \tag{5.5.2}$$

exists between each pair of ions (every possible combination); then, by summing over all ions in the infinite crystal we can obtain an expression for the binding energy per mole of NaCl.

For $N$ ions of $Na^+$ and $N$ ions of $Cl^-$, the equilibrium value of the lattice energy, $U(r_0)$, is

$$U(r_0) = -\left(\frac{1.74755Ne^2}{4\pi\epsilon_0 r_0}\right) + \left(\frac{1.74755Ne^2}{4\pi\epsilon_0 r_0}\right)\frac{1}{n}. \tag{5.5.3}$$

The coefficient 1.74755 is called the Madelung constant for sodium chloride; it arises from summing over all the ions. There is a discussion in Problems 5.26 and 5.27 of how the Madelung constant can be evaluated for a one-dimensional (imaginary) ionic crystal. In Equation (5.5.3), $e$ is the magnitude of the electron charge and $r_0$ is the separation between the nearest-neighbor $Na^+$ and $Cl^-$ ions at equilibrium. The separation distance $r_0$ equals one half the lattice parameter which can be measured by X-ray diffraction. One can also show from theoretical considerations that the exponent $n$ is related to the bulk modulus $K$ by

$$n = 1 + \frac{72\pi\epsilon_0 r_0^4 K}{1.74755e^2}. \tag{5.5.4}$$

The bulk modulus $K$ is defined in Problem 2.24. Using experimental values of $r_0$ and $K$, we find that $n \approx 10$ for many ionic solids. The first term in Equation (5.5.3) represents the coulombic attraction while the second (which is about a 10% contribution) represents the Born repulsion. Note that the repulsive term ($b/r^n$, where $n \approx 10$) in Equation (5.5.2) is a **short-range interaction.** Its contribution is negligible unless the ions are very close together. Table 5.5.1 gives some values of the lattice energy

Table 5.5.1. BONDING ENERGY RELATIVE TO ISOLATED IONS*

| Crystal | Theory (Mayer) | | Experiment | |
|---------|----------|------|----------|------|
|         | kcal/mole | eV | kcal/mole | eV |
| NaCl | 183.1 | 7.96 | 182.8 | 7.94 |
| KCl  | 165.4 | 7.19 | 164.4 | 7.14 |
| AgCl | 203   | 8.82 | 205.7 | 8.94 |
| TlCl | 167   | 7.26 | 170.1 | 7.39 |
| CuCl | 216   | 9.39 | 221.9 | 9.64 |

* 1 eV = 23.01 kcal/mole.

$U(r_0)$ calculated from Equations (5.5.3) and (5.5.4). Thus the binding energies per ion pair are in the neighborhood of 10 eV.

### Q. 5.5.3

A crystal structure which is an alternating array of positive and negative ions along three perpendicular directions is called (a)_____ _____. It is sometimes convenient to think of an ion as (b)_____ _____ with a well-defined (c)_____. Because the Born repulsive term falls off so rapidly with distance it is called a (d)_____ _____ interaction. By analogy we would call a coulombic attraction (or repulsion) a (e)_____. When experimental values of the (f)_____ are used to evaluate $r_0$ and $n$, the Born model leads to binding energies for NaCl in reasonably good agreement with experiment.

\* \* \*

## ANSWERS TO QUESTIONS

**5.5.1** (a) Ionic, covalent, and metallic; (b) 5; (c) 0.1.

**5.5.2** (a) Coulombic, (b) point, (c) overlap, (d) $-e^2/4\pi\epsilon_0 r + b/r^n$ (mks units), (e) 10, (f) force, (g) parabolic.

**5.5.3** (a) The sodium chloride type of crystal structure, (b) a rigid sphere, (c) radius, (d) short-range, (e) long-range interaction, (f) lattice parameter and bulk modulus.

### 5.6 COVALENT BONDING

Consider two hydrogen atoms which when brought closely together combine to form a hydrogen molecule. Why? In previous chemistry courses the student was told that this was due to the formation of an **electron pair bond** designated by H:H. The general picture is that the electrons are no longer localized around their respective protons but each is in the field of both protons. The detailed mathematical analysis of this is an involved quantum mechanical problem which is discussed in quantum chemistry books. A highly simplified mathematical discussion is considered next. We note that each hydrogen atom is electrically neutral. Hence when we bring them together, a first approximation of the change in energy can be to ignore any coulombic interactions and to consider only changes in kinetic energy. We recall that the kinetic energy of the electron in the hydrogen atom in the ground state is 13.6 eV. Let us consider that the hydrogen atom can be approximated by an electron moving in a potential $U = 0$ in a cubical box. Its energy in the lowest state is, from (5.2.8),

$$E_K = \frac{h^2}{8ma^2}(1^2 + 1^2 + 1^2) = \frac{3h^2}{8ma^2}.$$

We choose a value for $a$ such that $E_K = 13.6$ eV. Let us now consider the interaction of two such hydrogen "atoms" to form a hydrogen "molecule"

FIGURE 5.6.1. Model of interactions of two hydrogen atoms to form a hydrogen molecule.

as shown in Figure 5.6.1. The molecule is a box of dimensions $a$, $a$, and $2a$. The energy of each electron in the large box, $E_L$, is

$$E_L = \frac{h^2}{8m}\left(\frac{1^2}{a^2} + \frac{1^2}{a^2} + \frac{1^2}{(2a)^2}\right).$$

Since two electrons are involved the decrease in kinetic energy when the bond is formed is

$$\Delta E_K = 2(E_K - E_L) = \frac{1}{2}\left(\frac{h^2 3}{8ma^2}\right) = \frac{E_K}{2} = 6.8 \text{ eV}.$$

The experimental value is 4.7 eV. It is not surprising that we do not get the correct answer in this oversimplified model which completely ignores the potential energy due to the coulombic interaction of an electron with each of the protons and with the other electron and of the protons with each other. Precise calculations, while tedious, lead to the correct experimental result; such calculations show that the major contribution to the binding energy is indeed the decrease in kinetic energy as shown in our simple model. Note how very different this is from the ionic bond where the major contribution was due to the coulombic potential energy term.

The energy of the system is lowered because the electrons become delocalized; i.e., each electron is free to be associated with each atom in the bond. It is pointed out here that the bonding energy would be only 0.6 eV if the electrons had not been delocalized. Actually the average charge becomes concentrated more between the nuclei when the bond is formed and the two protons are pulled closer together; thus the electron pair concept of G. N. Lewis has substantiation from detailed quantum mechanical calculations. This is the **covalent bond**.

The outer shell electron configuration of sulfur is $3s^2 3p^4$. It thus needs two electrons to achieve the inert gas configuration of argon. In the compound $H_2S$, two $p$ bonds of sulfur are expected to be primarily involved to result in a molecule with a bond angle near 90 deg (actually 92°20′). Recall from Figure 5.4.2 that $p$ orbitals are at 90 deg to each other. A number of compounds form $p$-orbital-type bonds at 90 deg (or slightly more) to each other. Thus the bond angle in $H_2O$ (vapor) is 105 deg.

## Q. 5.6.1

The covalent bond in the hydrogen molecule is due to the fact that the electron from one atom can exchange with the electron from the other atom; we say the two electrons are (a) _____. The primary source of the bonding is a change in (b) _____. If sulfur had two more electrons it would have a full $n = 3$ shell. By combining with hydrogen, it forms $H_2S$ with an expected bond angle of (c) _____.

\* \* \*

BONDING IN CARBON COMPOUNDS. The covalent bond formed with carbon in many organic compounds and in diamond is of special interest. In these cases four bonds extend outward from the carbon atom at 109°28′. Carbon atoms, in the gas state, have the $1s^2 2s^2 2p^2$ electron configuration. The inner shell ($n = 1$) is full and very stable. The outer shell ($n = 2$) has four electrons and would require eight to be full.

Suppose that a $2s$ electron is excited to a $2p$ state. We then have a valence shell configuration of $2s2p^3$. These combine to form four equivalent **hybridized orbitals** called **$sp^3$ orbitals** directed toward the corners of a regular tetrahedron. (Four classical electrons placed on the surface of a sphere and allowed to move only on the surface of the sphere would position themselves as on the vertices of a regular tetrahedron so that their repulsive coulombic potential is a minimum.) Such bonds are sometimes called **tetrahedral bonds**. The simplest of compounds based on such bonds is methane $CH_4$.

Figures 5.6.2 and 5.6.3 illustrate how $sp^3$ orbitals form tetrahedral bonds in methane and ethane, respectively. We note that the carbons in all the compounds

$$H[-\underset{\underset{H}{|}}{\overset{\overset{H}{|}}{C}}-]_n H$$

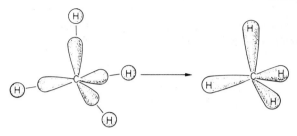

FIGURE 5.6.2. Formation of methane from carbon with four $sp^3$ hybrid orbitals and four hydrogens, each with one $1s$ orbital.

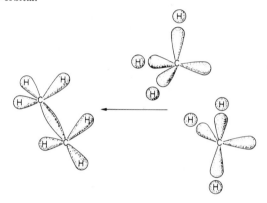

FIGURE 5.6.3. Formation of the ethane molecule.

also show such tetrahedral bonding. Such compounds include small molecules such as ethane

$$\begin{array}{c} \text{H} \quad \text{H} \\ | \quad | \\ \text{H}-\text{C}-\text{C}-\text{H} \\ | \quad | \\ \text{H} \quad \text{H} \end{array}$$

all the way to huge molecules such as polyethylene where $n$ is about 1000; the latter are called polymeric molecules. Examples of bond lengths and dissociation energies for bonds ordinarily found in polymeric materials are given in Table 5.6.1.

A carbon-carbon double bond means that four electrons are shared by two carbons.

### Q. 5.6.2

An element in a compound can have a different electron configuration than it does in the free atom. Thus in compounds of carbon, the

carbon usually forms (a)＿＿＿＿＿＿ called (b)＿＿＿＿＿＿.
Such electron clouds are directed outward toward the (c)＿＿＿＿＿＿.
The simplest molecule with this configuration is (d)＿＿＿＿＿＿.

\* \* \*

Table 5.6.1. BOND LENGTHS AND ENERGIES\*

| Bond | Bond Length (Å) | Dissociation Energy | |
|---|---|---|---|
| | | kcal/mole | eV |
| O—O | 1.32 | 35 | 1.5 |
| Si—Si | 2.35 | 42.5 | 1.85 |
| S—S | 1.9–2.1 | 64 | 2.8 |
| C—N | 1.47 | 73 | 3.2 |
| C—Cl | 1.77 | 81 | 3.5 |
| C—C | 1.54 | 83 | 3.6 |
| C—O | 1.46 | 86 | 3.7 |
| N—H | 1.01 | 93 | 4.0 |
| C—H | 1.10 | 99 | 4.3 |
| C—F | 1.32–1.39 | 103–123 | 4.5–5.3 |
| O—H | 0.96 | 111 | 4.8 |
| C=C | 1.34 | 146 | 6.3 |
| C—O | 1.21 | 179 | 7.8 |
| C≡N | 1.15 | 213 | 9.3 |

\* From J. A. Brydson, *Plastic Materials*, Van Nostrand Reinhold Company, New York (1966).

Covalent bonding in *solids* is the same as in *molecules*. The attractive forces arise from the concentration of electronic charge along the bonding directions joining nuclei. The distribution of electrons in a crystal structure can be determined relatively precisely by the degree to which they scatter X-rays. This has been used to define effectively the location of highly directed bonds in specific covalent crystals of the diamond type. Typical examples of crystals where bonding is essentially all covalent are diamond, silicon, germanium, and silicon carbide. Diamond, silicon, germanium, and gray tin all have the tetrahedral structure associated with $sp^3$ hybridized bonding found in simple hydrocarbon molecules such as methane. Silicon is an extremely important semiconductor. In diamond each carbon is tetrahedrally bonded to four other carbon atoms. This arrangement in solids is referred to as the **diamond structure** and will be discussed in the next chapter.

## Q. 5.6.3

A crystalline form of carbon known as (a)_____ also has tetrahedral bonding. The C—C bonds in this material have an energy of approximately (b)_____ eV. The atoms in the important semiconductor material (c)_____ are also tetrahedrally coordinated.

\* \* \*

## ANSWERS TO QUESTIONS

5.6.1 (a) Delocalized, (b) kinetic energy, (c) 90 deg.

5.6.2 (a) Four equivalent hybridized orbitals, (b) $sp^3$ orbitals, (c) vertices of a regular tetrahedron, (d) methane or $CH_4$.

5.6.3 (a) Diamond, (b) 4, (c) silicon.

## 5.7  BONDING IN METALS

Metals are characterized by having each atom surrounded equally by many other atoms. A metal atom in a crystal usually has 12 nearest neighbors, although in some cases it has only 8. This is to be compared with covalently bonded carbon in diamond which has 4 nearest neighbors, or with the sodium ion in NaCl which has 6 nearest-neighbor chloride ions.

**Valence electrons** are the electrons in an atom outside of the full shells; it is these electrons (and these alone) which are involved in bonding. We have already noted how they are involved in ionic bonding and covalent bonding. A metal, such as copper, which has one valence electron per atom (the electron configuration is $1s^2 2s^2 2p^6 3s^2 3p^6 3d^{10} 4s$) has the following structure: The $Cu^+$ ions are located at specific sites (as are the $Na^+$ ions of NaCl) but the valence electrons (one from each atom) are "free" to wander throughout the copper bar. This pool of electrons is called the **electron gas.** To a first approximation, it can be assumed that the motion of these electrons is independent of the presence of the other electrons or of the ions; i.e., each electron has a potential energy $U = 0$. In that case the energy of each of these **free electrons** is given by (5.2.8). Each electron in this pool must have a different set of four quantum numbers. At the absolute zero of temperature, $2N$ electrons would occupy all the possible energy states from the lowest to the $N$th state in increasing order (the factor of 2 enters because of the two-spin states). The highest energy state occupied is then called the **Fermi energy**; it is independent of the size of the copper bar (at finite temperatures, the Fermi energy is defined differently; however, it is virtually independent of temperature for metals).

Bonding in metal crystals is due primarily to the decrease in energy which results from the complete delocalization of the valence electrons. The heat of vaporization, $\Delta H_v$, of some metals, which is a good measure of the bonding energy, is shown in Table 5.7.1.

Table 5.7.1. COMPARISON OF COHESIVENESS IN METALS, $\Delta H_v$

|  | Molecule | | Crystal | |
| --- | --- | --- | --- | --- |
|  | kcal/mole | eV | kcal/mole | eV |
| Sodium | 17.6 | 0.76 | 26.2 | 1.13 |
| Copper | — | — | 81.7 | 3.55 |
| Iron | — | — | 96.5 | 4.19 |
| Tungsten | — | — | 203.0 | 8.82 |

### Q. 5.7.1

Electrons in an atom outside a completed shell or subshell are called (a)_____. The valence electrons in a metal are (b)_____. This pool of electrons is called (c)_____. In many cases these electrons can be considered to be completely (d)_____. The energy states of these free electrons at absolute zero temperature varies from almost (but not quite) zero to an energy called the (e)_____.

\* \* \*

## ANSWERS TO QUESTION

5.7.1 (a) Valence electrons, (b) "free" to wander through the metal, (c) an electron gas, (d) free electrons, (e) Fermi energy.

## 5.8 SECONDARY BONDS

There is a large group of solids called **molecular solids** which are held together primarily by forces between molecules. These forces were originally described by van der Waals to explain how molecular gases condense to liquids. Here the bonding is relatively weak and so the bonds are called **secondary bonds.** These forces actually arise from several different interactions: *permanent dipole* forces, *induced dipole* forces, and *dispersion* forces. Although the magnitude of these effects is relatively small, nevertheless, in many cases their contribution is often critical in defining the stability of solids. For example, they contribute to the condensation of the noble gases at low temperatures, to the equilibrium interionic distance in ionic crystals, and to the stability of many solids, including proteins, cellulose, and other polymeric materials. The **orientation effect** between permanent dipoles was suggested first by Keesom in 1912. If the centroid of the positive charges does not coincide with the centroid of the negative charges, the molecule is said to possess a **permanent dipole.**

The **dipole moment** is defined as

$$\mathbf{p} = q\mathbf{a}, \qquad (5.8.1)$$

where **a** describes the charge separation as shown in Figure 5.8.1. The direction of **p** is from the negative to the positive charge. For example, in

FIGURE 5.8.1. Electric dipole.

the hydrogen chloride molecule (in the gas phase) the hydrogen end of the molecule has a permanent positive charge, $\delta$, while the chloride end has a permanent negative charge, $-\delta$. Here $\delta$ is a fraction of the unit electron charge magnitude and in this case $\delta \approx e/6$. At absolute zero two such dipoles would align and they would attract each other as shown in Figure 5.8.2. (The attractive force is easy to calculate from coulomb's law by using it four times, once for each pair interaction.) However, at high temperatures the dipoles tend to be randomly oriented. (This is one of the important effects of temperature, namely, to cause randomization or disorientation.)

FIGURE 5.8.2. Dipoles fixed at a separation of $r$ from their centers but allowed to rotate. At equilibrium they are aligned as shown.

However, there will still be some attraction at high temperature. The potential energy interaction in mks units between two dipoles at a temperature $T$ can be shown to be (by statistical mechanics)

$$U_0 = -\frac{1}{8\pi^2\epsilon_0^2}\frac{p_1^2 p_2^2}{r^6}\frac{1}{k_B T}. \qquad (5.8.2)$$

Here $r$ is the distance between the molecules, $k_B$ is Boltzmann's constant, and $T$ is the absolute temperature.

The **induction effect** was pointed out by Debye who noted that the orientation effect (attraction of permanent dipoles) alone cannot account for the van der Waals cohesion since there is also an interaction term which is independent of temperature. The dipole is not rigid (but can be stretched or compressed), so that the electric field of one dipole changes the dipole moment of a neighboring one (*induces* an additional moment) in such a manner as to result in an additional attraction between the two dipoles. Again the potential interaction $U_I$ varies as $1/r^6$.

To explain the van der Waals interaction observed in gases which possess no permanent dipole moment such as argon, nitrogen, methane, etc., the existence of **dispersion forces** was postulated and proved to exist by F. London in 1930. These are transient or fluctuating electric moments induced between molecules having no permanent dipole moment which result in a significant attractive interaction. The dispersion energy $U_D$ also varies as $1/r^6$.

Table 5.8.1. COMPARISON OF THEORETICAL AND EXPERIMENTAL HEATS OF SUBLIMATION*

| Substance | Theoretical Values | | Experimental Values | |
|---|---|---|---|---|
| | kcal/mole | eV | kcal/mole | eV |
| Ne | 0.47 | 0.020 | 0.59 | 0.025 |
| Ar | 1.92 | 0.083 | 2.03 | 0.088 |
| Kr | 3.27 | 0.142 | 2.80 | 0.121 |
| $N_2$ | 1.64 | 0.071 | 1.86 | 0.080 |
| $O_2$ | 1.69 | 0.073 | 2.06 | 0.089 |
| CO | 1.86 | 0.080 | 2.09 | 0.090 |
| $CH_4$ | 2.42 | 0.105 | 2.70 | 0.117 |
| HCl | 3.94 | 0.171 | 5.05 | 0.219 |
| HBr | 4.45 | 0.193 | 5.52 | 0.239 |
| HI | 6.65 | 0.289 | 6.21 | 0.269 |

* F. London, *Transactions of the Faraday Society*, **33**, 8 (1937).

Table 5.8.1 gives the sublimation energies for solids in which the molecules are attracted to each other by one or more of the **van der Waals bonds**. It should be noted that the bonding energies of molecular crystals are of the order of 1 to 5 kcal/mole as contrasted to values of about 200 kcal/mole for ionic solids, for pure covalent solids, and high-melting metals. The molecules of cotton, silk, and other natural fibers are also bound to each other by secondary molecular forces. The contribution of the van der Waals bonding will be considered again in many instances when we consider structure in more detail.

## Q. 5.8.1

A dipole moment is (a)_____. The direction of **p** is (b)_____. All the types of van der Waals' bonding vary with separation $r$ of the dipoles according to (c)_____. Argon liquefies and solidifies at low temperature, although the atom has no permanent dipole; the origin of the binding forces is (d)_____.

\* \* \*

Mention should be made of **hydrogen bonding** which acts in a fashion similar to that of a van der Waals bond but is completely different in origin. We know that hydrogen can form only one normal covalent electron pair bond since it has only the single $1s$ orbital available for bond formation. However, since the proton is extremely small, its electrostatic field is intense, and bonding can occur due to the attraction of the positive proton for the electrons on the bonded atom. In some situations therefore a hydrogen atom can form a bond to two other atoms instead of to only one. It is an example of a particularly strong dipole-dipole interaction. It occurs in general between hydrogen and the electronegative elements N, O, and F of small atomic volumes. Hydrogen bonding, like the van der Waals bonding, is a secondary bonding effect and has many important applications.

The **hydrogen bond,** while not very strong, usually has a dissociation energy of about 5–10 kcal/mole (which is a strong secondary bond) but it is important in many structures such as water and organic compounds. The structure of ice is shown in Figure 5.8.3. The high freezing and boiling points of water are due to the hydrogen bonds.

It is interesting that hydrogen bonding also makes a significant contribution to the structure of proteins such as hair and muscle and to the structure of complex amino acids such as DNA (deoxyribonucleic acid). It is likewise very important in determining the structure of synthetic polymers such as nylon and plays a major role in the strength of nylon

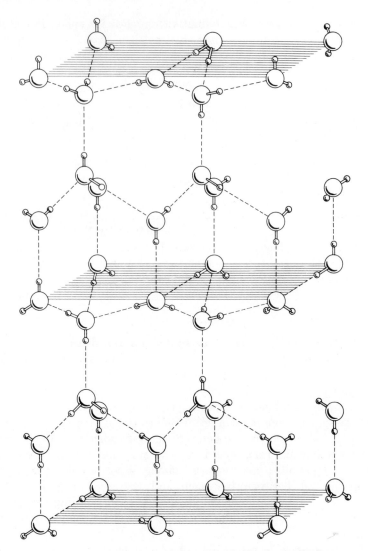

FIGURE 5.8.3. Arrangement of H$_2$O molecules in the ice crystal. The large spheres are the oxygens. Note that the orientation of the H$_2$O molecule as indicated is arbitrary. It is significant that every hydrogen atom is on a line between two oxygen atoms and is closer to one than to the other.

fibers. The structure of such polymeric materials will be discussed in Chapter 7.

## EXAMPLE 5.8.1

The bonding energy per mole of argon in the solid phase is, according to Table 5.8.1, given by 2.03 kcal/mole. Here the distance of closest approach of the atoms is 3.87 Å. Estimate the bonding energy per mole of gas at 1 atm and 0°C.

*Answer.* A mole of ideal argon gas under these conditions has a volume of 22.4 liters. It contains $6.02 \times 10^{23}$ argon atoms. Hence the

$$\text{volume/atom} = \frac{22.4 \times 10^3 \times 10^{24} \text{ Å}^3}{6.02 \times 10^{23}}$$

$$\approx 37 \times 10^3 \text{ Å}^3.$$

The distance between atoms is roughly the cube root of this or 33.4 Å. The binding energy is approximately determined by the attractive term $U \approx -a/r^6$. Then

$$\frac{U_{\text{gas}}}{U_{\text{solid}}} \approx \frac{r^6_{\text{solid}}}{r^6_{\text{gas}}} \approx \left(\frac{3.87}{33.4}\right)^6 \approx 10^{-6}.$$

Hence in the present case $U_{\text{gas}} \approx 2 \times 10^{-6}$ kcal/mole or $2 \times 10^{-3}$ cal/mole.

## EXAMPLE 5.8.2

The kinetic energy per mole of an ideal monatomic gas is $\frac{3}{2}RT$. Compare the potential energy of argon gas (see the previous example) with its kinetic energy at 0°C and at 1 atm.

*Answer.* Since $R = 1.987$ cal/mole °K we have

$$U_K = \tfrac{3}{2} \times 1.987 \times 273.2 = 814 \text{ cal.}$$

The potential energy is thus only a tiny fraction ($2 \times 10^{-3}/820 \approx 2 \times 10^{-6}$) of the kinetic energy.

In a truly perfect gas the potential interactions between the molecules is exactly zero. Argon at 1 atm and 0°C comes very close to this.

## Q. 5.8.2

The strongest secondary bond is the (a)_____ which is formed between a hydrogen atom (already covalently bonded to another atom) and a highly electronegative atom such as (b) (three). The high (c) (a property) of water is due to hydrogen bonding. At 1 atm and 300°K the potential energy of argon gas is only about (d)_____ that of the kinetic energy.

\* \* \*

## ANSWERS TO QUESTIONS

**5.8.1** (a) A pair of opposite charges of magnitude $q$ times the distance between them; (b) from the negative to the positive charge; (c) $1/r^6$; (d) dispersion forces; i.e., the atom has temporary dipoles due to charge fluctuations.

**5.8.2** (a) Hydrogen bond; (b) N, O, and F; (c) boiling or freezing point, (d) $10^{-6}$.

### 5.9 POLARIZATION AND MAGNETIZATION

ELECTRIC DIPOLES AND POLARIZATION. We noted in Section 5.8 that the molecule of hydrogen chloride has a permanent dipole. Let us suppose that we have a unit volume of gas. The dipoles would be randomly oriented at high temperatures so that the sum of all the dipole moments would equal zero. However, in the presence of an electrical field, these dipoles would tend to align (to overcome the randomizing effect of temperature). There would then be a net dipole moment per unit volume; this is the **polarization, P**, studied in Chapter 3. An electric field **E** causes a force on a charge $q$ equal to $\mathbf{F} = q\mathbf{E}$. This can be used to show that if the angle between **E** and **p** is $\theta$, the electric field will exert a torque

$$T = pE \sin \theta \qquad (5.9.1)$$

on the dipole, tending to line it up with the field **E**. In vector notation, $\mathbf{T} = \mathbf{p} \times \mathbf{E}$. Recall that a torque is generated by a force acting normal to a lever arm times the length of the lever arm. A unit often used for dipole moment is the **debye**; 1 D = $3.33 \times 10^{-30}$ C-m. Electrical fields also induce dipoles in atoms (or molecules); thus an argon atom would have a dipole moment in an electric field; this is illustrated in Figure 5.9.1; moreover, the dipole moment of an HCl molecule will be changed by the field. This process which involves only the motion of electrons can occur exceedingly rapidly and is independent of temperature, while those polarization processes which involve the rotation of molecules occur much more slowly. Both effects contribute to the polarization of HCl gas.

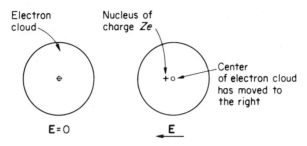

FIGURE 5.9.1. Induced electric dipole.

**EXAMPLE 5.9.1**

In Table 3.4.1 there are two values of dielectric constant for NaCl crystals, depending on the frequency. Explain.

*Answer.* The NaCl crystal (see Figure 5.5.2) has no permanent dipole moment. However, an electric field can induce a net moment and hence a polarization. First, it would cause the $Na^+$ and $Cl^-$ ions to move in opposite directions. Second, it would distort the spherical electron clouds of both the $Na^+$ and $Cl^-$ as in Figure 5.9.1. These are additive effects. At low frequencies both processes would occur; at high enough frequencies the former would not occur (the big ions move too slowly). Recall that $P = \epsilon_0 \chi E$ and $\chi = \kappa - 1$, where $\chi$ is the susceptibility and $\kappa$ is the dielectric constant.

Certain crystals such as $BaTiO_3$ at room temperature contain permanent dipoles just as do some molecules. These are the ferroelectrics.

**Q. 5.9.1**

A molecule such as HCl contains (a)_____ dipole. A molecule such as $CH_4$ contains (b)_____ dipole. When a gas such as HCl is placed in an electric field it shows a net polarization, i.e., $P > 0$; this is primarily due to (c)_____. At very high frequencies, the polarization of NaCl is due to (d)_____ while at low frequencies it is due to (e)_____. Some crystals, just like some molecules, have permanent electric dipoles; an example is (f)_____.

\* \* \*

**EXAMPLE 5.9.2**

Find the potential energy of a charge $-e$ which is a distance $r$ away from the center of a dipole $ea$. The dipole is directed toward the charge.

*Answer.* The potential energy for a charge $q_1$ and charge $q_2$ interaction is $q_1 q_2 / 4\pi\epsilon_0 r$. In the present case there are two such interactions. Thus the potential energy is $-e^2/4\pi\epsilon_0[r - (a/2)] + e^2/4\pi\epsilon_0[r + (a/2)]$. If $a \ll r$, then $1/(1 - a/2r) \doteq 1 + a/2r$, and the potential energy is $-e^2a/4\pi\epsilon_0 r^2 = -pe/4\pi\epsilon_0 r^2$. The electrical potential is

$$\frac{p}{4\pi\epsilon_0 r^2}$$

and the electric field of the dipole along a line extending in the direction of the dipole is

$$\frac{p}{2\pi\epsilon_0 r^3}$$

For the general description of the electric field surrounding the dipole see the text [Duckworth, Henry E., *Electricity and Magnetism*, Holt, Rinehart and Winston, New York (1961), p. 88.]

MAGNETIC DIPOLES. When a steady current $i$ flows in a wire it generates a magnetic field. If the wire forms a circular loop of area $A$ as shown in Figure 5.9.2, an external magnetic induction field **B** at an angle $\theta$ to the normal to the loop will exert a torque on the loop equal in magnitude to

$$T = iAB \sin \theta. \tag{5.9.2}$$

[See, e.g., N. H. Frank, *Introduction to Electricity and Optics*, McGraw-Hill Book Company, Inc., New York (1950), p. 115.] By analogy with (5.9.1) we are led to define a **magnetic dipole moment**

$$p_m = iA. \tag{5.9.3}$$

(In vector notation $\mathbf{p}_m = i\mathbf{A}$, where the direction of **A** is the normal to the loop as shown in Figure 5.9.2. Similarly $\mathbf{T} = \mathbf{p}_m \times \mathbf{B}$.)

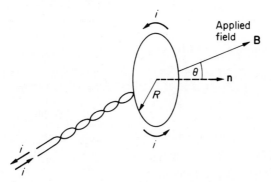

FIGURE 5.9.2. Magnetic dipole. The dashed line is normal to the circular current loop. The dipole moment is normal to the loop and equal in magnitude to $i\pi R^2$.

In the elementary Bohr theory of the atom, the ground state electron (with charge $-e$) passes a given point $v/2\pi r$ times per second. Hence the current is $-ev/2\pi r$ and the expected magnetic dipole moment is, from (5.9.3) and (5.1.8),

$$p_m = \frac{-evr}{2} = \frac{-emvr}{2m} = \frac{-e}{2m}\frac{h}{2\pi}. \tag{5.9.4}$$

The quantity

$$\beta = \frac{e}{2m}\frac{h}{2\pi} \tag{5.9.5}$$

is called the **Bohr magneton**. It is the unit magnetic dipole moment and has a value of $9.27 \times 10^{-24}$ A-m². We might rewrite (5.9.4) in the form

$$\mathbf{p}_m = \frac{-e}{2m}\mathbf{L}, \tag{5.9.6}$$

where $\mathbf{L}$ is the angular momentum. We know that the Bohr theory was inadequate in many respects (one of which is that the ground state electron actually has zero angular momentum). Actually modern quantum mechanics yields

$$L = \sqrt{l(l+1)}\,\frac{h}{2\pi}. \tag{5.9.7}$$

Recall from Section 5.3 that $l$ takes on all the integer values $n-1, n-2, \ldots, 0$. $l$ is called the angular momentum quantum number. Note that for the $n = 1$ state of the hydrogen atom, $l = 0$, so $L = 0$ and $p_m = 0$. It can also be shown that the total angular momentum of electrons in atoms with full shells is zero. Recall that we noted earlier that such atoms had electron clouds with spherical charge distributions. Therefore such atoms have no permanent magnetic moment. However, magnetic dipoles can be induced by the presence of a magnetic field. These induced dipoles are always antiparallel (i.e., in a direction opposite) to the field; hence the **magnetization** [m, the net magnetic dipole moment (the vector sum) per unit volume] is negative (diamagnetism); for an elementary discussion of this, see Section 14.1.

Atoms or molecules with a nonzero orbital angular momentum have permanent magnetic dipoles. The alignment of these in a magnetic field is analogous to the alignment of electric dipoles in an electric field; this leads to paramagnetism.

In **ferromagnetism** the magnetic dipoles are spontaneously (in the absence of a field) aligned; in **ferrimagnetism** part of the dipoles are spontaneously aligned, and the other dipoles are antiparallel but there is a net dipole.

**EXAMPLE 5.9.3**

The compound $Fe_3O_4$ has $Fe^{2+}$ ions with their dipole moments aligned and $Fe^{3+}$ ions with half their dipole moments in an opposite direction to the other half so the $Fe^{3+}$ ions do not contribute to the net magnetization. Calculate the saturation magnetization. The magnetic dipole moment of each $Fe^{2+}$ is $4\beta$ and there are $1.35 \times 10^{28}$ $Fe^{2+}$ ions/m³.

*Answer.*

$$M_S = 4(9.27 \times 10^{-24} \text{ A-m}^2)(1.35 \times 10^{28} \text{ m}^{-3})$$

$$= 0.5 \times 10^6 \text{ A/m}.$$

The experimental value is $0.48 \times 10^6$ A/m.

**Q. 5.9.2**

A current loop produces a magnetic dipole moment equal to (a) _____. The direction of this dipole moment is (b)_____ ____. A hydrogen atom in the $n = 1$ state would have (c)_____ magnetic moment due to orbital angular momentum.

* * *

ELECTRON SPIN. The silver atom has the electron configuration $(1s^2 2s^2 2p^6 3s^2 3p^6 3d^{10} 4s^2 4p^6 4d^{10})5s$. The full $n = 1$, $n = 2$, $n = 3$, and $n = 4$ shells have electron clouds with spherical symmetry as does the additional $5s$ electron ($n = 5$, $l = 0$). One would therefore be led to conclude that silver atoms in the vapor have no orbital angular momentum and hence no permanent magnetic dipole. However, when a collimated beam of silver atoms is passed through a magnetic field (normal to the initial path of the atoms) some are deflected in the direction of **B** and some in the opposite direction and of such a magnitude as to suggest that the electron has a magnetic dipole of $\pm \beta$. This is the famous Stern and Gerlach experiment: O. Stern and W. Gerlach, *Zeitschrift fuer Physik* **8**, 110, and **9**, 349 (1922). (Note that except for the $5s$ electron, the electrons would occur in pairs with one having a dipole $+\beta$, the other $-\beta$, resulting in no net dipole.) G. E. Uhlenbeck and S. Goudsmit, *Naturwissenschaft* **13**, 953 (1925) or *Nature* **117**, 264 (1926) as a result of their attempts to explain certain atomic spectra were led to conclude that the electron has an *intrinsic* dipole moment of $\pm \beta$ (independent of orbital motion). The Stern and Gerlach experiment agreed with this. This was the origin of the quantum number $m_s$.

## Q. 5.9.3

A hydrogen atom in the $n = 1$ state would have a net dipole moment due to (a)_____. A helium atom would have a permanent dipole moment of (b)_____. A sodium ion and a fluorine ion would have a dipole moment of (c)_____. NaF crystals (same structure as NaCl) would be (d) (diamagnetic, paramagnetic, or ferromagnetic).

\* \* \*

## ANSWERS TO QUESTIONS

**5.9.1** (a) A permanent, (b) no permanent, (c) partial orientation of the dipoles in the direction of the field, (d) electron cloud displacement, (e) both electron cloud displacement and ion motion, (f) $BaTiO_3$.

**5.9.2** (a) $iA$, (b) normal to the area, (c) zero.

**5.9.3** (a) Electron spin, (b) zero, (c) zero, (d) diamagnetic due to induced dipoles.

## REFERENCES

Sproull, R. L., *Modern Physics*, John Wiley & Sons, Inc., New York (1963). Sproull presents modern physics in a clear and exciting fashion.

Hill, T. L., *Matter and Equilibrium*, W. A. Benjamin, Inc., Reading, Mass. (1966). Chapter 3 on intermolecular forces is a good simple introduction.

Speakman, J. C., *Molecules*, McGraw-Hill Book Company, Inc., New York (1966). The structure of molecules is discussed from an elementary viewpoint.

Ryschkewitsch, G. E., *Chemical Bonding and the Geometry of Molecules*, Van Nostrand Reinhold Company, New York (1963). An elementary paperback which would make nice reading while waiting for a plane.

Wahl, A. C., "Chemistry by Computer," *Scientific American* (April 1970), p. 54. Illustrates computed electron structures in atoms and molecules.

## PROBLEMS

**5.1** Give, in electron volts, the binding energy of
   (a) Nuclear particles.
   (b) Inner electrons of the atom to the nucleus.

(c) Outer (valence) electrons to the ion.
(d) Chemical bonds.

The materials scientist is primarily concerned with which of the above binding energies?

**5.2** Give, in Ångstroms, the approximate size of
   (a) Protons.
   (b) Atoms.
   (c) The wavelength of optical light.
   (d) The length of a nylon polymer molecule.
   (e) The size of a crystal in steel.

**5.3** Calculate the electron speed in the Bohr orbit. Compare this with the speed of light in free space.

**5.4** Show that the coefficient of $1/n^2$ in (5.1.10) is $-13.6$ eV.

**5.5** Show that the Bohr radius is 0.529 Å.

**5.6** An electron has a kinetic energy of 3 eV. Show that its wavelength is 7 Å.

**5.7** Discuss the background which led to Schrödinger's equation.

**5.8** Discuss the physical significance of each of the conditions in (5.3.2).

**5.9** Why is the second ionization potential of the helium atom four times the first ionization potential of the hydrogen atom?

**5.10** Why is the first ionization potential of Li very small but the second ionization potential extremely large?

**5.11** Using the hydrogen-like approximation for wave functions in atoms, calculate the energy and wavelength of the photon emitted when an electron drops from an $L$ shell to a $K$ shell in copper. (*Note:* This is the source of the $K_\alpha$ line for copper, which is X-ray radiation widely used in diffraction studies. An electron is initially knocked out of the $K$ shell by bombarding the copper with high-energy electrons. Then an electron from an $L$ shell spontaneously falls into the lower-energy state with the emission of the X-ray radiation.) The experimental value is $\lambda = 1.54$ Å.

**5.12** X-rays are generated by bombarding a target material, such as molybdenum, with electrons which have been accelerated by an electric field. Approximately what potential (volts) would be necessary to knock a $K$ electron (i.e., a $1s$ electron) from molybdenum?

**5.13** (a) Discuss the assumption being made when it is stated that the electron configuration of carbon is $1s^2 2s^2 2p^2$.
   (b) Describe the nature of the electron distribution for each of these orbitals.

**5.14** (a) Give, in electron volts, the binding energies for
   1. Ionic bonding of NaCl.
   2. Metallic bonding of Fe.
   3. Covalent bonding of diamond.

(b) Give, in electron volts, the binding energies for secondary bonding of
  1. Argon by dispersion forces.
  2. $H_2O$ molecules in ice by hydrogen bonding.

**5.15** What is the essential feature of
  (a) The ionic bond in MgO?
  (b) The covalent bond in $H_2$?
  (c) The metallic bond in copper?
  (d) The van der Waals bond in solid argon?
  (e) The hydrogen bond in water?

**5.16** Define ionic radius. If the distance between the $Na^+$ center and the $Cl^-$ center in solid NaCl is 2.79 Å and the $Cl^-$ has a radius of 1.82 Å, what is the $Na^+$ radius?

**5.17** Describe the nature of the covalent bond in diamond.

**5.18** Why is it that water vapor in equilibrium with water at room temperature behaves as an ideal gas?

**5.19** Describe the origin of the Keesom, Debye, and London secondary bonds which are often grouped together and called van der Waals bonds.

# MORE INVOLVED PROBLEMS

**5.20** A particle moves along a circular path. Its position can be described by $x = r \cos \omega t$, and $y = r \sin \omega t$. At $t = 0$ it is located at $x = r$, $y = 0$. Compute $d^2x/dt^2$ and $d^2y/dt^2$. Show that at $t = 0$ the inward acceleration is $\omega^2 r$ or $v^2/r$. Note that $\omega$ is the angular velocity in radians per unit time.

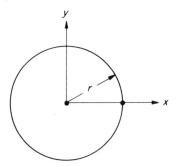

PROBLEM FIGURE 5.20.

**5.21** Write an essay on the production of characteristic X-rays. Among the many references which you might use are B. D. Cullity, *Elements of X-ray Diffraction*, Addison-Wesley Publishing Company, Inc., Reading, Mass. (1956).

**5.22** Show that (5.3.5) is a solution of (5.3.1) when $E$ is given by (5.3.3) with $n = 1$.

**5.23** Show that (5.3.5) satisfies all the conditions of (5.3.2).

**5.24** Assuming only dipole 2 in Figure 5.8.2 is allowed to rotate, what would be its orientation for the case of no attraction and for the case of maximum attraction?

**5.25** Show that the potential energy between a positive charge and an aligned dipole $p = qa$ which are separated by $r$ (where $r \gg a$) is given by

$$U = -\frac{qp}{4\pi\epsilon_0 r^2}.$$

We are using mks units in which the potential energy between two charges is

$$U = \frac{q_1 q_2}{4\pi\epsilon_0 r}$$

and the force is

$$F = \frac{q_1 q_2}{4\pi\epsilon_0 r^2}$$

and the field $E$ due to $q_1$ is

$$\frac{q_1}{4\pi\epsilon_0 r^2}.$$

(In cgs units the $4\pi\epsilon_0$ is absent.)

**5.26** Consider a one-dimensional prototype crystal of NaCl with the nearest-neighbor separation equal to $r$.

(a) Show why the coulomb attraction term of a given ion to all the other ions can be written as

PROBLEM FIGURE 5.26.

$$U_c = -\frac{2e^2}{4\pi\epsilon_0 r} + \frac{2e^2}{8\pi\epsilon_0 r} - \frac{2e^2}{12\pi\epsilon_0 r} + \frac{2e^2}{16\pi\epsilon_0 r} - \cdots$$

or

$$-\frac{2e^2}{4\pi\epsilon_0 r}\left[1 - \frac{1}{2} + \frac{1}{3} - \frac{1}{4} + \cdots\right].$$

(b) Show that the term in brackets is simply the Taylor series expansion of $\ln(1 + x)$ about $x = 0$ which is then evaluated for the case $x = 1$; i.e., the term in brackets equals $\ln 2$.

(c) Hence, show that

$$U_c = -\frac{1.38 e^2}{4\pi\epsilon_0 r} = -\frac{A e^2}{4\pi\epsilon_0 r}.$$

The coefficient $A$ is called the Madelung constant.

**5.27** In addition to the method used in Problem 5.26 there is an alternative way of obtaining Madelung's constant. This is very useful in three dimensions, but we shall illustrate it for the case of one dimension. In this technique, successively larger electrically neutral sections of the crystal are considered:

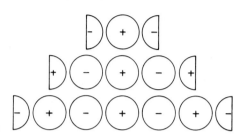

PROBLEM FIGURE 5.27.

$$U_1 = -\frac{e^2}{4\pi\epsilon_0 r}(1),$$

$$U_2 = -\frac{2e^2}{4\pi\epsilon_0 r} + \frac{e^2}{8\pi\epsilon_0 r} = -\frac{e^2}{4\pi\epsilon_0 r}\left(\frac{3}{2}\right),$$

$$U_3 = -\frac{2e^2}{4\pi\epsilon_0 r} + \frac{2e^2}{8\pi\epsilon_0 r} - \frac{e^2}{12\pi\epsilon_0 r} = -\frac{e^2}{4\pi\epsilon_0 r}\left(\frac{4}{3}\right).$$

(a) Show that

$$U_4 = -\frac{e^2}{4\pi\epsilon_0 r}(1.417) \quad \text{and} \quad U_5 = -\frac{e^2}{4\pi\epsilon_0 r}(1.367).$$

In this method, called Evjen's method, the terms in parentheses are approximations for Madelung's constant. Note that the $U_5$ approximation is within 1% of the answer given in Problem 5.26(c).

(b) Compare this with the answer obtained by truncating the series expression in Problem 5.26(a) at five terms.

(c) Give physical reasons why Evjen's method converges much more rapidly than the series of Problem 5.26(a).

**5.28** (a) Show that the total coulombic attraction of the $2N$ ions in Problem 5.26 ($N$ cations and $N$ anions) is

$$U_c = -\frac{1.38Ne^2}{4\pi\epsilon_0 r}.$$

(b) Assuming that we write the total repulsive potential in the form

$$U_R = \frac{B}{r^n},$$

we have

$$U = U_c + U_R = -\frac{1.38Ne^2}{4\pi\epsilon_0 r} + \frac{B}{r^n}.$$

**230** Binding in Atoms, Molecules, and Crystals

What conditions must apply to $U$ at equilibrium? At equilibrium $r$ has the value $r_0$.

(c) Show that

$$U_0 = U(r_0) = -\frac{1.38Ne^2}{4\pi\epsilon_0 r_0}\left(1 - \frac{1}{n}\right)$$

since

$$\frac{1.38Ne^2}{4\pi\epsilon_0 r_0^2} - \frac{nB}{r_0^{n+1}} = 0.$$

## SOPHISTICATED PROBLEMS

**5.29** Suppose an experimentalist has available apparatus for applying a tensile force $F$ to the one-dimensional crystal of Problems 5.26–5.28 and can measure $k_0$ as defined by

$$\lim_{r \to r_0} \frac{\partial F}{\partial r} = k_0.$$

Show that if $k_0$ and $r_0$ are known, then $B$ and $n$ in Problem 5.28(b) can be obtained.

**5.30** Show that if $U = -(NAe^2/4\pi\epsilon_0 r) + (B/r^n)$, that $\Delta U = U - U_0 \propto (r - r_0)^2$ for $|(r - r_0)/r_0| \ll 1$, if $U_0 = U(r_0)$ and $r_0$ is the value of $r$ at the minimum of $U$. In other words we want to show that near the minimum of the potential energy curve in Figure 5.5.1 the curve has a parabolic shape.

**5.31** The color of certain dye molecules which have alternating double and single carbon-carbon bonds can often be calculated from the simple one-dimensional particle in a box model. Given the polyene molecule (ion)

$$\text{H--N--C=C--C=C--C=N--H,}$$

with H atoms on each C and terminal N,

(a) Draw a resonance form of this molecule, i.e., a form in which the extra electron pairs (which form the double bond) move to a new but chemically equivalent position.

(b) The six $\pi$ electrons (the resonating electrons) may be considered as free electrons in a one-dimensional box of length

$$L = 7 \times 1.39 \text{ Å},$$

where we assume the ends of the box are $\frac{1}{2}$ bond; bond distance is taken as 1.39 Å. Recalling that the electrons have two possible spins, show that the energy of the highest occupied $\pi$ electron state is

$$E = \frac{h^2 n^2}{8mL^2}, \quad \text{where } n = 3.$$

(c) The first excited state corresponds to exciting a $\pi$ electron to $n = 4$ from $n = 3$. Calculate the change in energy for this transition.
(d) Calculate the wavelength of the absorbed light for this transition and compare with the experimental value of 4250 Å.

For a further discussion of calculating the color of dye molecules, see W. Kauzman, *Quantum Chemistry*, Academic Press, Inc., New York (1957), p. 675.

# Prologue

Many important materials are crystalline solids. In this chapter certain geometrical aspects of perfect crystals are introduced: *lattice, unit cell, basis,* and *lattice parameter*. The conventions used to describe planes and directions in crystals (*Miller indices* and *direction indices*) are explained. Six crystal structures are studied in detail because of their simplicity and frequency of occurrence in engineering materials. They are *body-centered cubic, face-centered cubic, diamond cubic, hexagonal closest packed, sodium chloride type,* and *zinc blende type*. Atoms and ions can be approximated as *rigid spheres* packed together to form crystals. One of the most important geometrical packings involves a *closest-packed* layer of spheres. Certain crystal structures such as the fcc crystal structure and the idealized hcp crystal structure can be constructed by stacking such closest-packed layers on top of each other in a specific *stacking sequence*. Another important packing concept is the *void*, i.e., the space between the rigid spheres. In metals such voids often can be occupied randomly by the smaller atoms, H, B, C, N, and O. Ionic crystals often can be considered to be derived from the above mentioned type of packing of anions with a fixed fraction of voids occupied by smaller cations. *Radius ratio rules* are available to help predict the structure of *ceramic* materials.

Crystals exhibit certain *symmetry* which can be described in terms of *symmetry operations*. The symmetry of a crystal determines the particular anisotropy of a given property.

Real crystals always have packing imperfections which are classified geometrically as *point, line, planar,* and *spatial* imperfections. Such imperfections are responsible for the structure-sensitive nature of properties.

Certain solids which do not have the periodic nature of crystals are called glasses. They have a *random network structure* and exhibit a *glass transition temperature*.

It is pointed out that crystals will diffract waves having wavelengths of the order of 1 Å. This applies to X-rays, electrons, and neutrons. Such diffraction leads to *Bragg's law* which provides a method for determining crystal structures and lattice parameters.

# 6

# ATOMIC ARRANGEMENTS

## 6.1 CRYSTALS AND LATTICES

Universals arise merely from our making use of one and the same idea in thinking of all individual objects between which there subsists a certain likeness; and when we comprehend all the objects represented by this idea under one name, this term likewise becomes universal.—René Descartes, *The Principles of Philosophy* from the translation by J. Vcitch entitled *The Meditations and Selections from The Principles of René Descartes.* Open Court Publishing Company, La Salle, Illinois (1950) p. 160.

Figure 6.1.1 is an example of a synthetically grown quartz crystal. Such quartz crystals occur in nature in various shapes and forms. In 1669 Niels Stensen (Steno) found that regardless of the difference in size and shape of such quartz crystals, the corresponding interfacial angles (in

FIGURE 6.1.1. Synthetically grown quartz crystal. Note the angles between the faces. (Courtesy of Sawyer Products Co., Cleveland, Ohio.)

different crystals) were always the same. (An interfacial angle is the angle between the normals to the faces.) Very careful measurements on materials other than quartz led to the same result. In 1784 Abbé Haüy proposed that this external form was due to the nature of the tiny building blocks of which the crystal was made. These little building blocks were cubes or other parallelepipeds as shown in Figure 6.1.2. When specific building blocks are stacked together in different ways, different crystal faces and hence different interfacial angles are developed as shown in Figure 6.1.3. This postulate also helped explain the cleavage of crystals: Thus NaCl cubes (common table

## 234  Atomic Arrangements

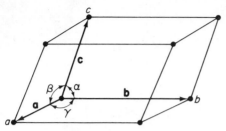

FIGURE 6.1.2. General building block of a crystal.

salt) fractured along planes parallel to the cube surface. The building blocks were postulated to be so small as to be invisible to the naked eye (as we now know, the edge of the block is often but a few Ångstroms). All the

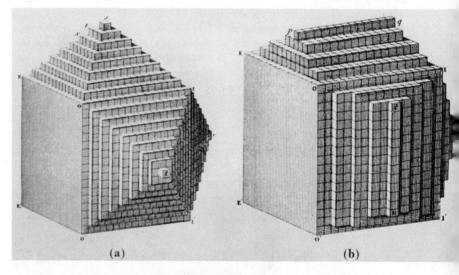

FIGURE 6.1.3. Relation of the external form of crystals to the stacking of the building blocks. Identical building blocks were used in (a) and (b) but different crystal faces and hence interfacial angles are developed. [From A. Haüy, from the atlas to the 1822 edition of his *Traite de cristallographie*, Paris (1822).]

external angles observed in all the various crystals were consistent with seven types of building blocks as shown in Table 6.1.1.

### Q. 6.1.1

Crystals, for example, quartz, although differing widely in shape and size always exhibit (a)_____. All the external angles observed in all the different crystals are consistent with Abbé Haüy's suggestions and (b)_____ types of building blocks are needed.

\* \* \*

**Table 6.1.1.** THE SEVEN CRYSTAL SYSTEMS

| | | |
|---|---|---|
| Triclinic | $a \neq b \neq c$ | $\alpha \neq \beta \neq \gamma \neq 90°$ |
| Monoclinic | $a \neq b \neq c$ | $\alpha = \gamma = 90° \neq \beta$ |
| Orthorhombic | $a \neq b \neq c$ | $\alpha = \beta = \gamma = 90°$ |
| Tetragonal | $a = b \neq c$ | $\alpha = \beta = \gamma = 90°$ |
| Rhombohedral | $a = b = c$ | $\alpha = \beta = \gamma \neq 90°$ |
| Hexagonal | $a = b \neq c$ | $\alpha = \beta = 90°, \gamma = 120°$ |
| Cubic | $a = b = c$ | $\alpha = \beta = \gamma = 90°$ |

LATTICES. To discuss the internal structure of crystals at the atomic level the concept of lattices is introduced. A **lattice** is a collection of points in a periodic arrangement (and therefore infinite in extent). A **plane lattice** is defined by two noncolinear *translations* and a **space lattice** by three noncoplanar translations. A line joining any two lattice points is a translation as indicated in Figure 6.1.4, where a plane lattice is depicted

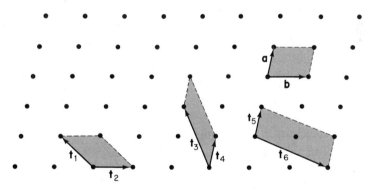

FIGURE 6.1.4. Plane lattice (portion of).

(it is useful to study plane lattices as these are simpler to visualize). These translations are, of course, vectors. For a plane lattice any two of these (noncolinear) vectors with a common origin define a **unit cell.** The unit cell is so called because the entire lattice can be derived by repeating this cell as a unit by means of the translations that serve as the unit cells edges. The unit cell is a **primitive cell** if all the lattice points in it are at vertices. In Figure 6.1.4 the cell $t_5$, $t_6$ is a nonprimitive unit cell; all the others shown are primitive cells. The choice of a unit cell is made on the basis of that which best represents the *symmetry* of the lattice (it is always possible to use a primitive cell). It is assumed at this point that the student has a feeling for what is meant by symmetry; Figure 6.1.5 will help clarify this feeling. The unit cell designated by $t_1$ and $t_2$ in Figure 6.1.5 *obviously* has

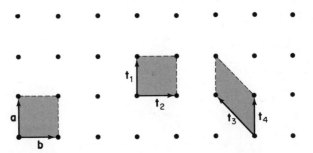

FIGURE 6.1.5. Plane lattice (portion of). Here $a = b$.

certain symmetry (clearly a square is more symmetric than is an arbitrary parallelogram). A general property of a lattice is that each point has the same identical surroundings as any other point. Figure 6.1.2 also can be used to describe the shape of the general three-dimensional cell. The quantities **a**, **b**, and **c** are called **lattice vectors** while $a$, $b$, $c$, $\alpha$, $\beta$, and $\gamma$ are called **lattice parameters.** We can consider the lattice as being built up from building blocks having the shape of the unit cell.

We might now ask, How many different types of lattices are possible? (We are not concerned with the size of a cell; thus a square lattice with the edge of the unit cell equal to either 3 Å or 12 m is just one type of lattice, namely, a square lattice.) We are only interested in determining those types of lattices in which a point has a different set of surroundings than does a point in another lattice. Frankenheim in 1842 deduced that there were 15 such lattices but in 1848 Bravais showed that two of those were equivalent. These 14 different lattices are now called **Bravais lattices.** Commonly used unit cells for four of the Bravais lattices are shown in Figure 6.1.6 (in

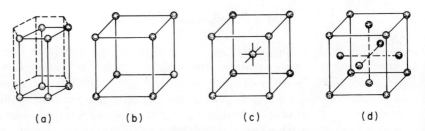

FIGURE 6.1.6. Commonly used unit cells for four of the most important Bravais lattices. (a) Hexagonal. (b) Simple cubic. (c) Body-centered cubic. (d) Face-centered cubic.

an elementary course, it is sufficient to consider only crystal structures based on these four lattices).

**EXAMPLE 6.1.1**

How many nearest-neighbor lattice points does a lattice point in the following lattices have: (a) simple cubic, (b) body-centered cubic, (c) face-centered cubic? *Note:* Each *nearest* neighbor is the same distance away from the given point.

*Answer.* (a) 6, (b) 8, (c) 12. Note that a lattice point in the simple cubic lattice has a different set of surroundings than a lattice point in the body-centered cubic lattice, although they both have cubic cells. This is why they are different Bravias lattices.

Note that in Figure 6.1.6, two possible choices for unit cells in the hexagonal lattice are shown, one primitive and one nonprimitive. Primitive cells could also be chosen for the unit cell of the body-centered lattice or for the unit cell of the face-centered cubic lattice.

**Q. 6.1.2**

A lattice is (a)_____
____. A lattice can be generated by translations of (b)_____. A unit cell is defined by three (c)_____ or by six (d)_____
____. Each lattice point in a specific Bravais lattice has (e)_____
_____. Altogether, there are (f)____ Bravais lattices. Nonprimitive cells are sometimes chosen because (g)____
_____.

\* \* \*

CRYSTAL STRUCTURE. A **crystal structure** is formed by the addition of a basis to every lattice point of the space lattice. By a **basis** we mean an assembly of atoms (such as a methane or $CH_4$ molecule) located at a lattice point which has the same composition, arrangement, and orientation as the assembly of atoms located at every other lattice point. Portions of two hypothetical two-dimensional crystals are shown in Figure 6.1.7. Here the lattice points are located at the intersection of the lines and the assembly of atoms (the basis) is as shown schematically. (A crystal structure is infinite in extent. Real crystals are not, although they contain about $10^7$ cells/cm of length.)

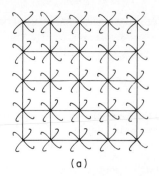 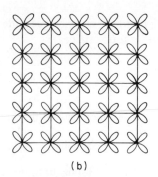

FIGURE 6.1.7. Square (two-dimensional) crystal structures with basis shown schematically. The crystal in (b) is more symmetrical than the one in (a).

## Q. 6.1.3

Every crystal structure consists of (a)_____. The crystal structures of Figure 6.1.7 exhibit (b) (the same or different) symmetry behavior.

* * *

## ANSWERS TO QUESTIONS

**6.1.1** (a) The same set of interfacial angles, (b) seven.

**6.1.2** (a) A collection of points in a periodic arrangement, (b) a unit cell, (c) lattice vectors, (d) lattice parameters, (e) the same surroundings as any other lattice point, (f) 14, (g) they exhibit obvious symmetry.

**6.1.3** (a) A lattice plus a basis, (b) different. In (b) a mirror placed along one of the lines shown would show a mirror image identical to the original image; this is not true of (a).

### 6.2 SOME SIMPLE CRYSTALS

This short section has one purpose: to familarize the student with four crystal structures containing only one kind of atom.

    1. BODY-CENTERED CUBIC (bcc). A cell of the **body-centered cubic** crystal structure is shown in Figure 6.2.1. Here an atom has been placed at each lattice point. This is not a primitive cell (although one can be chosen).

Two entire atoms are contained *within* this cell (the cube center atom and one eighth of each of the eight corner atoms). **Lattice coordinates** are used to describe the position of a point relative to the cell origin (which may

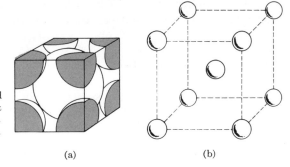

FIGURE 6.2.1. (a) bcc crystal unit cell. (b) Another view. (unit cell enclosed by the dotted line). Actually only one eighth of each of the spheres at the corners are within the cell.   (a)   (b)

be chosen to be any lattice point, since all are equivalent). Then position within the cell is measured in units of $a$ along **a**, of $b$ along **b**, and of $c$ along **c**. Thus the position of the atom in the body center of the cube is $\frac{1}{2}\frac{1}{2}\frac{1}{2}$ or, in general $uvw$, so a vector $r$ from the origin to the point $uvw$ is given by

$$\mathbf{r} = u\mathbf{a} + v\mathbf{b} + w\mathbf{c}. \qquad (6.2.1)$$

Molybdenum, tungsten, iron, and sodium have this crystal structure at room temperature and pressure.

2. FACE-CENTERED CUBIC (fcc). The **face-centered cubic** crystal cell is illustrated in Figure 6.2.2. Here an atom is located at each point of a

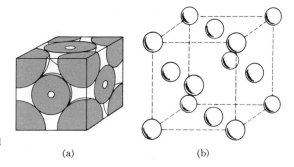

FIGURE 6.2.2. (a) fcc crystal unit cell. (b) Another view.   (a)   (b)

face-centered cubic lattice. Note that an atom is centered at each of the six faces. The unit cell contains four full atoms. Note that we would get the same structure by starting with a simple cubic lattice and adding a basis of four atoms, one each at positions 0 0 0, $\frac{1}{2}\frac{1}{2}0$, $\frac{1}{2}0\frac{1}{2}$, and $0\frac{1}{2}\frac{1}{2}$

(remember that the basis is added to *each* lattice point). Aluminum, copper, and nickel have this crystal structure.

3. DIAMOND CUBIC. The **diamond cubic** cell is illustrated in Figure 6.2.3. Here the unit cell contains eight full atoms. It can be con-

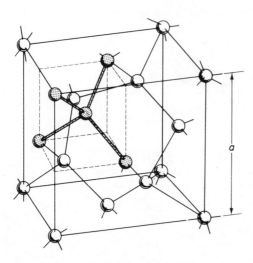

FIGURE 6.2.3. Crystal structure of diamond, showing the tetrahedral bond arrangement. [After W. Shockley, *Electrons and Holes in Semiconductors*, Van Nostrand Reinhold Company, New York (1950).]

sidered as two interlocking fcc crystals related by a translation of $\frac{1}{4}\frac{1}{4}\frac{1}{4}$. Thus the diamond structure is a face-centered cubic lattice with a basis of two atoms at positions 000 and $\frac{1}{4}\frac{1}{4}\frac{1}{4}$. Consider the eight cubic subcells of edge $a/2$ within the cell, of which the one outlined in Figure 6.2.3 contains an atom at its center. Half of these subcells have no atom at the center and they alternate in position with the half which do have an atom in the center. Diamond, silicon, and germanium have this structure.

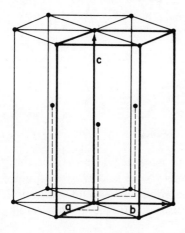

FIGURE 6.2.4. The hexagonal closest-packed structure.

4. HEXAGONAL CLOSEST-PACKED (hcp). The hcp crystal structure is formed from the simple hexagonal space lattice with a basis of two atoms at positions 000 and $\frac{2}{3}\frac{1}{3}\frac{1}{2}$. This is illustrated in Figure 6.2.4. Beryllium, zinc, and magnesium have this crystal structure.

Of the metals, more than two thirds commonly crystallize in the bcc, fcc, and hcp crystal structures which we have just described. Moreover, as we shall see in Section 6.4, the structures of many ionic compounds are closely related to these structures.

Some typical values of lattice parameters for crystalline forms of some of the elements are shown in Table 6.2.1. The determination of the crystal

**Table 6.2.1.** CELL DIMENSIONS OF SOME ELEMENTS AT ROOM TEMPERATURE (UNLESS NOTED)*

| Element | Structure | $a$ (Å) | $c$ (Å) |
|---|---|---|---|
| Aluminum | fcc | 4.05 | |
| Argon | fcc | 5.43 (20°K) | |
| Beryllium | hcp | 2.28 | 3.58 |
| Cadmium | hcp | 2.98 | 5.62 |
| Carbon | Diamond | 3.57 | |
| Copper | fcc | 3.61 | |
| Germanium | Diamond | 5.66 | |
| Iron | bcc | 2.87 | |
| Magnesium | hcp | 3.21 | 5.21 |
| Molybdenum | bcc | 3.15 | |
| Nickel | fcc | 3.52 | |
| Niobium | bcc | 3.30 | |
| Silicon | Diamond | 5.43 | |
| Sodium | bcc | 4.29 | |
| Titanium | hcp | 2.95 | 4.68 |
| Tungsten | bcc | 3.16 | |
| Zinc | hcp | 2.66 | 4.95 |

* From C. S. Barrett and T. B. Massalski, *Structure of Metals*, McGraw-Hill Book Company, Inc., New York (1966).

structure and the lattice parameters is carried out by X-ray diffraction; the first crystal structure determination was carried out by Bragg in 1913.

DENSITY OF CRYSTALS. We illustrate for the case of cubic crystals containing only one kind of atom how the density of the crystal is related to the lattice parameter:

$$\text{density of cell} = \frac{\text{mass of cell}}{\text{volume of cell}} = \rho \qquad (6.2.2)$$

$$= \frac{m}{v} = \frac{nm_a}{v}, \qquad (6.2.3)$$

where $m_a$ is the atomic mass and $n$ is the number of atoms per unit cell. If Avogadro's number of atoms $N_0$ has a mass $M$, then

$$m_a = \frac{M}{N_0} \qquad (6.2.4)$$

$$\rho = \frac{nM}{N_0 v} = \frac{nM}{N_0 a^3}. \qquad (6.2.5)$$

### Q. 6.2.1.

Most of the chemical elements crystallize in four crystal structures. These are (a)_____. The body-centered cubic crystal structure is a body-centered cubic lattice plus a basis of (b)_____ or a simple cubic lattice plus a basis of (c)_____. The diamond cubic crystal structure is a (d)_____ plus a basis of (e)_____. The cubic cell of volume $a^3$ contains (f)_____ atoms for the bcc, fcc, and diamond cubic crystals. The density of a crystal of a chemical element can be calculated if, in addition to the number of atoms per cell and the volume of the cell, the (g)_____ is known.

* * *

## ANSWERS TO QUESTIONS

**6.2.1** (a) bcc, fcc, hcp, and diamond cubic; (b) one atom at the lattice point; (c) one atom at 000, another at $\frac{1}{2}\frac{1}{2}\frac{1}{2}$; (d) face-centered cubic lattice, (e) one atom at 000, another at $\frac{1}{4}\frac{1}{4}\frac{1}{4}$; (f) two, four, and eight; (g) atomic mass.

### 6.3 CRYSTALLOGRAPHIC DIRECTIONS AND PLANES

DIRECTIONS. A **crystallographic direction** in a crystal is designated by the vector $u\mathbf{a} + v\mathbf{b} + w\mathbf{c}$, where $u$, $v$, and $w$ are whole numbers.

Since we have already decided upon our choice of **a**, **b**, and **c** for the unit cell we need only specify this direction by the ordered set enclosed in brackets [*uvw*]. For example, if the cell is simple cubic and the direction which we wish to indicate is from the origin along the body diagonal of the cube toward positive values of **a**, **b**, and **c**, we specify the direction by [111]. All lines parallel to it are also [111] directions.

To specify a negative component a bar is placed above the number, e.g., [11$\bar{2}$]. There are many times when we wish to talk about a certain type of direction such as any body diagonal in a cube rather than specifying a specific one. We designate such a family by carets ⟨111⟩; all the directions outward from the origin along cube diagonals are specified by [111], [$\bar{1}$11], [1$\bar{1}$1], [11$\bar{1}$], [$\bar{1}\bar{1}\bar{1}$], [1$\bar{1}\bar{1}$], [$\bar{1}$1$\bar{1}$], and [$\bar{1}\bar{1}$1]; the last four are antiparallel, respectively, to the first four.

## Q. 6.3.1

Directions along any cube edge are specified by (a)_____. Directions along any face diagonal are specified by (b)_____.

\* \* \*

MILLER INDICES. A plane which passes through a lattice point is called a **crystallographic plane.** Each crystallographic plane has an infinite number of planes parallel to it. It is, however, possible to designate the orientation of a crystallographic plane in a simple way by a set of three integers, called the **Miller indices** and designated by the ordered set enclosed in parentheses (*hkl*). This is illustrated with the help of Figure 6.3.1.

The procedure is as follows:

1. Determine the intercepts of the plane on the three crystal axes in units of $a$, $b$, and $c$.
2. Take the reciprocals of these numbers.
3. Clear fractions and reduce to smallest whole number set.

The intercepts along the **a**, **b**, and **c** axes in the present case are 2, 3, and 1, respectively. The reciprocals are $\frac{1}{2}$, $\frac{1}{3}$, and 1, respectively. These have a common denominator of 6; hence we multiply by 6 to obtain 3, 2, and 6, respectively.

The Miller indices 3, 2, and 6 are expressed by the ordered set enclosed in parentheses (326) or in general form (*hkl*). There will be a plane parallel to this plane which passes through the origin. It should be noted that the

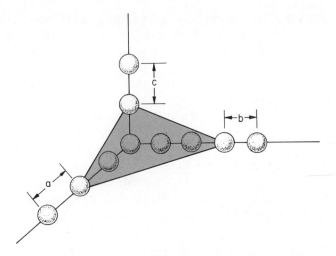

FIGURE 6.3.1. Intercepts of a plane on the three crystal axes. Only a few of the lattice points are shown.

set of Miller indices specifies not merely a single plane but the whole array of lattice planes parallel to it.

If a plane has a negative intercept such as is the case for the intercepts −3, 2, and 1, we indicate it as ($\bar{2}$36), i.e., with a bar over the negative intercept. Braces $\{hkl\}$ signify all the planes in a crystal which are equivalent; e.g., for a cubic crystal $\{100\}$ means all the cube faces (recall that the origin can be placed at *any* lattice point).

## Q. 6.3.2

A plane passing through a lattice point in a bcc crystal has intercepts $\frac{1}{2}$, ∞, ∞; its Miller indices are (a)_____. If a plane passing through a lattice point in a face-centered cubic crystal has intercepts 2, 2, and −1, its Miller indices are (b)_____. The plane with Miller indices ($\bar{1}\bar{1}\bar{1}$) (c) (is or is not) parallel to the plane with Miller indices (111). There are altogether how many types of $\{100\}$ planes in a cubic crystal? (d)_____.

\* \* \*

It should be noted that the distance, $d$, between two neighboring parallel planes can be determined and general mathematic expressions written out; this distance will depend on the lattice constants and the Miller indices. These distances are given in C. S. Barrett, and T. B. Massalski, *Structure of Metals*, McGraw-Hill Book Company, Inc., New York (1966).

The cubic case is particularly simple since the crystal axes form a Cartesian coordinate system. From his background in analytical geometry the student can show that for a *simple cubic* crystal

$$d = \frac{a}{\sqrt{h^2 + k^2 + l^2}}. \qquad (6.3.1)$$

*Note:* A **simple cubic** crystal contains an atom at each lattice point of the simple cubic lattice. It is not a common structure but is useful for illustration. Another interesting aspect of *cubic* crystals (only) is that the crystallographic direction $[hkl]$ is normal to a plane whose Miller indices are $(hkl)$. Because cubic crystals are particularly simple (and important) we often use them as illustrations. The Miller indices of some important planes in a

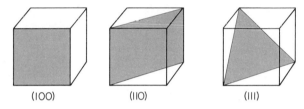

FIGURE 6.3.2. Important planes in a cubic crystal.

cubic crystal are shown in Figure 6.3.2. The $\{111\}$ planes are called the **octahedral planes** because eight different octahedral planes would form an octahedron.

In the hexagonal crystal we may choose a unit cell as shown by the heavy lines in Figure 6.3.3. Often, however, it is convenient to use a cell which is three times as large and is defined by four lattice vectors: $\mathbf{a}_1, \mathbf{a}_2, \mathbf{a}_3$, and $\mathbf{c}$. Here three of the crystal axes lie in the base of the hexagon (the **basal plane**). The intercepts of a crystal plane on these four axes leads to

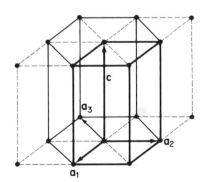

FIGURE 6.3.3. Hexagonal system.

the four Miller-Bravais indices ($hkil$). If ($hkl$) are the Miller indices of a specific plane, then the ($hkil$) are the **Miller-Bravais indices** of that same plane, where we must have $i = -(h+k)$; note that vector addition gives $\mathbf{a_1} = -(\mathbf{a_2} + \mathbf{a_3})$.

## Q. 6.3.3

The general equation of a straight line in terms of the Cartesian coordinates $x$ and $y$ is (a)_____ The intercepts of this straight line [in terms of the notation introduced in the answer to (a)] on the $x$ and $y$ axes are (b)_____. The slope of this line is (c)_____. The slope of a line perpendicular to the given line is (d)_____. If this perpendicular line goes through the origin, its equation is (e)_____. The two lines intersect at (f)_____. The distance from the origin to the point of intersection is (g)_____.

\* \* \*

### EXAMPLE 6.3.1

Which planes are the farthest apart in the simple cubic crystal?

*Answer.* The {100} planes are distance $a$ apart, where $a$ is the lattice parameter of the cubic crystal. Note that to specify the lattice parameters of a cubic crystal we need to specify only the length of an edge.

### EXAMPLE 6.3.2

Which planes are the farthest apart in a body-centered cubic crystal?

*Answer.* We note that Equation (6.3.1) applies to simple cubic crystals only. The answer in this case is the {110} planes which are a distance $a/\sqrt{2}$ apart. Consider Figure 6.2.1 to be infinite in extent. Note that the {100} type of planes are only $a/2$ apart. Because the body-centered cell is not a primitive cell, many of the planes {$hkl$} are closer than given by Equation (6.3.1).

## ANSWERS TO QUESTIONS

**6.3.1** (a) $\langle 100 \rangle$, (b) $\langle 110 \rangle$,

**6.3.2** (a) (100), (b) (11$\bar{2}$), (c) is, (d) three.

**6.3.3** (a) $hx + ky = A$; here $h$, $k$, and $A$ are *any* constants; (b) $A/h$ and $A/k$; (c) $-h/k$; (d) $k/h$, the negative reciprocal of the previous slope; (e) $y = (k/h)x$; (f) $x = Ah/(h^2 + k^2)$, $y = Ak/(h^2 + k^2)$; (g) $A/\sqrt{h^2 + k^2}$.

## 6.4 PACKING OF ATOMS IN CRYSTALS

To a first approximation atoms (or ions) can be considered to be rigid spheres. It is especially useful to investigate the packing in bcc, fcc, and hcp crystals in this context. Figure 6.4.1 shows two types of layers of

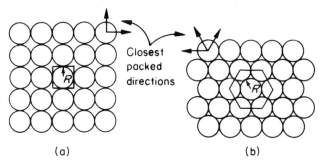

FIGURE 6.4.1. Portions of (a) cubic layers and (b) closest-packed layers.

spheres. In the **closest-packed layer** in Figure 6.4.1(b) each sphere has six other spheres as nearest neighbors, while in the cubic layer each sphere has only four spheres as nearest neighbors. The cubic layer could be the (100) plane of a simple cubic crystal (an atom at each lattice point of a simple cubic lattice) while the closest-packed layer could be the (001) plane of an hcp crystal. The **planar density** (atoms per area) is highest for the closest-packed layer where 90.7% of the area (of the plane which passes through the midpoint of the atoms in the layer) is occupied by atoms while on the cubic layer only 78.5% of the area is occupied. The directions along which the spheres touch are called **closest-packed directions.**

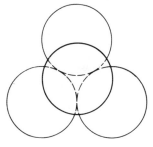

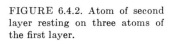

FIGURE 6.4.2. Atom of second layer resting on three atoms of the first layer.

We next consider the stacking of closest-packed layers on top of each other. An atom of the second layer rests in the "valley" formed by three atoms of the first layer as shown in Figure 6.4.2. Figure 6.4.3 (read the legend) illustrates the stacking of several layers on top of each other. Assume that we start with the $A$ layer and place a $B$ layer on top. The third layer might be an $A$ layer or a $C$ layer. Thus different stacking possibilities exist as illustrated in Figure 6.4.4. The stackings of most interest involve repeated sequences such as the stacking sequences $\cdots ABAB \cdots$ or $\cdots ABCABC \cdots$ as these were shown by W. Barlow in 1893 to lead to

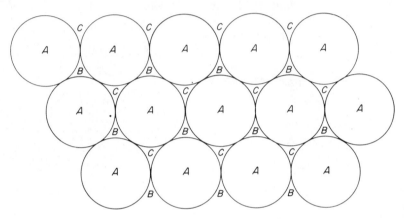

FIGURE 6.4.3. Packing of spheres. Consider the $A$ layer of spheres in place. The second closest-packed layer of spheres sits on top of the $A$ layer. Say that it is a $B$ layer. (*Note*: an atom of the $B$ layer then rests in the "valley" formed by three atoms of the first layer. This $B$ layer atom will have the three atoms below it, six around it, and three above it when the third layer is added.) Then the third closest-packed layer of spheres can be an $A$ layer or a $C$ layer.

FIGURE 6.4.4. Formation of two important stacking sequences. (a) Two layers. (b) Three layers of $ABA$. (c) Three layers of $ABC$.

ideal hcp and fcc crystal structures, respectively. The $\cdots ABAB \cdots$ sequence leads to an ideal hcp crystal (an ideal hcp crystal has $c/a = 2\sqrt{2}/\sqrt{3}$; real hcp crystals have slightly different $c/a$ values; see Table 6.2.1), as shown in Figure 6.4.5. In the ideal hcp crystal the closest-packed layer

FIGURE 6.4.5. Ideal hcp crystal structure formed from rigid spheres.

forms the basal plane. The $\cdots ABCABC \cdots$ sequence leads to the fcc crystal as shown in Figure 6.4.6; here the {111} planes, i.e., the octahedral planes, are the closest-packed layers. [Note in Figure 6.3.2 that the (111) plane intersects the cube face along face diagonals. Note in Figure 6.4.6 that the spheres are touching along the face diagonals which are the closest-packed directions of the closest-packed layer.]

It should be noted that in both fcc and ideal hcp crystal structures an atom has 12 nearest neighbors.

## Q. 6.4.1

In a closest-packed *layer* each atom has (a)____ atoms in contact with it. When closest-packed layers are packed on top of each other, regardless of the packing sequence each atom has (b)____ nearest neighbors. The $\cdots ABCABC \cdots$ sequence leads to a (c)____ crystal structure with the closest-packed layers corresponding to the (d)_____.

\* \* \*

VOIDS IN CRYSTALS. Many important classes of compounds such as oxides, for example, have one ion closest-packed (oxygen) and the other present in voids (the oxygen ion is quite large because of the two electrons added to the atom while a $Mg^{++}$ ion is small because it has lost two electrons). We shall discuss later the properties of alloys where the "foreign" atoms in

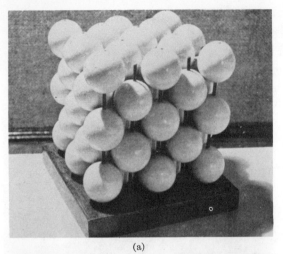
(a)

(b)

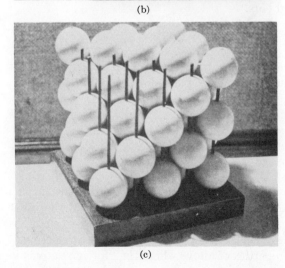
(c)

FIGURE 6.4.6. (a) fcc crystal structure formed from rigid spheres. (b) With one layer removed. (c) With two layers removed.

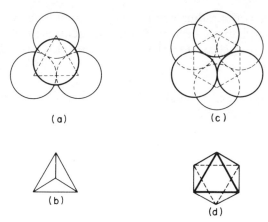

FIGURE 6.4.7. Voids in closest-packed structures. (a) Four atoms which enclose the tetrahedron. (b) Coordination void polyhedron joining the centers of the four atoms. (c) Six atoms which enclose the octahedral void. (d) Coordination polyhedron joining the center of the six atoms.

the voids between the closest-packed atoms influence important properties. The two kinds of voids that occur in closest packings are illustrated in Figure 6.4.7. If the void is surrounded by four spheres, it is called a **tetrahedral void,** and if surrounded by six spheres, an **octahedral void.** The number of spheres around a void is called the **coordination number** of the void and the space occupied by joining the centers of the spheres coordinating the void is called the **coordination polyhedron.** The coordination polyhedra for a tetrahedral void and an octahedral void are illustrated in Figure 6.4.8. (Note that these are both regular polyhedra.) An important

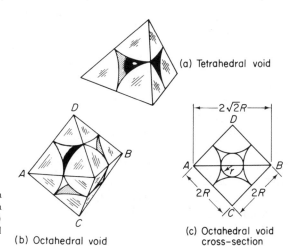

FIGURE 6.4.8. Coordination polyhedron with small atom in void. (a) Tetrahedral void. (b) Octahedral void. (c) Octahedral void cross section.

question is, What is the radius $r$ of the largest atom (sphere) which can be inserted in the void between atoms of radius $R$? This is illustrated in Figure 6.4.8 from which it is clear that the **void radius ratio** is $r/R = 0.414$ for octahedral voids. It can be shown that for tetrahedral voids the radius ratio is 0.225.

The next question we might ask is, Where are these voids located in the fcc crystal? Examination of the fcc crystal structure (see Figure 6.4.6) shows that octahedral voids are located at the centers of the cube edges (and at the body center). In other words, looking along a [100] direction, i.e., along a cube edge, there is a sequence of alternating atoms and voids.

**EXAMPLE 6.4.1**

There are four atoms in the fcc cell of Figure 6.2.2. How many octahedral voids are there per unit cell?

*Answer.* As is clear from the previous discussion there is one void for each atom so there are four octahedral voids per unit cell. There is another way to show this, namely, the cube has 12 edges and each of these voids is $\frac{1}{4}$ within the cube. There is also a void at the center of the cube. Thus there are four octahedral voids. The **void per atom ratio** is 1 for octahedral voids.

The center of the tetrahedral void is located at positions such as $\frac{1}{4} \frac{1}{4} \frac{1}{4}$ in the fcc crystal cell, i.e., at the center of the cubic subcell which has edges of length $a/2$ and a volume $a^3/8$ (Figure 1.1.2 is helpful in visualizing this, although there is an atom present in the center of the subcell shown; note how a regular tetrahedron is inscribed in a cube). There are eight such subcells per cell. Hence there are eight tetrahedral voids per fcc cell, or two tetrahedral voids per atom. The order of the stacking sequence does not change the void/atom ratio or the radius ratio.

**Q. 6.4.2**

An octahedral void in a closest-packed structure is surrounded by (a)____ spheres. The void/atom ratio for octahedral voids in closest-packed structures is (b)____. The void radius ratio is defined as (c)____

_____. The octahedral void in the fcc crystal structure is located (d)_____

_____. The tetrahedral void in the closest packed structure is surrounded by (e)____ spheres and has a void/atom ratio of (f)____.

\* \* \*

The body-centered cubic cell has a tetrahedral void centered at a position such as $\tfrac{1}{2}\,\tfrac{1}{4}\,0$ with a radius ratio of 0.292 and an octahedral void at the face center ($\tfrac{1}{2}\,\tfrac{1}{2}\,0$) with a radius ratio of 0.154. The results for radius ratios are summarized in Table 6.4.1. Note that the **atomic radius** $R$ can

**Table 6.4.1.** RADIUS RATIOS

| Crystal Voids | bcc* $r/R$ | fcc or Ideal hcp $r/R$ |
|---|---|---|
| Tetrahedral | 0.291 | 0.225 |
| Octahedral | 0.154 | 0.414 |

*Although the voids in the bcc structure have four and six neighbors, respectively, the coordination polyhedra are *not* regular.

be calculated from the lattice parameter. Thus in the fcc crystal, atoms touch along the face diagonal; hence $\sqrt{2}\,a = 4R$.

In the bcc crystal structure atoms touch along the body diagonal; hence $\sqrt{3}\,a = 4R$. In the diamond cubic structure $\sqrt{3}\,a = 8R$ (see Figure 1.1.2). These results are summarized in Table 6.4.2.

**Table 6.4.2.** ATOMIC RADIUS COMPUTED FROM LATTICE PARAMETER, $a$

| Structure | Atomic Radius |
|---|---|
| sc | $a/2$ |
| bcc | $\sqrt{3}\,a/4$ |
| fcc | $\sqrt{2}\,a/4$ |
| diamond cubic | $\sqrt{3}\,a/8$ |

### Q. 6.4.3

The atomic radius can be calculated from a knowledge of the (a)_____ and the (b)_____. Calculate the atomic radius of iron. It is (c)_____. Calculate the atomic radius of carbon in diamond. It is (d)_____.

\* \* \*

INTERSTITIAL IMPURITIES IN IRON. Iron has the bcc crystal structure below 910°C and the fcc crystal structure (with nearly the same atomic radius $R = 1.24$ Å) from 910 to 1390°C. (The existence of a given material in more than one equilibrium crystalline form depending on temperature, pressure, etc., is called **polymorphism.** It is a phenomenon common among the elements and compounds.) The carbon atom in diamond has a radius of $r = 0.78$ Å, from which we calculate $r/R = 0.63$ for the ratio of the carbon atom radius to the iron atom radius. While this is somewhat larger than $r/R = 0.414$ for octahedral voids in the fcc structure it is found experimentally that at 910°C about 1.3 wt. % of carbon (or 5.9 at. %) dissolves in fcc iron and that the carbon atoms are located in the octahedral void sites; carbon is called an **interstitial impurity.** However, in the bcc crystal of iron, the solubility at 910°C is nearly negligible, i.e., much smaller, as expected from Table 6.4.1. The heat treatment of steel depends on this difference in solubility of carbon in these two polymorphs of iron.

### Q. 6.4.4

Iron at atmospheric pressure has two crystalline forms depending on the temperature; this is called (a)_____. The solubility of carbon in the (b)____ structure which has the largest voids is (c)_____ than the solubility in the (d)____ structure.

\* \* \*

IONIC COMPOUNDS. The concepts of packing are extremely useful in studying ionic crystals. This will be illustrated for NaCl. The atom radius of sodium is 1.866 Å, but the radius of the sodium ion $Na^+$ is only 0.95 Å. Similarly the radius of the chlorine atom (estimated from the bond distance in covalent compounds) is 0.99 Å while the chlorine ion $Cl^-$ has a radius of 1.81 Å.

Consider $Cl^-$ ions to be spheres packed together in the face-centered cubic structure. (You should think of this both as a face-centered cubic lattice with $Cl^-$ ions at the lattice points and also as an $\cdots ABCABC \cdots$ stacking sequence of closest-packed layers.) Now visualize a smaller $Na^+$ ion in every octahedral void. Since the ratio of octahedral voids to $Cl^-$ ions is 1:1 the chemical formula is NaCl. Moreover, since the $\langle 100 \rangle$ directions consist of alternating spheres and voids this gives the **sodium chloride crystal structure** shown in Figure 5.5.2. This crystal structure is perhaps best illustrated by the model shown in Figure 6.4.9. The NaCl crystal structure is an fcc lattice with a basis (several choices are possible) of $Cl^-$ at 000 and $Na^+$ at 0 $\frac{1}{2}$ 0. The radius ratio in the present case is $r_{Na^+}/R_{Cl^-} = 0.95/1.81 = 0.525$; this is larger than the value 0.414 expected if the $Na^+$

FIGURE 6.4.9. Model of sodium chloride. [From A. Holden and P. Singer, *Crystals and Crystal Growing*, Doubleday & Company, Inc., New York (1960).]

ion exactly fitted in the closest-packed void and suggests that the Cl⁻ rigid spheres (which they are not) are slightly pushed apart.

**EXAMPLE 6.4.2**

Compute the lattice parameter of NaCl, based on the ionic radii.

*Answer.* The distance between the Na⁺ and Cl⁻ ion is $0.95 + 1.81 = 2.76$ Å; the lattice parameter is twice this distance or 5.52 Å in agreement with experiment.

NiO, MgO, and other important ceramic materials also have the NaCl-type structure. All the alkali halides have the NaCl-type structure at room temperature except CsCl. CsCl, at room temperature and pressure, has the structure obtained by placing a basis of a Cs⁺ at 000 and Cl⁻ at $\frac{1}{2} \frac{1}{2} \frac{1}{2}$ on a simple cubic lattice. This is referred to as the **CsCl crystal structure.** Many of the alkali halides have the NaCl-type structure at 1 atm but transform to the CsCl type at high pressures. NaCl itself requires a pressure of 141 kbars to transform (1 kbar = $10^9$ dynes/cm² = 14,500 psi = 980 atm).

The cubic form of zinc sulfide, called zinc blende, also can be visualized in terms of these packing concepts. Consider the large ions S²⁻ to be spheres packed in the fcc arrangement. Place a small $Zn^{2+}$ ion in every other tetrahedral void. Since there are two tetrahedral voids per large sphere the chemical formula will be ZnS. Each $Zn^{2+}$ appears to be tetrahedrally bonded to each $S^{2-}$; in fact it is true that there is considerable covalent bonding combined with ionic bonding in this case. The resultant crystal structure is called the **zinc blende structure.** Cubic ZnS is an fcc lattice with a basis of a $Zn^{2+}$ at 000 and $S^{2-}$ at $\frac{1}{4}\frac{1}{4}\frac{1}{4}$. A number of important

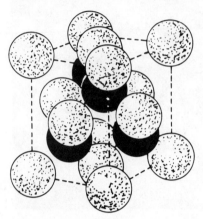

FIGURE 6.4.10. Unit cell of zinc blende structure.

semiconductor compounds such as GaAs, $\beta$-SiC, and InSb have this crystal structure. The unit cell is shown in Figure 6.4.10.

### Q. 6.4.5

The chlorine ion $Cl^-$ is (a) (much larger or smaller) than the chlorine atom. The sodium ion $Na^+$ is (b) (much larger or smaller) than the sodium atom. In terms of packing, NaCl may be considered to be (c)_____ _____ of (d)____ ions with (e)____ ions in (f)_____ voids. Depending on the pressure, NaCl may exhibit (g)_____.

\* \* \*

ZnS also exhibits a crystal structure called **wurtzite** which is based on an $\cdots ABAB \cdots$ packing of closest-packed layers of $S^{2-}$ ions with $Zn^{2+}$ ions in alternate tetrahedral voids. BeO and ZnO have this structure.

IONIC RADII. A. Lande in 1920 proposed a scheme of determining the size of the large ions by looking at compounds such as LiCl, LiBr, and LiI. In this case the $Li^+$ ion is expected to be sufficiently small so that the assumption that the Cl⁻ ions are actually touching (closest-packed) is justified. The radius of Cl⁻ can then be obtained from $\sqrt{2}\,a = 4R_{Cl^-}$, where $a$ is the lattice parameter for LiCl. The value of $r_{Na^+}$ could then be calculated from the lattice parameter for NaCl and the value of $4R_{Cl^-}$ since in this case the larger Na⁺ ion pushes the Cl⁻ ions apart. A number of similar schemes were used and on this basis a table of ionic radii was established by Linus Pauling and others. Selected values are shown in Table 6.4.3.

**Table 6.4.3.** PAULING* IONIC RADII, IN ÅNGSTROMS

|  |  |  | H⁻ | He | Li⁺ | Be²⁺ | B³⁺ | C⁴⁺ | N⁵⁺ | O⁶⁺ | F⁷⁺ |
|---|---|---|---|---|---|---|---|---|---|---|---|
|  |  |  | 2.05 | 0.92 | 0.59 | 0.43 | 0.34 | 0.29 | 0.25 | 0.22 | 0.19 |
| C⁴⁻ | N³⁻ | O²⁻ | F⁻ | Ne | Na⁺ | Mg²⁺ | Al³⁺ | Si⁴⁺ | P⁵⁺ | S⁶⁺ | Cl⁷⁺ |
| 4.14 | 2.47 | 1.76 | 1.36 | 1.12 | 0.95 | 0.82 | 0.72 | 0.65 | 0.59 | 0.53 | 0.49 |
| Si⁴⁻ | P³⁻ | S²⁻ | Cl⁻ | A | K⁺ | Ca²⁺ | Sc³⁺ | Ti⁴⁺ | V⁵⁺ | Cr⁶⁺ | Mn⁷⁺ |
| 3.84 | 2.79 | 2.19 | 1.81 | 1.54 | 1.33 | 1.18 | 1.06 | 0.96 | 0.88 | 0.81 | 0.75 |
|  |  |  |  |  | Cu⁺ | Zn²⁺ | Ga³⁺ | Ge⁴⁺ | As⁵⁺ | Se⁶⁺ | Br⁷⁺ |
|  |  |  |  |  | 0.96 | 0.88 | 0.81 | 0.76 | 0.71 | 0.66 | 0.62 |
| Ge⁴⁻ | As³⁻ | Se²⁻ | Br⁻ | Kr | Rb⁺ | Sr²⁺ | Y³⁺ | Zr⁴⁺ | Nb⁵⁺ | Mo⁶⁺ |  |
| 3.71 | 2.85 | 2.32 | 1.95 | 1.69 | 1.48 | 1.32 | 1.20 | 1.09 | 1.00 | 0.93 |  |
|  |  |  |  |  | Ag⁺ | Cd²⁺ | In³⁺ | Sn⁴⁺ | Sb⁵⁺ | Te⁶⁺ | I⁷⁺ |
|  |  |  |  |  | 1.26 | 1.14 | 1.04 | 0.96 | 0.89 | 0.82 | 0.77 |
| Sn⁴⁻ | Sb³⁻ | Te²⁻ | I⁻ | Xe | Cs⁺ | Ba²⁺ | La³⁺ | Ce⁴⁺ |  |  |  |
| 3.70 | 2.95 | 2.50 | 2.16 | 1.90 | 1.69 | 1.53 | 1.39 | 1.27 |  |  |  |
|  |  |  |  |  | Au⁺ | Hg²⁺ | Tl³⁺ | Pb⁴⁺ | Bi⁵⁺ |  |  |
|  |  |  |  |  | 1.37 | 1.25 | 1.15 | 1.06 | 0.98 |  |  |

* L. Pauling, *Journal of the American Chemical Society*, **69**, 542 (1947).

V. M. Goldschmidt in 1926 introduced the concept of minimum radius ratios and their effect on crystal structure. We noted previously that a tetrahedral void had a radius ratio $r/R = 0.225$ while for the octahedral void $r/R = 0.414$. Goldschmidt noted that an ion would *tend* to go into the tetrahedral void if $0.225 < r/R < 0.414$; i.e., it would tend to push the tetrahedrally coordinated atoms outward rather than fit loosely in the larger octahedral void. Thus the void ratio ($r/R = 0.225$ for the tetrahedral void) in fact is a **minimum radius ratio.** Table 6.4.4 gives the expected **coordination number** (number of nearest neighbors) of the atom in the void based on the radius ratio.

**Table 6.4.4.** MINIMUM RADIUS RATIOS

| Coordination Polyhedron | Coordination Number | Minimum Radius Ratio |
|---|---|---|
| Cubo-octahedron* | 12 | 1.000 |
| Cube | 8 | 0.732 |
| Square antiprism† | 8 | 0.645 |
| Octahedron | 6 | 0.414 |
| Tetrahedron | 4 | 0.225 |
| Triangle (planar) | 3 | 0.155 |

* The coordination polyhedron surrounding an *atom* in the fcc structure is an example.

† The square antiprism is formed as follows: Consider two squares of edge $L$ on parallel planes one directly above the other; rotate the top square by 45 deg; then join the corners of these two squares.

## EXAMPLE 6.4.3

What coordination is expected for NaCl?

*Answer.* From Table 6.4.3 the radius of $Na^+$ is 0.95 Å and $Cl^-$ is 1.81 Å. Therefore $r_{Na^+}/R_{Cl^-} = 0.525$. Hence because the radius ratio is greater than 0.414 but less than 0.645, we expect octahedral coordination as is the case. Note that this is also consistent with charge neutrality.

## EXAMPLE 6.4.4

What coordination is expected for ZnS?

*Answer.* Since $r/R = 0.88/2.19 = 0.40$ (which is greater than 0.225 but less than 0.414) we expect tetrahedral coordination as we have previously noted to be the case. This will be consistent with charge neutrality if only half the tetrahedral voids are occupied.

## EXAMPLE 6.4.5

What coordination is expected for $SiO_4^{4-}$?

*Answer.* The radius of $Si^{4+}$ is 0.65 Å while the radius of $O^{2-}$ is 1.76 Å. Hence the radius ratio is $r/R = 0.37$. Thus we expect tetrahedral coordination. Complex ions such as $CO_3^{2-}$, $SO_4^{2-}$, $SiO_4^{4-}$, etc., are ionic groups of great stability and hence their polyhedra serve as building blocks in more complicated arrays.

The coordination rules just discussed combined with the general requirement of electrical neutrality enable one to predict many possible crystal structures. These rules are known as **Pauling's rules.** We used these rules in their simplest form in discussing NaCl and ZnS. (There are many exceptions to the rules.) For a detailed discussion of these rules, see L. V. Azaroff, *Introduction to Solids*, McGraw-Hill Book Company, Inc., New York (1960), p. 83.

An understanding of these coordination rules is particularly important in predicting the structure of ceramic materials. **Ceramic materials** are commonly understood to be compounds of metallic and nonmetallic elements. Thus NiO, MgO, $Al_2O_3$, $BaTiO_3$, and $SiO_2$ are simple ceramic compounds. Ceramic compounds can also be very complicated and include clays, spinels, common window glass, etc. Many ceramic compounds are refractory materials (have very high melting points) as noted in Table 4.5.1. Ceramic materials have a wide range of application. They may be used as refractories (MgO), abrasives (alumina, $Al_2O_3$), electrical insulators (fired clay-alumina mixtures), ferroelectrics ($BaTiO_3$), piezoelectric transducers ($SiO_2$, quartz), magnets (spinels), building materials (concretes), surface finishes (vitreous enamel), water softeners, and molecular sieves (zeolites), etc. Silicates and silica compose a major fraction of our solid environment. Silicates are based on the coordination tetrahedron $SiO_4^{4-}$. Such diverse materials as clay, sand, mica, talc, asbestos, granite, and Portland cement are silicates. The silicates involve an incredible number of island, chain, sheet, and framework structures based on the tetrahedral $SiO_4^{4-}$ unit. Asbestos fibers, which have a double silicate chain, are very strong (750,000 psi, 50-kbar fracture strength). Asbestos fibers are often used in composite materials. Talc and clay are based on silica sheets. Quartz has a network structure. Problems 6.47–6.52 involve a programmed learning sequence on the silicates.

## Q. 6.4.6

According to Goldschmidt an ion of radius $r$ should be surrounded by (a)_____ oppositely charged ions of radius $R$ when $0.225 < r/R < 0.414$. The number of nearest neighbors is called the (b)_____. In the sodium chloride crystal structure the coordination polyhedron for the positive ion is the (c)_____. In the zinc blende structure the coordination polyhedron for the positive ion is the (d)_____. When the minimum radius ratio rules are combined with the requirement of electrical neutrality a set of rules for predicting possible crystal structures, called the (e)_____, are obtained. A ceramic material is defined as (f)_____. The silicates are based on the (g)_____ ion.

\* \* \*

THE STRUCTURE OF SIMPLE INTERSTITIAL COMPOUNDS. In many cases compounds are formed between transition metals and small atoms of nonmetals, H, B, C, N, and Si. Such compounds are called, respectively, hydrides, borides, carbides, nitrides, and silicides. They are called **interstitial compounds.** As a rule these compounds have many of the characteristics of metallic alloys such as opacity, high electrical and thermal conductivity, and metallic luster. Moreover, these compounds are often *extremely* hard, have very high melting temperatures (hence are called refractories), and are often chemically unreactive except toward strong oxidizing agents. Some typical melting points are shown in Table 6.4.5.

Table 6.4.5. MELTING POINTS OF SOME INTERSTITIAL COMPOUNDS

|       | M.P. (°K) |     | M.P. (°K) |
|-------|-----------|-----|-----------|
| TiC   | 3410      | TiN | 3220      |
| HfC   | 4160      | ZrN | 3255      |
| $W_2C$ | 3130     | TaN | 3630      |

The structure of many of these simple interstitial compounds can be described by simple packing concepts. The metal atoms are arranged in a certain array and the small nonmetal atoms fill a certain fixed proportion of the voids. The structures of some of the interstitial compounds which may be derived on this scheme are indicated in Table 6.4.6. (A similar scheme may be drawn up for hexagonal closest-packed interstitial com-

Table 6.4.6. INTERSTITIAL STRUCTURES DERIVED FROM CUBIC CLOSEST-PACKING OF METAL ATOMS*

| Nonmetal Atoms in | Proportion Occupied | Structure | Examples |
|-------------------|---------------------|-----------|----------|
| Octahedral holes  | All                 | Sodium chloride | TiC, TiN, ZrC, ZrN, UC, HfC, VC, NbC, TaC |
|                   | $\frac{1}{2}$       | —         | $W_2N$, $Mo_2N$ |
|                   | $\frac{1}{4}$       | —         | $Mn_4N$, $Fe_4N$ |
| Tetrahedral holes | All                 | Fluorite  | $TiH_2$ |
|                   | $\frac{1}{2}$       | Zinc blende | ZrH, TiH |
|                   | $\frac{1}{4}$       | —         | $Pd_2H$ |
|                   | $\frac{1}{8}$       | —         | $Zr_4H$ |

*After A. F. Wells, *Structural Inorganic Chemistry*, Oxford University Press, Inc., New York (1950).

pounds.) It is likely that there is considerable covalent bonding along with the "free" electron gas.

## Q. 6.4.7

Hafnium carbide has a sodium chloride crystal structure. In this case we can consider it as made up of a closest-packed arrangement of (a)_____ with (b)_____ in (c)_____ void positions. Interstitial compounds are interesting materials, especially those involving nitrogen and carbon, because (d)_____. Because of this they are very hard to fabricate.

\* \* \*

Because of the extraordinary behavior of many of these compounds, there is considerable research on them now.

## ANSWERS TO QUESTIONS

**6.4.1** (a) 6, (b) 12, (c) fcc, (d) octahedral planes or the {111} planes.

**6.4.2** (a) 6, (b) 1:1, (c) the ratio of the radius of the largest spheres which will fit into the void, divided by the radius of the spheres making up the structure, (d) at the center of the edge of the cubic unit cell and at its body center, (e) 4, (f) 2:1.

**6.4.3** (a) Crystal structure, (b) lattice parameter, (c) 1.24 Å from $R = \sqrt{3}a/4$ and Table 6.2.1., (d) 0.78 Å from $R = \sqrt{3}a/8$ and Table 6.2.1.

**6.4.4** (a) Polymorphism, (b) fcc, (c) much greater, (d) bcc.

**6.4.5** (a) Much larger, (b) much smaller, (c) an $\cdots$ABCABC$\cdots$ sequence of closest-packed layers, (d) $Cl^-$, (e) $Na^+$, (f) each of the octahedral, (g) polymorphism.

**6.4.6** (a) Four, (b) coordination number, (c) octahedron, (d) tetrahedron, (e) Pauling rules, (f) a compound of metallic and nonmetallic elements, (g) $SiO_4^{4-}$.

**6.4.7** (a) Hafnium, (b) carbon, (c) *all* the octahedral, (d) of their very high melting points.

## 6.5 SYMMETRY AND ITS RELATIONSHIP TO PROPERTIES

SYMMETRY. Symmetry is defined in terms of **symmetry operations** which are movements which transform the body into itself. The snowflake shown in Figure 6.5.1 exhibits an obvious symmetry. Consider the snowflake as a two-dimensional object. Any of the rotations (about an axis normal to the plane of the flake and through its center), by 60, 120, 180, −120, −60, 360 deg, produces a configuration identical to the original. When a certain operation such as rotation brings a body into superposition

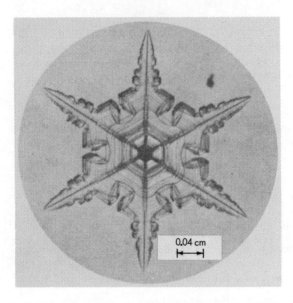

FIGURE 6.5.1. Snowflake. [From V. Nakaya, *Snow Crystals*, Harvard University Press, Cambridge, Mass. (1954).] Nakaya's book shows over 1400 plates of snowflakes and ice crystals.

with the original, we say that the body possesses this **symmetry element.** Thus the snowflake possesses the six symmetry elements just discussed. When a body can be rotated by $360/n$ into itself, it has an $n$-fold **rotation axis.**

### EXAMPLE 6.5.1

Other than the ±90-deg rotations and the 180- and 360-deg rotations, what other symmetry operations can be performed on the square; i.e., what symmetry elements does the square have?

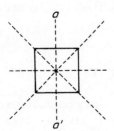

EXAMPLE FIGURE 6.5.1.

*Answer.* It contains **reflection planes** (or mirror planes) as shown by the dashed lines. Thus for the plane $aa'$ if all the points on the left are reflected through the plane $aa'$ to the right and vice versa, a new square will be obtained which superimposes on the original.

The square thus contains 8 symmetry elements. The snowflake discussed previously contains 12 symmetry elements. A regular pentagon would contain a five-fold rotation axis. Can crystals have five-fold symmetry? They cannot. A simple proof involves showing that it is impossible to fit equal-sized, regular pentagons together into an infinite array. Recall that crystals must have translational symmetry (certain translations transform the body into itself); this requirement limits the rotational symmetry to one-, two-, three-, four-, and six-fold axes.

To consider all the possible symmetry operations of a three-dimensonal crystal *about a point* (this excludes the translation operation or combined operations involving translation since these cause a movement of the point), it is also necessary to consider the **inversion operation** and the **rotoinversion operation.** The inversion operator takes every point from $\mathbf{r}$ to $-\mathbf{r}$. The rotoinversion operation is a combined operation of inversion followed by rotation. The operations about a point determine the external forms possible for crystals as well as the anisotropic nature of their properties.

## Q. 6.5.1

Symmetry is defined in terms of (a)_____
_____. An equilateral triangle contains (b)_____ symmetry elements. The building blocks of crystals never involve right prisms with a regular pentagonal base because (c)_____
_____. All the symmetry operations of a crystal about a point involve four types of operations. These are (d)_____
_____. Such operations are important because they determine the (e)_____ possible for a crystal as well as the (f)_____.

\* \* \*

PIEZOELECTRICITY AND SYMMETRY. Because three-dimensional crystals are more complicated than two-dimensional crystals (which are imaginary), it is often helpful to consider the behavior of a two-dimensional crystal as shown in Figure 6.5.2.

This crystal does not have an **inversion center** and such crystals are called **noncentrosymmetric crystals;** they do not possess an inversion element.

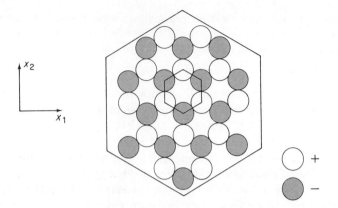

FIGURE 6.5.2. Two-dimensional ionic crystal.

Let us consider the ions to be rigid spheres which are just touching and let us apply a compressive stress in the $x_2$ direction as shown in Figure 6.5.3. After the compressive stress is applied ions 5 and 6 move inward, 1 and 2 move to the left, and 3 and 4 move to the right. From Figure 6.5.3

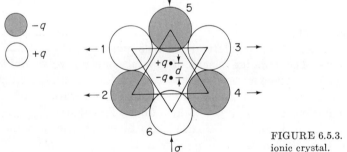

FIGURE 6.5.3. Stress applied to ionic crystal.

it can be seen that the centroids of the two triangles do not coincide with each other (as they did before the deformation), so that an electric dipole moment $p$ of magnitude $qd$ exists. Since a similar dipole occurs in each cell the crystal has become polarized (the polarization **P** is the sum of all the dipole moments per unit volume). The development of a polarization as a result of application of a stress is called the **piezoelectric effect.**

In a crystal which has a center of symmetry (possesses an inversion element) the charge centroids coincide when the crystal is deformed and there is no piezoelectric effect.

## Q. 6.5.2

If a crystal contains an inversion center it (a) (will or will not) be piezoelectric. NaCl (b) (*is or is not*) piezoelectric. Cubic ZnS (c) (is or is not) piezoelectric. One of the most important piezoelectric materials is quartz; quartz (d) (has or does not have) an inversion center.

\* \* \*

ANISOTROPY AND SYMMETRY. Suppose that the thermal expansion coefficient of molybdenum (bcc crystal structure) has been carefully measured along the [100] direction and you are asked to measure it along the [010] and [001] directions. "That is ridiculous," you say, "because anyone can see that these directions are identical from the symmetry viewpoint and hence the thermal expansion coefficient along all three cube axes is the same." That is correct. Suppose you are now asked to measure the thermal expansion coefficient along the [111] direction. Not being sure, you proceed to measure the thermal expansion coefficient and find the same value as for the $\langle 100 \rangle$ directions. In fact, had you been familiar with Cartesian tensors (an area of linear algebra) you could have readily shown that this also follows from the symmetry of the crystal (see the previously cited Ruoff, *Materials Science*, Chapter 7). Thus cubic crystals are isotropic with respect to thermal expansion coefficients. However, hexagonal crystals are not; they have a different thermal expansion coefficient along the $c$ axis, $\alpha_\parallel$, and perpendicular to the $c$ axis, $\alpha_\perp$, as can be shown experimentally or from symmetry considerations.

It can be shown from symmetry considerations that the thermal expansion coefficient along a line inclined at $\theta$ to the $c$ axis of the hcp crystal is

$$\alpha = \alpha_\perp \sin^2 \theta + \alpha_\parallel \cos^2 \theta. \tag{6.5.1}$$

Table 6.5.1 gives some properties for hexagonal crystals which exhibit the same type of anisotropy.

It should be pointed out that symmetry tells us only that $\alpha_\parallel \neq \alpha_\perp$ in general for hexagonal crystals; it says nothing of the magnitudes; to establish these quantities either requires direct measurement or a detailed knowledge of the bonding. This is illustrated for the graphite structure which is shown in Figure 6.5.4. It is a layer structure with each carbon in the layer being strongly bonded to three other carbons in the layer, the bond angles being 120 deg. (The bonds are $sp^2$ hybrids.) The fourth valence electron is (roughly) free to move *in the layer*. Chemists usually represent the bonding as consisting of alternating single and double bonds

**Table 6.5.1.** EXAMPLES OF ANIOSTROPIC PROPERTIES OF HEXAGONAL CRYSTALS*

| Crystal | Structure | Coeff. of Thermal Exp. ($10^{-6}$/°C) | | Electrical Resistivity ($10^{-8}\Omega$-m) | | Thermal Conductivity (W/cm-°C) | |
|---|---|---|---|---|---|---|---|
| | | $\|c$ | $\perp c$ | $\|c$ | $\perp c$ | $\|c$ | $\perp c$ |
| Mg | hcp | 27.0 | 25.4 | 3.78 | 4.53 | — | — |
| Zn | hcp | 63.9 | 14.1 | 6.05 | 5.83 | 1.24 | 1.24 |
| Cd | hcp | 52.6 | 21.4 | 8.36 | 6.87 | 0.83 | 1.04 |
| Hg | Hex. | 47.0 | 37.5 | 5.87 | 7.78 | 0.399 | 0.290 |

*After W. Boas and J. K. MacKenzie, *Progress in Metal Physics*, Vol. II, Pergamon Press, Inc., Elmsford, N.Y. (1949).

throughout the layer. Layers, however, are bonded to each other by weak van der Waals' forces. This difference in bonding in different directions explains the enormous anisotropy of many properties of graphite such as

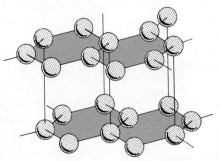

FIGURE 6.5.4. Graphite structure.

electrical conductivity. For example, it is a metal along the layers and a semiconductor normal to the layers. Moreover, it is a useful dry lubricant because the layers easily slip over each other. $MoS_2$ is a similar material.

## Q. 6.5.3

A casual examination of the symmetry nature of a bcc crystal such as tungsten reveals that it will have the same thermal conductivity along the [010] and [001] directions as along (a)_____. In fact it can be shown that this property is (b)_____ for cubic crystals.

In hexagonal crystals, the electrical resistivity is (c)_____ property. In fact for hexagonal crystals (d)_____ coefficients are needed to describe the electrical resistivity. Graphite can be considered a layer-like structure with (e)_____ in the layer. Hence it behaves as (f)_____ in directions parallel to the layers. However, it is (g)_____ normal to the layers. Because of their layer-like structure both graphite and (h)_____ are good solid lubricants.

* * *

## ANSWERS TO QUESTIONS

**6.5.1** (a) Symmetry operations which transform a body into itself, (b) six (three rotations and three reflections), (c) this would be incompatible with the translation symmetry, (d) rotations, reflections, inversion, and rotoinversions, (e) external form, (f) anisotropic nature of the properties.

**6.5.2** (a) Will not, (b) is not, (c) is, (d) does not have.

**6.5.3** (a) The [100] direction, (b) isotropic, (c) an anisotropic, (d) two, (e) free electrons, (f) a metal, (g) a semiconductor, (h) $MoS_2$ (molybdenum disulfide).

## 6.6 IMPERFECTIONS IN CRYSTALS

Let us distinguish more clearly between the *idealized* crystals with which we have been concerned so far and the *real* crystals upon which we base our measurements. We have already defined the former in terms of unit cells arranged in a periodic space lattice generated by repetition of three primitive lattice vectors in space (and hence infinite in extent).

Many properties of actual crystals such as cohesion (binding) and elastic coefficients can be accounted for quite adequately on the basis of our concept of an ideal crystal. There are other properties, however, which deviate seriously from such predictions, indicating that real crystals differ in some critical way from ideal crystals. This difference is attributed to **imperfections** which arise from deviations in arrangements of atoms from those arrangements typical of the ideal crystal.

**Packing imperfections** in crystalline structures are usually classified according to whether the disturbance is localized around a single lattice point, localized in a long narrow cylinder of related lattice points, localized around an internal or external surface, or spread throughout a sizable volume element. These are called, respectively, **point defects, line defects, plane defects,** and **space defects.**

SOLUTIONS. It is sometimes possible for small atoms to occupy in a random fashion some of the voids between atoms without causing a change in the basic crystal structure. Pure iron (near 1000°C) has the fcc crystal structure. As seen in Section 6.4 this iron (the solvent) can dissolve several atomic percent carbon (solute), without major change in crystal structure of the parent lattice; i.e., the iron atoms still occupy the lattice sites of an fcc lattice, although the lattice parameter is slightly changed. The resultant material is called an **interstitial solid solution.** It is a single phase; a **phase** is a structurally homogeneous part of a materials system.

### EXAMPLE 6.6.1

A beaker contains the compounds $H_2O$ and NaCl at room temperature. How many phases are present?

*Answer.* If all the NaCl is dissolved, there is only one phase, the solution. However if the NaCl is present in excess of the *solubility limit*, there will also be NaCl crystals present; then there are two phases, the NaCl crystals and the solution whose concentration is at the solubility limit.

Interstitial solutions in metals usually involve small atoms such as H, B, C, N, and sometimes O.

Atoms of one element can often replace atoms of another in a crystal. This results in a **substitutional solid solution.** Thus nickel added to copper yields a solution. Pure copper has an fcc crystal structure; when nickel is added the resulting alloy still consists of atoms placed at each lattice point of an fcc lattice, although some of the atoms at these points are nickel, the rest copper. If the different types of atoms are randomly distributed on the lattices sites, the solution is called a **random substitutional solid solution.** Semiconductors also form such solutions; e.g., some phosphorus dissolves in silicon and produces an $n$-type semiconductor. Likewise in the production of the ruby laser in ionic crystals such as $Al_2O_3$ some of the $Al^{3+}$ is replaced by some $Cr^{3+}$.

In some solid solutions the solubility is extensive. In the Cu–Ni system it is complete; i.e., since both copper and nickel have an fcc crystal structure and since the two types of atoms are very much alike [same valence, nearly the same atomic radius (see Table 6.2.1)], the solution can vary *continuously* from 0% Ni to 100% Ni. The lattice parameter will vary with concentration. In the Cu–Zn system, only about 40% of the copper atoms can be replaced by zinc atoms on the fcc lattice. Alloys of Cu–Zn with less than 40% zinc are called $\alpha$-brass. Brass alloys have many of the desirable properties of copper but are cheaper because zinc is cheaper than copper. The reason for the smaller solubility of **Zn** in fcc copper will be discussed further in Chapter 10.

Substitutional solid solutions at certain concentrations sometimes exist as **ordered solid solutions.** An alloy of composition $Cu_3Au$ will exist as a random solution of Au atoms and Cu atoms on an fcc lattice at high temperatures. However, at low temperature (room temperature), it is found that the Au atoms will occupy all the corners of the unit cells and the Cu atoms will occupy all the face centers; this is an example of an ordered solid solution. *High temperature favors disorder or randomness.*

### Q. 6.6.1

The four types of geometrical packing imperfections are (a)_____ _____. Boron in iron would be present as (b)_____ _____. A pearlitic steel consists of 1000 alternate layers of crystals of iron (with some carbon dissolved in it) and of crystals of the compound $Fe_3C$; there are (c)_____ phases present. Mo and W have the same valence. Their atom sizes are (d) (about the same or quite different). Therefore, (e)_____ solubility might be expected. FeO and MgO both have the NaCl-type crystal structure; they (f)(might or definitely could not) show complete solubility. A 50:50 Cu–Zn alloy is ordered at low temperature but disordered at high temperature; this happens because high temperature (g)_____.

\* \* \*

POINT DEFECTS IN PURE ELEMENTS AND COMPOUNDS. At a temperature near the melting point the stable structure for copper corresponds to one in which a significant fraction (about $10^{-4}$) of lattice positions is vacant. These defects are called **vacancies.** They were first postulated to exist by the great Russian scientist Frenkel in 1926. The equilibrium vacancy concentration (fraction of vacant sites) is given by

$$\frac{n}{N+n} = A e^{-E_f/k_B T}. \qquad (6.6.1)$$

Here $n$ is the number of vacancies, $N$ is the number of atoms, $E_f$ is the energy of formation of a vacancy, $k_B$ is the Boltzmann constant, $T$ is the absolute temperature, and $A$ is a parameter approximately equal to 1 in the present case. The **energy of formation of a vacancy** is the increase in energy of the crystal when an atom is taken from one of its regular sites and placed on the surface. It is thus a characteristic of the crystal. For copper $E_f = 1.17$ eV. Vacancies play a vital role in creep, in sintering, and in many other processes.

## Atomic Arrangements

**EXAMPLE 6.6.2**

Calculate the fraction of vacancies in copper just below the melting point of 1080°C.

*Answer.* The absolute temperature is 1353°K and $k_B = 8.5 \times 10^{-5}$ eV/°K. Hence

$$\frac{n}{N+n} \doteq \frac{n}{N} = e^{-1.17/8.62 \times 10^{-5} \times 1353} = e^{-10.0} = 10^{-10.0/2.3} = 10^{-4.3}.$$

**EXAMPLE 6.6.3**

If a metal has a fractional concentration of $10^{-4}$ vacancies just below $T_m$, what is the vacancy concentration at $T_m/2$?

*Answer.* If

$$e^{-E_f/k_B T_m} = 10^{-4},$$

then

$$e^{-2E_f/k_B T_m} = 10^{-8}.$$

Note how *very* rapidly concentration varies with temperature.

Vacancy concentrations in excess of the equilibrium concentration can be created by radiation (with neutrons, electrons, $\gamma$-rays) and by mechanical deformation.

Another common imperfection is the **self-interstitial** which is an atom not located at a lattice site in the lattice but squeezed between atoms. The energy of formation of the self-interstitial is considerably larger than that of a vacancy. A third important type of point defect occurs when an atom is removed from a lattice site and placed at an interstitial site. It is essentially a combination of the first two types and is called a **Frenkel defect.** Such defects are produced when metal crystals are bombarded with neutrons, electrons, and $\gamma$-rays. It is now possible to observe vacancies and interstitials in high melting point metals using field ion microscopy.

In pure ionic crystals of NaCl the dominant defect at high temperature is a **Schottky pair,** i.e., a vacant sodium site (with a net negative charge) and a vacant chlorine site (with a positive charge). The two defects are not necessarily adjacent to each other.) These vacancies occur in pairs to ensure charge neutrality.

One of the most important properties possessed by interstitial atoms or ions and vacancies is that they usually can migrate about the crystal relatively easily when assisted by thermal fluctuations; i.e., vibrating neighboring atoms may exchange with the vacancy and hence permit the *diffusion* of matter in solids.

At high temperatures ionic crystals become good electrical conductors because these charged defects can move rapidly and can therefore transport charge.

**Wustite** has the same lattice as NaCl. However, it has the chemical formula $Fe_{<1}O$; i.e., it deviates from the expected formula FeO. Such compounds are called **nonstoichiometric compounds.** In wustite certain of the positive ion ($Fe^{2+}$) lattice sites must therefore be empty. However, this leaves an overall deficit of positive charge (which is not allowed). For every $Fe^{2+}$ vacancy created, two $Fe^{2+}$ must be converted into the $Fe^{3+}$ state. Thus in the overall process

$$3Fe^{2+} + \tfrac{1}{2}O_2 \rightarrow 2Fe^{3+} + \boxed{v_c} + FeO. \tag{6.6.2}$$

As many as 14% of the normal cation sites can be empty in wustite.

### Q. 6.6.2

The equilibrium fraction of vacant lattice sites depends on temperature according to (a)_____. The equilibrium fractional concentration of vacancies near to the melting point is (b)_____. When stainless steel is bombarded with neutrons (c)_____ defects are produced. In pure NaCl at high temperatures, a pair of defects called (d)_____ is present. In certain nonstoichiometric compounds such as wustite, the vacancy concentration can be as large as (e)_____.

\* \* \*

LINE IMPERFECTIONS. **Dislocations** are line imperfections in crystals. There are many examples of their importance in crystals but one striking one is indicated in Table 6.6.1 in which the tensile strengths of a metal whisker (almost a perfect crystal) and an annealed single crystal are compared. An example of an edge dislocation is shown in Figure 6.6.1 for

Table 6.6.1. TENSILE STRENGTH VS. DISLOCATION DENSITY

| Type of Crystal | Effective Strength (psi) | Dislocation Density |
|---|---|---|
| Iron whisker | 1,000,000 | Nil |
| Ordinary crystal | 100–1000 | Approx. $10^8/cm^2$ |

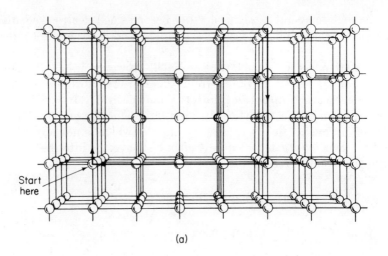

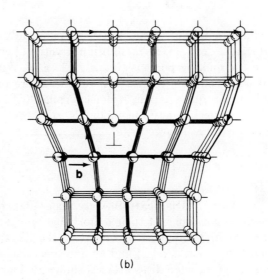

FIGURE 6.6.1. Description of a dislocation in terms of its Burgers vector. (a) A Burgers circuit in dislocation-free material. (b) The same Burgers circuit passing through dislocation-free material, but encircling a dislocation of unit Burgers vector, **b**. [From A. G. Guy, *Elements of Physical Metallurgy*, Addison-Wesley Publishing Company, Inc., Reading, Mass. (1959).]

the case of a simple cubic crystal. The symbol ⊥ is placed at the dislocation line. Note in part (b) how the atomic packing is disturbed in the immediate vicinity of the dislocation line. The atoms just below this dislocation line are in tension, and those just above in compression. Attention is drawn to the fact that far from the dislocation line the packing arrangement is correct, although the atoms there will be displaced slightly from their usual positions; i.e., there will be elastic strains. Energy is therefore needed to create a dislocation line. This energy is usually described in terms of the energy per atomic plane of length (about 2 Å or so) of the dislocation line. For an edge dislocation in copper this is about 8 eV/atomic plane.

A dislocation line is described by its direction (a unit vector parallel to the line) and by a vector known as its Burgers vector which is described next. Figure 6.6.1(a) shows a closed circuit of atom to atom steps. Suppose this circuit starts at the lower left corner of the loop. The circuit then consists of four steps upward, four to the right, four downward, and four to the left. If a similar circuit is made around a dislocated region (with the plane of the circuit perpendicular to the dislocation line), the circuit will not end at the starting point. The additional vector that must be added to make the circuit close is called the **Burgers vector, b**. See Figure 6.6.1(b). (Making the loop larger does not change this conclusion.) Notice that the Burgers vector is perpendicular to the dislocation line. This *defines* an **edge**

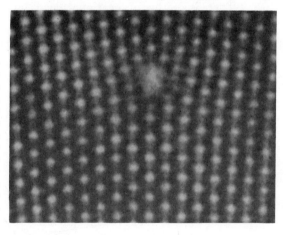

FIGURE 6.6.2. Electron micrograph moiré pattern of edge dislocation in cadmium. The thin cadmium crystal was grown on molybdenum sulfide. The slight mismatch between the two crystals produced the moiré effect, which makes possible the visualization of the pattern of an edge dislocation. (×1,400,000.) (Courtesy of W. Menter and D. W. Pashley, Tube Investments Research Laboratory, Saffron Walden, Essex, England.)

**dislocation**. We note that the Burgers vector equals a translation vector (or *repeat distance*) of the lattice (in the direction of the Burgers vector). We note an extra plane of atoms appears to have been inserted in the upper half of the crystal of Figure 6.6.1(b). (The word "edge" refers to the bottom edge of this extra plane.) Electron microscopy can be used to actually observe the displacement of atoms around an edge dislocation as shown in Figure 6.6.2.

Figure 6.6.3 shows how the edge dislocation can move under applied shear stresses. Since only tiny motions of the atoms are necessary, dislocation motion can take place very easily, particularly in metals where the bonding is long range. (In covalent materials, such as diamond, covalent bonds have to be broken and remade and hence a much higher stress is needed to move the dislocations.) In copper crystals of high purity and very

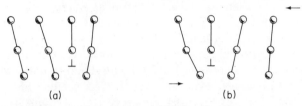

FIGURE 6.6.3. Motion of dislocation. (a) Original dislocation position. (b) Dislocation moved one repeat distance to the left under applied shear stresses.

low dislocation density, F. Young has found that shear stresses of only 1 psi will make the dislocations move. In a very pure single crystal of copper the dislocation moves easily because it is moving through an otherwise perfectly periodic atomic array. Anything which disrupts the perfect periodicity will make it more difficult for the dislocation to move.

There is another dislocation which is important in plastic deformation and in the growth of ordinary (nonperfect) crystals. This is the screw dislocation shown in Figure 6.6.4. The line $\overline{SO}$ is the dislocation line. Note that in this case the Burgers vector is parallel to the dislocation line: This

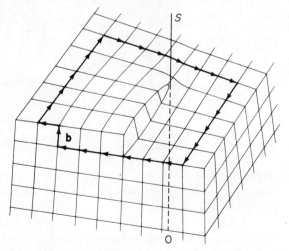

FIGURE 6.6.4. Burgers vector in simple cubic lattice for a screw dislocation.

is the defining characteristic of a **screw dislocation.** The motion of the screw dislocation (unlike the edge) is perpendicular to the Burgers vector. The crystal planes perpendicular to the dislocation line in Figure 6.6.4 actually form a spiral ramp (similar to the spiral ramp parking lot). Note that there is a ledge at the top of the crystal. The crystal grows by adding atoms from the gas or liquid; such atoms add at this ledge which therefore revolves around. This is one of the important mechanisms in the growth of imperfect crystals. See the growth spiral in Figure 6.6.5.

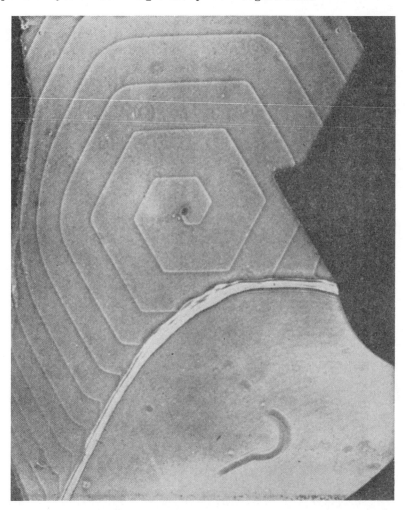

FIGURE 6.6.5. A regular hexagonal spiral on a SiC crystal originating from a group of screw dislocations at the center. ($\times$90.) [From A. R. Verma, *Crystal Growth and Dislocations*, Butterworth & Co. (Publishers) Ltd., London (1953).]

Figure 6.6.6 illustrates how the motion of a dislocation leads to plastic deformation. Actually, this is an idealized illustration since normally a crystal would contain $10^6$ cm/cm$^3$ of dislocation line when it crystallizes from the melt. When such materials are plastically deformed the dislocations multiply (dislocation multiplication, which is discussed in Chapter 12) and a highly deformed material may have a dislocation density of $10^{11}$ cm/cm$^3$, or $10^{11}$ cm$^{-2}$. It is much easier to form additional dislocation lines within a crystal than at the surface (as was illustrated in Figure 6.6.6).

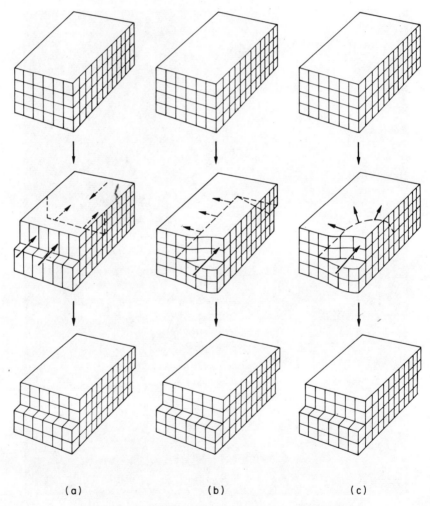

FIGURE 6.6.6. Plastic deformation by dislocation motion. (a) Edge. (b) Screw. (c) Mixed. [From L. H. Van Vlack, *Elements of Materials Science*, Addison-Wesley, Reading, Mass, (1964).]

Dislocation multiplication is illustrated in Figure 6.6.7 which shows the (111) face of a copper crystal whose original dislocation density was about 100 cm$^{-2}$. This crystal was first polished and etched. It was then indented by a diamond; next it was etched with chemicals which dissolved away the highly strained material around the dislocation more rapidly than

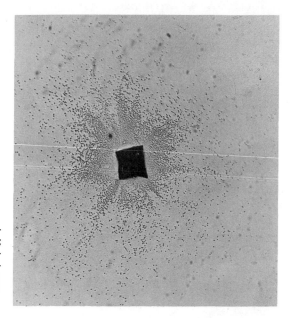

FIGURE 6.6.7. Etch pits associated with dislocations emerging from a copper crystal. The dislocations were formed by indentation of a (111) surface. (×75.) (Courtesy of B. F. Addis, Cornell University, Ithaca, N.Y.)

the good material. **Etch pit techniques** are one of the important ways of studying dislocations. Another important technique is **transmission electron microscopy.** Thin foils of material, $10^3$–$10^4$ Å thick, are transparent to electrons but the transmission through the highly strained material along the dislocation line is different from that through the good material. Dislocations in nickel are shown in Figure 1.2.5. An electron microscope is shown in Figure 6.6.8.

## Q. 6.6.3

Crystals without any dislocations would be (a)_____. If a dislocation is present in a metal or ionic crystal (where long-range attractive bonding dominates), it (b) (will or will not) move easily assuming that it does not have to cross other dislocation lines or precipitates. In an edge dislocation, the Burgers vector is (c)_____ to the dislocation line. In a screw dislocation, the Burgers vector is (d)_____ to the dislocation line. The plastic deformation of solids is due to (e)_____

_____. An ordinary crystal may have a dislocation density of about (f)_____, while in a heavily cold-worked crystal the density may be (g)_____.

* * *

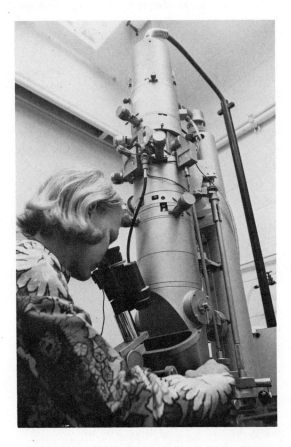

FIGURE 6.6.8. Student using a 100-kV electron microscope.

PLANAR IMPERFECTIONS. The **external surface** is a major defect of any crystal. The atoms in the surface have fewer nearest neighbors and hence their bonding requirements are not satisfied. A rough estimate is that only two thirds of their bonds are satisfied. Hence these atoms have a higher energy than do the internal atoms. This surface energy is about 1000–9000 ergs/cm$^2$ for solids. **Surface energy** is the work done in creating the new surface area. (Surface energy is negligible relative to the total binding energy unless the particle becomes very small, i.e., unless a significant fraction of the atoms are at the surface.) A similar situation arises in the case of liquids, e.g., water, where the surface energy is only about 80 ergs/cm$^2$; in this case a liquid droplet assumes a spherical shape (least surface area for a given volume). The process of powder metallurgy depends on a reduction in external surface energy. In powder metallurgy tiny particles of a material (a few microns in diameter) **sinter** together at high temperatures to form a solid object. The reason they sinter together is that in the process they eliminate surface energy; i.e., many external atoms become internal atoms and their energy is decreased.

When solids crystallize at the typical rates, the solid is not a single crystal but rather an assembly of a huge number of more or less randomly oriented crystals (grains) which meet each other at **grain boundaries.** The misorientation between two adjacent crystals (either tilt or twist or a combination of these) may be small (a few degrees or less for **small-angle grain boundaries**) or large (10 deg or more for **large-angle grain boundaries**). The simplest of grain boundaries is the **small-angle tilt boundary** shown in Figure 6.6.9. It is an array of equally spaced parallel

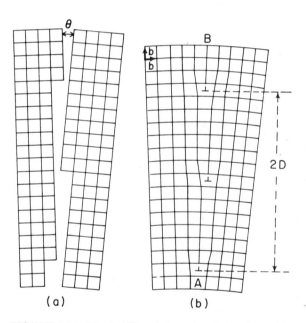

FIGURE 6.6.9. Simple tilt grain boundary lying along $AB$. The plane of the figure is parallel to a cube face and normal to the axis of relative rotation of the two grains. (a) Two grains have a common cube axis and an angular difference in orientation $\theta$. (b) The two grains are joined to form a bicrystal. The joining requires only elastic strain except where a plane of atoms ends on the boundary in an edge dislocation, denoted by the symbol ⊥. [After W. T. Read, *Dislocations in Crystals*, McGraw-Hill Book Company, Inc., New York (1953).]

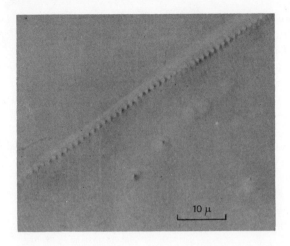

FIGURE 6.6.10. Etch pits that make up a grain boundary in germanium. [From F. L. Vogel, W. G. Pfann, H. E. Corey, and E. E. Thomas, *The Physical Review* **90**, 489 (1953).]

edge dislocations. Figure 6.6.10 shows etch pits at dislocations in such a boundary. The angle of misorientation between the two grains in Figure 6.6.9 is $\theta = 2\tan^{-1} b/2D \doteq b/D$, where $b$ is the Burgers vector and $D$ is the spacing between dislocations. This angle can also be measured using X-ray diffraction.

The misorientation of the large-angle grain boundary is too great for the dislocation model to apply. See Figure 6.6.11 which is a model of such a

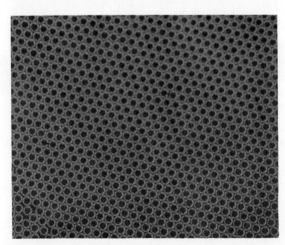

FIGURE 6.6.11. Bubble raft illustrating grain boundaries. Note that the boundaries are only a few bubbles wide. (Courtesy of W. R. Day, Jr., Mitron Research and Development Center.)

boundary. The boundary is only a few atoms wide as has been shown by field ion microscopy. The energies of grain boundaries are of the order of 100–1000 ergs/cm$^2$ for large-angle grain boundaries and 10–250 ergs/cm$^2$ for a 1-deg boundary. The equilibrium form of a crystalline material would have no grain boundaries.

One very striking feature of large-angle grain boundaries is their tendency to move through the crystal if the driving force and the temperature are sufficiently high. This results in *grain growth* and corresponding modifications in the mechanical and other properties of the material. Another important characteristic is the marked tendency of foreign atoms to "segregate" at these boundaries.

### Q. 6.6.4

One defect which every real crystal obviously always has is (a)_____. The energy associated with this defect is of the order of (b)_____. The success of powder metallurgy depends on a process called (c)_____ which takes place at high temperature; the reason this process occurs is (d)_____. A small-angle tilt boundary is a boundary between two crystals (e)_____. Large-angle grain boundaries (f) (would or would not) be present under equilibrium conditions.

\* \* \*

SPATIAL IMPERFECTIONS. An alloy of Al with 4% Cu after age-hardening heat treatment consists of approximately pure aluminum crystals with tiny particles very rich in copper distributed in the crystals.

In Section 1.4, we noted that large pores or bubbles are formed in reactor materials undergoing intense neutron irradiation. The radiation produces Frenkel pairs, and the fission products are often gases. The vacancies aggregate to form pores containing fission gases (**bubbles**).

Various defects such as large pores and bubbles and second-phase precipitates can occur during casting. It is common to have pores present after sintering, particularly of ceramic materials. This is illustrated in Figure 6.6.12 for sintered MgO which is white rather than transparent. It is opaque because of the presence of pores having a diameter of the order of a micron. The refractive index is 1.00 in the pore and 1.74 in the MgO itself; multiple refraction at the interfaces of the pores makes the material opaque.

Macroscopic defects are very important in determining many properties and will be studied in more detail later.

### Q. 6.6.5

Spatial imperfections in crystals may include (a)_____. Radiation damage often involves the formation of (b)_____. The opaqueness of many sintered ceramics is due to (c)_____. The fracture strength of sintered ceramics is often low because (d)_____.

\* \* \*

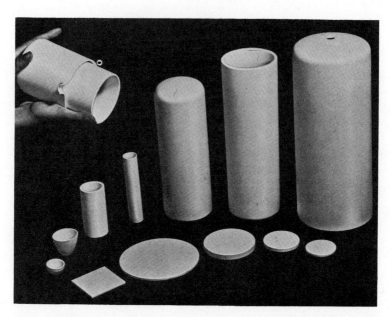

FIGURE 6.6.12. MgO crucibles and other shapes made by sintering.

## ANSWERS TO QUESTIONS

**6.6.1** (a) Point, line, planar, and spatial; (b) an interstitial impurity; (c) two; (d) about the same (see Table 6.2.1); (e) extended or even complete; (f) might; (g) causes randomness to be preferred.

**6.6.2** (a) $Ae^{-E_f/k_BT}$, (b) about $10^{-4}$, (c) Frenkel, (d) a Schottky defect, (e) several percent (14%).

**6.6.3** (a) Very strong, (b) will, (c) perpendicular, (d) parallel, (e) the motion of dislocations, (f) $10^6$ cm/cm$^3$, (g) $10^{11}$ cm$^{-2}$.

**6.6.4** (a) An external surface, (b) a few thousand ergs per square centimeter, (c) sintering, (d) it eliminates external surfaces and hence decreases the energy of the system, (e) consisting of a special array of edge dislocations, (f) would not.

**6.6.5** (a) Precipitates, pores, and bubbles; (b) bubbles; (c) porosity; (d) of the stress concentrations around the pores.

## 6.7 GLASSES

Glasses are not crystalline structures. To illustrate their structure, we compare crystalline $SiO_2$ with the glass form of $SiO_2$. Two of the crystalline forms of $SiO_2$ are shown in Figure 6.7.1 for purposes of illustration. There

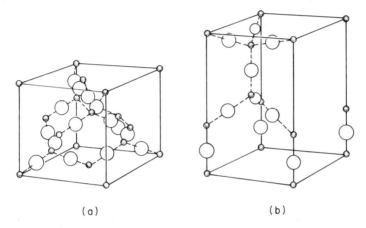

FIGURE 6.7.1. $SiO_2$ polymorphs. (a) Cristobalite. (b) Tridymite.

also are other polymorphs of which quartz is the most important. The cubic form is an fcc lattice plus a basis of an $SiO_4^{4-}$ group at 000 with a vertex nearly toward [111] and another $SiO_4^{4-}$ group at $\frac{1}{4}\frac{1}{4}\frac{1}{4}$ with a vertex nearly toward [$\bar{1}\bar{1}\bar{1}$]. From the symmetry viewpoint it is similar to the zinc blende structure of Figure 6.4.10. (Likewise tridymite is similar to the hexagonal wurtzite structure). The structure of the hexagonal tridymite viewed from along the $c$ axis is shown in Figure 6.7.2. The silicons form puckered rings of $SiO_4^{4-}$ units. Note that in each of these crystal structures an $SiO_4^{4-}$ unit is bonded to four others (by sharing oxygens).

When a crystalline form of $SiO_2$ is melted a viscous liquid is formed. The bonding in this liquid does not have the highly regular bonding characteristic of the crystalline solids. However, the bonding is quite strong for a liquid as evidenced by the high viscosity. [The viscosity coefficient of molten $SiO_2$ at the melting point of the cubic crystalline phase (the high-temperature polymorph) is $10^7$ times as large as the typical viscosity coefficient for liquid metals at their melting point.] If this liquid is not cooled *extremely* slowly, the "random" network structure becomes frozen in [recall

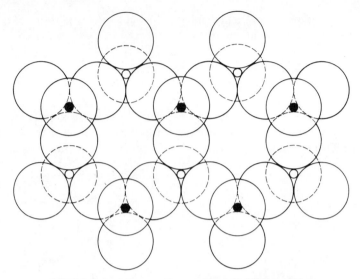

FIGURE 6.7.2. Silica structures. Note that the tetrahedra alternately point up and down. Similar layers are then directly bonded to each other. This is the SiO$_2$ polymorph, tridymite. The silicons in the tetrahedra pointing up (shown as solid circles) are above the plane of the paper, while the silicons in the tetrahedra pointing down (shown as open circles) are below the plane of the paper.

that the glass becomes more and more viscous as the temperature decreases; see Equation (2.3.4)]. The resulting structure is called **silica glass.** See Figure 6.7.3. Glass is essentially a random network. The **random network** may have rings of four, five, six, seven, and eight silica tetrahedra (SiO$_4^{4-}$ units) while the rings in the crystalline phase have only six tetrahedra.

### Q. 6.7.1

In crystalline networks of SiO$_2$, each (a)_____ tetrahedra is bonded to (b)_____ others. In the crystalline silica there are puckered rings containing (c)_____ tetrahedra. In silica glass, the rings contain (d)_____
____ tetrahedra. A glass is (e)_____.

\* \* \*

Figure 6.7.4 shows the volume change versus temperature for formation of a crystalline solid from the melt and for the formation of a glass from the melt.

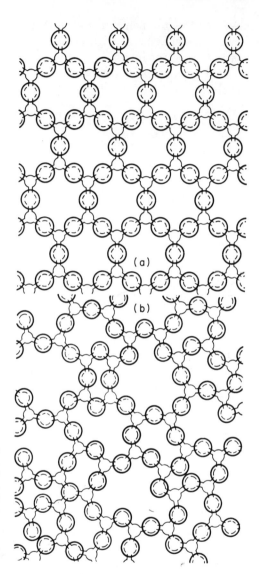

FIGURE 6.7.3. Schematic two-dimensional analogs, after W. H. Zachariasen, *Journal of the American Chemical Society* **54**, 3841 (1932), illustrating the difference between (a) the regularly repeating structure of a crystal and (b) the random network of a glass. The solid circles are oxygen atoms.

Upon very slow cooling, a crystalline phase forms. Note the large volume *discontinuity* at the melting point. Upon very rapid cooling (there is no time for crystallization), the viscosity increases rapidly; as the temperature decreases the contraction is rapid (but continuous) because more efficient packing is occurring. Below a certain temperature, $T_G$, the **glass transition temperature,** the packing arrangement is frozen in; further contraction is a result of smaller thermal vibrations, just as in the crystalline phase in this temperature range. A **glass** is a material with the thermal expansion behavior of Figure 6.7.4; i.e., $dV/dT$ is discontinuous but $V$ is continuous, as the liquid cools. Recall from Equation (4.1.4) that the volume thermal expansion coefficient is given by $\beta = d \ln V/dT$. Note that when the glassy state is reached, $\beta$ for the glass and for the crystal are nearly the same.

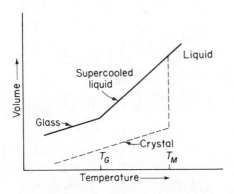

FIGURE 6.7.4. Volume changes upon cooling $SiO_2$ from the liquid.

Inorganic compounds which form glasses are $SiO_2$, $B_2O_3$, $GeO_2$, $BeF_2$, and a number of trioxides and pentoxides of the fifth periodic group. Organic compounds such as glycerine as well as rubber-like materials and many polymeric materials can exist as glasses at low temperatures.

**EXAMPLE 6.7.1**

Will the thermal expansion coefficient of glass be isotropic?

*Answer.* Yes. Since it is a random network, the average structure in each direction is the same. Thus it is isotropic with respect to *all* properties. As an example the velocity of light in all directions will be the same.

## Q. 6.7.2

At the melting point of cristobalite, the viscosity coefficient of the liquid silica is about (a)_____ times as large as the viscosity coefficient of liquid metals at the melting point. If such materials are not cooled extremely slowly, below the melting point, rather than getting (b)_____, (c)_____ is obtained. As the temperature is lowered still further, the freedom of motion of the silica units disappears at a temperature called (d)_____; this involves a *discontinuity* on (e)_____ versus temperature curve but no discontinuity (only a change of slope) on (f)_____ versus temperature curve.

\* \* \*

An important application of glasses is for coatings as illustrated in Figure 6.7.5. The clay-based ceramic insulators are often sintered aggregates of crystals with pores. Firing melts the surface layer, and upon cooling of the highly viscous liquid, a glass forms. The glass (glaze) is not porous.

FIGURE 6.7.5. Clay-based ceramics before and after firing.

Glasses have many engineering uses as containers, windows, and lenses and in combination with a variety of ceramics. The strongest bond in ordinary glass is the Si—O bond between silica tetrahedrons. Many commercial glasses are alloys; e.g., soda glass contains $Na_2O$ and $SiO_2$. The average bonding in this material is less strong than in pure silica glass so that soda glass has a much lower softening point. Pyrex contains boric oxide as an additive to $SiO_2$. In general in glasses containing additives some of the silicon atoms have been replaced by other atoms (such as boron). Ions which replace silicon atoms in this manner are called **network-forming ions** in contrast to metal ions which fill the voids in the network and are called **network-modifying ions** (such as sodium). The color, viscosity, density, and electrical properties of glasses can be modified considerably by additions of metal ions.

The **degree of crystallinity** (the fraction of the material which is crystalline) in glasses will increase slowly at temperatures sufficiently high to achieve mobility of atoms, but below that at which thermal softening occurs. This crystallization process is called **devitrification.** Although glasses are important materials, their physics and chemistry are not well understood. There is extensive research currently underway on the structure and properties of glasses.

### Q. 6.7.3

In addition to silica glasses, the inorganic compound (a)_____ forms a glass. Even rubber (b) (will or will not) form a glass at low temperatures. When a sodium atom attaches to an oxygen atom of a $SiO_4^{4-}$ tetrahedron only (c)_____ of the oxygen atoms of the latter are left to form a network. Hence the addition of $Na_2O$ to silica glass will (d)_____ _____. The slow crystallization of glass at temperatures below the softening point is called (e)_____.

\* \* \*

## ANSWERS TO QUESTIONS

**6.7.1** (a) $SiO_4^{4-}$; (b) four; (c) six $SiO_4^{4-}$; (d) various numbers of tetrahedra, usually from 4 to 8, including 5 and 7; (e) a random network structure.

**6.7.2** (a) $10^7$, (b) crystallization, (c) a supercooled liquid, (d) the glass transition temperature, (e) a volume thermal expansion coefficient, (f) a volume.

**6.7.3** (a) $B_2O_3$, etc.; (b) will; (c) three; (d) weaken the bonding and hence lower the softening temperature; this is of great commercial importance as it drastically reduces the cost of fabrication; (e) devitrification.

### 6.8 DIFFRACTION BY CRYSTALS

BRAGG'S LAW. It has been known for a long time that a ruled grating causes diffraction of light rays. In 1912 von Laue, a German scientist, suggested that if crystals were composed of regularly spaced atoms which would scatter X-rays, and if X-rays were electromagnetic waves with a wavelength of the same order as the atom spacing, then crystals should diffract X-rays. Experiments carried out under his direction on copper sulfate verified both of these assumptions. In 1913 Bragg obtained the crystal structure of NaCl. This was the first determination of a crystal structure.

Let us suppose that we have a parallel beam of copper $K_\alpha$ X-rays (wavelength $\lambda = 1.54$ Å). (The origin of such X-rays is discussed in Problem 5.11.) Figure 6.8.1 shows a set of parallel crystallographic planes (normal to the paper) represented by the lines $AA$ and $BB$. The line $LL_1$ represents the crest of an incoming wave; actually this wave will "strike"

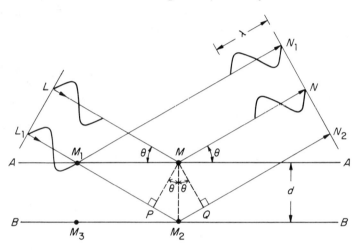

FIGURE 6.8.1. Bragg's law.

huge numbers of scattering centers, atoms, but only three are shown scattering here. The X-rays are scattered in all direction by the electrons in the atoms. We consider only one specific scattered ray from each atom. The distance $LM$ was chosen to be an integral number of wavelengths and $MN$ is also an integral number of wavelengths. We consider now the three path lengths: $LMN$, $L_1M_1N_1$, and $L_1M_2N_2$. The line $N_1NN_2$ has been assumed to coincide with the crest of the outgoing wave. We now proceed to prove that there is such a crest under certain conditions. Clearly, from the figures,

$$LMN = L_1M_1N_1 \qquad (6.8.1)$$

if the angle of incidence equals the angle of reflection (and we have chosen that case). (The point $M_1$ could be *any* atom on the $AA$ plane which scatters the incoming wave.) We note that

$$L_1M_2N_2 - LMN = PM_2Q. \qquad (6.8.2)$$

For $N_2$ to be at a crest the extra path difference $PM_2Q$ must equal an integral number of wavelengths, or

$$n\lambda = 2d \sin \theta, \tag{6.8.3}$$

where $n$ is an integer. A similar relation would apply to an incoming wave scattered from $M_3$. It would, moreover, apply between any neighboring pair of planes, e.g., $BB$ and $CC$, etc. Equation (6.8.3) is therefore the condition for mutual reinforcement of the scattered rays. It is called the **Bragg law** or the **Bragg equation.** It relates the **order of the reflection** $n$, the wavelength of the radiation $\lambda$, the distance between the planes $d$, and the angle of incidence $\theta$. $\theta$ is also called the **diffraction angle.**

BRAGG'S LAW AND SIMPLE CUBIC CRYSTALS. Although simple cubic crystals are not common (polonium has this structure at room temperature), we can easily analyze their diffraction pattern. If we combine Equation (6.8.3) with Equation (6.3.1) we have

$$n^2(h^2 + k^2 + l^2) = \frac{4a^2}{\lambda^2} \sin^2 \theta.$$

We choose to write this in the form

$$(H^2 + K^2 + L^2) = \frac{4a^2}{\lambda^2} \sin^2 \theta, \tag{6.8.4}$$

where $H = nh$, etc. We shall write the set $HKL$ without the parentheses which are used to describe Miller indices $(hkl)$; thus 200 is possible while (200) is not since Miller indices cannot have a common factor. The experimentalist measures a set of the angles $\theta$. To do so he may set up his experi-

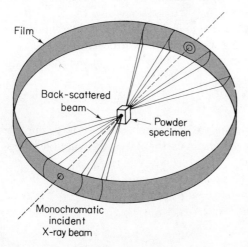

FIGURE 6.8.2. Sketch of Debye-Scherrer powder experiment using film.

ment as shown in Figure 6.8.2. This is known as the **powder technique** or the Debye-Scherrer method. An example of a powder film for a cubic metal is shown in Figure 6.8.3. The crystal structure and the lattice parameter may be obtained from such patterns. See B. D. Cullity, *Elements of X-ray Diffraction*, Addison-Wesley Publishing Company, Inc., Reading, Mass. (1956).

Suppose that the experimentalist obtains $\sin^2 \theta$ values as follows: 0.100, 0.201, 0.299, 0.401, 0.502, 0.599, 0.801, 0.900. We now show that this is consistent with a simple cubic crystal structure. We do not know the lattice parameter $a$. We *assume* $4a^2/\lambda^2 = 10$ in the present case. This is a guess! We then make Table 6.8.1. The data are consistent with Equation (6.8.4), except for small experimental errors which are to be expected. Assuming that the wavelength used was $\lambda = 2.00$ Å, we have $a^2 = 10$ Å$^2$ and $a = \sqrt{10}$ Å.

Table 6.8.1

| $\sin^2 \theta$ | $\frac{4a^2}{\lambda^2} \sin^2 \theta$ | $H^2 + K^2 + L^2$ | $HKL$ |
|---|---|---|---|
| 0.100 | 1.00 | 1 | 100 |
| 0.201 | 2.01 | 2 | 110 |
| 0.299 | 2.99 | 3 | 111 |
| 0.401 | 4.01 | 4 | 200 |
| 0.502 | 5.02 | 5 | 210 |
| 0.599 | 5.99 | 6 | 211 |
| — | — | — | — |
| 0.801 | 8.01 | 8 | 220 |
| 0.900 | 9.00 | 9 | 300 |

It can be shown that for a simple cubic crystal all diffraction lines, i.e., $HKL$, are possible; for a bcc crystal only those diffraction lines, i.e., $HKL$, for which the sum $H + K + L$ is even, have finite intensity; while for an fcc crystal only those diffraction lines, $HKL$, for which the $H$, $K$ and $L$ are all odd (e.g., 111) or all even (e.g., 220) are possible. Why such rules, called **selection rules** should exist is discussed in Problems 6.55 and 6.56. With this in mind we see that the data in Table 6.8.1 is consistent with a simple cubic crystal structure.

Suppose that instead of bombarding a huge number of randomly oriented single crystals with *monochromatic* radiation we bombard a single crystal with *polychromatic* radiation. The result we obtain (on a flat film) is shown in Figure 6.8.4. This is known as the **Laue method.** Note the symmetry. This method provides a means of orienting a single crystal, i.e., for experimentally determining the spatial orientation of the planes in the crystal.

292  *Atomic Arrangements*

## Q. 6.8.1

Waves will be diffracted from a grating if (a)_____. The waves scattered from the atoms in one plane of a crystal will be in phase with those scattered from an adjacent plane a distance $d$ away if and only if (b)_____. It is possible to obtain the (c)_____ of materials from the diffraction lines. Electrons and neutrons will also be diffracted from a crystal if (d)_____.

\* \* \*

FIGURE 6.8.3. X-ray diffraction lines.

FIGURE 6.8.4. Back reflection Laue patterns of tungsten, incoming X-ray normal to (111) plane.

# ANSWERS TO QUESTIONS

**6.8.1** (a) Their wavelength approximately equals the grating spacing; (b) $n\lambda = 2d \sin \theta$, where $n$ is an integer and $\theta$ is the angle between the plane and the incidence wave; (c) crystal structure and lattice parameters; (d) they have a wavelength approximately equal to the spacing between planes.

# REFERENCES

### Elementary

Holden, A., and Singer, P., *Crystals and Crystal Growing*, Doubleday & Company, Inc., Garden City, N.Y. (1960). This is an elementary paperback book. If the student has not already read it, he should. The student interested in the type of symmetry exhibited by a crystal about a point should refer to p. 284.

Bragg, Sir William, *Concerning the Nature of Things*, G. Bell & Sons Ltd., London (1948). Reprinted by Dover Publications, Inc., New York (1954).

### Packing

Azaroff, L., *Introduction to Solids*, McGraw-Hill Book Company, Inc., New York (1960). Chapters 3 and 4 give a good account of packing concepts.

Evans, R. C., *An Introduction to Crystal Chemistry*, Cambridge University Press, New York (1964).

Pauling, L., *Nature of the Chemical Bond*, Cornell University Press, Ithaca, N.Y. (1948). Pauling won the Nobel prize. You might get a perspective on how he thinks from reading this book. The reader is reminded of a statement made by Prof. John Wheeler at Princeton, "Tell us, great reader, that a book gives each of us the company of the great man of our choice."

Wells, A. F., *Structural Inorganic Chemistry*, Oxford University Press, Inc., New York (1962). This gives an extensive list of compounds in the index for which the structure and coordination is given in the text.

### Symmetry

Weyl, Hermann, *Symmetry*, Princeton University Press, Princeton, N.J. (1952). Illustrates symmetry in art, architecture, and nature as well as in crystals. This is a great book.

Jaswon, M. A., *Introduction to Mathematical Crystallography*, Longmans, Green & Co. Ltd., London (1965). Discusses the 32 point groups and the space groups. Requires an appreciation of matrix algebra. What Holden and Singer do pictorially is done by Jaswon with matrix theory. The reader could progress from Jaswon to a book utilizing group theory.

## Diffraction

Bragg, Sir Lawrence, "X-ray Crystallography," *Scientific American* (July 1968), p. 58. This is an excellent elementary discussion.

Bragg, W. H., *The Crystalline State*, Cornell University Press, Ithaca, N.Y. (1965). This book captures the excitement of the early years of X-ray analysis of crystals. The first correct structure determination is described by W. L. Bragg, *Proceedings of the Royal Society* (London) **A89**, 248 (1913).

Cullity, B. D., *Elements of X-ray Diffraction*, Addison-Wesley Publishing Company, Inc., Reading, Mass. (1956).

Bacon, G. E., *Neutron Diffraction*, Oxford University Press, Inc., New York (1962).

Gevers, R., "Electron Diffraction," in the book edited by Strumane et al., *Interaction of Radiation with Solids*, North-Holland Publishing Company, Amsterdam (1964).

Kittel, C., *Introduction to Solid State Physics*, John Wiley & Sons, Inc., New York (1968).

Cohen, J. B., *Diffraction Methods in Materials Science*, The Macmillan Company, New York (1966). For the instructor.

## Crystal Data

Wycoff, R. W. G., *Crystal Structures*, John Wiley & Sons, Inc. (Interscience Division), New York (1963). Look in this or the next reference for specific crystal structures.

*International Tables for X-ray Crystallography*, published for the International Union of Crystallography by Kynoch Press, Birmingham, England (1965).

## Imperfections

Simmons, R. O., and Balluffi, R. W., "Measurements of Equilibrium Vacancy Concentrations in Aluminum," *The Physical Review* **117**, 52 (1960).

Kröger, F., *The Chemistry of Imperfect Crystals*, John Wiley & Sons, Inc. (Interscience Division), New York (1964).

Hull, D., *Introduction to Dislocations*, Pergamon Press, Inc., Elmsford, N.Y. (1965). This monograph is a nice introduction to the subject. It has discussions of techniques for observing dislocations as well as of their properties.

Germer, L. H., "The Structure of Crystal Surfaces," *Scientific American* (March 1965), p. 32.

## Metals

Cottrell, A. H., "The Nature of Metals," *Scientific American* (September 1967), p. 90.

Barrett, C., and Massalski, T. B., *Structure of Metals*, McGraw-Hill Book Company, Inc., New York (1966). This is a comprehensive discussion.

## Ionic Crystals and Ceramics

Greenwood, N. N., *Ionic Crystals, Lattice Defects and Nonstoichiometry*, Butterworth & Co. (Publishers) Ltd., London (1968). This is an excellent treatment of packing concepts, binding, point defects, and nonstoichiometry in ionic materials.

Van Vlack, L. H., *Physical Ceramics for Engineers*, Addison-Wesley Publishing Company, Inc., Reading, Mass. (1964). This book contains both scientific and engineering aspects of ceramics.

Kingery, W. D., *Introduction to Ceramics*, John Wiley & Sons, Inc., New York (1960). This book is more advanced than the preceding one.

## Glasses

Charles, R. J., "The Nature of Glass," *Scientific American* (September 1967), p. 127.

Volf, M. B., *Technical Glasses*, Sir Isaac Pitman & Sons Ltd., London (1961).

# PROBLEMS

**6.1** Make a list of five *why* questions for each of the following:
(a) Crystals.
(b) Ionic crystals.
(c) Symmetry.
(d) Defects.
(e) Glass.
(f) Diffraction.
Example on defects: Why are vacancies present in equilibrium concentrations of about $10^{-4}$ near $T_m$ when it takes an energy $E_f$ to form each vacancy?

**6.2** There are three cubic lattices. Draw the unit cell for each. Imagine the cells to be moved slightly along a body diagonal. How many lattice points are then completely within the cell for each of three lattices?

**6.3** Comment on the following: The bcc crystal structure of molybdenum is a body-centered cubic lattice with a basis of one molybdenum atom at the origin or the bcc crystal structure of molybdenum is a simple cubic lattice with a basis of two atoms: Mo at 000 and Mo at $\frac{1}{2}\frac{1}{2}\frac{1}{2}$.

**6.4** What is the volume of a unit cell defined by **a**, **b**, and **c**?

**6.5** While we ordinarily think of the face-centered cubic crystal structure of aluminum as a face-centered cubic lattice with a basis of one aluminum atom centered at each lattice point, we could also form this crystal structure from a simple cubic lattice with a basis of four atoms: 000, $\frac{1}{2}\frac{1}{2}0$, $\frac{1}{2}0\frac{1}{2}$, $0\frac{1}{2}\frac{1}{2}$. Prove this.

**6.6** A NaCl crystal is a face-centered cubic lattice with a basis of Na$^+$ at 000 and Cl$^-$ at $\frac{1}{2}$00. When such a crystal grows from a water solution, cubes are formed. What can you say about the relative growth rates of the {100}, {110}, and {111} planes?

**6.7** A sodium chloride crystal grows as a cube from pure water solution, but the nature of growth is changed when borax is added to the solution and regular tetrahedrons are then formed.
(a) Describe the rate of growth of specific planes.
(b) What is a possible explanation of the effect of the borax?

**6.8** Copper has a face-centered cubic crystal structure with a lattice parameter of 3.62 Å. If the atoms are rigid spheres in contact along the ⟨110⟩ directions, calculate the sphere diameters.

**6.9** Lead has an fcc crystal structure with a lattice parameter of 4.95 Å and an atomic weight of 207.19. Calculate the density.

**6.10** Copper has an fcc structure with $a = 3.6154$ Å. Calculate its density.

**6.11** Molybdenum has a bcc structure with $a = 3.15$ Å. Calculate its density.

**6.12** Zinc has an hcp structure with $c = 4.95$ Å, $a = 2.66$ Å. Calculate its density.

**6.13** In the simple cubic crystal unit cell shown below, give the direction indices of

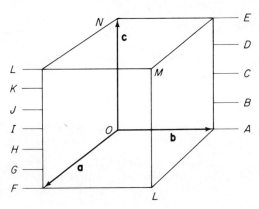

PROBLEM FIGURE 6.13.

(a) $G$ to $D$.
(b) $O$ to $L$.
(c) $F$ to $M$.

Give the Miller indices of the planes.
(d) $FAN$.
(e) $KFA$.
(f) $HAN$.

**6.14** (a) Define what is meant by a closest-packed layer.
(b) What is the stacking sequence for the fcc crystal structure and the hcp crystal structure?

**6.15** (a) Give the Miller indices for the closest-packed planes in the fcc crystal structure.
(b) Give the direction indices for the closest-packed directions in the fcc crystal structure.

**6.16** (a) Give the Miller indices for the closest-packed planes in the hcp crystal structure.
(b) What word is often used to describe these planes?

**6.17** (a) Give the $c/a$ ratio for four hcp metals.
(b) What would their $c/a$ ratio be for the ideal case where the atoms were really spheres?

**6.18** What is the ratio of tetrahedral voids and octahedral voids to atoms in the
(a) fcc crystal structure?
(b) hcp crystal structure?

**6.19** Along what direction (give the direction indices) do atoms touch in these crystals:
(a) fcc?
(b) hcp?
(c) bcc?

**6.20** Show that the expected crystal structure of MgO is the NaCl type.

**6.21** Why do you expect the $CO_3^{2-}$ complex to be a planar triangle?

**6.22** (a) Iron and NaCl both exhibits polymorphism. What are the polymorphs?
(b) Explain the difference between zinc blende and wurtzite.

**6.23** The compound $K_2SiF_6$ has a cubic crystal structure.
(a) Is this consistent (or nearly so) with concepts of packing?
(b) Show a unit cell.

**6.24** Would the expected crystal structure of BeO involve octahedral coordination or would it be one of the ZnS types?

**6.25** $Cr^{3+}$ has a radius of 0.67 Å. What is its expected coordination number in $Cr_2O_3$?

**6.26** (a) Compare the bonding in graphite and diamond.
(b) Compare the electrical conductivity in these materials.
(c) Which polymorph is the equilibrium one at room temperature and pressure?

**6.27** Define
   (a) Isotropic.
   (b) Anisotropic.
   Give some examples of anisotropic properties of crystals.
   (c) What is the maximum thermal expansion coefficient (depending on direction) in zinc?

**6.28** Discuss the symmetry elements of
   (a) An ordinary chair.
   (b) A Ping-Pong ball.
   (c) A hockey puck.

**6.29** Pb has an fcc crystal structure and Sn has a tetragonal crystal structure. Give reasons why you might expect some solubility but not continuous solubility of these elements in each other.

**6.30** Give an example of
   (a) A solid substitutional solution.
   (b) A solid interstitial solution.
   (c) An ordered solid solution.

**6.31** (a) What is the charge on a sodium ion vacancy in NaCl?
   (b) How does doping NaCl with $Ca^{2+}$ ions affect the concentration of sodium ion vacancies. Assume that the crystal remains electrically neutral.

**6.32** When a solid is bombarded with neutrons, large numbers of vacancies and self-interstitials are formed. Would you expect equal numbers of each? Explain.

**6.33** Give the dislocation density (centimeters per cubic centimeter) in copper for
   (a) Perfect crystal.
   (b) Whiskers.
   (c) Very good high-purity crystal.
   (d) Ordinary crystal.
   (e) A cold-rolled sheet of copper.

**6.34** It is much easier to plastically deform a crystal by passing dislocations through it than it would be to slip the top half of the crystal (as a rigid block) over the bottom half. In an analogous situation, it is easier to move a rug by sending a wave down it than to attempt to drag the rug. Discuss this analogy further.

**6.35** Give, in electron volts, for copper
   (a) The cohesive energy (81 kcal/mole).
   (b) The energy of vacancy formation.
   (c) The energy of dislocation line per centimeter of length.

**6.36** (a) What is the difference between a structure-sensitive and a structure-insensitive property?
   (b) In which class would elastic moduli and tensile yield strength be?

**6.37** A photon is a finite wave packet, and when such a wave interacts with a crystal, it interacts with a large number of scattering centers. Show that the condition for reinforcement of the scattered wave from all of such scattering centers is given by Bragg's law.

**6.38** Suppose we have the following set of $\sin^2 \theta$ for a cubic crystal: 0.100, 0.200, 0.300, 0.400, 0.500, 0.600, 0.700, 0.800. It is definitely either a simple cubic or bcc crystal. What is it?

**6.39** (a) Give the electron configuration of the $K^+$ ion and the $Cl^-$ ion.
(b) Although KCl has the NaCl-type structure, why does it appear, when studied by X-ray diffraction, to have a simple cubic structure?

**6.40** When a clay has been properly fired it has a shiny, nonporous surface. Why?

**6.41** A carefully cooled glass is free of stresses and is a random network. Will it be piezoelectric?

## MORE INVOLVED PROBLEMS

**6.42** Draw a face-centered cubic lattice (several cells) and pick out a primitive cell which could be used instead.

**6.43** Clearly sketch the (112) plane of a cubic crystal. Does a $\langle 110 \rangle$ direction lie in this plane? Which one? Prove that for *cubic* crystals a direction $[uvw]$ lies in a plane $(hkl)$ if $uh + vk + wl = 0$.

**6.44** Define Miller indices. Then show that for a simple cubic crystal the distance from the origin to the plane with Miller indices $(hkl)$ is given by (6.3.1).

**6.45** Suppose that 1 mole of water (18 cc) is broken into droplets having a diameter 31.7 Å.
(a) What is the total surface energy if the surface energy per unit area is $\gamma_{LV} = 72.75$ ergs/cm² at 20°C?
(b) How does this compare with the heat of vaporization of 1 mole of water?
(c) The molecules in these droplets are clearly not bonded as well as the molecules in bulk water. Would you expect these droplets to have a higher, lower, or the same vapor pressure as bulk water?

**6.46** Assume that there are about $2 \times 10^{15}$ copper atoms/cm² on the surface and that the volume per copper atom is $12 \times 10^{-24}$ cm³.
(a) Calculate the fraction of atoms in the surface layer versus radius, $r$, for particles having radii of $10^{-2}$, $10^{-3}$, $10^{-4}$, $10^{-5}$, $10^{-6}$, and $10^{-7}$ cm.
(b) For the fraction $f$, plot $\log_{10} f$ versus the $\log_{10} r$.
(c) For what value of $r$ does the surface energy become exceedingly important?
(d) One mole of copper is subdivided into spheres of radius $10^{-4}$ cm. Calculate the surface energy of the system and compare it to the cohesive energy of the solid which is 3.5 eV/atom.

Problems 6.47–6.52 comprise a learning sequence concerned with the structures of silicates.

**6.47** The compound forsterite, $Mg_2SiO_4$, has a melting point of 1888°C and is used as a refractory. Show that it has an **island structure;** i.e., a silicate unit is not directly attached to any other silicate unit; i.e., *no oxygen* of one tetrahedron is shared with another tetrahedron.

**6.48** Show that the infinitely long silicate **single chain** has the formula $SiO_3^{2-}$. Note that *two oxygens* from each tetrahedra are shared

PROBLEM FIGURE 6.48.

with *two* other tetrahedras. An $Mg^{2+}$ ion between neighboring chains can bond such chains together.

**6.49** By placing two single chains side by side, additional oxygens can be shared. Show that this leads to a **double chain structure** of formula $[SiO_{11/4}]^{3/2-}$.

**6.50** (a) Show that if each tetrahedra has three of its oxygens shared with three other tetrahedra and the base of all these tetrahedra lie in a plane, a **sheet structure** is obtained. This sheet can be considered to be a very large ion.
(b) What is its formula?

**6.51** If the oxygen at each of the four corners of every tetrahedron is shared with another tetrahedron, a **framework structure** is obtained. It has the formula $SiO_2$. Two of the $SiO_2$ polymorphs have structures similar to zinc blende and wurtzite (see Section 6.7). Discuss.

**6.52** Consider the compound $Mg_2SiO_4$. Assuming it is based on a closest packing of $O^{2-}$ ions, in what voids would we expect to find the $Mg^{2+}$ and $Si^{4+}$ ions and what fraction of such voids would be filled?

**6.53** Show that for electrons accelerated through small potential differences, $V$, the wavelength of an electron is

$$\lambda(\text{Ångstroms}) = \sqrt{\frac{150}{V(\text{volts})}}.$$

At low voltages ($\sim 10^2$ V) this expression is satisfactory while at high voltages ($\sim 10^5$ V) it is necessary to use relativistic expressions.

## SOPHISTICATED PROBLEMS

**6.54** (a) Show that in a bcc crystal, the distance between the (100) planes is only $a/2$.

(b) Show that the planes which are farthest apart in a bcc crystal are the (110) planes.

**6.55** Using the results of the previous problem show why the diffraction line for $HKL = 100$ has zero intensity. It would be possible, by proceeding in this fashion (although there is a simpler way to do it by using quantities called structure factors) to show that for a bcc crystal the intensity will be zero for $H + K + L =$ odd and nonzero otherwise. These are called **selection rules**.

**6.56** Show that in a fcc crystal structure the distance between (100) planes is not $a$ but $a/2$ and the distance between (110) planes is not $a/\sqrt{2}$ but $a/2\sqrt{2}$. Hence using the same approach as in Problem 6.55 show that the first diffraction line will be for $HKL = 111$, i.e., from a (111) plane. It can be shown that for fcc crystals that the intensity of diffracted lines will be nonzero only if each of the numbers in a set $HKL$ are even or if each is odd. Thus there will be a 111, a 200, and a 311 diffraction line; there will be no 100, 210, or 421 lines.

**6.57** (a) Using the results of Problems 6.55 and 6.56, list the first ten possible values of $H^2 + K^2 + L^2$ for the bcc structure. The fcc structure. Do the same for the simple cubic structure.
(b) How might this be used to analyze the following? A set of diffraction lines obtained with $CuK_\alpha$ radiation gave $\theta = 13.70, 15.89, 22.75, 26.91, 28.25, 33.15, 36.62, 37.60, 41.95$ deg for $\lambda = 1.54$ Å.

**6.58** Show that the only rotations consistent with the translational symmetry of the two-dimensional lattice are $n$-fold rotations where $n = 1, 2, 3, 4, 6$. *Hint:* Start with two parallel rows of points as shown:

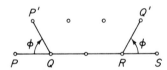

PROBLEM FIGURE 6.58.

**6.59** Show that there are five and only five two-dimensional Bravais lattices.

**6.60** (a) What are the symmetry elements of a regular tetrahedron?
(b) Does it contain a center of symmetry?
(c) Do diamond crystals have a center of symmetry?

**6.61** Give two examples where neutron diffraction yields structural determinations not possible with X-rays. Requires outside reading.

**6.62** Write an essay on the uses of X-ray diffraction. Include five applications. Requires outside reading.

**6.63** Write an essay on the uses of low-energy electron diffraction (LEED) in studying surfaces. Requires outside reading.

# Prologue

The solids described in the previous chapter could usually be considered as having been built from tiny building units such as copper atoms, placed together, to form copper metal. In the case of plastics and rubber the solid is made up of extremely large molecules, called *polymers*, which are studied in this chapter.

The most rapidly growing area of materials science is the study of polymers. Polymer molecules have many geometrical shapes and forms and some of the possibilities, ranging from *linear chains* to *three-dimensional networks*, are described in this chapter. The so-called linear chain is not always a "straight" chain but often takes on a *random configuration*. Polymer molecules in a material having a specific composition do not all have the same molecular weight. Rather, there is a *distribution of molecular weight*. There are two polymerization processes called *addition polymerization* and *condensation polymerization* and examples of each of these are given. The conditions which determine whether a polymer will form either a linear chain or a network are studied. Polyethylene, polypropylene, polytetrafluoroethylene, polyisoprene, nylon, silicone plastics and phenolics, and silicone rubber are used to illustrate the different structures. It is interesting (and commercially important) to note that polymers are sometimes crystals, sometimes amorphous, and sometimes part crystalline and part amorphous. One of the more interesting crystalline forms is the *folded-chain structure*. There is a brief introduction to the macromolecules in *living matter*, such as *cellulose* and the *proteins*.

# 7

# POLYMERS

## 7.1 INTRODUCTION

Organic polymers are formed by the combination of many small molecules. One of the simplest of these molecules is formed from ethylene. In ethylene the carbon-hydrogen bonds are single covalent bonds (one electron pair) while the carbon-carbon bond is a double covalent bond (two electron pairs). In the presence of heat, light, and an appropriate catalyst this molecule can be brought into an excited state as shown by the following reaction:

$$\begin{matrix} H & & H \\ & C::C & \\ H & & H \end{matrix} \xrightarrow[\text{catalyst}]{\text{heat}} \begin{matrix} H & & H \\ & \cdot C:C\cdot & \\ H & & H \end{matrix} \qquad (7.1.1)$$

If such excited molecules come in contact, it is possible to get addition of molecules (with single bonds between the carbons). When this process repeats itself, say 1000 times, we obtain a chain with a carbon backbone and a formula

$$H \left[ \begin{matrix} H & H \\ | & | \\ -C-C- \\ | & | \\ H & H \end{matrix} \right]_n H. \qquad (7.1.2)$$

This is called **polyethylene.** The building unit or the **mer** is ethylene. This process of polymerization is called **addition polymerization.** We have assumed that after $n$ ethylene units have been added, the reaction is terminated by the addition of a hydrogen on each end.

The carbon-carbon bonds in polyethylene are single bonds at 109 deg to each other. Rotation can readily take place about single bonds *unless* there are large side groups present which interfere with each other (in the present case the tiny hydrogens do not interfere). Because of this free rotation the polyethylene chain when dissolved in a solvent takes on many configurations.

Depending on the process used to make the polyethylene it may or may not have numerous **side chains;** these are short hydrocarbon chains attached to the longer chain. Polymers with side chains are called **branched polymers.** The possibility of packing nicely to form crystals is destroyed by the presence of these side chains. When side chains prevent good packing, the bonding between chains is weak and the softening point is low. Polymers without side chains are called **linear polymers.** The Zeigler catalyst makes the production of linear polyethylene possible. In this case the packing can be good so the bonding is stronger and the melting point is higher. As an example, baby bottles made from the latter material will maintain their shape when sterilized in boiling water while bottles made from branched polyethylene will collapse under their own weight under these conditions.

### Q. 7.1.1

It is possible to initiate a polymerization reaction in ethylene by (a)_____. The building block of a polymer molecule is called (b)_____. The process by which polyethylene forms is called (c)_____. If polyethylene is made by a special process, it has a negligible number of side chains and is called a (d)_____. The melting point of a polymer consisting of such molecules will be higher than for (e)_____.

\* \* \*

To develop a feeling for the nature of a polymer molecule a model of a polymer is shown in Figure 7.1.1. This is how large a molecule would be if the mers were 1 cm in diameter, rather than several Ångstroms. This figure shows the "molecule" at rest; if this molecule were present in a dilute *solution*, at room temperature, it would be thermally agitated and a more accurate presentation would be a motion picture which showed it twisting and writhing about.

The industrial use of polymers has grown and is growing at a rapid pace as shown in Figure 7.1.2. Polymers are used as plastics, adhesives, coatings, elastomers, fibers, etc. An understanding of them is important not only to appreciate technological systems, but also in the study of the life sciences.

Sec. 7.1                                                                                 Introduction     305

FIGURE 7.1.1. Polymer chain.

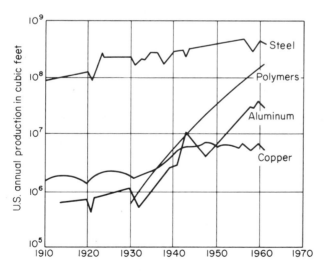

FIGURE 7.1.2. Production of polymeric materials. (Courtesy of T. G. Fox, Mellon Institute, Pittsburgh, Pennsylvania.)

## ANSWERS TO QUESTIONS

**7.1.1** (a) The presence of heat and a catalyst, (b) a mer, (c) addition polymerization, (d) linear polymer, (e) a polymer with branched molecules.

### 7.2 AN IDEALIZED RANDOM CHAIN

Let us consider a chain in which the bonds not only can rotate but can be freely bent to any angle (and not just restricted to 109 deg). A chain which has $N$ such links, each of length $l$, would, when completely stretched out, have a length $Nl$. However, if each link was added in a random direction, the **random chain configuration** of Figure 7.2.1 would be

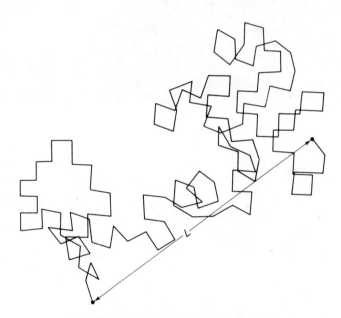

FIGURE 7.2.1. Random chain configuration.

obtained. The most likely distance between the ends of the chain is now only about $\sqrt{N}\, l$. (Problems involved with random chains are called random walk problems. For a discussion of random walk models, see the previously cited Ruoff, *Materials Science*, Section 17.3.) Many polymer molecules and particularly rubber-like materials tend to take on such random chain configurations instead of packing nicely to form crystalline materials.

## Q. 7.2.1

In the polymer chains based on the single bond carbon-carbon backbone, the bond angle is (a)_____ since the bonds are (b)___ orbitals which have (c)_____. A molecule in which the rotation about all bonds is completely random is called (d)_____ _____. Polymer molecules in dilute solutions often have such a structure. Moreover, polymer molecules in (e)_____ have such a configuration. The most probable distance between the ends of such a molecule which has a stretched out length $Nl$ is about (f)___.

\* \* \*

## ANSWERS TO QUESTIONS

**7.2.1** (a) 109 deg, (b) $sp^3$, (c) the tetrahedral configuration, (d) a random chain configuration, (e) rubber-like materials, (f) $\sqrt{N}l$.

### 7.3 DEGREE OF POLYMERIZATION

The **degree of polymerization** is the number of mers which have added together to form a polymer. It is directly related to the molecular weight, which for polymers is very high. It is not possible to speak of the actual molecular weight (as for $CH_4$) but rather of the **average molecular weight** since a polymeric material contains polymer molecules of varying degree of polymerization (chains of varying lengths).

The degree of polymerization can determine whether a polymer is a waxy substance or a plastic; it can strongly affect the softening and melting temperature and the degree of its solubility in specific solvents. Consider the boiling point of some straight chain hydrocarbons. Methane, $CH_4$, boils at 109°K; pentane, $C_5H_{12}$, boils at 309°K; while cetane, $C_{16}H_{34}$, boils at 561°K, and $C_{70}H_{142}$ does not boil at all but instead decomposes first. Decomposition means that the primary bonds which have an energy of about 4 eV are breaking. Although the molecules are held to each other by relatively weak secondary forces (bond energies of about 0.1 eV), there are so many of these secondary bridges in a long molecule that, in the aggregate, they can bind different molecules together more strongly than the covalent bond holds together a pair of atoms in one molecule. Figure 7.3.1 shows how the melting point of polymers increases with increasing molecular weight (and hence degree of polymerization). This is a typical behavior for a molecule. Some linear polymers which have particularly strong secondary bonds

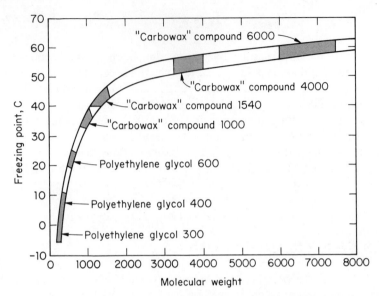

FIGURE 7.3.1. Melting point as a function of molecular weight of a specific polymer. [From A. X. Schmidt and C. A. Marlies, McGraw-Hill Book Company, Inc., New York (1948).]

between the molecules (hydrogen bridges in nylon) do not melt but decompose prior to melting. For these materials the tensile strength and resistance to solubility also increase as the degree of polymerization increases.

### Q. 7.3.1

If $n$ mers form a molecule the (a)_____ is $n$. Polymeric materials are not made of molecules of a fixed length but rather contain (b)_____. The melting point of polymers (c)_____ as $n$ increases. Some linear polymeric molecules (d)_____ before they melt because (e)_____

_____. The tensile strength and the viscosity of the melt (if the polymer melts) (f)_____ as $n$ increases.

\* \* \*

We digress briefly to study one method of molecular weight determination. The determination of the average molecular weight is not a simple matter as it is with compounds such as $CH_4$. Average molecular weights are determined by physical measurements such as the osmotic pressure of the polymer in a liquid solution, the viscosity of the polymer in solution, or by light scattered from the polymers in solution.

One of the simplest techniques (although often not the best) and the one first used to show that natural rubber is a polymer involved the use of osmotic pressure. Osmotic pressure was first observed by the Abbé Nollet in 1748 who placed "spirits of wine" in a cylinder which was then capped by an animal bladder and placed in water. He noted that the bladder swelled and sometimes burst. The bladder is a **semipermeable membrane,** which in this case allows water to pass through it but not alcohol. The pressure increase in the cylinder was caused by the increased amount of water which passed through the membrane. Flow of water eventually stops and an equilibrium pressure is reached; this is called **osmotic pressure.** In 1885 Van't Hoff showed from theoretical considerations that for *dilute* solutions the osmotic pressure $\pi$ is given by

$$\pi = CRT,$$

where $C$ is the moles of solute per liter of solution, $R$ is the universal gas constant, and $T$ is the absolute temperature. Hence if the osmotic pressure is measured for a solution containing a given weight of polymer, the molecular weight can be obtained.

**EXAMPLE 7.3.1**

If 10 g of a protein is dissolved in water to form 1 liter of solution at 7.0°C and the osmotic pressure is 8.7 mm Hg, what is the molecular weight?

*Answer.*

$$C = \frac{\pi}{RT} = \frac{8.7/760}{0.082 \times 280} \text{ (moles/l)}.$$

Hence

$$MW = \frac{10 \text{ g/liter}}{C} = 20{,}000 \text{ (g/mole)}.$$

Could you devise a simple experimental arrangement to measure osmotic pressure?

**Q. 7.3.2**

The pressure which builds up in a container which is sealed by a semipermeable membrane through which the solvent can pass but the solute cannot, when that container is placed in pure solvent, is called (a)_____. Water diffuses through cellulose acetate sheet but sodium and chlorine ions do not; the acetate sheet is called (b)_____

_____. The Van't Hoff relation for osmotic pressure is (c)_____; this expression holds only for (d)_____.

* * *

## ANSWERS TO QUESTIONS

**7.3.1** (a) Degree of polymerization, (b) a distribution of degree of polymerization, (c) increases, (d) decompose, (e) a primary bond in the chain is weaker than the sum of the secondary bonds between chains, (f) increase.

**7.3.2** (a) Osmotic pressure, (b) a semipermeable membrane, (c) $\pi = CRT$, (d) dilute solutions.

### 7.4 THE TOPOLOGY OF VINYL POLYMERS

There are a number of polymers formed from compounds much like ethylene, e.g.,

$$\begin{array}{c} H \\ \diagdown \\ H \end{array} C=C \begin{array}{c} H \\ \diagup \\ R \end{array}, \qquad (7.4.1)$$

where $R$ may be a halide, a benzene ring, etc. These are called **vinyl compounds**. Examples are shown in Table 7.4.1. Polyethylene is a plastic

Table 7.4.1. VINYL COMPOUNDS

| Name | —R |
|---|---|
| Ethylene | —H |
| Vinyl chloride | —Cl |
| Propylene | —CH$_3$ |
| Vinyl acetate | —O—C(=O)—CH$_3$ |
| Acrylonitrile | —C≡N |
| Styrene (vinyl benzene) | —C$_6$H$_5$ |

Sec. 7.4    The Topology of Vinyl Polymers    311

widely used for packaging, soil mulching between plants, temporary walls exterior to buildings under construction, baby bottles, containers, etc. Poly(vinyl acetate) is commonly used as the base for photographic materials; polyacrylonitrile when formed into fibers is called orlon; and polystyrene has many uses, one of which is for foamed plastics.

We note that for vinyl compounds there are a number of different topological arrangements possible depending on the manner in which mers are added. Thus, in vinyl chloride, chlorine atoms may be found on alternate carbons only (head to tail addition, written as $\cdots$ HTHT $\cdots$) and shown in Figure 7.4.1(a); chlorine may also be found alternately on two adjacent

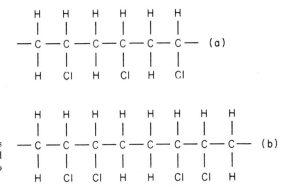

FIGURE 7.4.1. Different forms of poly(vinyl chloride). (a) Head to tail arrangement. (b) Head to head, tail to tail arrangement.

carbons and be absent from the next pair of adjacent carbons (head to head, tail to tail addition or $\cdots$ HHTT $\cdots$) as shown in Figure 7.4.1(b); or they may possibly involve other ordered or random sequences. This is a new topological feature which did not occur for straight chains of polyethylene. The possibility of packing closely would be strongly affected by such topology. The $\cdots$ HTHT $\cdots$ sequence is the most likely one to form, although vinyl polymers with other sequences have been made.

The $\cdots$ HTHT $\cdots$ arrangement itself can take on different configurations. To illustrate these configurations for polypropylene, we consider the zigzag carbon backbone to lie in a plane as shown in Figure 7.4.2. (In reality it does not necessarily lie in a plane, since there can be rotation about the carbon bonds.) A top view of this backbone is simply a straight line as in (b). Here (c), (d), and (e) show the top view with the side atoms added. There is a methyl group on alternate carbons. Three possible configurations are shown in Figure 7.4.2. These are called **stereoisomers** because they have a different topology. In the **isotactic** isomer the $CH_3$ groups are always on one side. In the **atactic** isomer the $CH_3$ groups are randomly located, while on the **syndiotactic** isomer the $CH_3$ groups are on alternate sides. [The reader should remember the actual zigzag nature of the chain; otherwise it appears that a simple rotation around the C—C

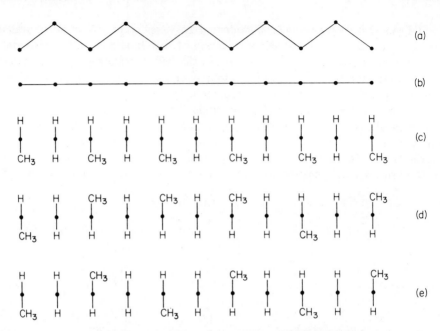

FIGURE 7.4.2. (a) Planar carbon backbone. (b) Top view of planar backbone. (c) Isotactic arrangement of $CH_3$. (d) Atactic (random) arrangement of $CH_3$. (e) Syndiotactic arrangement of $CH_3$.

bonds in (c) could give (d) or (e).] The atactic form is a waxy material at room temperature while the isotactic form is a hard plastic. This is because the chains can pack together more efficiently in the latter form and this leads to stronger secondary binding between molecules.

## Q. 7.4.1

The molecule $CH_2{=}CHR$ is called a (a)_____ compound. Three examples of R are (b)_____. A head to tail addition of polypropylene results in a different (c)_____ than a head to head, tail to tail addition. The head to tail molecule of polypropylene itself has many different possible (d)_____. Three of these are characterized by whether the $CH_3$ groups (also called the (e)_____ groups) are (f) _____
_____.

\* \* \*

A close-up view of the isotactic form of polypropylene is shown in Figure 7.4.3 as it would appear if it existed in the planar form. Because the

CH₃ groups are so close together, they would tend to overlap in the planar configuration as shown. Since it is more or less impossible for the two CH₃ groups to occupy the same space, they repel each other. This is called **steric hindrance.** Hence the chain twists out of the planar zigzag form

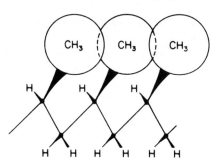

FIGURE 7.4.3. Isotatic polypropylene showing overlap of the CH₃ charge clouds in the planar zigzag arrangement.

by rotation about C—C bonds, in this case, by 120 deg. This repeated twist in the same direction gives a **helical chain** with three propylene units per complete turn.

These helical coils pack well and readily crystallize. Proteins often form similar but not identical coils. Steric hindrance tends to prevent the free rotation of the chain; this is called **hindered rotation.** Rotation is often still possible when aided by thermal fluctuations but stops when the temperature gets too low. A change from rubbery behavior to rigid plastic behavior (at the glass transition temperature) is associated with this elimination of free rotation. The glass transition temperature will be higher in materials with large steric hindrance.

### Q. 7.4.2

Because of (a)_____ the isotactic polypropylene molecule does not remain planar. Instead it forms (b)_____. A vinyl compound such as isotactic polystyrene will form a hard plastic at room temperature rather than a rubbery material because (c)_____
_____.

\* \* \*

## ANSWERS TO QUESTIONS

**7.4.1** (a) Vinyl; (b) —Cl, —CH₃, —C≡N, etc.; (c) stereoisomer; (d) stereoisomers; (e) methyl; (f) on the same side of the molecule (in the planar zigzag representation), alternately on opposite sides, or randomly located.

**7.4.2** (a) Steric hindrance, (b) a helical molecule (which happens to have three mers per complete turn), (c) the large side groups will hinder rotation and hence will cause **chain stiffening.**

## 7.5 OTHER ADDITION POLYMERS

POLYTETRAFLUOROETHYLENE. Another very important polymer is formed by addition from the compound

$$\begin{matrix} F & & F \\ & \diagdown & \diagup \\ & C=C & \\ & \diagup & \diagdown \\ F & & F \end{matrix} \quad . \tag{7.5.1}$$

This molecule is called tetrafluoroethylene. The resultant polymer has a carbon backbone with fluorine atoms attached and is called **polytetrafluoroethylene (PTFE)** or **Teflon**. It was discovered accidentally by R. J. Plunkett at du Pont laboratories while he was working with materials for refrigerants. Teflon is highly stable chemically. (Remember that fluorine itself is an extremely strong oxidizing agent and that concentrated sulfuric acid can be boiled in Teflon without ill effects.) Teflon has many other useful properties. Most materials do not wet it, i.e., do not stick to it. It therefore has a very low coefficient of friction and is often used as a sleeve bearing or in powder form as a lubricant.

1-4 ADDITION. Consider a molecule such as isoprene,

$$\begin{matrix} H & CH_3 & H & H \\ \diagdown & | & | & \diagup \\ C=C & —C=C & \\ \diagup & & & \diagdown \\ H & & & H \\ \phantom{H}\overset{①②}{\phantom{C=C}} & \overset{③④}{\phantom{C=C}} & \end{matrix} \quad . \tag{7.5.2}$$

This molecule will polymerize in the presence of heat and an appropriate catalyst by reaction at the ① and ④ carbon positions to form polyisoprene,

$$\begin{bmatrix} H & CH_3 & H & H \\ \diagdown & | & | & \diagup \\ —C—C=C—C— \\ \diagup & & & \diagdown \\ H & \overset{②}{\phantom{C}} & \overset{③}{\phantom{C}} & H \end{bmatrix}_n , \tag{7.5.3}$$

a long chain molecule in which double bonds are still present. This polymerization process is called **1-4 addition.** (Ordinary natural rubber is polyisoprene. It is interesting to note that Michael Faraday who is otherwise known for developments in electricity and magnetism was the first to show that rubber is polyisoprene. Polyisoprene is a product of the rubber tree. However, the same material was produced synthetically by Giulio Natta in 1955.)

In polyisoprene the carbons associated with the double bonds (②and ③) are still reactive points and can be used for **crosslinking** from one polymer chain to another, e.g., by sulfur atoms. (This is **vulcanization.**) All the polymer molecules can therefore be tied together at various points into a giant three-dimensional molecule. Prior to and after vulcanization the polyisoprene chains would be in a random chain configuration. In soft rubber bands only a few weight percent of sulfur is used and the crosslinks may be 100 units apart.

**EXAMPLE 7.5.1**

Calculate the extension ratio possible in rubber in which the number of chain links between crosslinks is 100.

*Answer.* The stretched out length of the portion of the molecule between crosslinks is $100L$. In the random configuration the actual length is about $\sqrt{100}\ L$. Hence the maximum extension ratio $l/l_0 = 10$. This means that the rubber could be stretched 900% before it would be necessary to stretch bonds.

Harder rubber materials such as tires and bowling balls contain more sulfur, and materials such as automobile battery cases contain as much as 40% sulfur by weight. Such materials are much stiffer and have much smaller extension ratios than soft rubber.

Soft rubber materials such as those in rubber bands are essentially liquids held together by the crosslinks. The chains between the crosslinks move quite readily over each other and are in constant thermal motion. Materials which can be stretched several hundred percent are called rubberlike materials or **elastomers.** Elastomers have stress-strain curves such as that shown in Figure 7.5.1. The Young's modulus of rubber is quite small (about $10^{-5}$ times that of steel). When rubber is stretched, the applied force tends to overcome the thermal motion of the random chains; this is the origin of the elasticity of elastomers. (You can think of the stretching of rubber as an analog of the pressurizing of gas. In both cases these processes are resisted by thermal agitation.)

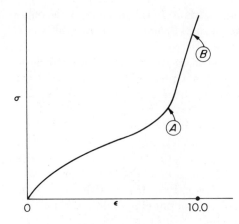

FIGURE 7.5.1. Stress-strain curve for an elastomer. Crystallization tends to occur at $A$ and bond stretching and bending at $B$.

## Q. 7.5.1

Teflon is resistant to strong oxidizing agents because (a)_____. The isoprene molecule is interesting because it has (b)_____ and after addition polymerization to form a linear chain it has (c)_____. These linear molecules can be (d)_____ by sulfur which forms (e)_____. The structure of elastomers could be described as (f)_____ held together by (g)_____ which impart (h)_____ rigidity to the material.

* * *

When the rubber is highly stretched, the molecules are aligned and even though the secondary bonding between them is quite small, crystallization will occur (even though the violent thermal motion of the chain segments prevents crystallization in the unstretched sample). This is illustrated in Figure 7.5.2. As the temperature is lowered the mobility of the chains decreases until a temperature, called the glass transition temperature, is reached (see Figure 6.7.4). At this temperature the mobility vanishes altogether and the configurations are frozen in place. The material is then rigid and brittle just as is ordinary window glass. The material called Plexiglas or Lucite [i.e., the polymer, poly(methyl methacrylate)] is a hard plastic near room temperature or below. (It is often used for aircraft windshields.) However, above about 120°C it is rubber-like.

## Q. 7.5.2

When an elastomer is highly stretched, a change in structure called (a)_____ often occurs. If the temperature of an elastomer is lowered sufficiently, (b)_____ will no longer be possible and the material will be (c)_____.

(a)

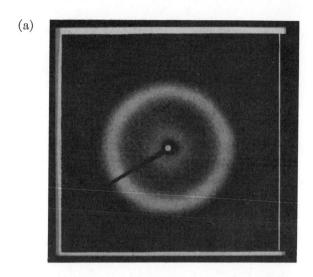

(b)

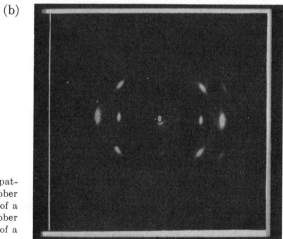

FIGURE 7.5.2. Laue X-ray pattern of (a) unstretched rubber (the pattern is characteristic of a liquid) and (b) stretched rubber (the pattern is characteristic of a crystal).

STEREOISOMERISM IN POLYISOPRENE. In natural rubber the methyl group and the hydrogen atom are on the same side of the C=C bonds as shown in Figure 7.5.3. This is called the *cis* **configuration** (same side). If the $CH_3$ and H are on opposite sides, the configuration is called the *trans* **configuration.** Molecules of the *trans* form, which is called **gutta percha,** pack well and readily crystallize. The *cis* configuration of polyisoprene does

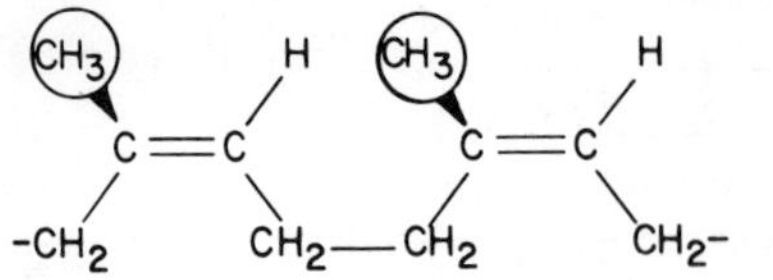

## 318  Polymers

$$\begin{array}{c} CH_3 \\ \diagdown \\ C=C \\ \diagup \\ -CH_2 \end{array} \begin{array}{c} H \\ \diagup \\ \\ \diagdown \\ CH_2-CH_2 \end{array} \begin{array}{c} CH_3 \\ \diagdown \\ C=C \\ \diagup \\ \end{array} \begin{array}{c} H \\ \diagup \\ \\ \diagdown \\ CH_2- \end{array}$$

FIGURE 7.5.3. *cis*-polyisoprene.

not pack well and instead of crystallizing forms a rubber-like material with a (nearly) random chain configuration called **natural rubber**.

POLYCHLOROPRENE. If instead of methyl ($CH_3$) side groups the isoprene molecule has chlorine atoms, the material is called **neoprene** (the building unit is called chloroprene). This material is chemically quite different from ordinary rubber. For example, ordinary rubber absorbs gasoline readily and swells. Without attempting to be quantitative, we recall the rules of chemistry, which say that like dissolves like. Rubber is a pure hydrocarbon and so is gasoline. Neoprene, however, contains the highly polar chlorine (which will have a net negative charge; recall that in HCl, the chlorine had a charge of $-e/6$). The polar neoprene does not absorb the nonpolar gasoline and in fact is widely used in service stations to dispense it.

## ANSWERS TO QUESTIONS

**7.5.1** (a) Of the fluorine atoms attached to the carbon backbone; fluorine itself is a stronger oxidizing agent than oxygen; (b) two double bonds with a single bond in between; (c) a remaining double bond; (d) crosslinked; (e) strong primary bonds between the chains; (f) a liquid; (g) crosslinks; (h) shear.

**7.5.2** (a) Crystallization; (b) free rotation about bonds; (c) a hard plastic; a glass transition will have occurred.

## 7.6 COPOLYMERS

It is often possible to combine two different mers into a polymer. Thus butadiene,

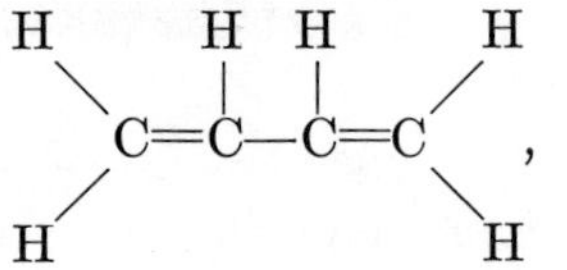

$$\begin{array}{c} H \\ \diagdown \\ \\ \diagup \\ H \end{array} C=C-C=C \begin{array}{c} H \\ \diagup \\ \\ \diagdown \\ H \end{array} \quad , \tag{7.6.1}$$

and styrene (see Table 7.4.1) can polymerize. The polymer which results from combining two different mers is known as a **copolymer**. Figure 7.6.1 shows schematically possible arrangements of such molecules. A **block**

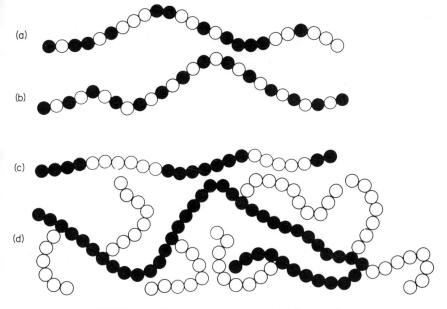

FIGURE 7.6.1. Copolymer arrangements. (a) A copolymer in which two different units are distributed randomly along the chain. (b) A copolymer in which the units alternate regularly. (c) A block copolymer. (d) A graft copolymer. [From J. Wulff et al., *The Structure and Properties of Materials*, John Wiley & Sons, Inc., New York (1964).]

**polymer** molecule consists of alternating segments of two polymers. A **graft polymer** consists of a backbone of one polymer with side chains of another.

## 7.7  CONDENSATION POLYMERS

NYLON.  **Organic acids** have the structure

$$R-\overset{\overset{\displaystyle O}{\|}}{C}-OH, \qquad (7.7.1)$$

where R is H, $CH_3$, $C_2H_5$, etc. **Amines** have the formula

$$R-N\begin{matrix}H\\ \\H\end{matrix}. \qquad (7.7.2)$$

Adipic acid is dibasic, i.e., it has two of the —COOH groups, namely,

$$HOOC—(CH_2)_4—COOH. \qquad (7.7.3)$$

Hexamethylenediamine has the formula

$$H_2N—(CH_2)_6—NH_2. \qquad (7.7.4)$$

These two compounds react, under the appropriate conditions, as shown:

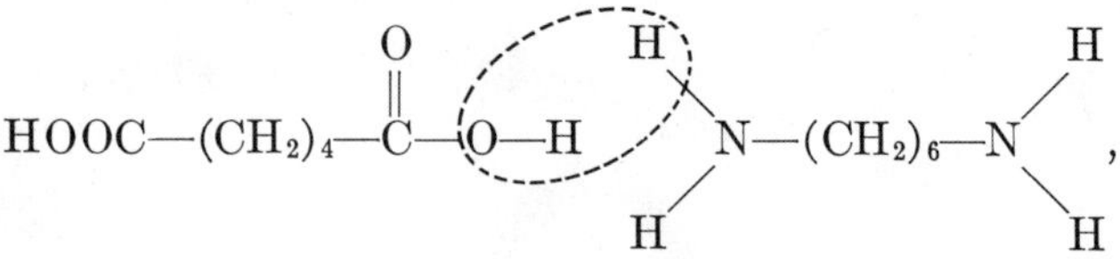

$$HOOC—(CH_2)_4—\overset{O}{\overset{\|}{C}}—\underbrace{O—H \quad\quad N}—(CH_2)_6—N\begin{smallmatrix}H\\ \\H\end{smallmatrix} , \qquad (7.7.5)$$

expelling $H_2O$ and forming a C—N bond between the two molecules:

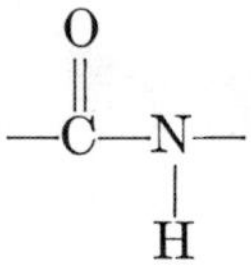

$$—\overset{O}{\overset{\|}{C}}—\underset{H}{N}— \qquad (7.7.6)$$

(the remaining hydrogen on the nitrogen in this new bond is not very reactive). This reaction can carry on with alternate mers to form a long chain. It is called **condensation polymerization** because of the by-product which in this case is $H_2O$. The by-product of these reactions is usually a small molecule such as $H_2O$, $NH_3$, HCl, etc. The polymer, since it is formed of two mers, is a copolymer. The specific molecule formed in the present case might be called polyhexamethylene-diamineadipate but it is usually called by its generic name, **nylon 66** (the 66 refers to the six carbons in each mer). Although there were earlier developments in the plastics industry, the synthesis of nylon by W. H. Carothers and its production by du Pont in 1939 signaled the beginning of the rapid growth of the plastics age.

Two of the important features of the nylon chain are the double-bonded oxygen and the hydrogen attached to the nitrogen. When two

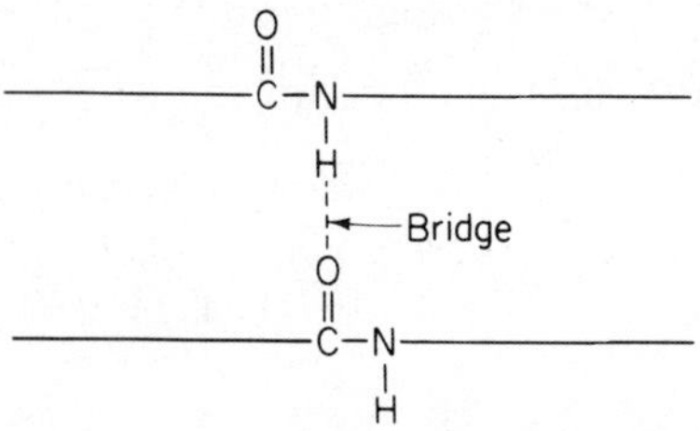

FIGURE 7.7.1. Hydrogen bridge in nylon.

nylon chains are brought together a *hydrogen-bonded bridge* forms as shown schematically in Figure 7.7.1. This is a particularly strong secondary bond.

When nylon fibers are made the molten nylon is forced out of pinholes and as it cools is stretched several hundred percent. This stretching along one direction causes alignment of molecules as shown in Figure 7.7.2. The

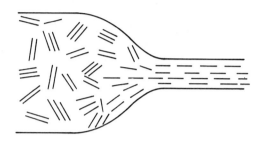

FIGURE 7.7.2. Schematic of alignment which occurs during stretching of a fiber.

resultant structure is very strong (see Table 1.3.5). If the nylon molecules are not aligned in this fashion, as in the case for bulk nylon, it is considerably weaker (see Table 1.3.5).

## EXAMPLE 7.7.1

Explain why nylon fibers are so sensitive to water. For example, they may be used in humidity-measuring devices. Nylon drapes will alternately stretch and shrink several inches in climates in which the humidity varies greatly.

*Answer.* Water is also a strong hydrogen bond bridge former. Water breaks many of the bridges in nylon and hydrogen-bridges itself to a nylon chain. This causes swelling.

Nylon is only one example of a large class of linear condensation polymers.

### Q. 7.7.1

Nylon 66 is formed by the addition of (a)_____ mers and is therefore a (b)_____. The reaction which leads to the nylon molecule is called (c)_____. A hydrogen attached to a nitrogen in one molecule is capable of forming a strong (d)_____ bond with an oxygen in another molecule. These lead to strong attractions between molecules in nylon. Drawing of nylon fibers produces alignment and strengths as high as (e)_____.

\* \* \*

SILICONES. In compounds, silicon has $sp^3$ orbitals. The compound

$$\text{HO}-\underset{\underset{Z}{|}}{\overset{\overset{R}{|}}{\text{Si}}}-\text{OH} \qquad (7.7.7)$$

(where Z is either R′ or OH, and R and R′ are alkyl groups such as $CH_3$) can polymerize by a condensation process to yield

$$-\underset{\underset{Z}{|}}{\overset{\overset{R}{|}}{\text{Si}}}-\text{O}-\underset{\underset{Z}{|}}{\overset{\overset{R}{|}}{\text{Si}}}-\text{O}-\underset{\underset{Z}{|}}{\overset{\overset{R}{|}}{\text{Si}}}-\text{O}-, \qquad (7.7.8)$$

a linear chain silicone if Z = R′. If Z = OH, then primary Si—O bonds can form between two or more of the chains. **Silicone plastics** are characterized by linear chains while crosslinked chains are found in **silicone rubber.** Note that the polymer backbone of silicones is a silicate-type structure (see, e.g., Section 6.7). The backbone structure is quite resistant to heat (relative to the C—C backbone) and so the silicones are often used because of their heat resistance. They are also good electrical insulators which are much more resistant to dielectric breakdown by electric arcing than are the carbon backbone molecules (which form electrical conducting carbon when arcing occurs). Rotation about silicon-oxygen bonds is freer than is rotation about the carbon-carbon bonds of natural rubber, so the glass transition temperature of silicone rubber is usually much lower. Hence they have good low-temperature properties; i.e., they do not become brittle until the temperature gets very low.

**EXAMPLE 7.7.2**

Although the Si—O bond length is not given in Chapter 5, the O—O and Si—Si bond lengths are. (a) Use this as a method to estimate the Si—O bond length. (b) What is another way to estimate the bond length? (c) Compare with the C—C bond distance. (d) Why is the rotation easier with the silicone backbone than with the carbon backbone?

*Answer.* (a) As an estimate, use the average, 1.83 Å. (b) It is also possible to use the sum of the ionic radii. Hence, from Table 6.4.3 a value of 2.41 Å is obtained. (c) The Si—O bond is longer. (d) Compare the steric hindrance of the two chains, assuming the side groups are the same. Note, however, that there are no side groups on the oxygen atoms in silicones.

Moreover, the silicon-oxygen bond is longer. Hence there is less steric hindrance with the silicon-oxygen backbone which means rotation is easier. Note how structure is related to properties.

**EXAMPLE 7.7.3**

Silicon rubber caps are stretched over steel containers. It is found that at temperatures of $-25°C$, the stress decreases as a function of time and the caps are no longer good seals and even fall off. Why? The glass transition temperature of this material is $-80°C$.

*Answer.* Recall from Figure 7.5.1 that rubber tends to crystallize when highly stretched. When this happens the thermal motion which ordinarily keeps the chains in a random configuration is drastically reduced. Since it is this thermal motion which resists the stretching force, the force drops when this motion ceases.

A silicone treatment for waterproofing is shown in Figure 7.7.3. The oxygen atoms of the chain are attracted to the polar substrate. Ordinarily the polar substrate such as glass, paper, or cloth would readily adsorb water.

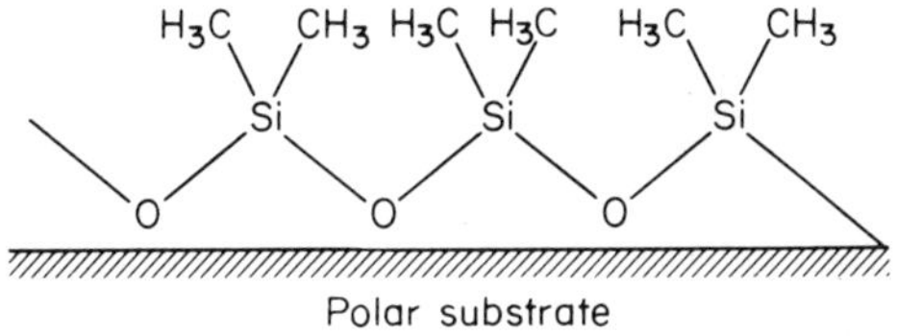

FIGURE 7.7.3. Silicone chain on a polar substrate.

However, the treated surface is now hydrocarbon in character (and hence repels water). This treatment also greatly increases the surface electrical resistivity of glass (an important dielectric) which is usually low because of the adsorbed water.

## Q. 7.7.2

The molecule

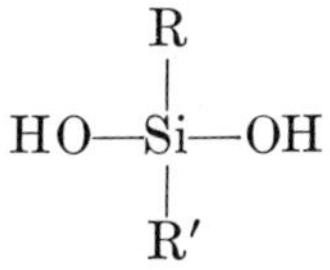

$$\text{HO—Si—OH} \quad \begin{array}{c} R \\ | \\ | \\ R' \end{array}$$

in a condensation reaction produces (a)_____. If R' in this molecule had been replaced by OH, the condensation reaction would produce (b)_____. Silicone rubber has a very low glass transition point compared to elastomers based on a carbon backbone because (c)_____. The (d)_____ represents a polar part of the silicone polymer which can attach to polar substrates.

* * *

## ANSWERS TO QUESTIONS

**7.7.1** (a) Two, (b) copolymer, (c) condensation polymerization, (d) hydrogen bridge or, (e) 7 kbars or 100,000 psi.

**7.7.2** (a) A linear molecule, (b) a three-dimensional molecule, (c) rotation about the bonds is freer inasmuch as the side groups are farther apart, (d) oxygen atom.

## 7.8 NETWORK POLYMERS

Phenol has the structure

$$\begin{array}{c} H \\ | \\ O \\ | \\ C \\ {}^*H{-}C{=}\phantom{C}\phantom{=}C{-}H^* \\ | \phantom{XXXXXX} | \\ H{-}C\phantom{=}\phantom{C}{=}C{-}H \\ C \\ | \\ H \\ * \end{array} \qquad (7.8.1)$$

Asterisks are placed on hydrogens which are reactive. Formaldehyde has the structure

$$\begin{array}{c} {}^*O^* \\ \| \\ C \\ / \phantom{X} \backslash \\ H \phantom{XX} H \end{array} \qquad (7.8.2)$$

Two asterisks are placed on the oxygen which is reactive and has a double bond. Under appropriate conditions of heat, etc., the molecules react as follows:

$$\text{(structure shown)} \tag{7.8.3}$$

with H$_2$O as a by-product. Here the rings with alternating double and single bonds are depicted simply as hexagons.

This reaction can proceed to form a long chain and because one of the mers, phenol in (7.8.1), has more than two reactive points there is also the possibility of reactions at the third reactive point on the phenol. Hence chains can be linked to other chains by primary bonds. The resultant structure is a three-dimensional framework. It is a noncrystalline giant single molecule. It is called a **phenol-formaldehyde plastic, a phenolic,** or by its generic name **Bakelite**. (If the light switches in your room or lecture hall are brown, they are very likely to be Bakelite. Each switch is a single molecule.)

Other important network polymers are the melamines and the epoxies. The epoxies are particularly useful as adhesives which have high strengths at relatively high temperatures (for plastics). Of course, the silicone rubber and the vulcanized polyisoprene rubber discussed previously are also crosslinked, network structures. The imido polymers are networks which have bulk strengths of 3–4 kbars ($\approx$ 50,000 psi) and retain relatively high strengths at high temperatures. The development of high-strength plastics which can be used at higher temperatures is an active area of research.

### Q. 7.8.1

To form a network in a condensation reaction one of the mers must have (a)_____. In the case of phenol-formaldehyde, phenol contains (b)_____ reactive points.

\* \* \*

## ANSWERS TO QUESTIONS

**7.8.1** (a) More than two reactive points, (b) three.

## 7.9 THERMOPLASTICS AND THERMOSETTING RESINS

Some polymeric materials such as polyethylene can be reversibly softened by heating and then conveniently formed by processes such as extrusion. These are called **thermoplastic** materials. The softening process here involves the breaking of *secondary* bonds.

Other polymeric materials once formed cannot be softened and reformed without actual decomposition. These are called **thermosetting** materials. The thermosetting materials have higher strengths at higher temperatures. An example is phenol-formaldehyde or Bakelite. Since this material is a three-dimensional network, *primary* covalent bonds would have to be broken. These softening characteristics are important properties both from the viewpoint of fabricating the materials and using them.

### Q. 7.9.1

A solid polymer which can be softened without degradation is called (a)_____. Network molecules which undergo permanent degradation before softening are called (b)_____.

\* \* \*

## ANSWERS TO QUESTIONS

**7.9.1** (a) A thermoplastic material, (b) thermosetting plastics.

## 7.10 CRYSTALLINITY IN POLYMERS

Polymeric materials range from highly crystalline, ordered arrangements of chains to noncrystalline random chain configurations. Because of the distribution of chain lengths in a linear chain polymer perfect crystals cannot be found. One concept of a polymer showing crystallinity is the **fringed micelle structure** illustrated in Figure 7.10.1. In this case an individual polymer chain may extend through several crystalline regions (each having a "diameter" of about 100 Å) and several amorphous regions.

It is also possible to grow single crystals of polyethylene from dilute solutions. Under these circumstances an individual chain is folded back and forth many times within the same crystal as shown in Figure 7.10.2 [A.

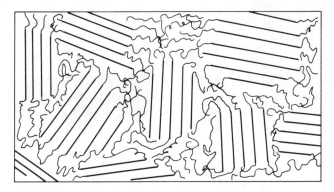

FIGURE 7.10.1. Diagram illustrating the proposed fringed micelle structure of semicrystalline polymers. [After P. J. Flory, *Principles of Polymer Chemistry*, Cornell University Press, Ithaca, N.Y. (1953), p. 49.]

Keller, Philosophical Magazine **2**, 1171 (1957)]. The fold length depends on the conditions present during crystallization. Numerous other polymers have since been grown from dilute solution in this **folded-chain structure.**

At the present time it is thought that the structure of partially crystalline polymers grown from the melt may be somewhere between the fringed micelle structure and the folded-chain structure. Figure 7.10.3 shows

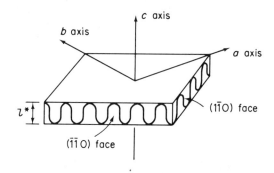

FIGURE 7.10.2. Folded-chain structure for polyethylene crystals.

an example of a structure grown from the melt. Here a crystal or crystals grow out from a point until they come into contact with similar structures and all the material is solidified. These structures are known as **spherulites.**

## Q. 7.10.1

(a) Single crystals of polymers grown from dilute solution tend to have _____. A solid polymer containing many crystals

also tend to have (b)_____
_____.

\* \* \*

FIGURE 7.10.3. Cross-sectional view of spherulites in isotatic polystryene (about halfway through the crystallization process) viewed under polarized light. (× 900.) (Courtesy of H. D. Keith, Bell Telephone Laboratories, Murray Hill, N.J.)

## ANSWERS TO QUESTIONS

**7.10.1** (a) The folded-chain structure, (b) amorphous regions between the crystals with molecules extending from one crystal to the other.

### 7.11 MACROMOLECULES IN LIVING MATTER

Cellulose in plants or cotton and also skin, hair, hoofs, toenails, muscle, and tendons in animals are all examples of organic polymers. These naturally occurring polymers are important to us all.

CELLULOSE. The building unit for cellulose can be considered to be glucose:

$$\text{(glucose ring structure)} \qquad (7.11.1)$$

The polymer cellulose is a linear chain of glucose molecules

$$\text{(cellulose chain structure)} \qquad (7.11.2)$$

with a degree of polymerization of 1800–3500.

The linear polymer chains of cellulose pack together surprisingly well and there is strong secondary bonding between the chains due to hydrogen bonding. Cellulose fibers are therefore very strong having a tensile stress in excess of 7 kbars ($\approx 100,000$ psi).

**EXAMPLE 7.11.1**

Explain the process of ironing a cotton shirt.

*Answer.* The shirt is first moistened. The water tends to break the hydrogen bonds between the cellulose chains, forming its own hydrogen bonds to the chains. This weakens the fiber and makes it much more flexible and plastically deformable. The hot iron shapes the cloth and causes the water to evaporate so new hydrogen bonds are formed between the chains.

Starch is formed from a different isomer of glucose and in addition is highly branched. Hence the packing is poor with the result that starch swells readily in water.

We note that each glucose unit in the cellulose chain has three —OH groups. A number of cellulose derivatives can be made by reactions with one or more of these —OH groups. This includes cellulose acetate (rayon) and nitrocellulose (gun cotton). The degree of polymerization decreases when these compounds are formed. One of the first commercial plastics was celluloid, which is nitrocellulose plasticized with camphor; it was first produced and sold by John Wesley Hyatt in 1868.

### Q. 7.11.1

The mer for cellulose is (a)_____, although the polymer is not manufactured commercially. Cellulose fibers are very strong because of the presence of (b)_____ atoms which are available for (c)_____. In the production of cellulose acetate one or more of the —OH groups are replaced by (d)_____ groups upon reaction with acetic anhydride. The resultant polymer is soluble in certain solvents and the viscous solutions can be extruded to form fibers, sheets, etc.

* * *

PROTEINS. $\alpha$-amino acids have the structure

$$\begin{array}{c} H \quad\quad H \quad O \\ \diagdown \quad | \quad\quad \| \\ N-C-C-OH, \\ \diagup \quad | \\ H \quad\quad R \end{array} \quad\quad (7.11.3)$$

where R, an alkyl group, may be H, $CH_3$, etc. Condensation reactions between two amino acids are possible, resulting in a **peptide linkage:**

$$\begin{array}{c} H \quad O \quad\quad H \quad O \\ | \quad \| \quad\quad | \quad \| \\ H_2N-C-C-N-C-C-OH. \\ | \quad\quad | \quad | \\ R \quad\quad H \quad R \end{array} \quad\quad (7.11.4)$$

Obviously the condensation process can continue with the same or other $\alpha$-amino acids to yield a **polypeptide chain.** A **protein** is a polypeptide chain with a specific order of amino acids. If the alkyl groups are small, the protein chain will tend to form the planar zigzag configuration (silk takes on this configuration). There is then excellent hydrogen bonding between the chains (intermolecular) as illustrated in Figure 7.11.1. Bulky side groups cause steric hindrance and lessen the possibility of strong hydrogen bonds between chains in the extended zigzag configuration. These polypeptide chains with large side groups twist into helices in which case there is strong

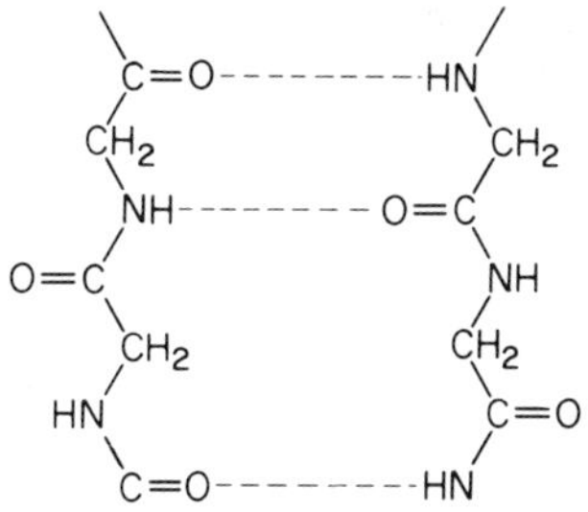

FIGURE 7.11.1. Hydrogen bonding between chains in polyglycine. About one half of the amino acids (on a mole basis) of silk is glycine.

intramolecular hydrogen bonding within the chains with the bulky R groups extending outward. Wool (hair) has such a structure. Moreover, one of its important mers is the di-$\alpha$-amino acid, cystine,

$$\text{HOOC}-\underset{\underset{H}{|}}{\overset{\overset{NH_2}{|}}{C}}-CH_2-S-S-CH_2-\underset{\underset{H}{|}}{\overset{\overset{NH_2}{|}}{C}}-\text{COOH}. \qquad (7.11.5)$$

This makes possible the formation of a primary bonded crosslink between chains (somewhat like the sulfur-sulfur crosslink in rubber). Dry wool, because of the strong hydrogen bonds, is a hard plastic; wet wool, on the other hand, can be stretched elastically about 80% and wool soaked in formic acid, which breaks hydrogen bridges, behaves in a rubber-like fashion.

## Q. 7.11.2

$\alpha$-amino acids have an organic acid group attached to a carbon and (a)_____ group attached to the same carbon. Condensation reactions involving $\alpha$-amino acids lead to (b)_____. A protein is defined as (c)_____. Although silk, which has small side groups forms (d)_____, most proteins tend to form (e)_____.

\* \* \*

**EXAMPLE 7.11.2**

Discuss the "home permanent."

*Answer.* If the S—S bonds in hair are broken by a suitable chemical agent the protein chains can readily slide over each other during the deformation process of curling (or straightening as the case may be). The S—S bonds can then be reformed in new positions which gives the hair fiber a new permanent configuration.

DNA. DNA, the "heredity" molecule, is a double helix of two sugar-phosphate backbones twisted onto each other. The details of how this structure was obtained are given in the book of the Nobel prize winner James D. Watson [*The Double Helix,* The New American Library, Inc., New York (1969)], who was responsible in part for determining its structure.

## ANSWERS TO QUESTIONS

7.11.1 (a) Glucose, (b) oxygen and hydrogen, (c) hydrogen bridges, (d) acetate.

7.11.2 (a) An amine or —$NH_2$, (b) polypeptide linkages, (c) a polypeptide chain with a specific sequence of amino acids, (d) the extended zigzag configuration, (e) helices.

## REFERENCES

Mark, H. F., "The Nature of Polymeric Materials," *Scientific American* (September 1967), p. 148. An hour of reading (plus perhaps more meditation) to establish contact with a lifetime of experiences in the polymer area is suggested.

Alfrey, T., and Gurnee, E. F., *Organic Polymers*, Prentice-Hall, Inc., Englewood Cliffs, N.J. (1967). This is a nice little monograph useful for structure studies and for study of viscoelasticity.

Geil, P. H., *Polymer Single Crystals*, John Wiley & Sons, Inc. (Interscience Division), New York (1963).

Frazer, A. H., *High Temperature Resistant Polymers*, John Wiley & Sons, Inc., New York (1968).

Brydson, J. A., *Plastic Materials*, Van Nostrand, Reinhold Company, New York (1966). The structure of molecules is nicely related to chemical, mechanical, electrical, and optical behavior. There is also extensive discussion of specific polymers.

Winding, C. C., and Hiatt, G. D., *Polymeric Materials*, McGraw-Hill Book Company, Inc., New York (1961). This is an interesting book concerned with the chemical production of polymers.

Rodriguez, R., *Principles of Polymer Systems*, McGraw-Hill Book Company, Inc., New York (1970). This book provides an excellent study of the chemistry of polymers and the physical principles of polymer behavior.

Rosen, S. L., *Fundamental Principles of Polymeric Materials*, Barnes and Noble, Inc., New York (1971). This book provides an appreciation of those fundamental principles of polymer science which are of practical significance.

# PROBLEMS

7.1 Make a list of ten "why" questions which you had about polymers. Example: Why does high temperature cause freer rotation?

7.2 (a) Give three examples of vinyl polymers.
(b) Why are vinyl compounds with different side groups of interest?

7.3 Discuss the different kinds of topology along a chain that is possible with polypropylene.

7.4 (a) What is meant by 1–4 addition?
(b) What feature of polyisoprene readily distinguishes it from polypropylene?

7.5 A polymer chain with 10,000 links achieved a random configuration in solution. If you were betting on the distance between the ends, what would be your choice?

7.6 Is it possible, in a direct fashion, to measure the molecular weight of propane? How?

7.7 What methods are used to measure the average molecular weight of polymers?

7.8 List five different polymers known from your everyday experience and suggest why each was used.

7.9 Ordinary vulcanized polyisoprene does not make a good gasoline hose. However, neoprene, a polymer whose mer is very much like isoprene, except that the —$CH_3$ side groups on the second carbon are replaced by —Cl, makes good gasoline hoses. Explain why.

7.10 What must be true of one of the mers if a condensation polymer is to form a framework structure?

7.11 (a) Give some reasons for the high tensile strength of nylon fibers.
(b) Why is bulk nylon not so strong?

7.12 Discuss the use of silicone for increasing the surface resistivity of glass.

**7.13** Give an example of a situation for which a thermoplastic might be preferable to a thermosetting material.

**7.14** (a) Why is it topologically impossible (as a rule) to have perfectly crystalline polymers?
(b) Discuss the two models of polymer crystallinity.

**7.15** Give some examples of macromolecules in plants and animals.

**7.16** Why is moist hair rubber-like?

## MORE INVOLVED PROBLEMS

**7.17** You are interested in developing a polymer for a container which is sunlight degradable but has a reasonable shelf life. (It appears easier to do than to civilize people.) What requirements do you have to meet?

**7.18** From what you know of the structure of rubber, sketch the stress-strain curve for a sample which is pulled at a constant speed until strain reaches 500% and is then released at the same speed.

**7.19** Devise a method for measuring the osmotic pressure of a solution.

**7.20** Devise a method for measuring the viscosity of a solution with a
(a) Low viscosity.
(b) High viscosity.

**7.21** The root mean squared length of a random one-dimensional chain (the links either add or subtract a length $l$ along a straight line when added) is given by

$$\sqrt{\overline{x^2}} = \sqrt{\overline{(l_1 + l_2 + l_3 + \cdots l_N)^2}}$$

If each $l_i$ is $\pm l$, show that

$$\sqrt{\overline{x^2}} = \sqrt{N}\, l.$$

## SOPHISTICATED PROBLEMS

**7.22** Write an essay on the measurement of viscosity, including the following types of instruments:
(a) Ostwald.
(b) Saybolt.
(c) Couette.

**7.23** Discuss quantitatively how the Newtonian viscosity coefficient is obtained experimentally. See, e.g., G. W. Scott-Blair, *A Survey of General and Applied Rheology*, Pitman Publishing Corporation, New York (1944).

**7.24** Discuss how measurements of viscosity coefficient can be used to obtain molecular weight. See, e.g., H. Tompa, *Polymer Solutions*, Academic Press, Inc., New York (1956).

# Prologue

In some applications single crystals are used in engineering practice. Thus microelectronic circuitry is built into a single silicon crystal, single piezoelectric quartz crystals are used as transducers and delay lines, while chromium-doped sapphire single crystals are used in the ruby laser. However, real engineering materials are usually not single crystals. Rather they are *polycrystalline aggregates*. In the simpler cases only one kind of crystal is present. The size and shape of the crystals which make up the aggregate can be studied by *metallography*. More complex materials involve mixtures, such as crystals of $Fe_3C$ mixed with crystals of bcc iron. These are known as *polyphase materials*. The size, shape, and orientation of these particles play a major role in determining the magnitude of many of the properties of a material. Such structures can be obtained by metallurgical processing and by mechanical means. Controlled structures obtained by mechanical means are called *composites*. Achieving controlled microstructures by either of these methods is an important part of materials science. From a practical viewpoint, this is where much of the payoff of materials research lies.

# 8

# MICRO- AND MACRO-STRUCTURE

## 8.1 SINGLE CRYSTALS

Single crystals are rare, which is one reason you see them in museums. They are not only rare in nature but they are not easily grown in the laboratory either; that is, special care has to be taken to get single crystals. Thus when a casting is made in a foundry, you can be rather sure that it will not be a single crystal.

Crystals may be **grown from solutions** which are supersaturated as shown in Figure 8.1.1. The solubility of the solute decreases with decreasing temperature, so that solute which is dissolved at the bottom of the vessel where the temperature is higher will be deposited on the seed crystal at the top where the temperature is less. The temperature difference causes convection currents which carry the dissolved materials upward. Large quartz crystals weighing several kilograms are grown this way, although in this case the procsss is carried out under hydrostatic pressure as well. Growth of such large crystals usually takes days. The quartz crystal of Figure 6.1.1 was grown this way.

Another method of growth called the **Bridgman method** does not involve a solvent. Rather the material is melted in a crucible which is slowly lowered from the furnace (perhaps 1 cm/hr) as shown in Figure 8.1.2.

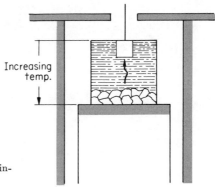

FIGURE 8.1.1. Growth of a single crystal from solution.

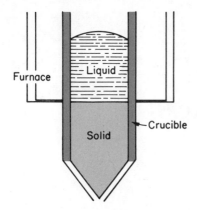

FIGURE 8.1.2. Bridgman method of growing single crystals. The crucible is lowered slowly from the furnace so that material at the tip solidifies first. The lowering continues slowly so that heat transfer takes place in a vertical direction.

A third method involves partially dipping a seed crystal into the melt and then slowly pulling the seed from the melt (perhaps 1 cm/hr). This is called **pulling from the melt.** Examples of crystals grown in this way are shown in Figure 8.1.3. Silicon crystals for microelectronic circuits are grown this way. High-purity NaCl crystals are also made by this process. Note the rather slow rates needed to form single crystals ($\approx 1$ cm/hr). This would surely not be a satisfactory rate for producing 100 million tons of material a year.

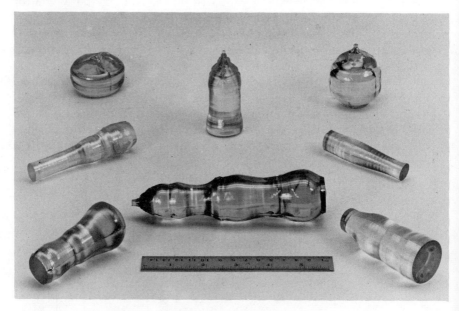

FIGURE 8.1.3. Single crystals grown by pulling from the melt. (Courtesy of the Isomet Corporation, Palisades Park, N.J.)

FIGURE 8.1.4. Sapphire whiskers (aluminum oxide).

The growth of crystals is an interesting and active area of research. One of the areas of single crystal growth of great interest is whisker growth. We noted earlier (see Table 6.6.1) that whiskers are often extremely strong. Hence they have great potential as reinforcements for weaker materials. An example of whisker growth is shown in Figure 8.1.4.

### Q. 8.1.1

When growing crystals from solution, a seed crystal is suspended in (a)_____ solution. Seed crystals are also used in growing crystals directly from the melt; the method used is called (b)_____. Another method of growing crystals called the (c)_____ involves lowering a pointed crucible containing the melt from a furnace. The growth of crystals is quite slow, a typical rate being about (d)_____.

\* \* \*

## ANSWERS TO QUESTIONS

**8.1.1** (a) A supersaturated, (b) pulling from the melt, (c) Bridgman technique, (d) 1 cm/hr.

## 8.2 THE REFLECTION MICROSCOPE

We have already noted that materials can be separated into crystalline and amorphous solids. These solids are distinguished from each other by the

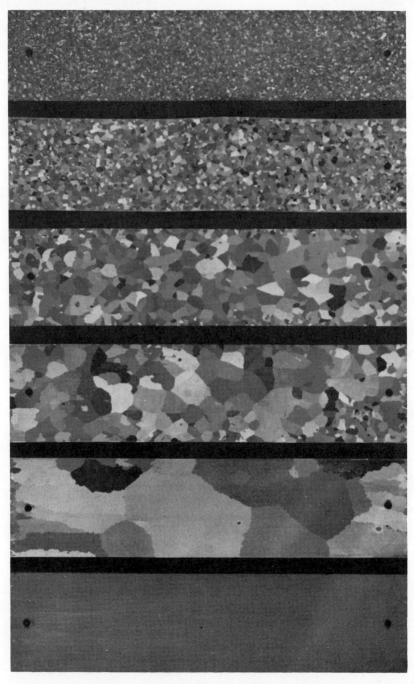

FIGURE 8.2.1. Micrograph of aluminum specimens. The crystals are much larger than would usually be found in commercial alloys. (×1.) (Courtesy of Stephen Sass, Cornell University, Ithaca, N.Y.)

Sec. 8.2                                        *The Reflection Microscope*     341

degree of order of the atomic arrangements, i.e., on the **atomic scale**. There are numerous techniques for studying structure on this scale: X-ray, electron and neutron diffraction, field ion microscopy, etc. In this chapter, we are concerned more with arrangements on a somewhat larger scale, e.g., on a scale where the optical microscope (with a resolution of about $10^{-4}$ cm) and the electron microscope (with a resolution of down to $10^{-7}$ cm) can be used for observation. The structure of materials on these levels is called **microstructure**. We shall also be concerned with different arrangements which are visible to the naked eye. This we call **macrostructure**.

Figure 8.2.1 shows the polished and etched surface of aluminum as seen in the (optical) reflection microscope (also called the metallurgical microscope), shown in Figure 8.2.2. The solid is a collection of single crys-

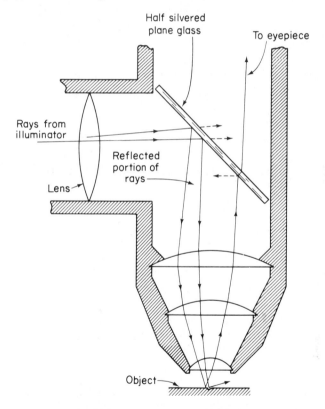

FIGURE 8.2.2. Path of light rays in a metallurgical microscope. The light from a bright point source enters through a lens at the left and is reflected downward onto the specimen. If the surface of the specimen is perpendicular to the axis of the microscope, it reflects the light to the eyepiece, with maximum intensity. If the surface is not perpendicular, the rays are deflected and produce less effect at the eyepiece. [From B. A. Rogers, *The Nature of Metals*, The M.I.T. Press, Cambridge, Mass. (1965).]

tals (of aluminum in this case) joined together at grain boundaries, and is called a **polycrystalline** material. Each crystal is essentially the same except for orientation, size, and shape. Because of their different orientations, different crystal planes coincide with the surface. (The surface is prepared by grinding, polishing with a wet slurry of very fine abrasive, and then etching. The different crystal planes react differently with chemical **etchants** which attack the surface.) This results in each individual crystalline surface showing a different reflectivity. The resultant photograph of a surface such as shown in Figure 8.2.1 is called a **micrograph.**

Figure 8.2.2 shows how a **reflection microscope** reveals the presence of a grain boundary after etching. We have noted in Chapter 6 that the atoms in a grain boundary are not packed as perfectly as those within the crystal; hence their bonding is not as good (this is the origin of the grain boundary energy). Because of this, grain boundaries tend to etch faster than the rest of the surface. Figures 1.2.10 and 1.2.11 show the presence of grain boundaries etched in this way.

The study of the surface of materials by use of the reflection microscope is called **metallography.** (Metallography originally referred to the study of metals only, but now refers to the study of all classes of materials.) Metallographic techniques can be used for various aspects of structural characterization: grain size and shape, precipitate distribution, slip bands formed during plastic deformation, cracks formed during fatigue, voids formed during creep, voids or bubbles formed during radiation damage, ferroelectric domain size and shape, magnetic domain size and shape, etc. Some examples will be considered now.

### Q. 8.2.1

The internal structure of materials at the level studied with the optical or electron microscope is called (a)_____. One of the ways of observing grain boundaries is to carefully polish a specimen and then to (b)____ it and observe it with the (c)_____ microscope. This process of examining microstructure is called (d)_____.

\* \* \*

FERROMAGNETIC DOMAINS. By placing a colloidal suspension of fine iron particles on a ferromagnetic material and observing these particles in a microscope, the experimentalist finds that the iron particles delineate certain regions as shown in Figure 1.2.9. This is called a **Bitter pattern.** Within each little region in Figure 1.2.9, the material is magnetized in a specific direction as shown by the arrows; this region is called a **ferromag-**

netic domain. Within a domain, the magnetic dipoles of the atoms are all aligned in one direction. A large piece of unmagnetized ferromagnetic material has a huge number of domains, but with different orientations, such that the net magnetization of the sample is zero. The use of the reflection microscope makes possible the observation of the behavior of these domains as shown in Figure 1.2.9. Note that the boundaries between the domains (**domain boundaries** or **domain walls**) move as the materials become magnetized. Based on this knowledge, made possible by the use of the reflection microscope, what should the materials scientist do to make a hard magnetic material? The answer is, prevent the motion of domain boundaries after the material is magnetized. Similarly, to make a soft magnetic material, it is necessary to make the motion of domain boundaries easy. The structural factors which control this domain wall motion are discussed in a later chapter.

## Q. 8.2.2

A region which has the same magnetization direction within a crystal is called (a)_____. Such regions can be made observable in the microscope by use of (b)_____. By observing the Bitter patterns during the application of magnetic fields, you would note that there is (c)_____. To make a hard magnetic material, you would want to (d)_____.

\* \* \*

NONPOROUS ALUMINUM OXIDE. Figure 8.2.3 is a micrograph of sintered aluminum oxide (the oxide powder is pressed together and then held at a high temperature). Note the porosity within the grains (it appears that there is much porosity in the grain boundaries; actually this is not the case, but, because the material is brittle, chipping occurs at the grain boundaries during polishing). The presence of these pores *within* the grains causes the material to be opaque as a result of light scattering at the pore surfaces. Otherwise, the material would be translucent (nearly transparent). What is apparent is that there are regions near the grain boundaries where there are few, if any, voids.

Based on this knowledge of structure, made possible by the reflection microscope, what should the materials scientist do to eliminate porosity and hence render the material translucent? One solution used by J. K. Burke and R. L. Coble: Keep the crystal size small [see R. L. Coble, Journal of Applied Physics **32**, 787 (1961)]. Usually during sintering, grain boundaries

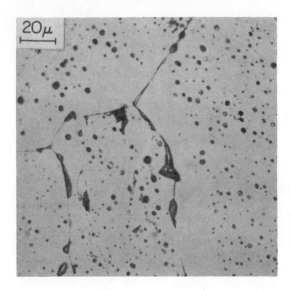

FIGURE 8.2.3. Sintered aluminum oxide. Note the pores within the solid and regions near the grain boundary without pores. The apparent big pores on the grain boundaries are not pores at all but are caused by chipping of the brittle material during polishing. (Courtesy of the General Electric Company, Schenectady, N.Y.)

move and crystal growth occurs. To prevent crystal growth an impurity, MgO, in this case, can be added in small amounts. This does not affect other properties. Figure 8.2.4 shows the resultant finer-grained structure. Note the absence of porosity. Once again the understanding of structure, made possible by the reflection microscope, plus the understanding of the fundamental properties of materials has led to the production of materials with the desired properties. This is molecular architecture.

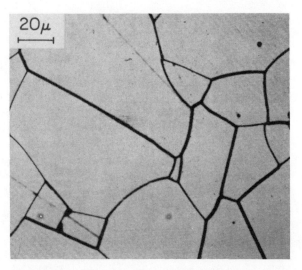

FIGURE 8.2.4. MgO-doped sintered $Al_2O_3$. The material is finer grained and shows essentially no porosity.

## Q. 8.2.3

Ordinary sintered alumina is opaque because of (a)_____
_____. Microscopic examination shows that near grain boundaries there are (b)_____. This suggests that the way to prepare nonporous material is to (c)_____
_____.

\* \* \*

FERROELECTRIC DOMAINS. X-ray diffraction studies show that barium titanate ($BaTiO_3$) has the cubic crystal structure shown in Figure 8.2.5 above 393°K. An examination of the radius ratios in Table 6.4.3 shows

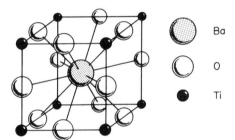

FIGURE 8.2.5. Crystal structure of $BaTiO_3$ above 393°K.

that $r_{Ti^{4+}}/R_{O^{2-}} = 0.55$, so that from Pauling's rules, we expect octahedral coordination of oxygen ions about the titanium ion, as is found. The barium ion is surrounded by 12 oxygen ions. When the temperature is lowered below 393°K the crystal structure changes to a tetragonal structure (the cell angles are still 90 deg but one length, $c$, is different from the remaining two) with a $c/a$ ratio = 1.04. We note that a cube edge consists of alternating $Ti^{4+}$ and $O^{2-}$ ions. When the transformation occurs (say the crystal stretches in the [001] direction), the $Ti^{4+}$ ions on this edge move upward by 0.06 Å while the $O^{2-}$ ions move downward by 0.09 Å (assuming that the $Ba^{2+}$ ion remains at the center of the tetragonal cell). [See G. Shirane, F. Jona, and R. Pepinsky, *Proceedings of the I.R.E.* **43**, 1738 (1955).] Assume this happens along all the edges in the [001] direction only. The crystal now contains numerous electrical dipoles along the [001] direction and is therefore polarized along this direction, i.e., along the $c$ axis. Earlier we had considered molecules such as the HCl molecule and the NaCl molecule (in the gas phase) which have permanent dipoles. The tetragonal crystal of $BaTiO_3$ also has a permanent dipole.

The actual transformation from cubic to tetragonal crystals is not a uniform transformation such as we have considered so far. Rather there are some regions of the original crystal which become polarized in the [100] direction, other regions which become polarized in the [$\bar{1}$00] direction, others in the [010] direction, etc. These regions are called **ferroelectric domains** and the boundaries between the domains are known as **domain walls**. Examples of ferroelectric domains obtained by microscopic examination are shown in Figure 8.2.6. The total polarization may be zero because it is

No field applied  +2000 volts/cm

Field removed  −2000 volts/cm

FIGURE 8.2.6. Ferroelectric domains made visible using polarized light.

equally likely that domains form along any of the $\langle 100 \rangle$ directions. When an electrical field is applied, the domains whose polarization are most nearly parallel to the electrical field have the lowest potential energy. Thus these domains grow at the expense of the others (which results in a lowering of potential energy). This involves the motion of domain walls, a process which does not take place reversibly; i.e., energy is dissipated when the domain walls move. Hence, a polarization versus electric field hysteresis loop as shown in Figure 3.5.1 is obtained.

### EXAMPLE 8.2.1

Would a single domain of tetragonal barium titanate be piezoelectric or not?

*Answer.* To be piezoelectric, a crystal must have no center of symmetry. Clearly, tetragonal $BaTiO_3$ has no center of symmetry (if it did, it could not be permanently polarized). Hence, it is piezoelectric, as are *all* ferroelectric materials.

### Q. 8.2.4

When $BaTiO_3$ is cooled below 393°K, it has a tetragonal crystal structure as can be shown to be the case by (a)_____. Because

of the manner in which the $O^{2-}$ and $Ti^{4+}$ ions move when the tetragonal structure is formed, (b)_____ are formed. However, the material shows (c)_____ when the transformation occurs in the absence of a field because it is equally likely that domains form along (d)_____.
Because energy is dissipated in ferroelectric domain wall motion, there is a (e)_____.

\* \* \*

## ANSWERS TO QUESTIONS

**8.2.1** (a) Microstructure, (b) etch, (c) reflection, (d) metallography.

**8.2.2** (a) A ferromagnetic domain, (b) colloidal suspensions of ferromagnetic materials, (c) motion of the domain wall boundaries, (d) pin the domain boundaries so that they cannot move or can move only with difficulty.

**8.2.3** (a) The presence of pores, (b) no pores, (c) prepare a fine-grained structure so that the center of a grain is also close to the boundary.

**8.2.4** (a) X-ray diffraction; (b) permanent electric dipoles; (c) no net polarization; (d) any of the [100], [$\bar{1}$00], [010], [0$\bar{1}$0], [001], and [00$\bar{1}$] directions; (e) polarization versus electric field hysteresis loop.

### 8.3 POLYCRYSTALLINE MATERIALS

While single crystals are often used in industry and research, the large bulk of commercial crystalline materials are polycrystalline. The reflection microscope can provide valuable data about such materials, e.g., the shapes of the **grains** (crystals in a polycrystalline aggregate) and their sizes. Grain boundaries are imperfections in crystalline materials and they increase the energy of the system above that of the single crystal. Under true equilibrium conditions grain boundaries would not be present in a single-phase crystalline material. Usually, however, as a result of the manner of solidification and other prior history grain boundaries are present; they often take up a **metastable** configuration in which it would be necessary for the material to pass through an intermediate state of higher energy before reaching true equilibrium (single crystal). An analog for a mechanical situation is shown in Figure 8.3.1. Such a metastable configuration for crystals which extend all the way through the specimen, in which we have assumed that all the grain boundaries have equal surface energy per unit area is shown in Figure 8.3.2; note that the angle between any two intersecting boundaries is 120 deg. Any small perturbation in the shape of some of these grains as

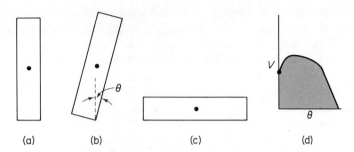

FIGURE 8.3.1. (a) Metastable equilibrium. (b) Intermediate state. (c) Stable equilibrium position. (d) Potential energy as a function of $\theta$. Note the maximum.

FIGURE 8.3.2. Grain boundaries extending through a sheet.

shown in Figure 8.3.3 would lead to an *increase* rather than a decrease in surface energy since it would lead to an increase in grain boundary area (prove this).

FIGURE 8.3.3. One grain attempts to grow at expense of neighboring grains. Dashed lines show the new grain boundaries.

The corresponding situation is more involved in the general bulk case because there is no regular polyhedron with plane faces which when packed together fills all space and which also meets the requirements of minimizing surface energy. The tetrakaidecahedrons (14 faces) of Figure 8.3.4 approxi-

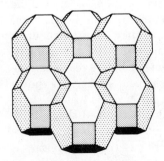

FIGURE 8.3.4. Stack of tetrakaidecahedrons. [From *Metal Interfaces*, American Society for Metals, Cleveland (1952).]

mately satisfy these conditions (the sphere, of course, has minimum surface area for volume but packed spheres do not fill space). The introduction of slight curvature in the faces allows the conditions for metastable equilibrium to be met.

Figure 8.3.5 shows grains of a Ti alloy which have a striking similarity to the "grains" of Figure 8.3.4.

FIGURE 8.3.5. These are individual grain groupings which parted from an arc-cast Ti alloy billet under the blows of a hammer. These fragments have preserved the true bounding facets of the individual grains which, belonging to a cast structure, are unusually large. Such perfect intercrystalline cleavage is rather rare and is usually associated with a thin intercrystalline film of a low-melting, liquid phase. ($\times \frac{1}{3}$.) [From W. Rostoker and J. R. Dvorak, *Interpretation of Metallographic Structure*, Academic Press, Inc., New York (1965).]

### Q. 8.3.1

Grain boundaries within crystals (a) __(are or are not)__ equilibrium structures. A polycrystalline aggregate has a (b) __(higher or lower)__ energy than the same quantity of single crystal. Grain boundaries within a polycrystalline aggregate often have (c)_____ configuration. We say that a book lying flat on a table has (d)_____ configuration while a book standing on end on a table has (e)_____ configuration.

\* \* \*

An important parameter in defining grains is the grain size. The **ASTM grain size** $N$ is defined as follows: the number of grains in a square inch of micrograph whose magnification is 100 (linear) is $2^{N-1}$. See Table 8.3.1. Grains which have nearly the same "diameter" $D$ in all directions are called **equiaxed** grains.

Many of the properties of materials are closely related to the grain size, shape, and orientation. For example, the creep of equiaxed fine-grained MgO at high temperatures and low stresses is proportional to $D^{-2}$, where $D$ is the grain diameter. Hence a large grain size would be desirable to reduce the creep rate. For other properties small grain sizes are best; the fracture stress of many polycrystalline materials which are brittle is

**Table 8.3.1. ASTM GRAIN SIZE***

| ASTM Number | Grains/mm$^2$ | Grains/mm$^3$ |
|---|---|---|
| −3 | 1 | 0.7 |
| −2 | 2 | 2 |
| −1 | 4 | 5.6 |
| 0 | 8 | 16 |
| 1 | 16 | 45 |
| 2 | 32 | 128 |
| 3 | 64 | 360 |
| 4 | 128 | 1,020 |
| 5 | 256 | 2,900 |
| 6 | 512 | 8,200 |
| 7 | 1,024 | 23,000 |
| 8 | 2,048 | 65,000 |
| 9 | 4,096 | 185,000 |
| 10 | 8,200 | 520,000 |
| 11 | 16,400 | 1,500,000 |
| 12 | 32,800 | 4,200,000 |

* From *Metals Handbook*, American Society for Metals, Cleveland (1948).

proportional to $D^{-1/2}$. Orientation of the grains can also be very important since many properties of single crystals are highly anisotropic. It is often desirable to have a specimen with **preferred orientation,** e.g., to have the $\langle 100 \rangle$ directions of crystals of cubic iron lie primarily along magnetic transformer sheet and also lie normal to the sheet.

### Q. 8.3.2

If the grains in a polycrystalline aggregate have the same cross section in all directions, they are called (a)_____. Assuming that the ASTM grain size of 12 is for the smallest grains found, what is the "diameter" of such a grain in microns? (b)_____. If the crystals in a polycrystalline aggregate tend to have a given axis lying along a specific direction (rather than randomly oriented) we say that the material exhibits (c)_____.

\* \* \*

CASTINGS. We have already mentioned that polycrystallinity can be a result of solidification. Figure 8.3.6 shows a macroetched casting. Thermal effects are very important in solidification because when a liquid freezes a large amount of heat (latent heat of solidification) is given off. This heat must be carried away either by conduction through the solid or by conduction and convection in the liquid. Many equiaxed crystals begin to grow at the surface of the mold (into which the liquid was poured) as shown

FIGURE 8.3.6. Typical ingot structure; a macroetched cross section through an iron-silicon ingot. [From Cecil H. Desch, *Metallography*, Longmans, Green & Co. Ltd., London (1937).]

in Figure 8.3.7, because the walls are initially cold and the liquid which comes in contact with these walls is rapidly supercooled. A large supercooling causes a rapid nucleation (initial formation) of crystals. (We shall study later on why this is so.)

The subsequent growth of these crystals is controlled by two factors: motion of the atoms (or molecules) from the liquid to an appropriate site on the crystal surface (which may be on the revolving step of a screw dis-

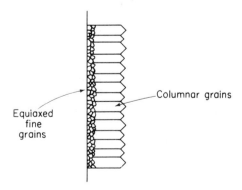

FIGURE 8.3.7. Crystals on surface of mold.

location) and the transfer of the heat released upon freezing. After the initial layers of equiaxed crystals are formed, the subsequent solidification is controlled by the transfer of heat through the equiaxed region to the walls. Because this heat transfer is perpendicular to the walls, growth of each grain is now inward and **columnar grains** are formed as shown in Figure 8.3.7.

The description of the solidification process in a casting is very important industrially and is often studied in detail in advanced materials science and engineering courses. One of the important processes studied there is the formation of **dendrites,** skeleton-like or Christmas tree-like crystals whose formation depends on the details of the rate of heat transfer from and material transfer to a growing branch. An example is shown in Figure 8.3.8.

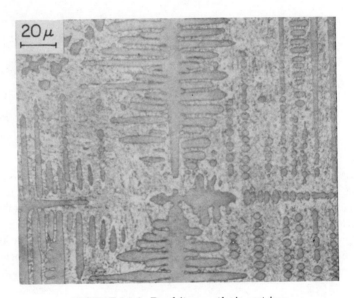

FIGURE 8.3.8. Dendrite growths in cast iron.

### Q. 8.3.3

To produce castings at a reasonable rate, molten material is poured into a mold which is at room temperature and allowed to cool. The liquid next to the walls is chilled very rapidly and forms (a)_____ _____. After this is formed (b)_____ grains are formed. The Christmas tree-like growths found in castings are called (c)_____.

\* \* \*

# ANSWERS TO QUESTIONS

**8.3.1** (a) Are not, (b) higher, (c) a metastable, (d) an equilibrium, (e) a metastable.

**8.3.2** (a) Equiaxed grains, (b) about 5 $\mu$, (c) preferred orientation.

**8.3.3** (a) A fine-grained equiaxed structure, (b) columnar, (c) dendrites.

## 8.4 POLYPHASE MATERIALS

Polycrystalline materials such as those discussed in the previous section may have been single-component materials such as pure aluminum or a multicomponent material (an **alloy**) such as 70:30 $\alpha$-brass (70 wt. % Cu–30 wt. % Zn) which is a substitutional solution of Zn in fcc Cu. Such materials are single-phase materials. However, the bulk of engineering materials consists of more than one solid phase.

A **phase** is the material in a region of space which in principle can be mechanically separated from other phases. Thus if we have pure $H_2O$, we may under certain conditions have the three phases present: liquid, solid, and gas. The student should not, however, confuse phase with the state of aggregation, meaning liquid, solid, or gas. Thus homogenized milk has two liquid phases, globules of fat suspended in skim milk. Note that a phase need not be pure; i.e., a phase may have several components.

Consider an alloy of iron containing 0.8 wt. % carbon at 850°C. Here only a single phase exists. The phase has the fcc crystal structure but with carbon randomly distributed in the octahedral voids. When this material is cooled below 723°C, say to 600°C, it forms a **lamellar structure** (a layer-like structure) of two phases as shown in Figure 8.4.1 and this structure persists at room temperature where the metallographic examination is made. One phase consists of bcc iron with only a tiny amount of dissolved carbon, while the other phase consists of the compound $Fe_3C$.

FIGURE 8.4.1. Lamellar structure in steel showing alternate layers of $\alpha$-iron (nearly pure bcc iron with some carbon dissolved) and the compound $Fe_3C$. (×1000.)

The yield stress of this steel increases as the spacing between the $Fe_3C$ layers decreases which is one reason for being concerned with the details of this microstructure. Methods to manipulate microstructure, in this case lamellae spacing, will be studied in later chapters. So will the reason that the yield stress is affected.

Another example of a lamellar structure is shown in Figure 1.2.13. This is the structure obtained when a liquid of 40 wt. % Pb and 60 wt. % Sn solidifies rapidly. An alloy of this composition is the common solder used for making solid electrical connections.

Polyphase materials are not restricted to crystalline solids. Figure 8.4.2 shows a glass which has separated into two phases. In general, the

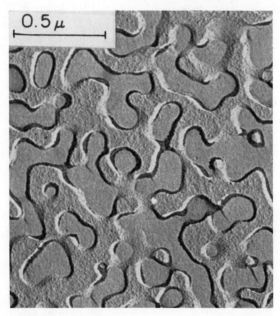

FIGURE 8.4.2. Replica electron micrograph of a thermally aged borosilicate glass showing the development of borate-rich and silica-rich phases. (Courtesy of F. E. Wagstaff and R. J. Charles, General Electric Company, Schenectady, N.Y.)

second phase may be present as rods, cubes, or spheres or in much more complicated configurations. The size, shape, orientation, and distribution of these second-phase particles are often *very* important in determining the magnitudes of various properties such as yield stress, coercive force of hard magnetic materials, and the current-carrying capacity of hard superconductors.

## Q. 8.4.1

When 0.8 wt. % carbon is added to iron, the resultant material is called (a)_____. This material has (b)____ component(s). At 850°C, all the carbon is present in interstitial positions in the fcc iron; there is (are) (c)⎯⎯ phase(s) present. When this material is cooled slowly to room temperature, bcc iron with some carbon dissolved is present and there are also crystals of Fe₃C; there are now (d)⎯⎯ phases present. The Fe₃C plates alternate with impure bcc α-iron plates to form (e)⎯⎯⎯⎯⎯⎯⎯⎯⎯⎯. The yield stress of the steel is affected by (f)⎯⎯⎯⎯⎯⎯⎯⎯⎯⎯.

\* \* \*

## ANSWERS TO QUESTIONS

**8.4.1** (a) An alloy, (b) two, (c) one, (d) two, (e) a lamellar structure, (f) the lamellar spacing.

### 8.5 COMPOSITE MATERIALS

**Composites** are made from two or more different phases and the distribution of these phases is controlled by *mechanical means* rather than only by heat treatment. The glass-reinforced plastic fishing rod is a composite. A cross section is shown schematically in Figure 8.5.1. (The polyphase mate-

FIGURE 8.5.1. Cross section of a filamentary composite.

rials of the previous section are essentially composites but the material distribution is controlled by thermal and chemical means.) By combining different materials in certain ways it is often possible to achieve a property which each of the individual materials did not possess (**synergism**). For example, fine individual glass fibers have a high tensile stiffness and a *very* high tensile strength. However, because of their small diameter their bending stiffness is very small. Likewise if the fishing rod were made of only epoxy plastic, it would have a low flexural stiffness. It could not be made of solid glass because bulk glass has a low tensile strength compared to that of fibers (see Table 1.3.5). However, when the fibers are placed in the epoxy plastic, the resultant structure has high tensile stiffness, high tensile strength, *and* high bending stiffness and bending strength. Such materials are called **glass-reinforced plastics (GRP)**.

**EXAMPLE 8.5.1**

A bundle of glass filaments is arranged to form a rod of circular cross section as shown in Figure 8.5.1. Let the filaments have radius $r$, the composite beam have radius $R$, and the volume fraction of filaments in the composite beam be $\frac{1}{2}$. Suppose there are 500,000 continuous filaments in the beam. Let $E_f = 10 \times 10^6$ psi for the filaments and $E_b = 5 \times 10^5$ psi for the plastic binder.
(a) Calculate Young's modulus for the rod.
(b) Calculate the radius $R$ of the rod in terms of the fiber radius $r$.
(c) When the rod is used as a fishing rod it is really a beam. The bending stiffness constant of a beam is directly proportional to its cross-sectional moment of inertia $I$ times its longitudinal Young's modulus $E$. For a circular beam $I = \pi R^4/4$. What is the product $EI$ for the bundle of fibers without binder? Assume the friction between the fibers is negligible.
(d) What is the product $EI$ for the composite?
(e) How much stiffer is the composite beam than the bundle?
(f) How much stiffer is it than a beam of the plastic alone? Assume the radius is the same.
(g) How much stronger is the composite in tension than the plastic alone? Assume the plastic has a fracture stress of 10,000 psi and the glass has a fracture stress of 500,000 psi.

*Answer.*

(a) $E_{\text{comp}} = \frac{1}{2}E_f + \frac{1}{2}E_b \simeq 5 \times 10^6$ psi. The reader should figure out why this is so.
(b) Since half the volume of the rod is fibers we have
$$\tfrac{1}{2}\pi R^2 = 500{,}000\,\pi r^2.$$
Hence $R = 1000r$.
(c) $500{,}000 \times (10 \times 10^6) \times (\pi r^4/4)$.
(d) $5 \times 10^6 \times (\pi R^4/4) = 5 \times 10^6 \times (\pi r^4/4) \times 10^{12}$.
(e) $10^6$ times.
(f) 10 times.
(g) 25.5 times.

In Example 8.5.1 the assumption is made that the plastic carries the shear stresses which prevent slippage between the fibers. Since bulk glass has a fracture stress of about 10,000 psi the composite is 25.5 times as strong, although it will deflect twice as much under a given load as a solid glass rod of the same dimension. In working with ultra-high-strength materials the designer is often faced with the problem of working with large deflections.

**Q. 8.5.1**

By combining materials in certain ways it is often possible to achieve properties which the individual material did not exhibit; we call this (a)_____. The integral structures obtained when materials are combined mechanically are called (b)_____. GRP stands for (c)_____. In GRP beams the tensile and compressive loads are carried by the (d)_____ while the shear stresses are carried by the (e)_____.

\* \* \*

BORON FIBER COMPOSITES. A cross section of a boron-fiber-reinforced composite with an aluminum matrix is shown in Figure 1.2.16. Because of the high Young's modulus of the boron filaments and their high tensile strength and low density, the resultant composite is extremely stiff and has a high strength to density ratio. This is illustrated in Figure 8.5.2 where

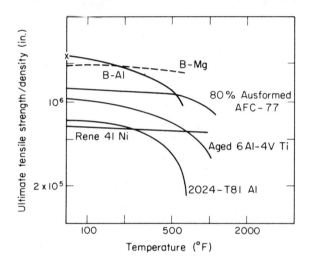

FIGURE 8.5.2. Comparison of strength to density ratio versus temperature for two composites and four alloys.

this ratio for the composite is compared with those of alloys of aluminum, nickel, titanium, and steel. Such composite materials because of their high stiffness to density and strength to density ratios are being used at an increasing rate in aircraft (the economy of reducing structural weight was discussed in Section 1.4). The boron filaments are made by the decomposition of boron halides on a tiny hot tungsten wire. Under appropriate conditions this results in very strong and stiff boron filaments. (Bulk boron is far inferior.) One of the problems associated with the making of this composite (with an aluminum matrix) is that boron dissolves in aluminum. The fiber is therefore coated with a layer of silicon carbide. Note that the fiber itself is a composite! For a further discussion of these composites, see *Metal Matrix Composites*, American Society for Testing and Materials, Philadelphia (1968).

Composites can also be made of whiskers (short, fine, perfect, or nearly perfect single crystals) dispersed in an appropriate matrix. The problem with whiskers is that they have to be very fine to be strong (see Figure 8.5.3). There is a problem with growing them fast enough and a problem of placing them in the matrix. It is hoped that in the future fine whiskers can be grown continuously. This is an active area of research.

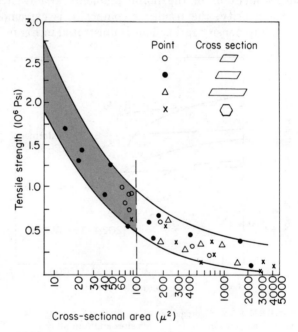

FIGURE 8.5.3. Tensile strength versus cross-sectional area of $\alpha$-$Al_2O_3$ (sapphire) whiskers. (Courtesy of E. Scala, Cornell University, Ithaca, N.Y.)

## Q. 8.5.2

Boron filaments have a Young's modulus of (a)____ kbars. They can be made in continuous filaments with strengths of (b)____ kbars. When boron is placed in an aluminum matrix it is necessary to (c)_____
_____.

\* \* \*

LAMINATES. Cotton-cloth-reinforced phenolics have been in use for many years. The cloth greatly increases the toughness of the material. Many composites are made by alternately adding layers of woven fabric and impregnating it with plastic or in some cases metal; these layer-like structures are called **laminates.** The fabrics include cotton, asbestos, glass, and metal. An example of a reinforced plastic car body is shown in Figure 8.5.4. One of the current production processes for this body involves

FIGURE 8.5.4. Racing car with reinforced-plastic body.

**prepregs,** fabric which has been impregnated with partially cured plastic. These sheets can be pressed together and then cured at high temperature without the need for further addition of plastic binder, and this results in substantial reductions in cost.

HONEYCOMBS. Figure 8.5.5 shows typical **honeycomb** and honeycomb panels. The panels are stiffer in bending strength per unit weight than other materials. In general bending loads are carried by the face sheets while the smaller shear loads are carried by the core material. The analysis of the stiffness of such panels is carried out in strength of materials or in elasticity courses. Corrugated cardboard is based on the same principles.

## Q. 8.5.3

A reinforced plastic which consists of properly arranged layers of fabric within a matrix is called (a)_____. A fabric impregnated with partially cured plastic so that these need only to be pressed together and cured is called a (b)_____. From the strength to density viewpoint, the strongest beams and panels are (c)_____.

\* \* \*

FIGURE 8.5.5. Aluminum and paper honeycomb and honeycomb panels.

CONCRETE. Concrete is a composite of rocks (coarse aggregate), sand (fine aggregate), hydrated Portland cement, and, in most cases, voids. Figure 1.2.17 is a photograph of a cross section of concrete. Portland cement is a mixture of calcium aluminum silicates such as

| | |
|---|---|
| tricalcium silicate | $3CaO \cdot SiO_2$, |
| dicalcium silicate | $2CaO \cdot SiO_2$, |
| tricalcium aluminate | $3CaO \cdot Al_2O_3$. |

When mixed with water the fine cement particles form a suspension; the water is absorbed into these particles forming in time a rigid gel which is composed of a hydrate having approximately the formula $Ca_3Si_2O_7 \cdot 3H_2O$. A **gel** is a mixture of a solid and liquid (or gas) which behaves mechanically like a solid (e.g., Jello is a gel). One of the important properties of a wet concrete mix is the viscosity. Ordinarily much more water than is needed for the hydration is added to the concrete mix to decrease its viscosity so that the concrete will fill the mold. Hence when the hydration is more or less complete (a week or a month later) there is excess water which either escapes, leaving behind voids, or remains trapped in tiny capillaries. In either case it weakens the concrete. It should be noted that the water in a hydrate is bonded by secondary bonds only (as is water absorbed by hair).

The strength of concrete will be affected by the completeness of the hydration, the aggregate size and distribution, the void volume, and the excess water used. Concrete is usually weak in tension but fairly strong in compression so that designers often consider the tensile strength to be zero.

Prestressed concrete is another example of a composite structure. Here roughened steel rods under tension are placed in a mold which is filled with the concrete mix. When the concrete is cured, the external forces holding the rods in tension are released and the rods therefore place the concrete in compression. The rods are placed near the bottom side of a beam which is being used as a horizontal beam. When the beam is loaded there would ordinarily be a tensile stress in the concrete (on one side) which would cause fracture. However, in the prestressed beam the concrete is in compression initially and hence the beam can undergo considerable loading before the tensile side of the beam actually has a tensile rather than a compressive stress.

## Q. 8.5.4

Concrete is a composite of (a)_____.

When Portland cement is mixed with water it forms a (b)____. Ordinarily concrete has voids or retained water because (c)_____.

\* \* \*

COATINGS. In all cases where a coating is a permanent, integral part of the structure, the material can rightfully be called a composite. Certain examples have been mentioned previously. Thus in Chapter 7, silicone on glass is a composite which has a higher surface electrical resistivity than untreated glass. Coatings are used for chemical, optical, mechanical, and electrical purposes.

In chemical applications, the purpose of the coating usually is to isolate the reactive structure from a chemical environment. Thus a chemical reactor may be internally coated with neoprene. Hydrofoils are coated with neoprene rubber, which can withstand the stresses of vibration, impact with small foreign bodies, etc. Graphite may be coated with silicon carbide to greatly increase the oxidation resistance of rocket nozzles. A layer of silicon vapor deposited on molybdenum reacts with the substrate to produce molybdenum silicide, which resists oxidation (molybdenum, like graphite, forms a volatile oxide). Teflon is coated on the inside of cooking utensils because of its nonstickiness. Porcelain is coated and baked onto steel for such applications as kitchen ranges. Aluminum is anodized, a special electrochemical treatment, which creates a hard durable layer of aluminum oxide on the aluminum surface.

Optical coatings are used for the reflection, transmission, and filtering of light. Vacuum-deposited aluminum on glass is used as a front surface mirror, which then is usually coated with hard quartz to increase durability. Filters can be made by coating glass with multilayer thin film coatings, with the thickness of the film related to the wavelength of the light being rejected.

Transparent, electrically conductive coatings are used on aircraft windshields for heating for purposes of deicing, etc. A thin layer of copper is clad to aluminum wire for CATV cable; because of the high frequency the current is conducted only near the surface (the skin effect) so the interior material serves only as a structural material.

Coatings are applied for mechanical purposes such as to reduce abrasion, erosion, cavitation, etc. Thus piping in sewer plants may have a porcelain interior because it chemically isolates the pipe and is resistant to abrasion. Many of the plating processes used by engineers are for the purposes of decreasing wear.

OTHER COMPOSITES. Fine particles of tungsten carbide, which are extremely hard, are mixed with about 6% cobalt powder and sintered at high temperatures to obtain sintered tungsten carbide for use as cutters in machining, rollers, etc. Ordinary grinding wheels are composites of an

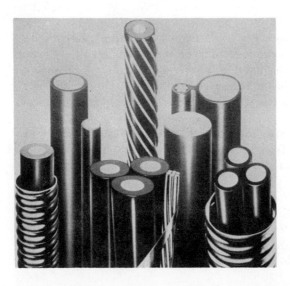

FIGURE 8.5.6. Polyethylene-coated sodium cable.

abrasive with a binder, plastic or metallic. Walls used for portable housing units are often made of thin aluminum sheets epoxied on to polyurethane foam. The presence of the latter provides excellent thermal insulation. The composite has high structural rigidity which the thin aluminum sheets or the foam do not have separately. Plywood is, of course, a composite also as are reinforced and prestressed concrete.

Composites are made for electrical as well as for mechanical applications. Sodium metal, which is very reactive chemically, is enclosed in polyethylene and used as an underground electrical cable as illustrated in Figure 8.5.6. $Nb_3Sn$ is deposited on copper to make superconducting ribbons. Liquid lead is forced under pressure into porous glass fibers to make a composite which at cryogenic temperatures is a synthetic hard superconductor. Microelectronic circuits are made from silicon which can be oxidized to form a layer of an electrical insulator, $SiO_2$; this can be etched away in various places by hydrofluoric acid, and phosphorus can be diffused into the silicon (initially a $p$-type material, say) to make $p$-$n$ or $p$-$n$-$p$ junctions. Aluminum or another metal can be deposited at specified places to provide microconductors between points, etc. Hence a microelectronic circuit is a tailored composite.

## ANSWERS TO QUESTIONS

**8.5.1** (a) A synergistic effect, (b) composites, (c) glass-reinforced plastic, (d) glass fibers, (e) plastic matrix.

**8.5.2** (a) 4000, i.e., $4 \times 10^{12}$ dynes/cm² from Table 1.3.3; (b) 30 (see Table 1.3.5); (c) coat the filaments with silicon carbide.

**8.5.3** (a) A laminate, (b) prepreg, (c) honeycomb panels.

**8.5.4** (a) Coarse aggregate, fine aggregate or sand, hydrated Portland cements, and voids or retained free water; (b) gel; (c) excess water is added to the concrete mix to decrease its viscosity.

## 8.6 QUANTITATIVE MICROSCOPY

The quantitative examination of microstructure is called **quantitative microscopy.** We have already noted in previous sections the need for quantitative data. The measurement of the grain size in a single-phase material is an example of quantitative microscopy. The study of dislocation distributions in a pure copper foil is another example. The study of the size and shape of precipitate particles in an alloy is yet another example of quantitative microscopy. We shall give a brief illustration of quantitative microscopy for the case in which spherical precipitates (called the $\alpha$ phase here) are randomly distributed in a uniform matrix. The sample is sliced along an arbitrary plane, polished, etched, and observed in the microscope. For our purposes, we assume that a photograph of the observed image is made and that a square grid is placed upon this photograph as in

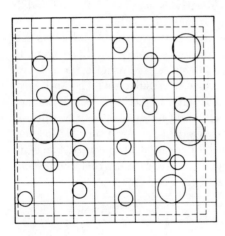

FIGURE 8.6.1. Square grid superimposed on a micrograph of a planar section of an alloy with spherical precipitate particles in a uniform matrix. Dashed line shows the edge of the photograph.

Figure 8.6.1. It can be shown in the present case [see R. T. Howard and M. Cohen, *Transactions of AIME* **172**, 413 (1947)] that

$$\frac{V_\alpha}{V} = \frac{A_\alpha}{A} = \frac{N_\alpha}{N} = \frac{L_\alpha}{L} \tag{8.6.1}$$

if a sufficiently large sample is chosen. Here $V_\alpha/V$ is the volume fraction of the spheres, $A_\alpha/A$ is the area fraction of circles on the planar area, $N_\alpha/N$ is the fraction of grid intersections lying within circles, and $L_\alpha/L$ is the fraction of grid line length lying in circles. In the present example for that portion of the sample shown $A_\alpha/A = 0.136$, $N_\alpha/N = 0.136$, and $L_\alpha/L = 0.130$. It is the purpose of quantitative microscopy to develop relations such as those of (8.6.1) which are applicable in specific cases and then to develop convenient experimental means of measuring the desired quantities. Thus in the present case, counting of points (grid intersection points) is probably the most convenient.

For further discussions of quantitative microscopy the reader may study Rostoker and Dvorak or Dehoff and Rhines (see the References). For a discussion of the quantitative analysis of dislocation distributions, see D. G. Brandon and Y. Komem, *Metallography* **3**, 111 (1970).

# REFERENCES

### Crystal Growth

Laudise, R. A., *The Growth of Single Crystals*, Prentice-Hall, Inc., Englewood Cliffs, N.J. (1970).

### Polycrystals

Smith, C. S., "Some Elementary Principles of Polycrystalline Microstructures," *The Metallurgical Reviews* 9, 1 (1964).

### Classical Metallography

Kehl, G. L., *The Principles of Metallographic Laboratory Practice*, McGraw-Hill Book Company, Inc., New York (1949). This book is a classic in the field.

Rostoker, W., and Dvorak, J. R., *Interpretation of Metallographic Structures*, Academic Press, Inc., New York (1965). This is an excellent introductory book which includes a chapter of quantitative metallography.

Dehoff, R. T., and Rhines, F. N., *Quantitative Microscopy*, McGraw-Hill Book Company, Inc., New York (1968). This book contains a thorough discussion of the analytical and experimental techniques of quantitative metallography.

### Modern Microscopy

Cosslett, V. E., *Modern Microscopy*, Cornell University Press, Ithaca, N.Y. (1966). This is an elementary book and can be easily read by the student. Includes electron microscopy.

Smallman, R. E., and Ashbee, K. H. G., *Modern Metallography*, Pergamon Press, Inc., Elmsford, N.Y. (1966). This monograph is at an intermediate level and discusses many of the techniques for studying microstructure.

Brandon, D. G., *Modern Techniques in Metallography*, Butterworth & Co. (Publishers), Ltd., London (1966). A more advanced discussion for the instructor.

### Composites

Kelly, A., "The Nature of Composite Materials," *Scientific American* (September 1967), p. 160.

Kelly, A., *Strong Solids*, Oxford University Press, Inc., New York (1966). Chapters 5 and 6 give a good introductory account of fiber reinforcement of materials.

**Whiskers**

Levitt, A. P., Ed., *Whisker Technology*, John Wiley & Sons, Inc., New York (1970).

**Concrete**

Brunauer, S., and Copeland, L. E., "The Chemistry of Concrete," *Scientific American* (April 1964), p. 80.

Neville, A., *Properties of Concrete*, John Wiley & Sons, Inc., New York (1970). This book gives a thorough discussion of the structure and behavior of concrete.

## PROBLEMS

**8.1** Give five "why" questions concerned with polycrystalline aggregates and five concerned with composites.

**8.2** Explain why etching takes place preferentially at grain boundaries.

**8.3** (a) Would grain boundaries be present under equilibrium conditions in a single-phase material?
(b) Explain what is meant by metastable grain boundary configurations.

**8.4** Describe briefly the various physical principles used in the design of the reflection microscope.

**8.5** A metal has an ASTM grain size of 8. Show that there are 65,000 grains/mm$^3$.

**8.6** What three features of grains in polycrystalline materials might determine properties?

**8.7** (a) Describe how grain size affects high-temperature creep of MgO.
(b) Describe how grain size affects brittle fracture.

**8.8** Describe two ways of growing NaCl crystals and note the temperature of each of these processes.

**8.9** How is yield strength related to the lamellar spacing of $\alpha$-Fe and $Fe_3C$ in steel?

**8.10** Define a composite and distinguish between the polyphase material of Figure 8.4.1 and the composite of Figure 1.2.16.

**8.11** Give an example of synergism in composites.

**8.12** Why are boron filament composites used in aircraft and space applications?

**8.13** (a) What is concrete?
(b) Why does concrete ordinarily contain considerable nonhydrated water when "completely" cured?
(c) What factors affect the strength of concrete?

**8.14** List ten composites. Include at least two natural composites.

# MORE INVOLVED PROBLEMS

**8.15** A second phase precipitates in a solid as spherical particles all of the same size. The specimen is sliced, polished, and etched, and photomicrographs are made with a linear magnification of 500. Discuss quantitatively how you would measure the sphere diameter.

**8.16** Consider Problem 8.15 for the case where there is a distribution of sphere diameters. Find an expression for the distribution function in terms of the measured diameter of the circles.

**8.17** An important parameter in a material is the grain size (because properties are often strongly affected by it). Suppose an industry produces metal with equiaxed grains. Design transparent overlays that could be used with an image projected from a microscope to quickly estimate the ASTM grain size number.

# SOPHISTICATED PROBLEMS

**8.18** Three grain boundaries meet at a line as in Figure 8.3.2. Show that if their surface free energies are the same, then they must meet at 120 deg.

**8.19** An optical scanner is available for scanning reflected light intensity from a point. The scanner can be programmed (controlled by a digital computer) to scan continuously along a line $x$ = constant or $y$ = constant. Write a program which will solve Problem 8.15.

A book by R. E. Smallman and K. H. G. Ashbee, *Modern Metallography*, Pergamon Press, Inc., Elmsford, N.Y. (1966), can be used as a starting reference for the following problems.

**8.20** Describe the limit of resolution of the reflection microscope.

**8.21** Describe the high-temperature microscope.

**8.22** Describe the use of the interference microscope in studying surface topography.

**8.23** Describe the electron microscope.

# Prologue

Temperature is atom motion which in turn affects the behavior of materials. The study of the effect of temperature fields is known as *thermodynamics*. We begin this chapter with a study of the motion of atoms in a dilute gas. The *average kinetic energy* of such an assembly of gas molecules (at a constant temperature) can be used to define the *ideal gas temperature scale*. The *ideal gas law* illustrates how atom motion is related to equilibrium behavior. The concept of *mean free path* between *collisions* is introduced and used to develop *Maxwell's theory of viscosity*, which illustrates how atom motion is related to kinetic behavior. The concepts of *internal energy*, *entropy*, and *Gibb's free energy* are introduced and used to derive the expression for the equilibrium concentration of vacancies in a crystal. The reason nature abhors a perfectly pure material just as it does a vacuum is discussed. The *barometric formula* is derived. The *Boltzmann distribution formula* is discussed and used to develop the *Einstein model* of the heat capacity of solids.

Interstitial atoms, such as carbon in iron, jump from one site to another due to thermal energy. Atoms located at regular lattice sites move through the lattice by interchanging with vacancies. In both cases the atoms move over a *potential energy barrier*. The height of this barrier is characterized by a quantity called the *activation energy*. The number of jumps per second is described by the *Arrhenius equation*. The process of motion of atoms through a crystal from regions of high concentration to low concentration is known as *diffusion*. Diffusion is described by *Fick's law* which involves a *diffusion coefficient*. We briefly study the mathematics of diffusion and the *root mean square diffusion distance*. We derive an expression for the diffusion coefficient based on the *random walk model*.

The nature of the diffusion coefficient of *gases*, *liquids*, and *solids* is examined. We study diffusion of carbon in steel, which involves the *error function* and the *carburization depth*. This is followed by a description of diffusion of vacancies in solids, *self-diffusion* in metals and ionic crystals, and diffusion in polymers. The diffusion processes used in the making of microelectronic circuits are studied in detail.

The kinetics of the *nucleation process*, which is often the first step in the occurrence of a phase transformation, is described.

# 9

# EQUILIBRIUM AND KINETICS

## 9.1 ATOM MOTION AND TEMPERATURE

> When I say of motion that it is the genus of which heat is a species I would be understood to mean, not that heat generates motion or that motion generates heat (though both are true in certain cases) but that heat itself, its essence and quiddity, is motion and nothing else . . . . Heat is a motion of expansion, not uniformly of the whole body together, but in the smaller part of it . . . the body acquires a motion alternative, perpetually quivering, striving, and struggling, and initiated by repercussion, whence springs the fury of fire and heat.—Francis Bacon (1561–1626).

Atoms are in constant random motion. Energy is associated with this motion. In the case of an ideal gas the energy associated with the random motion can be conveniently measured.

An **ideal gas** is a gas in which there are no potential energy interactions between the molecules. Real gases at low pressures approximate ideal gases. The classical mechanics expression for the kinetic energy of an atom, such as an argon atom, is $\frac{1}{2}mv^2$. Thus the total kinetic energy, $U_K$, of a system of $N_0$ argon atoms is given by

$$U_K = \sum_{i=1}^{N_0} \tfrac{1}{2}mv_i^2, \tag{9.1.1}$$

where $v_i$ is the magnitude of the velocity of each atom. The velocity is a vector

$$\mathbf{v} = [v_x, v_y, v_z].$$

The square of the magnitude of the velocity is given by

$$v^2 = v_x^2 + v_y^2 + v_z^2. \tag{9.1.2}$$

It was shown by Boyle in 1660 that gases at low pressure (say, air at 1 atm or less) obey the relationship

$$PV = \text{constant} \tag{9.1.3}$$

when the temperature is constant. This is now known as Boyle's law. In 1738, Daniel Bernoulli showed from theoretical considerations that

$$PV = \tfrac{2}{3} U_K. \tag{9.1.4}$$

His derivation proceeds as follows:

Consider a gas particle whose $x$ component of velocity is $v_x$ moving in a cubic box of edge $a$. When the particle strikes the wall of the box there is a change of the magnitude of the $x$ component of momentum equal to $2mv_x$. The time required to make a round trip in the $x$ direction is $2a/v_x$ so that the rate of momentum transfer by one particle to one wall is $2mv_x(v_x/2a) = mv_x^2/a$. The rate of change of the $x$ component of momentum of $N_0$ particles (a mole of particles) is therefore

$$\frac{dp_x}{dt} = \frac{m}{a} \sum_{i=1}^{N_0} v_{x_i}^2.$$

But by Newton's equation $F_x = dp_x/dt$.

Hence

$$F_x = \frac{m}{a} \sum_{i=1}^{N_0} v_{x_i}^2.$$

But pressure, or stress on the wall normal to the $x$ direction, is just force per area, $P = F_x/a^2$. Hence

$$PV = m \sum_{i=1}^{N_0} v_{x_i}^2$$

since $V = a^3$. Let us rewrite this in the form

$$PV = N_0 m \frac{\sum_{i=1}^{N_0} v_{x_i}^2}{N_0} = N_0 m \overline{v_x^2},$$

where $\overline{v_x^2}$ is the average value of $v_x^2$ for all the particles. Since the motion of the particles is random and since there are a large number of particles, the average values of $v_x^2$, $v_y^2$, and $v_z^2$ are therefore equal. This being the case, the average value of $v^2$ is $3v_x^2$ by (9.1.2). Hence we have

$$PV = N_0 \frac{m}{3}\overline{v^2} = \frac{2}{3}N_0 \frac{\overline{mv^2}}{2} = \frac{2}{3}U_K. \quad (9.1.5)$$

If a mole of gas is placed in a container and the product $PV$ is measured, then $U_K$ is obtained. For example, for argon gas at 0°C,

$$U_K = 3510 \text{ J/mole} = 814 \text{ cal/mole}.$$

In the above derivation, it was assumed that (1) the volume of the particles is a negligibly small fraction of the volume occupied by the gas, (2) no appreciable forces act between the particles except during collisions, and (3) collisions are elastic and of negligible duration compared to the time spent between collisions.

**EXAMPLE 9.1.1**

Calculate the fraction of volume actually occupied by the argon atoms themselves at 1 atm of pressure and 0°C if the argon atom has a diameter of 2.86 Å and a mole of gas occupies 22.4 liters.

*Answer.* $\frac{4}{3}\pi(2.86/2)^3 \times 10^{-24} \times 6.02 \times 10^{23} \approx 7$ cm³. Hence the fraction equals $7/22,400 \approx 3 \times 10^{-4}$.

In Examples 5.8.1 and 5.8.2 it is shown that the potential energy of argon per mole at 1 atm and 0°C is only of the order of $10^{-3}$ cal/mole, which is only about $10^{-6}$ that of the kinetic energy. Hence the potential energy interactions and therefore the forces between the atoms on the average are negligible. (It is of interest to note that Newton had suggested, incorrectly, that the pressure of a gas was due to the repulsive forces between the atoms.)

The quantity $(\overline{v^2})^{1/2}$ is called the **root mean squared velocity** and is designated by $v_{\text{rms}}$. The average of the magnitude of the velocity $v$ differs only slightly from $v_{\text{rms}}$. It is shown (in texts on the kinetic theory of gases) that

$$\bar{v} = \left(\frac{8}{3\pi}\right)^{1/2} v_{\text{rms}}. \quad (9.1.6)$$

Examples of the average magnitudes of the velocities are shown in Table 9.1.1. Note that between collisions the molecular velocities are of the same

**Table 9.1.1** AVERAGE MAGNITUDES OF MOLECULAR VELOCITIES AT 0°C

| Gas | $v$ (m/sec) |
|---|---|
| Hg | 170.0 |
| Ar | 380.8 |
| $O_2$ | 425.1 |
| $N_2$ | 454.2 |
| He | 1204.0 |
| $H_2$ | 1692.0 |

order of magnitude as the velocity of a rifle bullet. In Section 9.2, the distance traveled between collisions will be calculated.

At low pressures all these materials very closely follow Boyle's law, and all have the same translational kinetic energy, $U_K$, at the same temperature.

## Q. 9.1.1

According to Boyle, $PV =$ (a)_____. Daniel Bernoulli in 1738 showed that $PV =$ (b)____. His derivation begins by considering the rate of (c)_____ to the walls. The velocity squared of a particle is related to its velocity components $v_x$, $v_y$, and $v_z$ according to (d)_____. The average squared velocity, $\overline{v^2}$, is related to the velocity components by (e)_____.

\* \* \*

There are experiments, called molecular beam experiments, which directly measure the velocity distribution of atoms or molecules in a gas. See R. C. Miller and P. Kusch, *The Physical Review* **99**, 1314 (1955). These experiments have been used to verify the prediction made by Maxwell in 1860 that the components of velocity, such as $v_x$, have a random or Gaussian distribution, i.e., the fraction $df$ having velocity components between $v_x$ and $v_x + dv_x$ is given by

$$df = \frac{b^{1/2}}{\sqrt{\pi}} e^{-bv_x^2} dv_x, \tag{9.1.7}$$

where $b$ is a constant to be determined. [Note that $\int_{-\infty}^{\infty} (b^{1/2}/\sqrt{\pi}) e^{-bv_x^2} dv_x = 1$. In this equation the integrand is the "bell-shaped" curve of probability theory.]

It should also be noted that since these molecular beam experiments provide a direct measurement of the velocities, the kinetic energy of the assembly of particles given by (9.1.1) is directly measured.

For a further discussion of velocity distributions in gases, see W. J. Moore, *Physical Chemistry*, Prentice-Hall, Inc., Englewood Cliffs, N.J. (1972), Chap. 4.

THERMOMETRY. The earliest "thermometers" were based on thermal expansion coefficients. One of the earliest consisted of water in a glass bulb and stem and was used by the French physician Jean Rey in 1631 to study the change in fever of his patients. A calibration based on two "points" was introduced by Ferdinand II in 1641; he marked the thermometer at the liquid levels for the "coldest winter cold" and the "hottest summer heat." The distance between these two points was equally divided. The idea of more definite fixed points was introduced in 1688 by Dalence. He set the melting point of snow at $-10°$ and the melting point of butter at $+10°$. Then in 1694 Rinaldi took the boiling point of water as the upper fixed point with the melting point of ice as the lower fixed point. (To be precisely fixed points, the pressure should be specified, at, say, 1 atm.)

Charles in 1787 and Gay-Lussac in 1802 used the expansion of a gas at constant pressure to define a temperature scale. Thus the freezing point of water was set at $0°$ and the boiling point of water at $100°$ and the volume expansion of the gas was used to mark the 100 divisions in between (Centigrade scale). For example, Gay-Lussac in 1808 found that $V_{100°}/V_{0°} = 367/267$, so that he set $1°C$ equal to a fractional change in volume of $1/267$ relative to the volume at $0°C$. The presently measured value is $1/273.16$ instead of $1/267$.

The **ideal gas temperature scale** is assigned a value of $273.16°K$ at the ice point and $373.16°K$ at the boiling point of water. The symbol $T$ is used for this scale. For conditions of constant pressure

$$\frac{V(T_2)}{V(T_1)} = \frac{T_2}{T_1}.$$

The **ideal gas law** (for one mole of gas),

$$PV = RT, \qquad (9.1.8)$$

includes both Boyle's law and Charles' law. Here $R$ is the proportionality constant, called the **universal gas constant**; $R = k_B N_0$, where $N_0$ is Avogadro's number and $k_B$ is Boltzmann's constant (whose value is fixed by the size of the division on the temperature scale). Since $PV$ is proportional to the kinetic energy, it follows that the kinetic energy per mole, $U_K$, is directly proportional to the temperature. In fact from (9.1.4) and (9.1.8)

$$U_K = \tfrac{3}{2}RT. \tag{9.1.9}$$

Thus we are led to the conclusion that the kinetic energy of the translational motion of the molecules is directly proportional to the temperature. Why the ideal gas temperature scale should be an absolute temperature scale, i.e., why $-273.16°C$ should be the ultimate zero of temperature, is not proved here. The proof of that depends on a thorough understanding of thermodynamics.

We note that the ideal gas law for a mole of gas is an **equation of state,** i.e., the volume is a uniquely determined function of the fields $P$ and $T$, i.e., $V = V(P, T)$. The material may be taken through various $P$, $V$, and $T$ cycles but when it is returned to the original $P$ and $T$ (assuming it is still the same material and has not decomposed) it will have the same volume as before under equilibrium conditions. We say that $V$ is a **state function** since it depends only on the field variables $P$ and $T$ (in this case) and not on the path (i.e., the sequence of pressures and temperature) needed to reach a fixed $V$.

You can now show [using (9.1.5) and (9.1.9)] that the root mean square velocity of an atom in an ideal gas is

$$v_{\text{rms}} = \sqrt{\frac{3k_B T}{m}}. \tag{9.1.10}$$

Starting with Equation (9.1.7), the reader, with considerable effort can show that the average magnitude of the velocity is

$$\bar{v} = \sqrt{\frac{8}{\pi}\frac{k_B T}{m}}; \tag{9.1.11}$$

see Problems 9.38–9.40.

DEVIATIONS FROM THE IDEAL GAS LAW. In 1873 van der Waals proposed the equation of state which now bears his name:

$$\left(P + \frac{a}{V^2}\right)(V - b) = RT. \tag{9.1.12}$$

The term $a/V^2$ can be attributed to the attractive forces between the molecules; $b$ is called the **excluded volume** and its origin can be explained as follows. The centers of two atoms regarded as rigid spheres cannot approach each other more closely than $d$, where $d$ is the atom diameter. Hence because of the presence of the first atom, the center of the second atom is excluded from a volume $\frac{4}{3}\pi d^3$. Hence the excluded volume in (9.1.12) is $\frac{2}{3}\pi d^3 N_0$. At high temperatures and modest pressures the intermolecular force term becomes negligible and therefore $P(V - b) = RT$; i.e.,

$$PV = RT + bP. \qquad (9.1.13)$$

The measurement of $b$ provides a measure of the atom diameter. In this way the argon atom is found to have a diameter of 2.86 Å while the helium atom has a diameter of 2.00 Å. (Recall that in Chapter 5 we mentioned that Bohr in 1912 knew the atomic radius should be about 1 Å; this was one source of his knowledge.)

### Q. 9.1.2

Inasmuch as $PV = \frac{2}{3} U_K = RT$ for an ideal gas, $v_{rms} =$ (a)_____. The mercury thermometer is based on what material property? (b)_____. The ideal gas thermometer is based on what material property? (c)_____. The reason that 0°K corresponds to an odd number such as $-273.16°C$ is (d)_____. At high temperature and moderate pressure the equation of state of gases takes the form $PV = RT + bP$. The quantity $b$, called the (e)_____, can be used to measure (f)_____.

\* \* \*

## ANSWERS TO QUESTIONS

**9.1.1** (a) Constant, (b) $\frac{2}{3} U_K$, (c) momentum transfer, (d) $v^2 = v_x^2 + v_y^2 + v_z^2$, (e) $\overline{v^2} = \overline{3v_x^2} = \overline{3v_y^2} = \overline{3v_z^2}$.

**9.1.2** (a) $\sqrt{3RT/M}$, where $M$ is the molecular weight (g/mole); (b) thermal expansion (at constant pressure); (c) thermal expansion (at constant pressure); (d) due to the fact that 0°C is fixed by the freezing point of water and 100°C by the boiling point of water; (e) excluded volume; (f) atom or molecule diameters.

## 9.2 KINETICS IN AN IDEAL GAS

MEAN FREE PATH. In a dilute gas an atom (or molecule) moves freely through space until it collides with another atom (or molecule) and then it moves off in some other direction.

Let $\lambda$ be the **mean free path** (average distance a particle travels between collisions). It can be shown that the mean free path is given by

$$\lambda = \frac{1}{\sqrt{2}\,\pi n d^2}, \qquad (9.2.1)$$

where $n$ is the number of molecules per unit volume and $d$ is the molecular collision diameter. An approximate derivation follows. When a test molecule of diameter $d$ moves along as shown in Figure 9.2.1, the centers of other

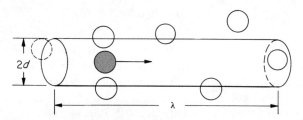

FIGURE 9.2.1. Molecule (shown darkened) moves through cylinder without colliding.

molecules must be at least a distance $d$ from the center of the test molecule in order that there be no collision. Hence the test particle sweeps a cylinder of diameter $2d$, area $\pi d^2$, and moves through a volume $\pi d^2 \lambda$ before it meets a molecule in its path and undergoes a collision. Note that there is one molecule per volume of $\pi d^2 \lambda$. Hence the number of molecules per unit volume is

$$n = \frac{1}{\pi d^2 \lambda},$$

which leads to (9.2.1) except for the factor $1/\sqrt{2}$. An exact derivation accounts for the fact that all the molecules are moving and gives the extra factor $1/\sqrt{2}$ in (9.2.1). For a rigorous derivation of (9.2.1) see E. H. Kennard, *Kinetic Theory of Gases*, McGraw-Hill Book Company, Inc., New York (1938).

**EXAMPLE 9.2.1**

Calculate the molar volume of an ideal gas at 0°C and 1 atm.

*Answer.* $V = RT/P = 0.0823/1 \times 273.16 = 22.4$ liters. Thus $n$ under these conditions is given by

$$n = \frac{6.02 \times 10^{23}}{22{,}400} = 2.69 \times 10^{19} \text{ cm}^{-3}.$$

It is easy to estimate the values of $\lambda$ from (9.2.1) based on the assumption that atoms have a diameter of a few Ångstroms; at 0°C and 1 atm, $\lambda \approx 10^3$ Å. Some values of $\lambda$ are given in Table 9.2.1.

**Table 9.2.1.** MEAN FREE PATHS AT 0°C, 1 ATM

| Gas | $\lambda$ (Å) |
|---|---|
| Ar | 635 |
| $O_2$ | 647 |
| $N_2$ | 600 |
| He | 1798 |
| $H_2$ | 1123 |

VISCOSITY OF GASES. Gases, liquids under certain conditions, and even solids under special circumstances exhibit linear viscous behavior.

It is our intention in this section to derive for gases **Newton's equation of viscosity**.

$$\tau = -\eta \frac{\partial v}{\partial x}, \tag{9.2.2}$$

i.e., the shear stress causing the flow, $\tau$, equals the negative of the velocity gradient normal to the flow direction, $-\partial v/\partial x$, times the viscosity coefficient, $\eta$. In so doing we shall also obtain an expression for $\eta$ for gases. The viscosity coefficient of gases is an important parameter in the design of airplanes, air bearings, etc. Consider Figure 9.2.2, which shows two planes (on which a number of gas molecules are located) whose normal is $x_3$, and which are a distance $\lambda$ apart, where $\lambda$ is the mean free path of the gas molecules (distance between collisions). The molecules in the lower plane have a *net* velocity $v$ superimposed on their random motion while the molecules in the upper plane have a *net* velocity of $v + (\partial v/\partial x_3)\lambda$ superimposed on their random motion. As a result of their random motion molecules move from one plane to the other plane where they have a collision and transfer momentum. (A parallel can be drawn with two trains traveling side by side at different velocities in which people jump back and forth in equal numbers. Those that jump from the slower moving train to the faster tend to slow the faster train down and vice versa.) When an atom of mass $m$ moves from the upper layer and interchanges with an atom

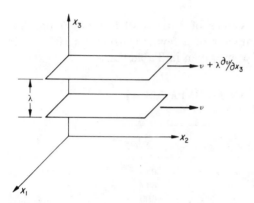

FIGURE 9.2.2. Viscosity of gases.

from the lower layer there is net momentum transfer $-m\lambda\,(\partial v/\partial x_3)$ to the upper layer. The fraction of molecules moving up or down is $\frac{1}{3}$. The number moving up or down across a *unit area* per second is $\frac{1}{3}n\bar{v}$, where $n$ is the number of molecules per unit volume and $\bar{v}$ is the average magnitude of their velocity (due to random motion). Hence the momentum change of unit area of the upper layer per second is

$$\frac{dp_2}{dt} = -\frac{1}{3}n\bar{v}m\lambda\frac{\partial v}{\partial x_3}.$$

Newton's equation is $\mathbf{F} = d\mathbf{p}/dt$ so the component of force in the $x_2$ direction is $F_2 = dp_2/dt$ and since we are dealing with unit area, $dp_2/dt = \tau$, the shear stress acting in the $x_2$ direction, which causes the flow. Therefore

$$\tau = -\frac{1}{3}n\bar{v}m\lambda\frac{\partial v}{\partial x_3}. \tag{9.2.3}$$

It follows from (9.2.1) to (9.2.3) that

$$\eta = \frac{1}{3}nm\bar{v}\lambda = \frac{m\bar{v}}{3\sqrt{2}\,\pi d^2}. \tag{9.2.4}$$

Sec. 9.2    Kinetics in an Ideal Gas    379

This equation was first derived by James Clerk Maxwell, one of the greatest scientists. It predicts that the viscosity coefficient of an ideal gas does not depend on pressure (recall that $\bar{v} = \sqrt{8RT/\pi M}$); this was an unexpected prediction.

$\eta$ provides a convenient measure of molecular diameters which may be compared with values obtained by other techniques. These results are illustrated in Table 9.2.2. Values of $\eta$ are given in Table 2.3.1.

**Table 9.2.2.** MOLECULAR DIAMETERS IN ÅNGSTROMS

| Molecule | From Viscosity* | From the van der Waals $b$† | From Closest Packing of the Solidified Material‡ |
|---|---|---|---|
| Ar | 2.86 | 2.86 | 3.83 |
| $O_2$ | 2.96 | 2.90 | 3.75 |
| $N_2$ | 3.16 | 3.14 | 4.00 |

\* Calculated from measured viscosity coefficients and Equation (9.2.4).
† See Section 9.1.
‡ From packing of spheres in crystals as in Chapter 6.

The concept of mean free path is also used in calculating values of the thermal conductivity, not only of gases but of liquids and solids as well, and is also used in calculating values of the electrical conductivity of a metal or semiconductor.

### Q. 9.2.1

The mean free path between collisions in an ideal gas is (a) <u>(directly or inversely)</u> proportional to the number of molecules per unit volume. Newton's equation of viscosity is (b)_____. One of the most surprising predictions of Maxwell's equation for the viscosity coefficient of a gas is (c)_____. The viscosity coefficient of a monatomic gas can be used to measure (d)_____. The concept of the (e)_____ is also used in calculating the thermal conductivity of gases, liquids, and solids.

\* \* \*

Although the ideal gas law, $PV = RT$, owes its origin to the motion of atoms, no net motion is involved and this is called **equilibrium behavior**.

However, the Maxwell-Newtonian viscosity equation,

$$\tau = -\frac{m\bar{v}}{3\sqrt{2}\,\pi d^2}\frac{\partial \mathscr{v}}{\partial x}$$

which also owes its existence to the motion of atoms, does involve a net motion (or displacement) of atoms and is therefore an example of **kinetic behavior**.

## ANSWERS TO QUESTIONS

9.2.1 (a) Inversely, (b) $\tau = -\eta(\partial \mathscr{v}/\partial x)$, (c) the fact that the viscosity coefficient is independent of pressure, (d) atomic diameters, (e) mean free path.

## 9.3 INTERNAL ENERGY

The **internal energy**, $U$, of a system is the sum of the kinetic energy and the potential energy of the atoms independent of the motion of the system as a whole.

**EXAMPLE 9.3.1**

What is the internal energy per mole of an ideal monatomic gas?

*Answer.* The potential energy is zero and the translational kinetic energy is given by $U_K$ in Equation (9.1.9). It equals $3RT/2$.

**EXAMPLE 9.3.2**

Discuss qualitatively the internal energy of nitrogen gas (diatomic molecules) at low pressure near room temperature.

*Answer.* Nitrogen under these conditions approximates an ideal gas since the potential energy interactions between the molecules are very small. The molecules are in constant motion so that there is translational kinetic energy, rotational kinetic energy, and vibrational energy (which would be both kinetic and potential energy since the total energy of a classical harmonic oscillator is $\frac{1}{2}mv^2 + \frac{1}{2}kx^2$, where $x$ is the displacement and $v$ is the velocity of the particle). The sum of all these energies is the internal energy.

The various molecular motions of a diatomic molecule are illustrated in Figure 9.3.1. The internal energy of a gas of diatomic molecules includes a contribution of $3RT/2$ due to translation, $RT$ due to rotation, and a contribution due to vibration which is nearly zero at room temperature but

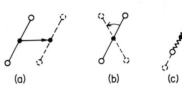

FIGURE 9.3.1. Motion of nitrogen molecule. Black dot shows the center of mass. (a) Translation of center of mass. (b) Rotation about center of mass. (c) Vibration about center of mass.

increases to $RT$ at several thousand degrees Kelvin. The latter contribution is discussed in Section 9.7.

**EXAMPLE 9.3.3**

Discuss the internal energy of a copper crystal.

*Answer.* If the widely separated copper atoms are assigned zero potential energy, then the bonding energy represents a potential energy term. We might evaluate this bonding energy for the static case (atoms not vibrating). However, the atoms in a crystal vibrate about their lattice positions. If there are $N_0$ atoms, there are a total of $3N_0$ modes of vibrations (each atom can potentially vibrate in $x$, $y$, or $z$ directions) and each of these vibrations can be approximated by a harmonic oscillator. The average internal energy associated with a harmonic oscillator is discussed in Section 9.7.

**Q. 9.3.1**

The internal energy of an ideal monatomic gas is due entirely to (a)_____. Its value per mole is $U =$ (b)____. The molar heat capacity at constant volume (see Section 4.3) of an ideal monatomic gas is (c)____. The internal energy of an ideal diatomic gas contains contributions due to (d)_____.

\* \* \*

## ANSWERS TO QUESTIONS

**9.3.1** (a) Translational motion, i.e., kinetic energy of translation; (b) $\tfrac{3}{2}RT$; (c) $\tfrac{3}{2}R$, since $C_V = (\partial U/\partial T)_V$; (d) translation, rotation, and vibration of the molecule.

## 9.4 RANDOMNESS AND ENTROPY

Consider a system consisting of two tanks of gas at the same pressure and temperature as in Figure 9.4.1. One tank contains helium, the other neon. Assume $P = 1$ atm, $T = 300°K$. Then each gas behaves (very nearly) as an ideal gas: The internal energy per mole of each of these monotomic gases

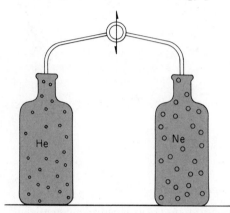

FIGURE 9.4.1. Two tanks of gas.

is simply $\frac{3}{2}RT$, which equals the kinetic energy (the potential energy is zero). This means that the gases in each tank have the same value of internal energy per mole and the same molar volume.

Suppose the valve between the two tanks is now opened. The two gases will eventually form a solution as illustrated in Figure 9.4.2. (The

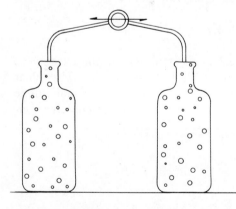

FIGURE 9.4.2. Solution of He and Ne gases.

solution is often erroneously called a mixture and the process of interdiffusion is called mixing.) There is no change in the total pressure $P$ or the temperature $T$ (pressure gauges and temperature sensors show no change). The solution still obeys the relationship $PV = RT$. The internal energy $U$

of the total system is unchanged; clearly no work was done (no displacements in the environment); moreover no heat was added or subtracted. Although $U$ and $P$ and $T$ remained constant, there was a profound change in the system! What did change? Why did the change occur?

**EXAMPLE 9.4.1**

Air contains about 4 parts of $N_2$ to 1 part of $O_2$. Some manufacturers go to a great deal of trouble (*work*) to separate these two components. They then sell the oxygen gas for such applications as oxyacetylene welding. The nitrogen gas is also sold and, after being liquefied, may be used to maintain a constant temperature of 77°K. In either case the purchaser pays for the cost of the original separation.

What is lost if the separated gases both at the same $P$ and $T$ and in 4:1 volume ratio are allowed to diffuse together to form air?

*Answer.* The work done in the separation.

The fundamental change which occurred in mixing the gases in Example 9.4.1 was the overall increase in randomness of the total system.

No student would deny that the two gases of Figure 9.4.1 would diffuse together to form a solution as in Figure 9.4.2. (It seems to be accepted as a fact of life, like taxes and, perhaps, death.) There must therefore be a fundamental belief that *nature strives to attain maximum randomness* (at least in the case of gases at room temperature).

To measure this randomness we define a quantity $\omega$ which is the total number of **distinct configurations** (or arrangements) of the system.

As an example consider the number of ways of placing different colored balls one to a box in boxes fixed in a row. If there are three balls $A$, $B$, and $C$ and three boxes, the arrangements are

$$ABC \quad BAC \quad CAB$$
$$ACB \quad BCA \quad CBA;$$

i.e., there are 3! or 6 arrangements.

If there are three balls $A$, $B$, and $A$ and three boxes, then the distinct arrangements are

$$ABA \quad BAA \quad AAB,$$

i.e., 3!/2!. The reason for the 2! in the denominator is that interchanging the two $A$ balls does not give a distinct configuration.

If all three balls are $A$, $A$, and $A$, then there is only one configuration, i.e., 3!/3!. Now if $N + n$ *different* kinds of particles (atoms) were placed on $N + n$ lattice sites (as in a crystal), there would be $(N + n)!$

configurations. However, if $N$ of the particles were the same kind of atoms, there would be only $(N + n)!/N!$ distinct configurations; if in addition $n$ of the particles were all atoms of another kind, then the number of distinct configuration is, in this case,

$$\omega = \frac{(N + n)!}{N!n!}.$$

Ludwig Boltzmann in 1896 showed that a useful measure of the randomness is

$$S = k_B \ln \omega, \qquad (9.4.1)$$

where $k_B$ is a universal constant which we now call Boltzmann's constant. The quantity $S$ is called the **entropy**. Equation (9.4.1) is one of the fundamental equations of science.

### Q. 9.4.1

When two ideal gases such as He and Ne at the same low pressure and at 300°K, say, are allowed to mix together, there is no change in (a)_____. There is, however, a change in the (b)_____ of the system. One measure of the randomness of a system is the number of (c)_____ possible. The quantity called the (d)_____ is a useful measure of randomness.

* * *

From a historical viewpoint, the concept of entropy had been introduced on a macroscopic scale following the work of Carnot. It was Boltzmann who gave entropy meaning on the atomic scale.

Let us now compute the change in entropy of a perfect crystal of $N$ atoms, to which $n$ vacancies are then added. The entropy of the perfect crystal is zero since the number of configurations is 1. The entropy of the crystal with vacancies is

$$S = k_B \ln \frac{(N + n)!}{N!n!} \qquad (9.4.2)$$

and this also equals the change in entropy in the present case.

We now use an approximation, namely, $\ln y! \doteq y \ln y - y$, if $y$ is very large; this is called Stirling's approximation. Hence the change in entropy (from the perfect crystal where $S = 0$ to the crystal with vacancies) can be written in the approximate form

$$\Delta S \doteq -(N+n)k_B \left[ \frac{N}{N+n} \ln\left(\frac{N}{N+n}\right) + \frac{n}{N+n} \ln \frac{n}{N+n} \right]; \quad (9.4.3)$$

this follows from (9.4.2) and Stirling's approximation. This is called the **entropy of mixing.** Note that the term in square brackets in (9.4.3) is *always* negative (since the logarithm of a fraction of one is negative) except for $n = 0$ where it equals zero (to show the latter, you need to know that $\lim_{x \to 0} x \ln x = 0$). Hence the entropy of mixing is always positive.

Note that $N/(N+n)$ represents the mole fraction of atoms in the crystal while $n/(N+n)$ represents the mole fraction of vacancies. If instead of atoms and vacancies being mixed, $N$ atoms of one kind had been mixed with $n$ atoms of another kind, as in forming the gas solution of Figure 9.4.2, the entropy of mixing would also be given by (9.4.3). Note that the rate at which the entropy of mixing changes (i.e., $\partial[\Delta S]/\partial n$) is the greatest when one solute atom is added to the pure solvent (to show this, find the partial derivative and then find its maximum). For very small concentrations, $n \ll N$, (9.4.3) becomes

$$\Delta S \doteq -Nk_B \left[ \frac{n}{N+n} \ln \frac{n}{N+n} \right], \quad n \ll N. \quad (9.4.4)$$

### Q. 9.4.2

The entropy of mixing is always (a)_____. The change of entropy per vacancy added to a crystal is the largest for what concentration? (b)_____. In terms of mole fractions the entropy of mixing of 1 mole total of ideal gas with a mole fraction $X_1$ of component 1 and $X_2$ of component 2 is (c)_____.

\* \* \*

## ANSWERS TO QUESTIONS

9.4.1 (a) Internal energy, (b) randomness, (c) distinct configurations, (d) entropy.

9.4.2 (a) Positive, (b) the first vacancy added, (c) $-R \sum_{i=1}^{2} X_i \ln X_i$, where $R$ is the gas constant.

## 9.5 EQUILIBRIUM IN CHEMICAL SYSTEMS

We state without proof that a chemical system will be in equilibrium under conditions of constant pressure and temperature if the quantity

$$G = U + PV - TS \qquad (9.5.1)$$

is a minimum. Here $U$ is the internal energy and $S$ is the entropy. We call $G$ the **Gibbs free energy.** The proof of this statement is carried out in thermodynamics. A corollary of this statement is that a chemical reaction or process will occur if at constant pressure and temperature it results in a decrease in the Gibbs free energy.

We apply this principle to the process of adding $n$ vacancies to a perfect crystal of copper containing $N$ atoms. Recall that when a vacancy is created an atom is taken from the interior and placed on the surface; this requires an energy $E_f$, called the **energy of formation of a vacancy,** and the volume of the crystal increases by $V_f$, called the **volume of formation of a vacancy.** In a perfect crystal of copper the $N$ atoms occupy the $N$ lattice sites of an fcc lattice. When $n$ vacancies are present within the crystal containing $N$ atoms, there are then $n + N$ lattice sites since the vacancies occupy $n$ lattice sites and the atoms occupy $N$ lattice sites. Thus the change in Gibbs free energy when $n$ vacancies are created is

$$\Delta G = n(E_f + PV_f) - T \Delta S, \qquad (9.5.2)$$

where $\Delta S$ is given by (9.4.3).

Figure 9.5.1 shows curves of the various energy terms versus the number of vacancies. The slope of the $\Delta U + P \Delta V$ vs. $n$ curve is $E_f + PV_f$,

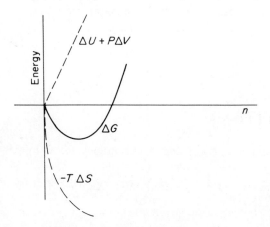

FIGURE 9.5.1. Free energy terms versus number of vacancies.

while the initial slope of the $-T\,\Delta S$ vs. $n$ curve is $-\infty$ [according to the approximate expression (9.4.3)]; according to (9.4.2) it is $-k_B T \ln (N+1)$, which is, at temperatures higher than about $T_M/3$, greater in magnitude than $E_f + PV_f$. Hence at high temperatures, $\Delta G$ is negative for small values of $n$. Note that $\Delta G$ has a minimum. This corresponds to $\partial(\Delta G)/\partial n = 0$; it can be shown that this minimum occurs when

$$\frac{n}{N+n} = e^{-(E_f + PV_f)/k_B T}. \tag{9.5.3}$$

In other words vacancies can be added to the crystal so long as the Gibbs free energy decreases.

At atmospheric pressure the $PV_f$ term is negligible relative to the $E_f$ term so that

$$\frac{n}{N+n} = e^{-E_f/k_B T}. \tag{9.5.4}$$

This equation will often be written in the form

$$\frac{n}{N+n} = e^{-E_f/RT}. \tag{9.5.5}$$

In that case $E_f$ is the energy required to create a mole of vacancies. Typical units then used for $E_f$ are kilocalories per mole. Some values of $E_f$ for metals are shown in Table 9.5.1.

Table 9.5.1. ENERGIES OF VACANCY FORMATION

| Metal | kcal/mole | eV |
|---|---|---|
| Al | 15 | 0.7 |
| Au | 21.6 | 0.94 |
| Ag | 25.1 | 1.09 |
| Cu | 26.5 | 1.15 |

IMPURITIES IN DILUTE SOLUTIONS. Consider now an impurity atom which is added to a solid; suppose that the internal energy of the solid increases by $\Delta U$ when the atom is placed in solution. At the absolute zero of temperature, there would be no solubility in this system; note that the

$T \, \Delta S$ term would be zero. However, at high temperatures, there is solubility. Why? Because the addition of a certain concentration of impurities results in a decrease in the Gibbs free energy in spite of the fact that it results in an increase in the internal energy. The problem is completely analogous to the problem of the vacancies just discussed. The mixing of the solute and solvent, resulting in a solution, causes a large increase in entropy. This entropy of mixing is responsible for the solubility. In general the change in the Gibbs free energy at zero pressure is given by

$$\Delta G = \Delta U - T \, \Delta S.$$

At sufficiently low solute concentrations and moderate temperatures the $T \, \Delta S$ term will always be greater than the $\Delta U$ term and there will be solubility, since the formation of the solution leads to a decrease in $\Delta G$. As an example, consider a situation in which the internal energy change is 2 kcal/mole of solute added and a mole fraction of $10^{-6}$ impurity is present in 1 mole of solution at 300°K. Then $\Delta U = 2 \times 10^{-6}$ kcal/mole. $\Delta S$ can be calculated from (9.4.4) with $n/(N + n) = 10^{-6}$ and $N_0 k_B = R$. The result is $T \, \Delta S = 8.0 \times 10^{-6}$ kcal/mole. Clearly $T \, \Delta S > \Delta U$, so $\Delta G < 0$ and solubility occurs. Nature abhors a perfectly pure material just as it does a vacuum.

MIXING OF IDEAL GASES. In Section 9.4, it was noted that two ideal gases at the same pressure and temperature would diffuse together and that work would be needed to separate them. The change in the Gibbs free energy where $N$ atoms of neon and $n$ atoms of helium are mixed at constant temperature and pressure is given by

$$\Delta G = -T \, \Delta S,$$

where $\Delta S$ is given by Equation (9.4.3). The internal energy change is zero since this is entirely determined by the temperature of the ideal gas. If $N = n = N_0$, then by (9.4.3)

$$\Delta G = 2N_0 k_B T [\tfrac{1}{2} \ln \tfrac{1}{2} + \tfrac{1}{2} \ln \tfrac{1}{2}]$$
$$= -2N_0 k_B T \ln 2.$$

The mixture is more stable by the amount of this energy lowering. The minimum work which must be done to separate these gases is $2N_0 k_B T \ln 2$ even if it is done by a process which is 100% efficient.

## Q. 9.5.1

A system is in chemical equilibrium under conditions of constant pressure and temperature if (a)_____. Vacancies are present in a solid even though their presence increases the (b)_____ because of the (c)_____. Tin dissolves in solid lead at high temperatures because of the (d)_____.

\* \* \*

## ANSWERS TO QUESTIONS

**9.5.1** (a) The Gibbs free energy is at a minimum, (b) internal energy, (c) entropy of mixing term in the Gibbs free energy, (d) entropy of mixing.

### 9.6 THE BAROMETRIC FORMULA

What happens to the pressure of air in the atmosphere with increasing altitude? Consider a small element of volume of air located between height $x$ and $x + dx$. Assume that the cross-sectional area (normal to the vertical) is *unity*. Then the volume of air in the element is $dx$. The gravitational pull on the element is $\rho g\, dx$, where $\rho$ is the density of the air and $g$ is the gravitational acceleration. Figure 9.6.1 shows the forces acting in the $x$ direction on the air in the element of unit cross-sectional area. A balance of the forces

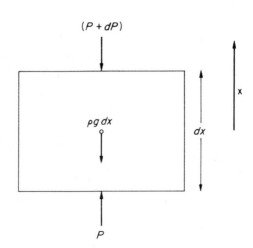

FIGURE 9.6.1. Forces acting on a small volume element of air with unit cross-sectional area normal to the $x$ direction.

shown in Figure 9.6.1 gives

$$P = P + dP + \rho g\, dx$$

or

$$dP = -\rho g\, dx. \qquad (9.6.1)$$

The density equals the mass per mole $M$ (the molecular weight) divided by the volume per mole $V$. It is also assumed that the gas can be treated as an ideal gas. Hence

$$\rho = \frac{M}{V} = \frac{MP}{RT}$$

and (9.6.1) can be written as

$$\frac{dP}{P} = -\frac{Mg\, dx}{RT}.$$

If $P = P_0$ at $x = 0$, then by integration,

$$\ln \frac{P}{P_0} = \frac{-Mgx}{RT}$$

or

$$\frac{P}{P_0} = e^{-Mgx/RT}. \qquad (9.6.2)$$

This is the **barometric formula**. The design of altimeters is based on this formula. It can also be written in the form

$$\frac{n}{n_0} = e^{-Mgx/RT} = e^{-mgx/k_B T}, \qquad (9.6.3)$$

where $n_0$ is the number of atoms per unit volume at $x = 0$, $n$ is the number per unit volume at height $x$, and $m$ is the mass per atom. The quantity $mgx$ represents the *potential energy*, P.E., of a gas atom at a height $x$.

Suppose that the total number of molecules in a column, of unit cross section, extending to the edge of the atmosphere (call it $x = \infty$) is $N$. Then we have

$$N = \int_0^\infty n\, dx = n_0 \int_0^\infty e^{-mgx/k_B T}\, dx,$$

which, combined with (9.6.3), leads to

$$\frac{n}{N} = \frac{e^{-\text{P.E.}/k_B T}}{\int_0^\infty e^{-\text{P.E.}/k_B T}\, dx},$$

where P.E. represents the potential energy per atom. This is a physically significant quantity because $n\,dx/N$ is the fraction of molecules with potential energy between $mgx$ and $mg(x+dx)$.

If we assume that a similar equation holds with the potential energy replaced by the total energy $U$ and consider that there are discrete rather than continuous energy states $u_i$ available, we have for the fraction $f_i$ of particles with energy $u_i$,

$$f_i = \frac{n_i}{N} = \frac{n_i}{\sum_i n_i} = \frac{e^{-u_i/k_B T}}{\sum_i e^{-u_i/k_B T}}. \tag{9.6.4}$$

This is the famous **Boltzmann distribution** formula.

### EXAMPLE 9.6.1

*Suppose* a molecule has two energy levels, namely, a ground state $u_0 = 0$, and one excited state $u_1 = 1$ eV. What fraction of such molecules will be in the excited state under *equilibrium conditions* at room temperature? Let $n_0$ and $n_1$ be the number in the ground state and excited state, respectively.

*Answer.*

$$f_1 = \frac{n_1}{n_0 + n_1} = \frac{e^{-u_1/k_B T}}{1 + e^{-u_1/k_B T}}.$$

At room temperature $k_B T \approx 1/40$ eV. Hence we can write

$$f_1 = \frac{n_1}{n_0 + n_1} \doteq e^{-u_1/k_B T} = e^{-40} = 10^{-40/2.3}.$$

### Q. 9.6.1

The pressure of a column of gas varies with the height of the column according to (a)_____. The average value of the potential energy per atom is given by (b)_____. One of the fundamental equations of science is Boltzmann's equation, $f_i = n_i/N = $ (c)_____.

\* \* \*

If there are only two states $u_0 = 0$ and $u$, and $u \gg k_B T$, then the fraction $f$ of particles with energy $u$ is by the Boltzmann formula (9.6.4),

$$f \doteq e^{-u/k_B T}. \tag{9.6.5}$$

This form of the Boltzmann equation is used on several future occasions in the text.

What precisely does the Boltzmann expression do for us? It enables us to compute the fraction $f_i$ of particles in each energy state (specified by $i$) at a given temperature if we know ahead of time exactly what the value of the energy $u_i$ is for each of these states. Once you have $f_i$, then it is a simple matter to get the average energy $\bar{u}$ of the particles at a given temperature because

$$\bar{u} = \sum_{i=1}^{N_0} f_i u_i. \qquad (9.6.6)$$

Equation (9.6.6) therefore leads to the molar internal energy $U = N_0 \bar{u}$. Other important thermodynamic quantities such as the entropy also follow from a knowledge of the $f_i$ as is shown in texts on statistical thermodynamics.

## ANSWERS TO QUESTIONS

**9.6.1** (a) $P/P_0 = e^{-Mgx/RT}$; (b) $\overline{\text{P.E.}} = \int_0^\infty (\text{P.E.})(n/N)\,dx$; since P.E. $= mgx$, it can be shown that $\overline{\text{P.E.}} = k_B T$; (c) $e^{-u_i/k_B T} \Big/ \sum_i e^{-u_i/k_B T}$.

### 9.7 ATOM VIBRATIONS

DIATOMIC MOLECULES. Consider a diatomic molecule. For small displacements $x$ of the atoms from their equilibrium separation distance in the molecule, the force tending to restore the atoms to the equilibrium position is $-kx$. The equation of motion is

$$m \frac{d^2 x}{dt^2} = -kx.$$

The motion is simple harmonic motion $x = x_0 \cos \omega t$, where $\omega = \sqrt{k/m}$. Here the angular frequency $\omega$ is measured in radians per second; if the frequency $\nu$ in cycles per second is desired, we have

$$\nu = \frac{1}{2\pi} \sqrt{\frac{k}{m}}. \qquad (9.7.1)$$

The force which stretches this linear spring is $kx$ and the work done in stretching it from $x = 0$ to $x$ is $\int_0^x kx\,dx = kx^2/2$ so that the potential energy of the system is

$$\frac{kx^2}{2}. \qquad (9.7.2)$$

Hence from the viewpoint of classical mechanics the total energy of the harmonic oscillator is given by

$$\frac{1}{2}m\left(\frac{dx}{dt}\right)^2 + \frac{1}{2}kx^2 = \frac{kx_0^2}{2}. \qquad (9.7.3)$$

Hence the energy has any value from zero up. We might expect $x_0 = 0$ at the absolute zero of temperature.

In quantum mechanics, the Schrödinger's equation for the one-dimensional harmonic oscillator is

$$\frac{h^2}{8\pi^2 m}\frac{d^2\psi}{dx^2} + \left(u - \frac{1}{2}kx^2\right)\psi = 0, \qquad (9.7.4)$$

where we have used the symbol $u$ for the unknown total energy of the oscillator. We now have to solve an eigenvalue problem. It can be shown (see texts on quantum mechanics) that this differential equation has a nontrivial solution if and only if $u$ is given by

$$u_i = (i + \tfrac{1}{2})h\nu, \qquad i = 0, 1, 2, \ldots, \qquad (9.7.5)$$

where $\nu$ is given by Equation (9.7.1) in which $m$ is the reduced mass, $m_1 m_2/m_1 + m_2$, and $m_1$ and $m_2$ are the atom masses. Thus at the absolute zero of temperature, there is still a finite energy of vibration. When heat is added to a system of such vibrators, some of them are excited into higher energy states. If there were $N_0$ identical vibrators, and the value of $i$ were known for each, then the total energy of the system could be calculated. It is an impossible task to know specifically the energy level of each oscillator. But the fraction of oscillators in a certain state can be calculated from the Boltzmann expression (9.6.4). Hence the average energy is given by

$$\bar{u} = \sum_i u_i f_i; \qquad (9.7.6)$$

where $u_i$ is given by (9.7.5) and $f_i$ is given by (9.6.4).

After some considerable analysis (see Problems 9.41 and 9.42) involving no approximations, it can be shown that

$$\bar{u} = \frac{h\nu}{2} + \frac{h\nu \, e^{-h\nu/k_BT}}{1 - e^{-h\nu/k_BT}}. \qquad (9.7.7)$$

The fundamental frequency of vibration, $\nu$, can be measured by spectroscopy. Thus if the harmonic oscillator is in a ground state and is excited to the first excited state, its energy will be increased by $h\nu$ according to (9.7.5) as the result of absorption of a photon whose energy is also equal to $h\nu$. The measured frequency of the light absorbed equals the oscillator frequency. Some typical frequencies for diatomic molecules are given in Table 9.7.1.

Table 9.7.1. VIBRATIONAL FREQUENCIES

| Molecule | Fundamental Frequency (Hz) | $\frac{h\nu}{k_B}$ (°K) |
|---|---|---|
| HCl | $8.97 \times 10^{13}$ | 4310 |
| $N_2$ | $7.08 \times 10^{13}$ | 3400 |
| NaCl | $1.14 \times 10^{13}$ | 548 |
| $O_2$ | $4.74 \times 10^{13}$ | 2280 |

The vibrational energy of a mole of diatomic molecules is $U = N_0 \bar{u}$. At high temperatures where $h\nu \ll k_BT$, $\bar{u} \doteq k_BT$ so $U \doteq RT$. To prove this use the fact that $e^{-x} \doteq 1 - x + (x^2/2)$ when $x \ll 1$. At low temperatures where $h\nu \gg k_BT$, $\bar{u} \doteq h\nu/2$ (the zero point energy). The temperature for which $h\nu = k_BT$ is given in the third column of Table 9.7.1.

## Q. 9.7.1

The total energy of a simple harmonic oscillator is according to classical mechanics (a)_____ and according to quantum mechanics (b)_____. The average energy of an oscillator can be calculated from (c)_____. At high temperatures where $k_BT \gg h\nu$, the average energy of an oscillator is (d)_____.

\* \* \*

VIBRATIONS IN SOLIDS. A solid consisting of $N_0$ atoms has $3N_0$ oscillators (each atom in a solid might be described by a potential energy $k_1 x_1^2/2 + k_2 x_2^2/2 + k_3 x_3^2/2$). To calculate the vibrational energy associated with a solid, Einstein assumed that all $3N_0$ oscillators had the same frequency, which, like the NaCl molecule of Table 9.7.1, is about $10^{13}$ Hz. This is not the case, but is a convenient approximation. Then the internal energy of the solid per mole is

$$U = 3N_0 \bar{u},$$

where $\bar{u}$ is given by (9.7.7) and the molar specific heat is

$$C_V = \left(\frac{\partial U}{\partial T}\right)_V = 3N_0 k_B \frac{x^2}{2(\cosh x - 1)}, \qquad (9.7.8)$$

where $x = h\nu/k_B T$.

The value of $\nu$ is such that $x = 1$ usually occurs on the range of temperatures $100 < T < 300°K$. The temperature for which $x = 1$ is called the Einstein temperature, $\theta_E$. $\theta_E$ has values of about $T_M/5$ (roughly). If $T \gg \theta_E$, then $x \ll 1$ and in this case, called the high-temperature limit,

$$C_V \doteq 3N_0 k_B = 3R. \qquad (9.7.9)$$

This is the result noted experimentally by Dulong and Petit in 1819.

**EXAMPLE 9.7.1**

(a) Using the fact that $\cosh x = 1 + x^2/2! + (x^4/4!) + \cdots$, show that (9.7.9) follows from (9.7.8) for $x \ll 1$. (b) Show that $C_V = 0$ for $T = 0$.

*Answer.* (a) $(\cosh x - 1) \doteq x^2/2$, where higher order terms in $x$ are neglected because $x \ll 1$. Therefore from (9.7.8)

$$C_V \doteq 3N_0 k_B = 3R.$$

(b) As $T \to 0$, $x \to \infty$ in (9.7.8). It is necessary to use l'Hospital's rule to show that $C_V \to 0$.

The frequency which is chosen in the Einstein model to make the data agree with experimental results (except at very low temperatures) is called the **Einstein frequency**. It has a value of about $10^{13}$ Hz.

Debye has developed a model for specific heat which takes account of the fact that there is a distribution of vibrational frequencies. His theory is in close agreement with experiments even near the absolute zero of temperature where the Einstein relation fails.

### Q. 9.7.2

According to the Einstein model of a solid of $N_0$ atoms, there are (a)____ vibrations having (b)_____. This model leads to a heat capacity per mole of atoms at high temperatures of (c)____. The Einstein frequency of a solid is about (d)____ Hz.

\* \* \*

## ANSWERS TO QUESTIONS

**9.7.1** (a) $u = \frac{1}{2}mv^2 + \frac{1}{2}kx^2$; (b) $u_i = (i + \frac{1}{2})h\nu$, where $i = 0, 1, 2, \cdots$; (c) $\bar{u} = \sum_i u_i f_i$;

(d) $k_B T$; this can be computed from (9.7.7).

**9.7.2** (a) $3N_0$, (b) the same frequency, (c) $3R$, (d) $10^{13}$.

## 9.8 KINETICS OF REACTIONS

Consider the process of vacancy motion in a solid. Suppose that a vacancy is at the 000 position in an fcc crystal such as copper. Consider the process by which it moves to a $\frac{1}{2}\frac{1}{2}0$ position. Let us move it slowly along the [110] direction, i.e., from 000 to $\frac{1}{2}\frac{1}{2}0$, and calculate the potential energy of the system along the way. We would find that the potential energy of the system varied with displacement as shown in Figure 9.8.1. There is an energy peak which has to be crossed if the vacancy is to change positions. The height of this peak is called the **activation energy for motion,** $E_m$, and the peak itself is called the **activation energy barrier.** The probability of finding the vacancy at the top of the peak, called the **activated state,** is by the Boltzmann expression (9.6.5)

$$f = e^{-E_m/k_B T}. \tag{9.8.1}$$

This is simply the fraction of the time the vacancy is in the activated state. Since an atom is in the process of exchanging with the vacancy, it is also the fraction of time during which that atom is in the activated state.

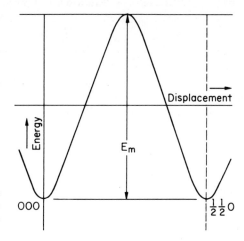

FIGURE 9.8.1. Energy of a crystal plotted versus distance as a vacancy moves from 000 to $\frac{1}{2}\frac{1}{2}0$ in copper.

Atoms in crystals are vibrating. The high energy vibrations have frequencies, $\nu$, of about $10^{13}$ sec$^{-1}$. The atom at 000 would usually tend to have displacements which would keep it near the minimum of the potential energy curve. However, for a fraction, $f$, of these vibrations the atom would move to the activated state due to thermal energy where it could then pass over the activation energy barrier to the new position. The jumps per second for such a process is therefore given by

$$\Gamma = \nu f = \nu e^{-E_m/k_B T}, \tag{9.8.2}$$

where $\nu$ is the vibrational frequency along the reaction path. (It can be assumed to be roughly equal to the Einstein frequency.)

**EXAMPLE 9.8.1**

(a) Assuming that aluminum has $\nu = 0.8 \times 10^{13}$ sec$^{-1}$ and $E_m = 0.6$ eV, calculate the number of vacancy jumps per second at 300°K. (b) Calculate the number of vacancy jumps in the crystals when $T = T_m = 943°$K.

*Answer.* (a)
$$\Gamma = 8 \times 10^{12} \times 10^{-0.6/2.3 \times 8.62 \times 10^{-5} \times 300}$$
$$= 570 \text{ jumps/sec.}$$

(b) Jumps/sec $= 4.7 \times 10^9$.

Some values of the activation energy for vacancy motion are given in Table 9.8.1.

**Table 9.8.1. ENERGIES FOR VACANCY MOTION**

| Metals | kcal/mole | eV |
|---|---|---|
| Al | 14 | 0.6 |
| Au | 20 | 0.87 |
| Ag | 19 | 0.83 |
| Cu | 23 | 1.0 |

GENERALIZATION. The above discussion applies not only to vacancy motion but to chemical reactions in general. In most cases the activation energy barrier is not symmetric and the products have a lower energy than the reactants.

From a historical viewpoint the chemist Svante Arrhenius in 1889 had noted empirically that chemical reactions proceed at a rate which depends exponentially on the negative reciprocal of absolute temperature: Rate = $Ae^{-B/T}$. This is called the **Arrhenius equation.** The chemical physicist Henry Eyring has in recent years developed the reaction rate theory much further but these developments are not discussed here.

### Q. 9.8.1

Chemical reactions, including the motion of vacancies, often involve overcoming an energy peak, the top of which is called the (a)_____. The jump frequency is given by the product of (b)_____ _____. The overall rate of many chemical reactions depends on the absolute temperature according to (c)_____, an expression known as the (d)_____.

\* \* \*

In discussing the behavior of polymer molecules, hindered rotation was described. In this case there is an activation barrier for rotation, $Q_r$. Rotation takes place only when aided by thermal fluctuations. The number of rotational jumps per unit time made at a specific bond is proportional to

$$e^{-Q_r/k_B T},$$

where $Q_r$ is of the order of 0.01 eV. If the material is to have a large elongation (either elastic or plastic) it is necessary that segments of the molecules can move over each other. It is found that the segments which move have a length of about 20 bonds. For this segment to be mobile, it must be flexible. To be flexible, rotation about each of the 20 bonds is necessary. The probability that each of the 20 bonds rotates, so that the segment is flexible, is proportional to $e^{-20Q_r/k_B T}$.

**EXAMPLE 9.8.2**

Given that $20Q_r = 0.4$ eV, compare the flexibility of a chain at 200°K to that at 400°K.

*Answer.* The relative flexibility is $e^{-[0.4/8.62 \times 10^{-5}][(1/200)-(1/400)]} \approx 10^{-5}$. Thus a chain which is barely flexible at 400°K will clearly be inflexible at 200°K.

# ANSWERS TO QUESTIONS

**9.8.1** (a) Activated state, (b) the attempt frequency $\nu$ and the probability of success of an attempt $f = e^{-E_m/k_B T}$, (c) rate $= Ae^{-B/T}$, (d) Arrhenius equation.

## 9.9 INTRODUCTION TO DIFFUSION

If a tank of argon gas is connected to a tank of helium gas at the same pressure and temperature, the two gases will diffuse into each other until a uniform composition is reached throughout. Similarly two miscible liquids such as water and alcohol, if placed in contact with each other, will diffuse into each other until a uniform composition is reached throughout. Likewise if a block of copper and a block of nickel are clamped together and held at 1000°C for a long time, they will diffuse into each other and eventually a uniform composition will be reached throughout (copper and nickel form a continuous solid substitutional solution). From the macroscopic viewpoint the only difference between these processes is the rate: Diffusion is fastest in the gas and slowest in the solid.

MACROSCOPIC VIEWPOINT. In 1855 Adolph Fick proposed empirically that diffusion is described (in the one-dimensional case) by the following equation:

$$J_x = -D \frac{\partial C}{\partial x}. \qquad (9.9.1)$$

Here $J_x$ is the flux in the $x$ direction (quantity of material crossing a unit area, whose normal is in the $x$ direction, in unit time). $C$ is the concentration of the diffusing species and $x$ measures distance, so that $\partial C/\partial x$ is a gradient. $D$ is the diffusion coefficient. This equation is now known as **Fick's law** or more properly as **Fick's first equation.** Note that it is a direct analog of Fourier's heat conduction equation and Ohm's electrical conduction equation. If $C$ is given in mole fraction, $x$ in centimeters, and $t$ in seconds and flux is measured in terms of cubic centimeters per square centimeter-second, then the unit of $D$ is square centimeters per second. This is the most common unit. According to (9.9.1) atoms will diffuse from regions of high concentration to regions of low concentration (in multicomponent systems there can be exceptions to this rule but that is a more involved topic which need not concern us here). There are often cases in which $D$ does not vary with $C$ in which case (9.9.1) becomes a *linear* relation.

### Q. 9.9.1

The macroscopic law of diffusion in the $x$ direction is (a)_____. This is known as (b)_____. A commonly used unit for the diffusion coefficient is (c)_____.

\* \* \*

STEADY-STATE DIFFUSION. In some problems involving diffusion the concentration of the diffusing atom does not vary with time although it varies with position. These are called **steady-state diffusion** problems. As an example hydrogen gas can be purified by passing it through a palladium foil. Hydrogen dissolves to a considerable extent in palladium (the molecule dissociates and the hydrogen in the palladium is present as an atom). The solubility of hydrogen increases as the pressure increases so if there is a high gas pressure on one side of the foil the concentration in the palladium on that side will be high while if there is a low gas pressure on the other side the concentration in the palladium on that side will be low as shown in Figure 9.9.1. It can be safely assumed (although not proved here) that the rate at which the hydrogen enters or leaves the foil is very fast compared to the rate of diffusion through it. (*Note:* Other gases which might be present in the hydrogen as impurities dissolve only slightly if at all in the palladium and diffuse much more slowly than the tiny hydrogen atoms which can readily move from one interstitial position to another interstitial position.) The hydrogen flow rate can then be written in terms of Fick's law [see Equation (9.9.1) and Figure 9.9.1]:

$$Q = J_x A = \frac{D(C_2 - C_1)A}{L}. \tag{9.9.2}$$

Sec. 9.9                                Introduction to Diffusion    401

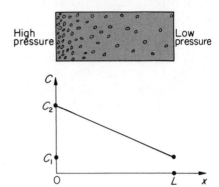

FIGURE 9.9.1. Steady-state flow
of hydrogen through palladium.

Here $A$ is the foil area, $C_2$ is the dissolved hydrogen concentration on the
high-pressure side and $C_1$ on the low-pressure side, while $L$ is the foil thickness. Note that in this case the concentration does not vary with time; this
is why the process is called a steady-state process.

## Q. 9.9.2

In steady-state diffusion the concentration does not (a)_____.
In a one-dimensional steady-state diffusion process, the concentration versus
distance curve is a (b)_____.

\* \* \*

TIME-DEPENDENT DIFFUSION IN ONE DIMENSION. Consider diffusion
in a volume element of material lying between $x$ and $x + dx$ and having
unit cross-sectional area. From a materials rate balance, involving the flux
in the $x$ direction into the element at $x$, the flux out in the $x$ direction at
$x + dx$, and the rate of accumulation, and also by use of Fick's law, the
equation

$$\frac{\partial}{\partial x}\left(D\frac{\partial C}{\partial x}\right) = \frac{\partial C}{\partial t}$$

is derived.

If $D$ does not vary with $C$ (or $x$), we obtain the linear differential
equation

$$D\frac{\partial^2 C}{\partial x^2} = \frac{\partial C}{\partial t}, \qquad (9.9.3)$$

often known as Fick's second equation or the **parabolic diffusion equa-**

**tion.** In the present case $C$ is a function of both $x$ and $t$. Applications of this equation are considered next.

THIN LAYER OF TRACER ATOMS ON A THICK SLAB. Suppose a very thin layer of radioactive silver is plated on a very thick slab of ordinary silver. Let the quantity of radioactive tracer per unit area be $\alpha$. If the slab is now heated to a high temperature and held for a long time (days), the radioactive silver will diffuse into the silver. The concentration $C^*$ of radioactive silver varies according to

$$C^*(x, t) = \frac{\alpha}{\sqrt{\pi D t}} e^{-x^2/4Dt}. \qquad (9.9.4)$$

You can verify that this is a solution of (9.9.3) by direct substitution. It can also be shown experimentally that this is the correct variation with distance by slicing successive thin layers off the surface of a cooled specimen and measuring their concentration. A graph of $\ln C^*$ vs. $x^2$ is found to be a straight line with slope $-1/4Dt$. Since $t$, the time spent at the high temperature, is known, the value of $D$ can be obtained. If the experiment is repeated at various temperatures, then the temperature dependence of $D$ can be obtained. It is found that

$$D = D_0 e^{-Q_D/RT}. \qquad (9.9.5)$$

Both $D_0$ and $Q_D$ are parameters of a material. The quantity $Q_D$ is called the **activation energy for tracer diffusion** and $D_0$ is called the **preexponential diffusion factor.** The process just studied is also called **self-diffusion** since silver is diffusing in silver.

**EXAMPLE 9.9.1**

How is a radioactive Ag* atom in Ag the same as an Ag atom and how is it different?

*Answer.* Both have identical electron clouds so the bonding is the same. However, the nuclei have different masses so that the vibrational frequencies are slightly different (since $\nu = (1/2\pi)\sqrt{k/m}$). Hence the internal energy is slightly different.

### Q. 9.9.3

If $D$ does not vary with $C$, Fick's second law is (a)_____. It is found that the diffusion coefficient varies with temperature according to (b)_____.

\* \* \*

MEAN SQUARED DIFFUSION DISTANCE. If in Equation (9.9.4), $\alpha$ is set equal to 1, then $C^*(x, t)\,dx$ represents the fraction of tracer material between $x$ and $x + dx$. Therefore the mean squared position (the average value of $x^2$) is given by

$$\overline{x^2} = \int_0^\infty x^2 C^*(x, t)\,dx = 2Dt. \qquad (9.9.6)$$

Hence for a one-dimensional diffusion problem the **root mean squared diffusion distance** is defined by

$$L = \sqrt{\overline{x^2}} = \sqrt{2Dt}. \qquad (9.9.7)$$

This equation is one of the most important equations in the study of diffusion. Because of its simplicity it is very useful to use in making estimates of how far an atom (or molecule) diffuses in a time $t$. Engineers and scientists must develop the ability to make rapid estimates!

### EXAMPLE 9.9.2

Estimate the diffusion distance of radioactive silver in solid silver at the melting point assuming that $D = 10^{-8}$ cm²/sec and $t = 10$ days.

*Answer.* Since 10 days $\approx 10^6$ sec we have from (9.9.7)

$$L \approx 0.14 \text{ cm.}$$

### EXAMPLE 9.9.3

Given that $D_0 \approx 1$ cm²/sec for Ag, calculate $D$ at $T_m/3$ and use this to estimate $L$ for $t = 10$ days at $T_m/3$.

*Answer.* By (9.9.5) we have at $T = T_m$

$$D = D_0 e^{-Q_D/RT_m} \approx 10^{-8} \text{ cm}^2/\text{sec},$$

and we have at $T_m/3$

$$D = e^{-3Q_D/RT_m} = [e^{-Q_D/RT_m}]^3 = 10^{-24} \text{ cm}^2/\text{sec}$$

and by (9.9.7)

$$L \approx 1.4 \times 10^{-9} \text{ cm} \approx 0.$$

THE RANDOM WALK. Diffusion in a crystal can be described in terms of a random walk. The simplest random walk is the one-dimensional case.

Here a particle steps from one position to another a distance $\pm l$ along the $x$ axis. The probability that a given step is in the positive $x$ direction equals the probability that it is in the negative $x$ direction. Hence the position after $n$ steps (each step being described by a vector $l_i$) is

$$x = l_1 + l_2 + l_3 + \cdots + l_N$$
$$= (\pm l) + (\pm l) + (\pm)l + \cdots (\pm l).$$

Thus for a large number of steps, $N$, the average value of $x$, $\bar{x} = 0$, but

$$\overline{x^2} = \overline{\{l_1 + l_2 + l_3 + \cdots + l_N\}^2} = \overline{[l_1^2 + l_2^2 + l_3^2 + \cdots + l_N^2]} + \overline{(2l_1l_2 + 2l_1l_3 + \cdots + 2l_1l_N)} + \cdots.$$

(The last quantity, in parentheses, and similar quantities like it are zero since each term in this quantity may have the value $-2l^2$ or $2l^2$ with equal probability.)

We then have

$$\overline{x^2} = Nl^2$$

since only the term in square brackets contributes so

$$L = \sqrt{\overline{x^2}} = \sqrt{N}\, l. \tag{9.9.8}$$

From (9.9.7) and (9.9.8) it follows that

$$D = \frac{N}{t}\frac{l^2}{2} = \frac{1}{2}\Gamma l^2, \tag{9.9.9}$$

since $N/t$ is the number of jumps per second, i.e., $\Gamma$. Note that the diffusion coefficient equals one half of a jump frequency times a jump distance squared.

### Q. 9.9.4

The root mean squared diffusion distance for one-dimensional diffusion is in terms of the diffusion coefficient (a)_____ and in terms of the number of jumps in a random walk (b)_____. The diffusion coefficient in terms of the atom jump frequency is (c)_____.

\* \* \*

In a liquid or a solid $\Gamma$ depends exponentially on the negative reciprocal of temperature. This explains why the diffusion coefficient in these materials varies with temperature according to (9.9.5). The self-diffusion coefficient in solids is often about $10^{-8}$ cm$^2$/sec at the melting point.

The diffusion coefficient for simple organic liquids at the melting point is about $10^{-4}$ to $10^{-5}$ cm$^2$/sec. Numerous values are given in S. Glasstone, K. J. Laidler, and H. Eyring, *The Theory of Rate Processes*, McGraw-Hill Book Company, Inc., New York (1941).

DIFFUSION IN GASES. If Equation (9.9.9) were used as a starting point for discussing diffusion in gases, and $\lambda$ (the mean free path) were considered the jump distance, then

$$D = \frac{1}{2}\left(\frac{\bar{v}}{\lambda}\right)\lambda^2 = \frac{1}{2}\bar{v}\lambda,$$

where $\bar{v}$ is the average magnitude of the velocity. This assumption is not quite right and the effective jump distance is only $2\lambda/3$ as shown in texts on the kinetic theory of gases. Then

$$D = \tfrac{1}{3}\lambda\bar{v}. \tag{9.9.10}$$

### EXAMPLE 9.9.4

Calculate the self-diffusion coefficient for argon at 0°C and at 1 atm.

*Answer.* From Table 9.2.1, $\lambda = 6.35 \times 10^{-6}$ cm, and from Table 9.1.1, $\bar{v} = 3.8 \times 10^4$ cm/sec. Hence $D = 0.08$ cm$^2$/sec.

### Q. 9.9.5

The diffusion coefficient in many solids just below the melting point is of the order of (a)_____ while the diffusion coefficient in liquids at the melting point is of the order of (b)_____ and the diffusion coefficient in gases at the boiling point is of the order of (c)_____.

\* \* \*

## ANSWERS TO QUESTIONS

**9.9.1** (a) $J_x = -D(\partial c/\partial x)$, (b) Fick's law, (c) square centimeters per second.

**9.9.2** (a) Change with time, (b) straight line.

**9.9.3** (a) $D(\partial^2 c/\partial x^2) = \partial c/\partial t$, (b) $D = D_0 e^{-Q_D/RT}$.

**9.9.4.** (a) $L = \sqrt{2Dt}$, (b) $L = \sqrt{N}\, l$, (c) $D = \tfrac{1}{2}\Gamma l^2$.

**9.9.5** (a) $10^{-8}$ cm²/sec, (b) $10^{-4}$ cm²/sec, (c) $10^{-1}$ cm²/sec.

## 9.10 SPECIAL CASES OF DIFFUSION

INTERSTITIAL IMPURITY DIFFUSION. The diffusion of carbon in steel is an example of interstitial diffusion. In the carburization of steel, carbon is diffused inward from the surface to produce a steel with higher carbon content near the surface; upon quenching, such a steel attains a high hardness near the surface. For the diffusion of carbon in bcc iron it can be shown that the pre-exponential factor in Equation (9.9.5) is given by

$$D_0 = \frac{\nu a^2}{6},$$

where $\nu$ is the vibrational frequency and $a$ is the lattice parameter.

Table 9.10.1 shows some results of $D_0$ and $Q_D$ [defined by (9.9.5)] for solute diffusion.

**Table 9.10.1.** INTERSTITIAL DIFFUSION DATA

| Solvent | Solute | $D_0$ (cm²/sec) | $Q_D$ (kcal/mole) | Ref. |
|---|---|---|---|---|
| Fe bcc | C | $2.0 \times 10^{-2}$ | 20.1 | Wert |
| Fe fcc | C | 0.21 | 33.8 | Guy |
| Fe fcc | N | $1.07 \times 10^{-1}$ | 34.0 | Smithells |
| Fe fcc | S | $4.8 \times 10^{-6}$ | 23.4 | Barrer |
| Fe fcc | P | $4.5 \times 10^{-2}$ | 43.3 | Barrer |
| Al fcc | H | $3.4 \times 10^{2}$ | 24.6 | Smithells |
| Fe bcc | H | $1.6 \times 10^{-2}$ | 9.2 | Barrer |
| Ni fcc | H | $2.04 \times 10^{-3}$ | 8.7 | Barrer |
| Pd fcc | H | $1.5 \times 10^{-2}$ | 6.8 | Smithells |

**EXAMPLE 9.10.1**

How many jumps per second does a carbon atom make in iron (a) at room temperature? (b) At 1800°K assuming the structure is bcc and the data for bcc apply?

*Answer.* (a) Jumps per second = $\Gamma = \nu e^{-Q_D/RT}$. We can solve for $\nu$ since we have $D_0 = 2.0 \times 10^{-2}$ cm²/sec from Table 9.10.1 and $a = 2.87 \times 10^{-8}$ cm from Table 6.2.1. We have

$$\nu = \frac{6D_0}{a^2} \approx 1.5 \times 10^{14} \text{ sec}^{-1}.$$

Hence

$$\text{jumps/sec} = 1.5 \times 10^{14} \, 10^{-20,100/2.3 \times 1.987 \times 300}$$

$$\approx \tfrac{1}{2} \text{ jump/sec}.$$

(b)

$$\text{Jumps/sec} = 1.5 \times 10^{14} \, 10^{-20,100/2.3 \times 1.987 \times 1800}$$

$$\approx 5 \times 10^{11} \text{ jumps/sec}.$$

DIFFUSION INTO A SEMI-INFINITE SLAB WITH CONCENTRATION FIXED AT THE SURFACE. We are often concerned with one-dimensional diffusion of an impurity into a solid in which the initial impurity concentration is

$$C(x, t) = C_0 \quad \text{for all } x > 0 \text{ at } t = 0 \tag{9.10.1}$$

and in which the concentration at the surface is very rapidly (assume instantaneously) brought to a fixed concentration $C_s$ so that

$$C(x, t) = C_s \quad \text{for } x = 0 \text{ at } t > 0. \tag{9.10.2}$$

We are therefore interested in solving Equation (9.9.3) subject to (9.10.1) and (9.10.2). The solution, after rearranging certain terms, is

$$\frac{C(x, t) - C_0}{C_s - C_0} = 1 - \text{erf}\left(\frac{x}{2\sqrt{Dt}}\right). \tag{9.10.3}$$

Here we have the function erf $y$, which is called the **error function of** $y$, the Gaussian error function, or the normalized probability integral,

$$\text{erf } y = \frac{2}{\sqrt{\pi}} \int_0^y e^{-z^2} \, dz.$$

This function, like many other well-known functions such as $\sin \theta$, $\cos \theta$, etc., is tabulated. Table 9.10.2 gives a few results.

## EXAMPLE 9.10.2

Suppose a 0.2 wt. % carbon steel is carburized at 950°C; this means the steel is placed in carbon which dissolves at the surface and diffuses inward. Assume the surface carbon content under these conditions reaches 1.4 wt. %. For diffusion at a

**Table 9.10.2.** THE ERROR FUNCTION

| $y$ | erf $y$ | $y$ | erf $y$ |
|---|---|---|---|
| 0.0 | 0 | 0.8 | 0.742 |
| 0.1 | 0.113 | 0.9 | 0.797 |
| 0.2 | 0.223 | 1.0 | 0.843 |
| 0.3 | 0.329 | 1.2 | 0.910 |
| 0.4 | 0.428 | 1.4 | 0.952 |
| 0.5 | 0.521 | 1.6 | 0.976 |
| 0.6 | 0.604 | 1.8 | 0.989 |
| 0.7 | 0.679 | 2.0 | 0.995 |

temperature $T$ for a period of time $t$, what is the depth at which the increased carbon content equals one half the increased carbon content at the surface? This is often the definition of the **carburization depth**. The question is really, What is $x/2\sqrt{Dt} = y$ when

$$\frac{C(x, t) - C_0}{C_s - C_0} = \frac{1}{2}?$$

*Answer.* Now erf $y = \frac{1}{2}$ when $y \approx \frac{1}{2}$. Hence the answer is $x \approx \sqrt{Dt}$. If it is desired in the application that carburization occurs to a depth of 0.005 cm, then the product $Dt$ can be calculated. Since $D$ can be calculated from the data in Table 9.10.1, $t$ can then be computed.

DIFFUSION OF VACANCIES IN METALS. It can be shown that the diffusion coefficient of vacancies in fcc metals is given by

$$D_v = \Gamma a^2 = \nu a^2 e^{-E_m/k_B T}. \tag{9.10.4}$$

Here $\Gamma$ is the jump frequency, $a$ is the lattice parameter, $\nu$ is the vibrational frequency, and $E_m$ is the activation energy of motion. (See Ruoff, *Materials Science*, previously cited, Chapter 18.)

The activation energies for motion, $E_m$, listed in Table 9.8.1 can be measured as follows. At temperatures approaching the melting point, the equilibrium vacancy concentrations of these metals as given by (9.5.4) is about $10^{-3}$ to $10^{-4}$ mole fraction. If a thin wire is rapidly quenched to a low temperature (50,000°K/sec), most of these vacancies are quenched in; i.e., they do not have time to escape to sinks (which are discussed later). At the low temperature, the vacancy concentration under equilibrium conditions is zero. Hence the wire is highly supersaturated with vacancies. At the low temperature (room temperature for copper) these vacancies slowly diffuse to sinks (as discussed in the next paragraph) and the concentration decreases. Since the presence of each vacancy causes a slight increase in the electrical resistivity of the material, the decrease in the vacancy concentration can be followed by very careful resistance measurements. By studying the rate of decay of vacancy concentrations at different temperatures, $E_m$ can be measured.

Surfaces, grain boundaries, and dislocations are as a rule good sources and sinks for vacancies. Vacancies can be readily created or destroyed at jogs on dislocations. **Jogs** are positions where the dislocation line changes from one slip plane to another. Figure 9.10.1 shows the extra plane of atoms inserted to create an edge dislocation (see also Figure 6.6.1).

If a vacancy interchanges with the shaded atom, the vacancy is destroyed and the jog moves in the negative $x_3$ direction in Figure 9.10.1(b).

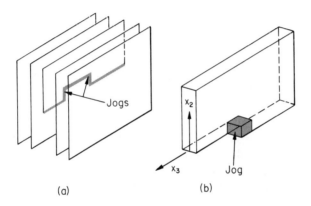

FIGURE 9.10.1. (a) Jogs on an edge dislocation. (b) Extra half-plane of atoms with dislocation line at bottom and jog on dislocation line.

This is called **dislocation climb.** Note that by repeated operations of this sort the edge dislocation line climbs upward (in the positive $x_2$ direction) and a portion of the extra plane of atoms is removed. This type of dislocation motion is different from the motion of slip or glide discussed in Chapter 6. Dislocation climb is an important process in high-temperature creep, in annealing of strain-hardened materials, and in the rapid drop of yield stress near $T_m/2$.

### Q. 9.10.1

If a solid which has been cooled very slowly from near $T_m$ to a low temperature is rapidly heated, vacancies will be (a)_____ at (b)_____ _____. If the specimen is held for a long enough time at the high temperatures, (c)_____ concentration of vacancies will be formed. A segment of a dislocation which joins a dislocation line on one slip plane to the dislocation line on a nearby slip plane is called a (d)____. When an edge dislocation acts as a vacancy source (or sink), (e)_____ occurs.

* * *

SELF-DIFFUSION IN CERTAIN ELEMENTS. It has been established by E. O. Kirkendall, R. W. Balluffi, and others that self-diffusion in metals occurs by a vacancy mechanism. An atom will move only if there is a vacancy next to it; the probability of this is $n/(N + n)$, where $n$ is the number of vacancies and $N$ the number of atoms in the crystal. The self-diffusion coefficient is therefore the product of this probability times the vacancy diffusion coefficient $D_v$; i.e.,

$$D = \frac{n}{N + n} D_v. \qquad (9.10.5)$$

Because $n/(N + n)$ is given by Equation (9.5.4) and $D_v$ by Equation (9.10.4)

$$D = \nu a^2 e^{-(E_f + E_m)/k_B T} = D_0 e^{-Q_D/k_B T}; \qquad (9.10.6)$$

i.e., in this case $D_0 = \nu a^2$ and $Q_D = E_f + E_m$. The self-diffusion coefficient can, except for a small correction factor (the correlation factor), be measured by studying the diffusion of radioactive isotopes. Some examples of self-diffusion coefficients are given in Table 9.10.3.

Table 9.10.3. SELF-DIFFUSION DATA IN PURE METALS*

|  | $Q_D$ (kcal/mole) | $D_0$ (cm²/sec) |
|---|---|---|
| Aluminum | 28.75 | 0.035 |
| Copper | 49.56 | 0.62 |
| α-iron (paramagnetic) | 57.3 | 2.0 |
| Lead | 26.06 | 1.37 |
| Nickel | 68.0 | 1.9 |
| Silicon | 110.0 | 1800. |
| Silver | 44.27 | 0.44 |
| Sodium | 10.09 | 0.145 |
| Tungsten | 153.0 | 42.8 |

* From N. L. Petersen, in *Solid State Physics*, Vol. 22, ed. by F. Seitz, D. Turnbull, and H. Ehrenreich, Academic Press, Inc., New York (1968).

Sec. 9.10    Special Cases of Diffusion    411

## Q. 9.10.2

(a) The self-diffusion coefficient in metals is given by the product of _____. The activation energy for self-diffusion is the sum (b)_____.

\* \* \*

DIFFUSION IN IONIC CRYSTALS. In crystals such as NaCl, there are both sodium ion vacancies and chlorine ion vacancies as was discussed in Section 6.6. These are present in equal concentrations under equilibrium conditions in the pure crystal. The creation of these defects can be expressed in terms of a chemical reaction:

$$\text{perfect crystal} + \text{space} \rightarrow \text{Na}_v + \text{Cl}_v.$$

The product of the concentrations of these defects is the equilibrium constant $K_{eq}$ for the reaction,

$$[\text{Na}_v][\text{Cl}_v] = K_{eq} = K_0 e^{-E_f/k_B T}. \tag{9.10.7}$$

This equation could also be derived by methods similar to those used to derive Equation (9.5.4). Note that the concentrations are highest at high temperatures. Here $E_f$ represents the energy required to create a separate sodium ion vacancy plus a separate chlorine ion vacancy. When $Cd^{2+}$ ions are present, a sodium vacancy must be created to maintain charge neutrality; also, for each chlorine ion vacancy created, there must also be a sodium ion vacancy. Hence the concentration of sodium ion vacancies is

$$[\text{Na}_v] = [Cd^{2+}] + [\text{Cl}_v]. \tag{9.10.8}$$

At high temperatures, NaCl becomes an electrical conductor because of the rapid exchange of the sodium ion vacancies with $Na^+$ ions. If $[Cd^{2+}] \ll [\text{Cl}_v]$, then $[\text{Na}_v] \doteq [\text{Cl}_v]$ and the material is called an **intrinsic** conductor; note from (9.10.7) and (9.10.8) that then

$$[\text{Na}_v] = K_0^{1/2} e^{-E_f/2k_B T}. \tag{9.10.9}$$

However, if the impurity concentration is such that $[Cd^{2+}] \gg [\text{Cl}_v]$, then $[\text{Na}_v] \doteq [Cd^{2+}]$ and the sodium vacancy concentration depends only on the $Cd^{2+}$ concentration; i.e., it does not depend on temperature. Such a material is called an **extrinsic** conductor. For a vacancy mechanism the diffusion coefficient of sodium is the product of the sodium vacancy concentration times the vacancy diffusion coefficient [as in (9.10.5)]. The electrical conductivity is directly proportional to the self-diffusion coefficient, $D$, of the fast-moving sodium ions. In fact

$$\sigma = \frac{Dq^2}{k_B T}, \tag{9.10.10}$$

where $q$ is the charge of the ion. This is the **Nernst relation.** The derivation involves the use of thermodynamics. See the previously cited Ruoff, *Materials Science*, Sect. 18.13. A material may be an intrinsic conductor at high temperatures but an extrinsic conductor at lower temperatures as

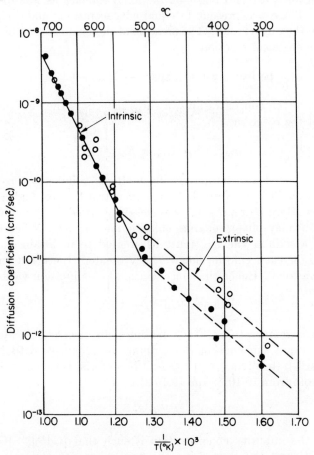

FIGURE 9.10.2. Logarithm of the tracer diffusion coefficient versus reciprocal of the temperature as measured by radioactive sodium (open circles) and as calculated from conductivity (full circles) for sodium chloride. [From D. Mapother, H. N. Crooks, and R. Maurer, *Journal of Chemical Physics* **18**, 1231 (1950).]

illustrated in Figure 9.10.2. Do you see why the slope at low temperatures is $-E_m/k_B$, while the slope at high temperatures is $-[(E_f/2) + E_m]/k_B$?

### Q. 9.10.3

The concentration of sodium ion vacancies in sodium chloride can be increased by increasing either the (a)_____ or the (b)_____. The ac electrical conductivity in an ionic crystal is proportional to the (c)_____ _____.

\* \* \*

DIFFUSION IN POLYMERIC MATERIALS. Diffusion occurs when small molecules (soluble in the polymer under the conditions of the experiment) pass through the "voids" between the polymer molecules. The rate of diffusion will depend to a large extent on the size of the small molecules and the size and mobility of the voids. The latter depend on the state of the polymer, i.e., whether it is crystalline, glassy, or rubber-like. The diffusion rate will be high in the rubber-like state because the packing is poor in the amorphous state and because the highly flexible chain segments can easily move out of the way of a diffusing molecule. The rate will be lower in the same material when it is in the glassy state (below the glass transition temperature for the polymer—small molecule solution) because there is a reduction in volume (see Figure 6.7.4) and a decrease in the flexibility of the chain. Finally, the molecular packing in the crystalline material is still better and the chain flexibility is also low, so diffusion will be the slowest of all when the polymer is in the crystalline state.

The permeation of molecules through a polymer, i.e., steady-state diffusion, depends on two factors, namely, the solubility of the solute molecules in the polymer and the diffusion coefficient. The solubility depends on chemical factors. In the simplest form the rule of solubility is that "like tends to dissolve like." Thus nonpolar materials such as rubber dissolve nonpolar gasoline (octane, $C_8H_{16}$). Highly polar materials such as nylon dissolve water. This results in swelling of the polymers.

### EXAMPLE 9.10.3

Rubber gloves of thickness $L$ are attached to a chamber containing pure argon (crystals of sodium are growing in the chamber). Show that the flow of oxygen through the gloves is determined by the coefficient

$$p = DC_s, \tag{9.10.11}$$

where $D$ is the diffusion coefficient and $C_s$ is the solubility limit of the oxygen in the rubber at a given pressure.

*Answer.* For steady-state diffusion in one direction, Fick's law is

$$J = -D\frac{dC}{dx};$$

$J$ is a constant and $dC/dx = -(C_s - 0)/L$. We have assumed that the oxygen pressure inside the chamber is virtually zero, so the dissolved oxygen at the inner face is zero. Hence

$$J = \frac{DC_s}{L} = \frac{p}{L}.$$

If the solubility $C_s$ were directly proportional to the gas pressure (Henry's law), then, by Fick's law, the flux rate would be directly proportional to the pressure gradient; i.e.,

$$J = -k_p \frac{dP}{dx}. \tag{9.10.12}$$

This is **Darcy's equation.** Both $p$ and $k_p$ are called permeability coefficients in the literature. Some values of $k_p$ are given in Table 9.10.4. (Recall that 1 bar = $10^6$ dynes/cm$^2$ = 0.98 atm.)

**Table 9.10.4.** PERMEABILITY DATA AT 30°C, ($10^{-8}$ cm$^2$/sec-bar)

| POLYMER | GAS | | | Nature of Polymer |
|---|---|---|---|---|
| | $N_2$ | $O_2$ | $CO_2$ | |
| Poly(vinylidene chloride) (Saran) | 0.007 | 0.041 | 0.21 | Crystalline |
| Polytetrafluoroethylene | 0.02 | 0.075 | 0.54 | Crystalline |
| Nylon | 0.07 | 0.28 | 1.2 | Crystalline |
| Poly(vinyl chloride) | 0.03 | 0.90 | 7.5 | Partially crystalline |
| Cellulose acetate | 2.1 | 5.9 | 51 | Glassy |
| Polyethylene (high density) | 2.1 | 7.9 | 27 | Crystalline |
| Polyethylene (low density) | 14 | 41 | 264 | Partially crystalline |
| Butyl rubber | 2.3 | 9.7 | 39 | Rubber-like |
| Polybutadiene | 48 | 143 | 55 | Rubber-like |
| Natural rubber | 60 | 175 | 982 | Rubber-like |

## Q. 9.10.4

In general diffusion is fastest in polymeric materials in the (a)_____ _____ state, slower in the (b)_____ state, and slowest of all in the (c)_____ state. Diffusion of pressurized fluids through polymeric materials obeys (d)_____ equation.

\* \* \*

## ANSWERS TO QUESTIONS

**9.10.1** (a) Created; (b) dislocations, grain boundaries, and surfaces; (c) an equilibrium; (d) jog; (e) dislocation climb.

**9.10.2** (a) The vacancy concentration times the diffusion coefficient for vacancies; (b) $E_f + E_m$.

**9.10.3** (a) Temperature, (b) divalent cation concentration, (c) diffusion coefficient of the faster-moving ion.

**9.10.4** (a) Rubber-like, (b) glassy, (c) crystalline, (d) Darcy's.

### 9.11 APPLICATIONS OF DIFFUSION THEORY

MICROELECTRONIC CIRCUITS. The production of such circuits was discussed in Section 1.1. We consider here, quantitatively, two aspects of the problem:

1. The oxidation process which yields the oxide film.
2. The diffusion of phosphorus into silicon.

The oxidation process can be described on the basis of Figure 9.11.1. The assumption is made (1) that once the oxide film has become several

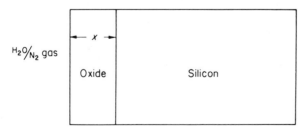

FIGURE 9.11.1. Oxidation of silicon.

atomic layers thick a steady-state concentration of silicon is reached at the gas interface. This means that a fixed concentration difference $\Delta C$ exists across the oxide layer of thickness $x$. We assume (2) that a steady-state concentration exists within the oxide layer as well. The flux of the small silicon atoms through the oxide is proportional to the overall gradient $\Delta C/x$ and hence, by Fick's law, equals $D\,\Delta C/x$; the flux is also proportional to $dx/dt$. Hence we have

$$\frac{dx}{dt} \propto \frac{D\,\Delta C}{x},$$

which by integration gives

$$x^2 \propto Dt.$$

Hence

$$x^2 = Kt, \qquad (9.11.1)$$

where $K$ is a constant determined by temperature and the gas pressure. This is an example of the **parabolic growth law** which is characteristic

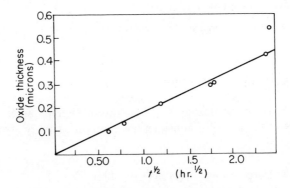

FIGURE 9.11.2. Oxide thickness on silicon versus the square root of oxidation time. The films were grown in wet $N_2$ at 1250°C with a gas flow of 1.5 liters/min. [From R. B. Allen, H. Bernstein, and A. D. Kurtz, *Journal of Applied Physics* **31**, 334 (1960).]

of the rate of adherent oxide film formation and is illustrated in Figure 9.11.2.

The oxidation of metals is similar and is an important area of materials research. See, e.g., Carl Wagner, "Diffusion and High Temperature Oxidation of Metals" in *Atom Movements*, American Society for Metals, Cleveland, Ohio (1951), p. 153. The study of the formation of protective films on materials by chemical reaction and diffusion is also an important area of materials research. As an example molybdenum, which at high temperatures forms an oxide which readily vaporizes, can be protected by formation of a nitride at the surface.

### Q. 9.11.1

When an adherent oxide film forms on a metal, the rate of formation is determined by the (a)_____. The thickness versus time relation is (b)_____. This is called the (c)_____ law.

\* \* \*

Let us now etch a hole in the oxide layer as shown in Figure 9.11.3. Hydrofluoric acid readily etches $SiO_2$ but not Si. We can mask areas which we do not want etched by covering them with beeswax.

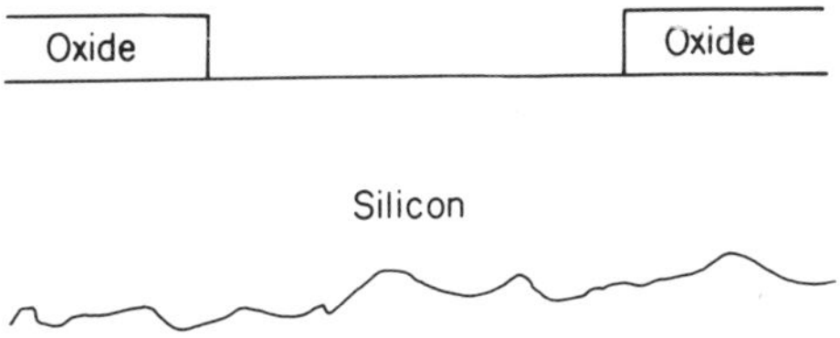

FIGURE 9.11.3. Hole in oxide layer on silicon.

Let us next consider the diffusion of the phosphorus into the silicon at the hole. Here a gas with controlled concentration of $P_2O_5$ is passed over the surface which rapidly (instantaneously) becomes saturated with phosphorus; i.e., the surface concentration of phosphorus reaches a certain value, say $C_s$. (The $P_2O_5$ decomposes at the surface.) We assume one-dimensional diffusion. The solution is, from (9.10.3),

$$C_p(x, t) = C_s \left\{ 1 - \text{erf}\left[\frac{x}{2(D_p t)^{1/2}}\right] \right\}. \tag{9.11.2}$$

The values of $D_p$ were measured by C. S. Fuller and J. A. Ditzenberger, *Journal of Applied Physics* **27**, 544 (1956). The result is

$$D_p = 10.5 e^{-85,000 \text{ cal/mole}/RT} \text{ cm}^2/\text{sec}. \tag{9.11.3}$$

At 1200°C this gives a diffusion coefficient of about $3 \times 10^{-12}$ cm²/sec.

The diffusion of phosphorus in the silicon oxide layer has been studied by R. B. Allen et al., *Journal of Applied Physics* **31**, 334 (1960). Their value for the diffusion coefficient is

$$D = 3.9 \times 10^{-12} e^{-32,000 \text{ cal/mole}/RT} \text{ cm}^2/\text{sec}. \tag{9.11.4}$$

At 1200°C this gives $D \approx 7 \times 10^{-17}$ cm²/sec. Thus if the oxide coating is of sufficient thickness, it will certainly mask the silicon below it from phosphorus.

To obtain a properly operating electronic device, we want to obtain a phosphorus concentration near to $C_s$ at a depth of $x = 10^{-3}$ cm but a concentration very close to zero at twice this depth. A good estimate of this diffusion depth is $x = 2\sqrt{D_p t}$; note that $1 - \text{erf } 1 = 0.157$ and $1 - \text{erf } 2 = 0.005$. Hence we set $10^{-3} = 2\sqrt{D_p t}$ and solve for $t$, using the value of $D_p$ at 1200°C. The result is $t = 83,000$ sec, nearly a day. Of course the designer of microelectronic circuits is interested in the exact nature of the concentration profile, i.e., Equation (9.11.2).

Let us ask how thick an oxide layer would be needed to essentially completely mask the silicon from the phosphorus. We want the ratio of the concentration of the phosphorus at the oxide-silicon interface to the phosphorus concentration at the external surface to be only a tiny fraction. We let the thickness of the oxide film be

$$x = 4\sqrt{Dt}.$$

This will ensure that the required ratio is 0.005. Since $D \approx 7 \times 10^{-17}$ cm²/sec and the diffusion time is 83,000 sec, the oxide thickness must be $1 \times 10^{-5}$ cm.

**Table 9.11.1.** SOME DIFFUSION-CONTROLLED PROCESSES

Semiconductor doping by diffusion
Oxidation
Homogenization
Graphitization
Age precipitation hardening
Eutectic and eutectoid lamellar formation
Vacuum melting and casting
Carburization, etc.
Annealing of radiation damage
Sintering
Creep
Recovery
Recrystallization
Osmosis
Reverse osmosis for desalinization
Passage of nerve impulses
Charge transport in batteries and fuel cells
Purification using molecular sieves

The reader interested in microelectronics technology should see W. R. Runyan, *Silicon Semiconductor Technology*, McGraw-Hill Book Company, Inc., New York (1965).

A partial list of processes in which diffusion plays a vital role is given in Table 9.11.1.

## ANSWERS TO QUESTIONS

**9.11.1** (a) Rate at which the metal ion diffuses through the oxide, (b) $x^2 = Kt$, (c) parabolic growth.

## 9.12 NUCLEATION

A process called nucleation is usually, but not always, the first step in the formation of a new phase.

### EXAMPLE 9.12.1

Suppose we have pure water vapor but no liquid or solid $H_2O$ and no other phases present in a large region at 300°K. What are the conditions (pressure) under which liquid water droplets will form?

*Answer.* Our first conclusion would be that if the water vapor pressure is increased to a pressure slightly above the equilibrium vapor pressure of water, liquid water droplets should form. We would be wrong. They do not and they should not. Why?

When we measure the so-called equilibrium vapor pressure, $p_0$, of water we use a large body of water as in a liter flask. Here we have a surface with essentially zero curvature. Moreover, a *negligible* fraction of the molecules is at the surface. Consequently, the equilibrium vapor pressure $p_0$ is a measure of how well the molecules are bound in the *interior* of the liquid.

Suppose that you have a little spherical droplet containing only 157 water molecules. Now a sizable fraction of the molecules are at the surface where they have fewer nearest neighbors than normal for the bulk. Hence the average molecule in this droplet is less effectively bound to other molecules than a molecule in the bulk of a large body of water; hence the vapor pressure is higher. The vapor pressure $p$ for a droplet of radius $r^*$ can in fact be derived from thermodynamics. It is found that

$$\ln \frac{p}{p_0} = \frac{2\gamma \bar{V}_l}{RTr^*}. \tag{9.12.1}$$

This is called the **Kelvin equation.** Here $\bar{V}_l$ is the molar volume of the liquid and $\gamma$ is the surface energy. For $r^* = 10.4$ Å, which corresponds to a droplet with 157 molecules at 300°K, using the fact that $\gamma = 72$ ergs/cm², we find that $p/p_0 = 2.7$. At 300°K, $p_0 = 0.0349$ atm.

A droplet with a larger radius would have a lower vapor pressure [see (9.12.1)]. Hence, if there is a droplet with a radius of 15 Å and $p/p_0 = 2.7$, that droplet would tend to grow.

However, if there is a droplet with a radius of 7 Å and $p/p_0 = 2.7$, that droplet would tend to evaporate.

Therein lies a dilemma if droplets are to begin to form from a vapor even though the pressure is greater than the bulk equilibrium pressure, $p_0$. Clearly the droplet must be small before it can become large. But unless the droplet reaches a certain critical radius $r^*$ depending on the vapor pressure, it will tend to evaporate rather than grow. From another viewpoint, consider Figure 9.12.1. Once the droplet has reached a size at which the

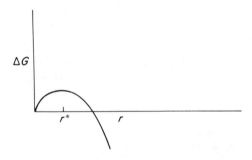

FIGURE 9.12.1. Free energy of formation of a droplet.

Gibbs free energy is a maximum (the critical nucleus), further growth (the addition of one more molecule) results in a decrease in Gibbs free energy. It can be shown (see the previously cited Ruoff, *Materials Science* Section 18.14) that the maximum value of the Gibbs free energy of formation of the droplet, $\Delta G^*$, is

$$\Delta G^* = \frac{4}{3}\pi\gamma r^{*2}.$$

The rate of formation of supercritical nuclei is called **nucleation.** Its rate is calculated as follows. The concentration $n^*$ (number per unit volume) of critical size droplets is

$$n^* = n e^{-\Delta G^*/k_B T}$$

where $n$ is the concentration of single water molecules (the number of single molecules per unit volume). This droplet becomes supercritical, if one more molecule is added. It can be shown from the kinetic theory of gases that the number of gas particles striking unit area per unit time is $\frac{1}{4}n\bar{v}$ where $\bar{v}$ is the average speed. The area in the present case is the surface area of the spherical droplet, $4\pi r^{*2}$. Therefore the time required for one molecule to hit the droplet is

$$\frac{1}{n\bar{v}\pi r^{*2}}.$$

Hence, the number of nuclei formed per unit time is

$$I = \pi n^2 \bar{v} r^{*2} e^{-\Delta G^*/k_B T}$$

or

$$I = \pi n^2 \bar{v} r^{*2} e^{-(4/3)\pi \gamma r^{*2}/k_B T}. \tag{9.12.2}$$

Since $r^*$ is a known function of the pressure [see (9.12.1)] and since the ideal gas law is $p = nk_B T$, the nucleation rate is given in terms of the pressure. The pre-exponential factor in (9.12.2) equals about $10^{28}$ nuclei/cm³-sec for $p/p_0 = 2.7$ and is nearly independent of pressure.

Table 9.12.1 gives results for $I$ depending on the ratio $p/p_0$, which is called the **supersaturation ratio.** Note that increasing the supersaturation ratio by 100% (from $p/p_0 = 2$ to 4) changes $I$ by $10^{54}$. This is a *strong* dependence!

**Table 9.12.1.** RATE OF NUCLEATION AT 300°K

| $p/p_0$ | $I$ (nuclei/cm³-sec) | $t^*$ |
|---|---|---|
| 2 | $10^{-43}$ | $\sim 10^{35}$ years |
| 3 | 1 | 1 sec |
| 4 | $10^{11}$ | $10^{-11}$ sec |

* $t$ = time for the appearance of one nucleus in a volume of 1 cm³.

OTHER EXAMPLES OF NUCLEATION. The nucleation of the water droplet as just described is called **homogeneous nucleation.** It may happen that dust (or some other particle) is present. Absorbed layers of water on this dust particle would effectively make a water droplet available for further condensation. This is called **heterogeneous nucleation.** This process can take place at much lower supersaturations.

The formation of solid crystals from the melt (e.g., ice in water) is another example of nucleation. The motion of the molecules in the water to the crystal involves diffusion in the water or diffusion on the surface to kinks in ledges on the surface.

The formation of a precipitate such as $CuAl_2$ in aluminum is also a nucleation process. This differs from the water droplet nucleation in at least three ways. First, the mechanism of aggregation of $CuAl_2$ or its equivalent concentration fluctuation involves the diffusion of Cu into and Al out of a critical radius region. Second, the surface energy may vary with direction and with the concentration and the concentration gradient. Third, because solids support static shear stress elastic strains may be introduced. The details of analysis in this case are therefore more involved than with the single component liquid-vapor system.

Finally, there are two-dimensional and one-dimensional nucleation processes. Consider a perfect crystal. Further growth of the crystal requires that a disc of atoms of a critical size form on a surface. This requires nucleation. Consider also the addition of another polymer molecule to a folded chain structure. This also requires nucleation.

In general, the rate of homogeneous nucleation depends on the product of the probability of formation of a critical nucleus which is given by

$$n^* = n e^{-\Delta G^*/k_B T}, \qquad (9.12.3)$$

where $\Delta G^*$ is the critical Gibbs free energy and the rate at which a molecule is added to such a nucleus. The greater the supersaturation (or the supercooling) is, the smaller $\Delta G^*$ will be.

In the formation of a solid nucleus from a liquid, $\Delta G^*$ can be shown to be equal to

$$\Delta G^* = \frac{16\pi \gamma^3 T_0^2}{3\Delta H_{\text{vol}}^2 (T_0 - T)^2}, \qquad (9.12.4)$$

where $T_0 = T_m$ in the present case and $\Delta H_{\text{vol}}$ is the enthalpy change (in the bulk) per unit volume during freezing.

**EXAMPLE 9.12.2**

Assume that the rate of nuclei formation from the melt equals

$$I = A e^{-\Delta G^*/k_B T},$$

where $A$ does not vary with temperature. Is there a temperature for which $I$ has a maximum if $\Delta G^*$ is given by (9.12.4)?

*Answer.* Yes. $dI/dT = 0$ when $T = T_0/2$.

In general the rate of homogeneous nucleation varies with temperature as shown in Figure 9.12.2. It depends strongly on the supercooling, $\Delta T = T_0 - T$. The nucleation of a solid precipitate in a solid follows similar rules.

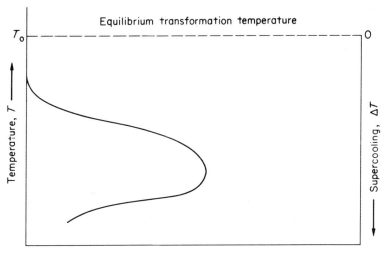

FIGURE 9.12.2. Nucleation rate for homogeneous nucleation as a function of temperature. $T_0 - T = \Delta T$ represents the supercooling.

### Q. 9.12.1

The equilibrium vapor pressure of a droplet is (a) (more or less) than the equilibrium vapor pressure of the bulk. The equation which describes the vapor pressure as a function of droplet radius is called the (b)_____. To form a water droplet in the presence of a supersaturated vapor and *nothing* else, it is necessary that the (c)_____ of the system first increase before it can decrease. To form a droplet under these conditions, a process known as (d)_____ must occur. The rate of homogeneous nucleation varies (e) (slowly or rapidly) with the supersaturation ratio.

\* \* \*

## ANSWERS TO QUESTIONS

**9.12.1** (a) More, (b) Kelvin equation, (c) Gibbs free energy, (d) nucleation, (e) incredibly rapidly.

# REFERENCES

### Equilibrium

Mark, M., *Thermodynamics, An Auto-Instructional Text*, Prentice-Hall, Inc., Englewood Cliffs, N.J. (1967). A programmed learning approach.

Sears, F. W., *Thermodynamics*, Addison-Wesley Publishing Company, Inc., Reading, Mass. (1953). A good discussion of the fundamentals of thermodynamics, kinetic gas theory, and the elements of statistical mechanics.

Fast, J. D., *Entropy*, McGraw-Hill Book Company, Inc., New York (1962).

### Random Walk

Kac, M., "Probability," *Scientific American* (September 1964). An excellent elementary discussion of the random walk.

### Diffusion

Shewmon, P. G., *Diffusion in Solids*, McGraw-Hill Book Company, Inc., New York (1963). A good discussion of the mechanics of diffusion.

Jost, W., *Diffusion in Solids, Liquids, Gases*, Academic Press, Inc., New York (1952).

Peterson, N. L., "Diffusion in Metals," in *Solid State Physics*, Vol. 22, ed. by F. Seitz, D. Turnbull, and H. Ehrenreich, Academic Press, Inc., New York (1968), p. 409.

Runyan, W. R., *Silicon Semiconductor Technology*, McGraw-Hill Book Company, Inc., New York (1965). Has an excellent discussion of various aspects of diffusion in silicon.

### Nucleation and Transformations

Turnbull, D., "The Undercooling of Liquids," *Scientific American* (January 1965), p. 38.

Fine, M. E., *An Introduction to Phase Transformations in Condensed Systems*, The Macmillan Company, New York (1964).

# PROBLEMS

**9.1** (a) Can it be shown directly by experiment that

$$PV = \tfrac{2}{3} U_k$$

for a dilute argon gas?

(b) If so, explain how.

**9.2** Why is it true that the internal energy of an ideal monatomic gas is given by its translational kinetic energy?

**9.3** (a) Derive the volume thermal expansion coefficient (at constant pressure) for a gas having the equation of state of (9.1.13).
(b) Then discuss the experiments which you would perform to obtain the value of the absolute zero of temperature ($-273.16°$K). (This does not mean that you attempt to reach $0°$K experimentally.)

**9.4** Derive the expression for $v_{rms}$ for an ideal gas.

**9.5** Derive Maxwell's expression for the coefficient of viscosity of an ideal gas.

**9.6** (a) About how large are argon atoms?
(b) About how far apart are they on the average at 1 atm pressure and $0°$C?
(c) About what is their mean free path at 1 atm and $0°$C?

**9.7** Discuss (quantitatively if possible) why argon would or would not be an ideal gas at 1000 atm and $0°$C.

**9.8** A copper ball is dropped in a vacuum. It has a mass $M$ and a velocity of 3000 cm/sec. The temperature is $0°$C. Differentiate between internal energy of the ball and its rigid body or external energy.

**9.9** Two ideal gases are mixed together, and their entropy of mixing is $\Delta S_{mix}$. What is the minimum work that has to be done (assuming 100% efficiency) to separate these two gases at temperature $T$?

**9.10** One of the most profound discoveries in all of science was the finding that the ordinary energy terms such as kinetic energy and potential energy were not sufficient to define the equilibrium of a chemical system. Discuss.

**9.11** When $n$ vacancies are added to a solid of $N$ atoms at low pressure (effectively zero) the internal energy of the system increases by $nE_f$. If this is so, why is it still true that such vacancies can be present under equilibrium conditions at high temperatures?

**9.12** Explain why Sn dissolves in Pb at high temperatures although it does not at low temperatures.

**9.13** Why do ordered solutions become disordered at high temperatures?

**9.14** Why do materials vaporize at high temperatures?

**9.15** Describe the possibility of an altimeter based on the barometric formula.

**9.16** Suppose that individual particles in a system of particles have only two energy levels, $u_0 = 0$ and $u_1 = \frac{1}{40}$ eV.
(a) Compute the fraction in each energy state at $300°$K.
(b) Compute the average energy of a particle.

**9.17** Show that
$$x = x_0 \cos \sqrt{\frac{k}{m}} t$$

is a solution of the differential equation of motion for the simple harmonic oscillator.

**9.18** Outline (without going through the details) the steps for Einstein's theory of the specific heat of solids.

**9.19** Discuss the origin of the equation

$$\Gamma = \nu e^{-Q_a/RT}$$

for the specific reaction rate between a hydrogen molecule and a deuterium atom:

$$H_2 + D \rightarrow HD + H.$$

**9.20** A good engineer or scientist is often capable of taking limited data and using them to calculate (almost instantaneously) good approximate values (factor of 2, order of magnitude, etc.). What two diffusion equations would the engineer have memorized?

**9.21** Derive the expression for the diffusion coefficient for diffusion in a one-dimensional lattice:

$$D = \tfrac{1}{2}\Gamma l^2.$$

**9.22** Is the variation of $D$ with temperature for an ideal gas more or less rapid than the variation of $D$ with temperature for a solid?

**9.23** Three processes which can be used for desalinization of sea water are distillation, freezing, and reverse osmosis. If the salt content is reduced from a fixed initial concentration to a fixed final concentration by any of the three processes, the energy required if the processes are carried out reversibly (at maximum efficiency) is the same for all. However, when carried out at a finite rate, additional energy is required. In the case of reverse osmosis, for what is the additional energy required?

**9.24** Is a $p$-$n$ junction a thermodynamically stable solid or is it metastable? Why?

**9.25** Give a process in ionic crystals which is controlled by the diffusion rate of
(a) The fastest-moving ion. Why?
(b) The slowest-moving ion. Why?

**9.26** Explain how surfaces, grain boundaries, and dislocations (edge and screw) can act as sources or sinks for vacancies.

**9.27** A vacancy is created on a dislocation line by dislocation climb. It then makes 10,000 jumps through the good crystal. What is the most probable distance from the origin?

**9.28** Calculate the time it would take at 1750°F to diffuse carbon into iron to a penetration depth 0.005 in.

**9.29** Discuss the role which an adherent oxide film plays in oxidation resistance.

**9.30** The oxidation of a certain iron alloy boiler tube follows the law

$$X^2 = Kt,$$

where $X$ is the thickness of the oxide and

$$K = e^{-17,500/T(°K)} \text{ cm}^2/\text{sec}.$$

If $T = 600°C$, calculate how much time has to elapse before an oxide layer of 0.1 cm forms on the surface.

**9.31** You have an assignment to measure the self-diffusion coefficient of copper. Describe in detail (with equations) how you would do this. Include the size of the sample, etc.

**9.32** Silicon is somewhat transparent in the infrared. In what application is it desirable to have a *p-n* junction near to the surface?

**9.33** A new steel being marketed for automotive applications consists of a low alloy steel core and a high Cr surface layer diffused into the steel for increased corrosive resistance. Calculate how long a heat treatment would be required at 1000°C to diffuse Cr far enough into the steel so that the composition is 18% Cr at 0.020 cm below the surface and 100% Cr at the surface. Assume $D_0 = 0.47$ cm²/sec and $Q_D = 79.3$ kcal/mole for diffusion of chromium in iron.

**9.34** Suppose a large number of vacancies ($10^6$, say) are placed side by side within a circle of radius $r$ within a copper crystal on the (111) plane.
(a) What is the result?
(b) Under what sort of physical conditions might this be achieved?

**9.35** The kinetics of the growth of precipitates from a supersaturated solution of Al–4 wt. % Cu is $10^8$ times as fast as expected from lattice diffusion under equilibrium conditions. Give a possible explanation.

## MORE INVOLVED PROBLEMS

**9.36** Explain quantitatively the origin of $a$ and $b$ in van der Waals' equation of state.

**9.37** On the basis of (9.1.7) find the fraction $dg$ of molecules simultaneously having velocity components between $v_x$ and $v_x + dv_x$, $v_y$ and $v_y + dv_y$, and $v_z$ and $v_z + dv_z$.

**9.38** Show from Problem 9.37 that the probability of the *velocity* having magnitude between $v$ and $v + dv$ is

$$dF = \frac{b^{3/2}}{\pi^{3/2}} e^{-bv^2} 4\pi v^2 \, dv.$$

**9.39** (a) Explain why the average value of $v^2$ is given by

$$\overline{v^2} = \int_{v=0}^{v=\infty} v^2 \, dF(v).$$

(b) Use this to show that $\overline{v^2} = 3/2b$. Since you have shown earlier that $\overline{v^2} = 3k_BT/m$, the value of $b$ is now determined.

**9.40** Using the result of Problem 9.38, what is
(a) The average magnitude of the velocity?
(b) The most probable magnitude of the velocity?

**9.41** (a) Show that

$$\frac{1}{1-x} = \sum_{i=0}^{\infty} x^i \quad \text{for } -1 < x < 1.$$

(b) Use this to show that

$$\sum_{i=0}^{\infty} e^{-ih\nu/k_BT} = \frac{1}{1 - e^{-h\nu/k_BT}}.$$

**9.42** (a) Using (a) from Problem 9.41 show that

$$\frac{x}{(1-x)^2} = \sum_{i=0}^{\infty} ix^i.$$

(b) Use this to evaluate

$$\sum_{i=0}^{\infty} ih\nu e^{-ih\nu/k_BT}.$$

**9.43** (a) What is the flux of atoms of H (hydrogen) across a foil of Pd $1.5 \times 10^{-2}$ cm thick at $750°K$, if the concentration of H on the high-pressure side of the foil is $1.1 \times 10^{22}$ atoms/cm$^3$ and on the low-pressure side the concentration is only $0.1 \times 10^{22}$ atoms/cm$^3$.

(b) How long would you have to run this system, assuming the foil has a cross section of 1 cm$^2$ to produce 1 mole of pure hydrogen? *Note:* Assume that the dissolved hydrogen concentration is fixed in time at each side of the foil. It will, in fact, be determined by the pressure according to

$$C(\text{H}) = K(P_{\text{H}_2})^{1/2}.$$

This is an example of the *assumption of local equilibrium*. In this case, the assumption is made that the transport of the gas to the surface and the decomposition and dissolution of the gas is very rapid compared to the rate of diffusion through the solid, so that the concentration at the surface is the same as if no diffusion were occurring.

**9.44** The process of reverse osmosis, where water is squeezed out of a salt solution by applying pressure of several hundred pounds per square inch, has great potential for desalting ocean water. See U. Merten, *Desalinization by Reverse Osmosis*, The M.I.T. Press, Cambridge, Mass. (1966). Discuss this process in as much quantitative detail as possible.

**9.45** When a hot gold wire is cooled very rapidly, there is a supersaturation of vacancies. However, it is usually assumed that in the good crystal material near grain boundaries, free surfaces, and dislocations the concentration very quickly reaches the equilibrium concentration. Criticize this assumption of *local equilibrium vacancy concentration*.

**9.46** A piece of Pb and another piece of radioactive Pb* are welded together as shown.

PROBLEM FIGURE 9.46.

After a sufficiently long time at high temperature (but below $T_m$) the Pb* is uniformly distributed through the couple.
(a) Are Pb and Pb* similar chemically?
(b) Do Pb and Pb* crystals have the same vibrational internal energy?
(c) What energetically causes the diffusion to occur?

**9.47** Hydrogen can readily be purified by passing it through Pd. Invent similar separation or purification procedures for He and $O_2$.

# SOPHISTICATED PROBLEMS

**9.48** (a) What is the Doppler shift?
(b) A mercury atom at 100°C in an excited state emits radiation with a wavelength of 5461 Å. If the atom is moving with a velocity of magnitude $\bar{v}$ away from the observer, what is the wavelength?
(c) Discuss Doppler broadening in general.

**9.49** A gas is placed in a centrifuge and rotated at an angular velocity $\omega$ for a long time. Calculate the gas pressure distribution versus radius $r$, assuming the chamber has an inner radius $a$ and an outer radius $b$. Suggest values for $a$, $b$, and $\omega$ which would make this a suitable device for bringing about enrichment of $^{235}UF_6$ in a gas solution containing predominantly $^{238}UF_6$.

**9.50** (a) Calculate the vacancy concentration in gold at $T_m$ and $T_m/4$.
(b) Comment on the measurability of the vacancy concentration in an 0.005-in. diameter gold wire 30 in. long if vacancies contribute a resistivity change of 1.5 $\mu\Omega$-cm/at. % at $T_m$.
(c) Do the same if the wire is quenched from $T_m$ to $T_m/4$ and no vacancies are lost.

# Prologue

The size, distribution, and composition of the different phases in an engineering material can have profound effects on its properties. A map of pressure versus temperature can be used to show which phases are present under equilibrium conditions in a one-component system. A map of temperature versus composition can be used to show which phases are present under equilibrium conditions in a two-component system at constant pressure. The use of these *phase diagrams* and the application of the *lever rule* is described. Some of the more common types of phase diagrams involving *continuous solubility* and *limited solubility* are discussed. Most transformations which occur in the real world take place under *nonequilibrium conditions* and phase diagrams give us a hint of what will result. The *lamellar structure* which occurs during a nonequilibrium eutectic transformation is described. So is *coring*. A nonequilibrium transformation forms the basis of the *heat treatment* process known as *age precipitation hardening*.

We also describe some of the many microstructures possible in steel and cast iron because of the nonequilibrium processes which occur upon cooling in the Fe–C system. The *eutectoid transformation* and the *martensitic transformation* are introduced.

The *distribution coefficient* is used to describe *segregation* and *zone refining*.

The student should note several ways of manipulating microstructure which are discussed in this chapter; these are examples of molecular architecture.

# 10

# PHASE DIAGRAMS

## 10.1 INTRODUCTION

We begin with a study of equilibria of a pure substance. Then, equilibrium in a system in which composition may be a variable is discussed.

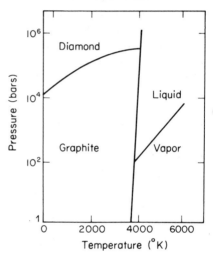

FIGURE 10.1.1. Carbon $P$-$T$ equilibrium diagram.

A ONE-COMPONENT SYSTEM. Figure 10.1.1 is the **phase diagram** for a single component, carbon. It is a map in $P$-$T$ space (pressure and temperature space) of the various equilibrium phases. Recall that a phase is the material in a region of real space which, in principle, can be mechanically separated from other phases. Each phase has a characteristic structure which distinguishes it from another phase. As an example, graphite, which has a hexagonal crystal structure, is a different phase than diamond, which has a diamond cubic crystal structure.

It is of interest to point out that it can be shown from thermodynamics that the slope $dP/dT$ of the dividing line between the liquid and vapor phases is given by

$$\frac{dP}{dT} = \frac{\Delta H_v}{T\,\Delta V_v}, \tag{10.1.1}$$

where $\Delta H_v$ is the heat absorbed by a mole of the material at constant pressure (enthalpy) and $\Delta V_v$ is the volume change when a mole of liquid is transformed to vapor at temperature $T$ and pressure $P$. The quantities $\Delta H_v$ and $\Delta V_v = V_g - V_l$ also can be measured independently of the determination of the $P$-$T$ curve. Thus, a phase diagram contains quantitative information in addition to a mapping of the phase regions in $P$-$T$ space.

Note that in the single-component phase diagram, there is but a single phase within a given region of $P$-$T$ space, although two phases are in equilibrium along certain lines. The equilibrium phase boundary between graphite and diamond has been a subject of research for several hundred years, but its determination had to await the development of modern high-pressure technology. The ability to make diamonds, the hardest known material, of industrial quality, has obvious commercial implications in this age when many very hard materials of great potential use can be machined only by use of a harder substance to wear away unwanted material.

To grow diamond crystals from pure carbon at a reasonable rate, it would be necessary to have molten carbon present to assure rapid atom motion. This would require a very high temperature (4000°K or so). This would then require a very high pressure (140 kbars or so) since the growth would take place along the line between diamonds and liquid. This combination of pressure and temperature is not easy to attain. Consequently, materials scientists took the following approach. Why not find a solvent for carbon such that the solvent would be in the liquid state at a much lower temperature? Since carbon would diffuse rapidly in this liquid, diamond growth could occur at a lower temperature. Moreover, as is clear from Figure 10.1.1, if the temperature is reduced, diamond can form at a reduced pressure. Nickel is used as such a solvent.

A TWO-COMPONENT SYSTEM. Consider now the two-component system of nickel and carbon whose phase diagram is shown in Figure 10.1.2. This is a map in $T$-$x$ space (where $x$ is the mole fraction of carbon) at a constant pressure of 54 kbars. Because this system has two-components (nickel and carbon) it is called a **binary system** and the diagram is called a **binary phase diagram.** Note that it describes equilibrium only at a fixed pressure. If pressure were varied, a three-dimensional representation would be needed. Alternatively, a series of $T$-$x$ diagrams at different constant pressures might be used. In Figure 10.1.2, the phases are labeled as follows: $l$, liquid; $g$, graphite; $d$, diamond; $\alpha$, an interstitial solid solution of carbon in the fcc crystals of nickel.

In the case of binary systems, the phase regions sometimes contain one phase and sometimes two phases. In fact, at a constant temperature, if there is more than one phase as composition varies, the phase regions will have alternately 1, 2, 1, ..., 2, 1 phases/region. Consider the situation at

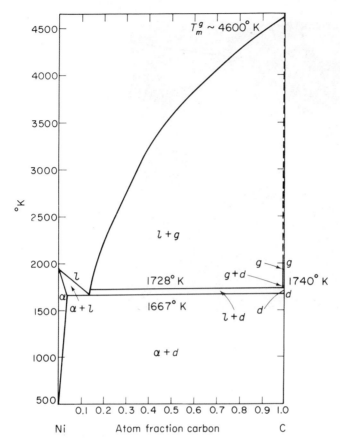

FIGURE 10.1.2. Nickel-carbon phase diagram at 54 kbars. [From H. M. Strong and R. E. Hanneman, *Journal of Chemical Physics* **46**, 3668 (1967).]

1700°K beginning at $x = 0$, i.e., at 0%C. There is first the $\alpha$ region which may vary from pure nickel to a few percent carbon but in any case is a single phase. At higher carbon compositions, there is the $\alpha + l$ region (solid particles of $\alpha$ in liquid, sometimes called a slush region). At still higher carbon compositions, there is the $l$ region, followed by the $l + d$ region (solid diamonds in the liquid) and finally diamonds. The growth of diamonds from nickel solution depends on the presence of the narrow $l + d$ (liquid plus diamond) region from 1667 to 1728°K. Thus diamonds can form there from the liquid phase at modest temperatures (compared to the 4000°K needed with pure carbon), just as salt crystals grow from water solution at room temperature. Figure 10.1.3 shows a diamond grown in this fashion. Note the well-developed {111} planes. (How could you show that they were {111} planes?)

## 434  Phase Diagrams

FIGURE 10.1.3. Diamond grown from nickel solvent. (Photographed by C. C. Chao in author's laboratory.)

### Q. 10.1.1

A phase diagram of a one-component system is a map of thermodynamic variables, such as $P$ and $T$, showing (a)_____. The atomic arrangements in one phase are (b)_____ the atomic arrangements of another phase. Phase diagrams contain certain quantitative information. Thus the slope of the line separating liquid and vapor is (c)_____. The growth of diamonds from molten carbon requires extremely high pressures and temperatures. More modest conditions can be used if (d)_____.
Is Figure 3.10.1 a phase diagram? (e)_____. A two-component phase diagram involving $T$ and $x$ at constant $P$ has regions containing (f)_____.

* * *

## ANSWERS TO QUESTIONS

**10.1.1** (a) Regions (of $P$-$T$ space in this case) in which a given phase is the equilibrium form, (b) different from, (c) $\Delta H_v / T \Delta V_v$, (d) the carbon is dissolved in a solvent such as liquid nickel, (e) yes, (f) one or two phases.

## 10.2 BINARY SYSTEMS

A number of useful materials are binary systems. For instance, age precipitation hardenable aluminum alloys are essentially based on Al with 4 wt. % Cu, although usually commercial aluminum alloys contain other alloying elements also. Steel is essentially Fe with C, although commercial alloys contain various other elements. Ordinary solder is a Pb and Sn alloy. $p$-type silicon is silicon with boron or some other trivalent element while $n$-type silicon is silicon with phosphorus or some other pentavalent element added. High field superconductive wire is often made of an alloy of niobium and zirconium. Hard magnetic materials are made of samarium and cobalt. Clearly, binary systems form the basis for an interesting array of materials. These systems can be studied in a systematic fashion. The systems discussed here are assumed to be at a pressure of 1 atm. In more advanced courses ternary and other more complicated multicomponent systems are discussed.

COMPLETE SOLUBILITY IN BOTH SOLID AND LIQUID. The phase diagram of a typical binary system showing complete solubility in both the solid and liquid phases is illustrated in Figure 10.2.1. Here the components are the compounds NiO and MgO. Note that at 2900°C a single liquid, of any concentration from 0% to 100% MgO, is possible. The two components are completely soluble in each other in the liquid phase. Solids of both pure NiO and pure MgO have the NaCl-type crystal structure. The solid solution phase has $Ni^{2+}$ and $Mg^{2+}$ placed (more or less) randomly on the positive ion sites of the NaCl-type crystal structure. Hence, the solid

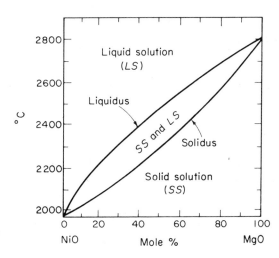

FIGURE 10.2.1. MgO–NiO system.

has only one type of crystal structure for compositions of 0% to 100% MgO, although the lattice parameter is a function of composition. The single solid phase is a substitutional solid solution.

There are three regions of interest in Figure 10.2.1. Above the **liquidus** line, a single liquid solution is found (single phase). Below the **solidus** line, a single solid solution is found (one phase). The third region is the two-phase region between the liquidus and the solidus. Here, particles of solid solution will be suspended (assuming the gravitational field is negligible) in the liquid. Examples of other systems showing the general form of Figure 10.2.1 are given in Table 10.2.1.

**Table 10.2.1.** SYSTEMS SHOWING COMPLETE SOLUBILITY

| Elements | Compounds |
|---|---|
| Au–Pt | CoO–CoS |
| Cu–Ni | CoO–MgO |
| Mo–W | FeO–MgO |
| Ag–Au | CoS–FeS |
| Ag–Pd | $ThO_2$–$UO_2$ |
| Au–Pd | $PbCl_2$–$PbBr_2$ |
| Pt–Rh | $\alpha$-$Al_2O_3$–$Cr_2O_3$ |
| Pt–Ir | $2MgO \cdot SiO_2$–$2FeO \cdot SiO_2$ |

The major important difference between the metallic systems (under elements in Table 10.2.1) and the ceramic systems (under compounds) is the *rate* at which equilibrium is achieved. In the metals this is fairly rapid, while in the ceramics it is often *very* slow. The fluidity coefficient of liquid metals at the melting point is about 50 poise$^{-1}$, while for liquid silica at the melting point of cristobalite this is $10^{-7}$ poise$^{-1}$. [Fluidity, defined in Equation (2.3.3), is a measure of the ability of atoms, molecules, or complexes (such as $SiO_4^{4-}$) to move.] The ratio of nearly $10^9$ represents an extreme situation, and in this case silica tends to form glass when solidifying from the melt rather than forming an equilibrium crystal structure.

**Composition** means the mole fraction or percentage (or the weight fraction) of the component. The components in the binary system whose phase diagram is shown in Figure 10.2.1 are NiO and MgO.

If an alloy whose composition is 80 mol. % MgO and 20 mol. % NiO is heated, it will exist (see Figure 10.2.1) as a single phase (crystalline with NaCl-type structure) up to 2600°C. Between 2600 and 2700°C, it will exist as two phases, and above 2700°C there will be only liquid present.

## Q. 10.2.1

Above a line (on a two-component $T$-$x$ phase diagram at constant pressure) called the (a)_____ there is only liquid solution, while below a line called the (b)_____ only solid solution is found. Ceramic systems often have phase diagrams which are analogous to metallic systems; however, they differ in the time needed to establish (c)_____. 25 mol. % NiO is mixed with 75% mol. % MgO and melted and cooled; the resultant alloy has a (d)_____ of 25 mol. % NiO and 75 mol. % MgO.

\* \* \*

Let us now examine the behavior at 2600°C as the composition of the alloy varies from pure NiO to pure MgO. With pure NiO, there is only a single liquid phase; as MgO is added there is still only a single phase until a composition of 64 mol. % MgO has been achieved. At this point, the liquid (at 2600°C) has dissolved the maximum amount of MgO; it has reached its solubility limit. It cannot dissolve any more MgO. If more MgO is added to the system to bring its overall composition to 75 mol. % MgO, two phases will be present: liquid solution and solid solution.

**EXAMPLE 10.2.1**

What will be the composition of the liquid solution at 2600°C for an alloy of 25 mol. % NiO–75 mol. % MgO?

*Answer.* The liquid will be 64 mol. % MgO. It cannot be more than this because that is the solubility limit at this temperature. It cannot be less because then only a single phase would be present and the overall composition could not be 75% MgO.

As additional MgO is added at 2600°C and the net composition of the alloy reaches 80% MgO, the solidus line is reached. Beyond 80%, there is only one phase present at 2600°C, namely, the solid solution. The two-phase region (enclosed by the solidus and liquidus lines) extends from 64% MgO to 80% MgO at 2600°C. The liquid phase in this two-phase region always has a composition of 64% MgO at 2600°C, and the solid phase in this two-phase region always has a composition of 80% MgO at 2600°C regardless of whether the alloy composition is 64.4% MgO, 70% MgO, or 79.4% MgO. An alloy with overall composition of 65% MgO at 2600°C consists of two phases. It is mostly liquid with a composition of 64% MgO but there is some solid phase with a composition of 80% MgO. The fraction which is liquid can be calculated from the phase diagram by using a mass balance as illustrated in the next example.

## EXAMPLE 10.2.2

An alloy of 24 mol. % NiO, 76 mol. % MgO is heated to 2600°C. What is the mole fraction of liquid present?

*Answer.* Let $f_l$ be this unknown fraction. Then $1 - f_l$ is the mole fraction $f_s$ of solid solution. The MgO in the liquid plus the MgO in the solid solution must equal the total MgO. Let us suppose we start with 1 mole of alloy. Then we have 0.76 moles of MgO altogether. Hence, the MgO in the liquid plus the MgO in the solid phase equals the total MgO, or

$$0.64 f_l + 0.80(1 - f_l) = 0.76, \tag{10.2.1}$$

which can be rearranged to give

$$f_l = \frac{0.80 - 0.76}{0.80 - 0.64} = 0.25. \tag{10.2.2}$$

Hence, there are 0.25 moles of liquid whose composition is 64% MgO and 0.75 moles of solid solution whose composition is 80% MgO.

Figure 10.2.2 is used as a basis for illustrating the calculation of the previous example in another way. A line at constant temperature (2600°C in this case) is drawn across the two-phase region. This is called the **tie line**. The composition at the liquidus side of the tie line gives the composition of the liquid in the *entire* two-phase region at 2600°C. Likewise, the composition at the solidus side of the tie line gives the composition of the solid solution in the *entire* two-phase region at 2600°C. The mass balance

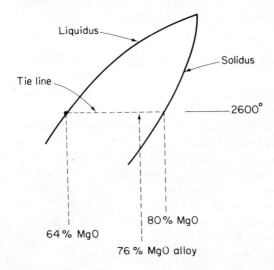

FIGURE 10.2.2. Expanded view of the two-phase region of Figure 10.2.1.

which is described by Equation (10.2.2) leads to a simple rule called the lever rule. Since $f_l + f_s = 1$, Equation (10.2.1) can be rewritten as

$$0.64 f_l + 0.80 f_s = 0.76(f_l + f_s),$$

or as

$$f_l(0.76 - 0.64) = f_s(0.80 - 0.76). \qquad (10.2.3)$$

This suggests the following mechanical balance which is an application of the principle of the lever.

Consider a fulcrum placed at the given alloy composition and the tie line to be a lever as in Figure 10.2.3 and the fraction of each phase to be a

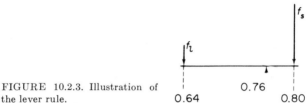

FIGURE 10.2.3. Illustration of the lever rule.

force. Then a balance of moments (force times a lever arm) leads to Equation (10.2.3) which, since $f_l = 1 - f_l$, can be written as

$$f_l = \frac{0.80 - 0.76}{0.80 - 0.64} = 0.25.$$

This illustrates the **lever rule,** which states: In a two-phase region the fraction of one phase equals the length of the lever opposite this phase divided by the length of the tie line.

### Q. 10.2.2

A tie line is a line drawn at (a)_____ across a (b)_____. The composition of the two phases in a two-phase region at a given temperature $T$ is (c)_____ _____. A phase diagram not only tells us which phases are present in a two-phase region but it also tells us (d)_____. To obtain the relative proportions of two phases at a given point $(T, x)$, we use the (e)_____, which can be derived by using (f)_____.

\* \* \*

To form a continuous solid solution, both components must have the same crystal structure. Moreover, in the case of elements, the radii of the atoms (and hence the lattice parameter) must be within about 14%. Similar rules apply for the lattice parameters of compounds such as CoO and MgO.

Figure 10.2.4 shows a system with continuous solubility in both the liquid and solid phases but with a coincident minimum in the liquidus and

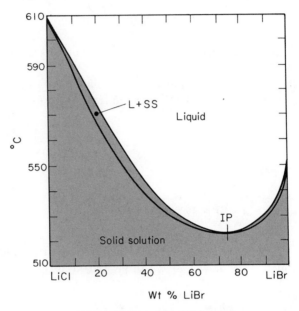

FIGURE 10.2.4. LiBr–LiCl system.

solidus, called an **indifferent point.** Other systems showing an indifferent point (labeled I.P. in Figure 10.2.4) are shown in Table 10.2.2.

It is often a characteristic of binary systems with an indifferent point that they tend to form stable ordered structures at low temperatures even

Table 10.2.2. SYSTEMS SHOWING COMPLETE SOLUBILITY WITH A COINCIDENT MINIMUM IN THE SOLIDUS AND LIQUIDUS CURVES

| Elements | Compounds |
|---|---|
| Ni–Pd<br>Cr–Mo | $KNO_2$–$NaNO_2$<br><br>$UF_4$–$ZrF_4$<br>$CaO \cdot SiO_2$–$SrO \cdot SiO_2$ |

though their behavior at higher temperatures is characterized by the curve in Figure 10.2.4. This is illustrated in Figure 10.2.5 where the phase diagram of the gold-copper system is shown. The solid $\alpha'$ and $\alpha''$ solutions are based on the fcc lattice. Their specific crystal structures are discussed just prior to Question 6.6.1. Figure 10.2.5 shows both atomic percentage and weight

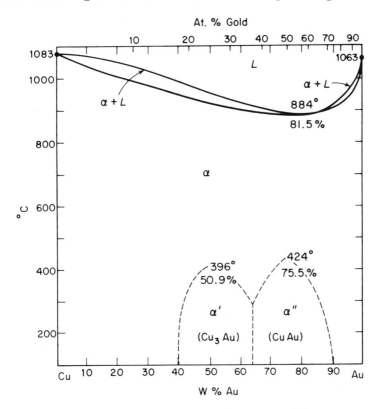

FIGURE 10.2.5. Cu–Au system. Ordered CuAu and Cu$_3$Au both consist of an ordered arrangement of Cu and Au atoms on an fcc lattice.

percentage for the gold-copper system. Atomic percentage is related to weight percentage according to

$$\text{At. \% Au} = \frac{\text{Wt. \% Au}/M_{Au}}{(\text{Wt. \% Au}/M_{Au}) + (\text{Wt. \% Cu}/M_{Cu})} \times 100\%. \quad (10.2.4)$$

Here, $M_{Au}$ and $M_{Cu}$ are the molecular weights of gold and copper, respectively. When dealing with compounds, we relate the mol. % to wt. % in the same way.

Other systems showing complete solubility and ordering phenomena are LiCl–NaCl, Cd–Mg, Cu–Pd, Cu–Pt, and Ni–Pt. Moreover, the Cr–Fe and V–Fe systems are similar to the above, although they show added effects as well.

## Q. 10.2.3

A coincident minimum in the liquidus and solidus is called (a)_____. A necessary condition for materials to have continuous solubility is (b)_____. Metals will not show extended solubility if the atomic radii differ by more than (c)_____. Binary systems with an indifferent point often exhibit (d)_____ at low temperatures.

\* \* \*

COMPLETE SOLUBILITY IN THE LIQUID BUT ONLY SLIGHT SOLUBILITY IN THE SOLID PHASE. Because long-range disorder is a characteristic of liquids, it is often true that complete solubility is possible in the liquid phase even under conditions in which there is little solubility in the solid phase. This is illustrated in Figure 10.2.6. Continuous solubility in the solid state is not

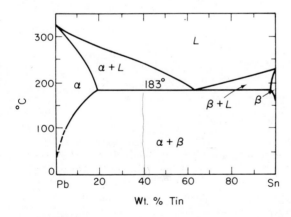

FIGURE 10.2.6. Pb–Sn diagram.

expected in this case inasmuch as lead has an fcc crystal structure while tin has a tetragonal crystal structure at high temperatures and a diamond cubic structure at low temperatures.

Note that Figure 10.2.6 should be considered to be a modification of Figure 10.2.4. The phase labeled $\alpha$ is a substitutional solid solution of Sn in Pb and it has an fcc crystal structure. The phase labeled $\beta$ is a substitutional solid solution of Pb in Sn and it has a tetragonal crystal structure.

Note that at any constant temperature the number of phases present in a given region alternates from one to two to one, etc.

A point worthy of particular notice is the point at 61.9 wt. % Sn and 183°C. Here three phases coexist: solid $\alpha$ of 19.2% Sn in Pb, solid $\beta$ of 2.5% Pb in Sn, and liquid with 61.9% Sn. This point (defined by both the temperature and the composition) is called a eutectic point. By definition a **eutectic point** is a point (in temperature and composition space) involving a dynamic equilibrium:

$$\text{liquid} \underset{\text{heating}}{\overset{\text{cooling}}{\rightleftarrows}} \text{solid } \alpha + \text{solid } \beta \qquad (10.2.5)$$

A transformation from a liquid to two solids is called a **eutectic transformation** and the temperature at which this occurs is called the **eutectic temperature** (183°C in the present case). The line at the eutectic temperature is called the **eutectic line.** An alloy of this composition (61.9 wt. % Sn in the present example) is called a **eutectic alloy.**

### EXAMPLE 10.2.3

Suppose you have an alloy with an overall composition of 40 wt. % Sn. Calculate the wt. % $\alpha$ and the wt. % liquid at $183 + \epsilon$°C, where $\epsilon$ is small.

*Answer.* Draw the tie line in the two-phase region $\alpha + L$ at $183 + \epsilon$°C. Then by the lever rule

$$\text{wt. } \% \ \alpha = \frac{61.9 - 40}{61.9 - 19.2} \times 100\% = 51.3\%$$

$$\text{wt. } \% \ L = 100\% - 51.3\% = 48.7\%.$$

Note that these calculations are for *equilibrium* conditions; under actual continuous cooling conditions, equilibrium is not attained due to the slowness of diffusion in the liquid and the even slower diffusion in the solid.

### EXAMPLE 10.2.4

For the 40 wt. % Sn alloy, calculate the wt. % $\alpha$ at $183 - \epsilon$°C.

*Answer.* Now the tie line is in the two-phase region $\alpha + \beta$. Hence

$$\text{wt. } \% \ \alpha = \frac{97.5 - 40}{97.5 - 19.2} \times 100\% = 73\%.$$

**EXAMPLE 10.2.5**

Why use a eutectic Pb–Sn solder instead of some other composition?

*Answer.* Because it melts at a lower temperature and is hence more convenient to use then either pure lead or pure tin.

Table 10.2.3 gives some further examples of eutectic systems. We note that although LiF and LiBr have the same crystal structure, there is a large variation in cell parameter; these are 4.0279 Å and 5.501 Å, respectively, and hence differ by 37% or 27% depending on which compound is the solvent. Hence, little solubility is expected in the solid state.

**Table 10.2.3.** SIMPLE EUTECTIC SYSTEMS WITH LIMITED SOLUBILITY

| Elements | Compounds |
|---|---|
| Bi–Cd | LiF–LiBr |
|  | CaO–MgO |
| Ag–Cu | FeO–Na$_2$O·2SiO$_2$ |
| Ag–Pb | Al$_2$O$_3$–Ca$_3$(PO$_4$)$_2$ |
| Al–Si | KF–LiF |
| Au–Be | KF–NaF |
| Cd–Pb | KF–BaF$_2$ |
| Cd–Zn | MgF$_2$–BeF$_2$ |
| Pb–Sb | LiCl–LiCO$_3$ |
|  | CaMgSi$_2$O$_6$–2MgO·SiO$_2$ |
| Sn–Zn | CaAl$_2$Si$_2$O$_8$–SiO$_2$ |
|  | Ethylene glycol–water |

**Q. 10.2.4**

A eutectic point at constant $T$, $x$, and $P$ involves an equilibrium between (a)_____. When a single liquid transforms at a single temperature into two solids, the transformation is called (b)_____. When a eutectic liquid solidifies, the solid contains (c)___ phases.

\* \* \*

COMPOUND FORMATION. The formation of compounds by two components which themselves are compounds in this case is illustrated in Figure 10.2.7. Here the new compound shows a **congruent melting point;** i.e.,

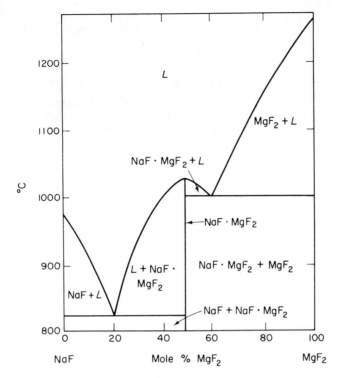

FIGURE 10.2.7. NaF–MgF$_2$ system.

the solid compound NaF·MgF$_2$ changes to a liquid of identical composition at 1030°C. Compound formation between the elements or between salts is extremely common. Some systems showing compound formation are listed in Table 10.2.4. Thus in the Al–Ca system we find the stable compound Al$_2$Ca. Note that the phase diagram of Figure 10.2.7 could be considered to

Table 10.2.4. SYSTEMS SHOWING COMPOUND FORMATION

| Elements | Salts |
| --- | --- |
| Al–Ca | BaO–TiO$_2$ |
| Ba–Pb | CaO–TiO$_2$ |
| Fe–B | CaO–ZrO$_2$ |
| Mg–Ca | CaO·SiO$_2$–CaO·Al$_2$O$_3$ |
| Cu–P | Li$_2$CO$_3$–K$_2$CO$_3$ |
| Mg–Si | Li$_2$SO$_4$–K$_2$SO$_4$ |
| Mg–Sn | PbO–PbSO$_4$ |
|  | NaF–AlF$_3$ |

be two separate phase diagrams, one for the system $NaF-NaF \cdot MgF_2$ and another for the system $NaF \cdot MgF_2-MgF_2$.

OTHER CHARACTERISTICS. We shall list, at this time, a number of important transformation points: (1) inversion point, (2) congruent melting point, (3) indifferent point, (4) eutectic point, (5) eutectoid point, (6) peritectic point, and (7) peritectoid point. The **inversion point** is a transformation point such as that between two polymorphs, e.g., bcc iron and fcc iron at 910°C. We now define the others.

$$\textbf{eutectoid:} \quad \text{solid A} \underset{\text{heating}}{\overset{\text{cooling}}{\rightleftarrows}} \text{solid B} + \text{solid C} \quad (10.2.6)$$

$$\textbf{peritectic:} \quad \text{solid A} + \text{liquid} \underset{\text{heating}}{\overset{\text{cooling}}{\rightleftarrows}} \text{solid B} \quad (10.2.7)$$

$$\textbf{peritectoid:} \quad \text{solid A} + \text{solid B} \underset{\text{heating}}{\overset{\text{cooling}}{\rightleftarrows}} \text{solid C}. \quad (10.2.8)$$

In all the transformations given by (10.2.5)–(10.2.8) we shall be particularly interested in the kinetics of the process, because, in general, at normal cooling rates the reactions are often so slow that we do *not* obtain an equilibrium structure. These points are all called **invariant points** because it is impossible to vary either $T$ or $x$ without changing the number of phases.

### Q. 10.2.5

Components in a binary system often form (a)_____ with a fixed melting point. A eutectoid transformation involves a transformation from (b)_____.

\* \* \*

## ANSWERS TO QUESTIONS

**10.2.1** (a) Liquidus, (b) solidus, (c) equilibria, (d) composition.

**10.2.2** (a) Constant temperature, (b) two-phase region, (c) the compositions at the ends of the tie line, (d) the fraction of each phase, (e) lever rule, (f) a mass balance.

**10.2.3** (a) An indifferent point, (b) that they must have the same structure on the atomic scale, (c) 14%, (d) ordered structures.

**10.2.4** (a) A single liquid and two solid phases, (b) a eutectic transformation, (c) two.

**10.2.5** (a) Compounds, (b) one solid phase to a second and a third solid phase.

## 10.3 NONEQUILIBRIUM TRANSFORMATIONS

EUTECTIC TRANSFORMATIONS. From the viewpoint of equilibrium itself the eutectic transformation is not very interesting, but from the kinetic viewpoint such a transformation is of great interest. Note that if a lead-tin eutectic alloy (see Figure 10.2.6) is cooled sufficiently slowly through the eutectic transformation temperature (or more simply the eutectic temperature) the melt would yield two distinct crystals, one of $\alpha$ and one of $\beta$ plus an interface between them. (If tiny crystals of $\alpha$ and $\beta$ were formed, then considerable interfacial energy would be involved because of the large interfacial area between the particles and the system would not be at equilibrium.) To obtain two large crystals would mean that nearly all the Pb atoms would have to transport themselves to one side of the crucible and nearly all the Sn atoms to the other. We cannot expect this to happen, if the melt is cooled rapidly; that is, we will *not* get an equilibrium structure unless the alloy is cooled extremely slowly.

**EXAMPLE 10.3.1**

Assuming that two crystals (one of $\alpha$ and one of $\beta$ phase) form from a eutectic Pb–Sn melt and the average diffusion distance is 1 cm, estimate the diffusion time necessary at $183 - \epsilon$°C. The diffusion coefficient in the liquid near the melting point is $D \simeq 10^{-5}$ cm²/sec.

*Answer.* Using $L = \sqrt{2Dt} = 1$ cm [see Equation (9.9.7)], we have $t = 5 \times 10^4$ sec $\simeq$ 14 hr. This would be the time which must be spent just below the eutectic temperature but before freezing is complete. In the making of an actual solder joint, the time spent in the temperature interval just below 183°C but prior to the completion of freezing is probably only a few seconds. Thus in reality the possible diffusion length is only about $10^{-2}$ cm.

In the time in which the liquid ordinarily solidifies, the atoms could only move a certain distance $S$ and hence we get a **lamellar structure** (alternating layers of $\alpha$ and $\beta$ with repeat distance $S$ which equals the combined thickness of one $\alpha$ layer and one $\beta$ layer). This is called a **eutectic structure** and a micrograph of a typical eutectic structure is shown in Figure 1.2.13. The spacing $S$ decreases as the rate of cooling increases.

Mechanical properties such as yield stress are often strongly dependent on the lamellar spacing.

Figure 10.3.1 illustrates what happens if a liquid of eutectic composition is cooled unidirectionally by lowering a crucible at a velocity $v$ from a furnace. We shall assume that the temperature is constant across the specimen (perpendicular to the motion of the crucible). There will be a slight supercooling, $\Delta T$; i.e., the eutectic transformation temperature $T_E$ will be above the liquid-solid interface which is at a temperature $T_E - \Delta T$.

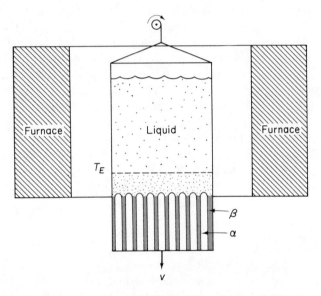

FIGURE 10.3.1. Crucible containing eutectic Pb–Sn liquid alloy is lowered from furnace at velocity $v$.

It is in this narrow liquid region that Pb atoms are moving toward the $\alpha$ phase and Sn atoms are moving toward the $\beta$ phase. *Note:* The diffusion of atoms in liquids is rapid compared to the diffusion in solids (at the same temperature) and in fact the latter diffusion rate can often be considered negligible with respect to the former; moreover, the diffusion rate drops off very rapidly as temperature decreases, it is only about $10^{-8}$ as fast at half the melting point as at the melting point in the solid metal.

It is found that the spacing $S$ varies with velocity $v$ according to

$$S^2 v = \text{constant}, \tag{10.3.1}$$

as shown in Figure 10.3.2.

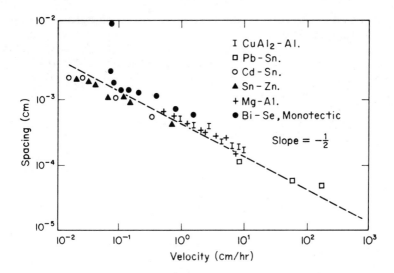

FIGURE 10.3.2. The $S^2v$ relationship for eutectic transformation. (Courtesy of C. Y. Li and H. Weart, Cornell University, Ithaca, N.Y., and University of Missouri at Rolla.)

## Q. 10.3.1

When a liquid of the eutectic composition freezes (a)_____. The solidification of a large quantity of a eutectic alloy under *equilibrium* conditions necessarily involves (b)_____. If the liquid freezes rapidly, there is not (c)_____; consequently a structure called the (d)_____ forms. This structure involves alternate layers of the two phases and is therefore called a (e)_____. The spacing between successive layers (f) (increases or decreases) as the velocity of cooling increases.

\* \* \*

**EXAMPLE 10.3.2**

A 40 wt. % Sn alloy of Pb–Sn is cooled rapidly. Discuss the microstructure which is obtained. *Note:* An equilibrium structure is not obtained.

*Answer.* From Example 10.2.3 we note that the wt. % liquid = 48.7% at $183 + \epsilon°C$. This liquid has a composition arbitrarily close to 61.9% Sn, i.e., the eutectic composition. When this remaining eutectic liquid freezes (at $T_E - \Delta T$) it forms a eutectic structure, i.e., the lamellar structure or the "thumbprint" structure of $\alpha$ and $\beta$ layers. We also note that at $183 + \epsilon°C$ the wt. % $\alpha$ = 51.3%. This remains as separated $\alpha$ phase when freezing occurs. The microstructure of the cold alloy consists of regions of $\alpha$ and regions of lamellar eutectic. There is 51.3 wt. % $\alpha$ and 48.7 wt. % of lamellar eutectic. *Note:* Actually, these ratios would be slightly changed as would the composition of the $\alpha$ and eutectic because the solid forms under nonequilibrium conditions, i.e., at $T_E - \Delta T$ and not at $T_E$. The eutectic, of course, also contains $\alpha$ phase since it consists of alternate $\alpha$ and $\beta$ phases.

The use of uniaxial solidification where the temperature gradient is unidirectional can result in a eutectic structure which is very similar to a filamentary composite. The Al–Al$_3$Ni phase diagram has a eutectic point. When this eutectic alloy is cooled from the liquid, rod-like Al$_3$Ni forms in an Al matrix. When uniaxially solidified, the rods are all lined up in one direction. A cross section of such a solidified alloy is shown in Figure 10.3.3. The rods are many times longer than their diameter. A goal of current re-

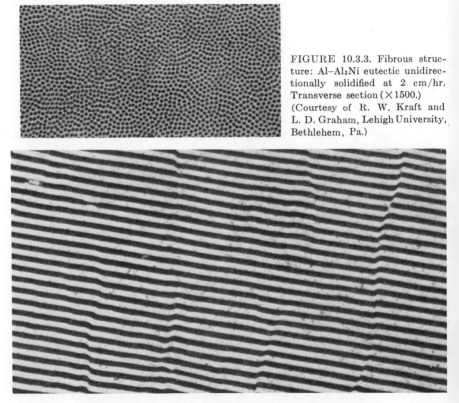

FIGURE 10.3.3. Fibrous structure: Al–Al$_3$Ni eutectic unidirectionally solidified at 2 cm/hr. Transverse section ($\times$1500.) (Courtesy of R. W. Kraft and L. D. Graham, Lehigh University, Bethlehem, Pa.)

FIGURE 10.3.4. Lamellar structure of Al–CuAl$_2$ eutectic unidirectionally solidified at 1 cm/hr. Etchant 20% HNO$_3$, 80% H$_2$O. Dark phase is CuAl$_2$. ($\times$1000.) Transverse section. (Courtesy of R. W. Kraft and L. D. Graham, Lehigh University, Bethlehem, Pa.)

search is to make the rods continuous for any desired length of a manufactured part. The unidirectional eutectic would then have the same structure as a filamentary composite such as the glass-reinforced plastics discussed in Chapter 8. The possibility of growing eutectics uniaxially so that materials will have superior properties in certain directions is discussed in the paper by R. W. Kraft, "Controlled Eutectics," *Scientific American* (February 1967), p. 86. A uniaxially solidified lamellar structure is shown in Figure 10.3.4.

## Q. 10.3.2

When an alloy of 80 wt. % Sn and 20 wt. % Pb is frozen its microstructure consists of (a)_____. When the eutectic alloy of the Al–$Al_3Ni$ solution cools it forms a matrix with (b)_____ $Al_3Ni$; if this alloy is cooled uniaxially it forms a structure roughly similar to a (c)_____. The structure in Figure 10.3.4 contains defects in the lamellar structure itself which reminds us of a defect studied earlier; this is the (d)_____.

\* \* \*

SOLIDIFICATION IN A SYSTEM SHOWING CONTINUOUS SOLUBILITY. Figure 10.3.5 shows the phase diagram of the Ni–Cu system. Consider an

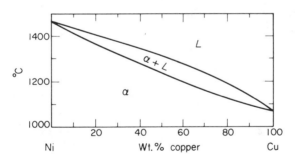

FIGURE 10.3.5. Ni–Cu system.

alloy with 22 wt. % Cu at 1500°C. This is a single-phase liquid until cooled to 1400°C; then $\alpha$ phase begins to form. The $\alpha$ phase is a continuous solid solution of Cu in nickel with the fcc crystal structure. At 1400°C the tie line extends from 14 wt. % Cu on the solidus line to 22 wt. % Cu on the liquidus line. Therefore, the first tiny particles of $\alpha$ which form upon cooling to just below 1400°C have a composition of 14 wt. % Cu. Consequently the liquid which remains has a composition above 22 wt. % Cu. As the cooling proceeds the particles will grow. The material subsequently added to the outside of the particle will have a higher and higher copper content. If there is no diffusion within the solid (or a negligible amount of diffusion), there will therefore be a variation of composition from the inside of the particle to the outside. This is known as **coring.** Recall that the average overall composition when solidification is complete must be 22 wt. % Cu; because the composition inside is only 14%, the outside composition of this particle must be considerably above 22%.

An example of a cored structure is shown in Figure 10.3.6. In coring, the center of the particle is richer in the higher melting component and has a

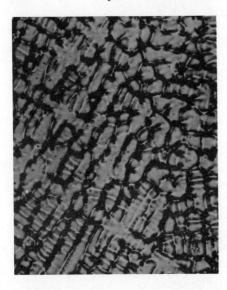

FIGURE 10.3.6. Micrograph of a cored 30 wt. % Ni–70 wt. % Cu alloy. (×75.)

higher melting point than the outside of the particle. Likewise the outside of the particle (where the copper content exceeds 22 wt. % in the previously discussed case) will have a melting point considerably *below* that of the homogeneous α-alloy of 22 wt. % Cu. Therefore, a bar of cored material would melt locally and hence be mechanically weak at a much lower temperature than the homogeneous solution.

To prevent or eliminate coring,

1. Cool *very* slowly (so that diffusion in the solid state will occur and the coring will not occur).
2. Cool very rapidly to obtain coring on a *fine* scale and then anneal at a high temperature (so that solid state diffusion can speed up homogenization).

In Section 10.6, the one-dimensional solidification of a binary alloy with a single solid phase will be described quantitatively. In this case, the

Sec. 10.4                                         Age Precipitation Hardening    453

actual variation of composition with distance upon solidification can be obtained.

### Q. 10.3.3

When solidification involves cooling through a two-phase region, with one phase being a liquid, the other solid, a phenomenon known as (a) _____, occurs. In such a case, the center of the particle has a (b) _____ melting point than the outside. In the case of a 50:50 alloy in the Cu–Ni system, assuming there is no diffusion in the solid state, the last drop to freeze would have a melting point (c) _____.
Coring can be prevented by cooling extremely slowly because (d) _____
_____.
Coring can be eliminated by cooling rapidly to form (e) _____
_____
which can then be readily homogenized at a high temperature.

\* \* \*

## ANSWERS TO QUESTIONS

**10.3.1** (a) Two phases are formed, (b) diffusion over long distances, (c) time for diffusion over long distances, (d) eutectic structure, (e) lamellar structure, (f) decreases.

**10.3.2** (a) $\beta$ phase plus eutectic regions, (b) rod-like, (c) filamentary composite, (d) edge dislocation.

**10.3.3** (a) Coring; (b) higher; (c) equal to the melting point of copper; (d) then the diffusion distance, $\sqrt{2Dt}$ [see Equation (9.9.7)] is much greater than the particle radius; here $D$ is the diffusion coefficient in the solid; (e) an extremely fine-cored structure as a result of the rapid nucleation which occurs at high supercooling.

### 10.4 AGE PRECIPITATION HARDENING

Many commercial alloys depend on a heat treatment known as age precipitation hardening to greatly enhance their mechanical properties. We have already noted this in Section 1.4 where we pointed out the importance of this process in making aluminum alloys for the aircraft industry. These aluminum alloys basically contain about 4 wt. % copper plus small percentages of other elements (which we shall ignore for the sake of simplicity).

The portion of the Al–Cu phase diagram relevant to our discussion is shown in Figure 10.4.1.

We consider an alloy with 4 wt. % Cu at 550°C. If held at that temperature for a long time (so diffusion in the solid can occur), a single homogeneous phase called the $\kappa$ phase will form. This process is called **solutionizing**. The $\kappa$ phase is a substitutional solid solution of Cu in Al. It has an fcc crystal structure. If the homogenized alloy is cooled very

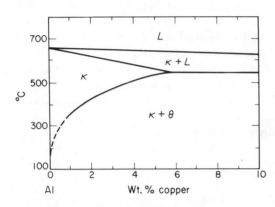

FIGURE 10.4.1. Phase diagram for aluminum-rich end of the aluminum-copper system. The pure $\theta$ phase (not shown) contains about 46 wt. % Cu at room temperature and corresponds to the compound $CuAl_2$.

slowly so that equilibrium always exists, then at room temperature there will be two phases: $\kappa$ phase (nearly pure Al) and $\theta$ phase (which is the compound $CuAl_2$). During such slow cooling, large crystals of $\kappa$ and large crystals of $CuAl_2$ form. Clearly, solid state diffusion of both copper and aluminum is necessary for this to occur.

**EXAMPLE 10.4.1**

If the $\kappa$ phase containing 4 wt. % Cu–96 wt. % Al is cooled from 550°C *very* rapidly (**quenched**) so that there is no time for much diffusion, what do we get at room temperature?

*Answer.* Cold $\kappa$ phase.

We could also add that at room temperature this quenched $\kappa$ phase is supersaturated with copper (it contains 4 % Cu). The $\kappa$ phase would contain only a tiny fraction of 1% of Cu at room temperature if the alloy had been cooled very slowly under equilibrium conditions. The supersaturated $\kappa$ phase is mechanically weak (has a low yield stress).

Thus we see from Example 10.4.1 that certain structures can be "frozen in." Such structures are nonequilibrium structures. If the supersaturated $\kappa$ phase is heated up somewhat, say to 200°C, it is still greatly supersaturated and supercooled (about 200°C). However, at this temperature atom motion is possible and tiny particles (Cu-rich precipitates) nucleate and begin to grow. This treatment is called **aging**. Because of the large supercooling, many tiny particles are nucleated rather than one larger particle; i.e., the nucleation rate is very high (see Example 9.12.2 and Figure 9.12.2).

**EXAMPLE 10.4.2**

(a) A bottle of Seven-Up is placed in a freezer and frozen. What happens? (b) A bottle of Seven-Up is placed in a freezer and cooled nearly to the point where freezing begins. It is then removed from the freezer and opened. What happens?

*Answer.* (a) You might get one piece of ice forming initially, and as cooling continues slowly, this piece of ice grows larger. (b) First, the pressure is removed. The effect of the $CO_2$ pressure was to lower the freezing point. When the pressure is removed, the freezing point is raised. Hence the liquid is supercooled rapidly (compare with the quench which supersaturates the $\kappa$ phase). *Millions* of tiny ice particles are instantaneously nucleated (so it appears) because of the large supercooling. Large supersaturation or supercooling favors nucleation of large numbers of particles.

The age precipitation hardening process in the aluminum-copper system was discovered in 1906 by Alfred Wilm. We now know that when the Cu-rich particles first form they are coherent with the $\kappa$ lattice. A **coherent precipitate** is a region which has different composition than the matrix, but it is part of the same lattice; i.e., there are no dislocations at the interface between the Cu-rich and the Cu-poor regions. Con-

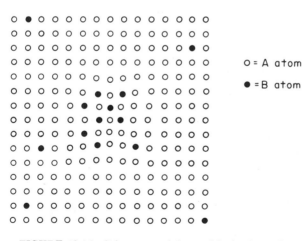

FIGURE 10.4.2. Coherent precipitate with elastic strains. Clustering takes place prior to formation of the coherent precipitate: The eventual precipitate has a different crystal structure and will then be incoherent with the lattice.

siderable elastic strains, however, may be associated with the formation of such precipitates as is shown in Figure 10.4.2. The size and shape of these coherent precipitates can be studied even when they are very small in diameter ($\sim$10–20 Å) by X-ray scattering [J. B. Cohen, *Diffraction Methods in Materials Science*, The Macmillan Company, New York (1966)] and by electron transmission microscopy [R. E. Smallman and K. H. G. Ashbee, *Modern Metallography*, Pergamon Press, Inc., Elmsford, N.Y. (1966)], which also reveals details of the strain field around the precipitate particle. As the particles grow larger, they eventually become noncoherent with the $\kappa$ lattice (grain boundaries appear between the particle and the matrix) and approach, eventually, the equilibrium concentration of $\theta$, i.e., $CuAl_2$.

**EXAMPLE 10.4.3**

Of what possible use are the tiny Cu-rich precipitate particles formed during aging of a 4% Cu–96% Al alloy?

*Answer.* They act as barriers for the motion of dislocations and hence greatly increase the yield strength and the hardness when present in sufficient numbers and optimum size. See Table 1.3.5. This process of strengthening is discussed further in Chapter 12. It is of enormous commercial importance.

Alloys heat-treated to obtain fine precipitates are called **age precipitation hardened alloys.** The three-stage heat treatment process consists of

1. Solutionizing.
2. Quenching.
3. Aging.

If the age-hardened alloy is heated for too long a time at high temperatures, the larger precipitate particles grow at the expense of the smaller and eventually there are only a few large precipitate particles essentially in equilibrium with the matrix. This is called **overaging.** The resultant alloy is mechanically weak.

We note from Figure 10.4.1 that the $\kappa$ phase has a decreasing solubility of Cu in Al as temperature decreases. This type of variation of solubility on temperature is one of the important requirements of all age precipitation hardenable alloys. The Ni–Ti system has the same general form at nickel-rich composition as does the Al–Cu system at the aluminum-rich end. Figure 10.4.3 shows an electron transmission micrograph of precipitates in nickel. Note that at this stage of precipitation there is a definite orientation relationship between the precipitate and the matrix. Upon aging, the precipitate will eventually grow into stable second phase particles of $Ni_3Ti$.

Sec. 10.4                                              Age Precipitation Hardening    457

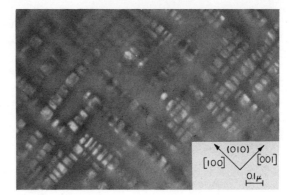

FIGURE 10.4.3. Precipitate in Ni–Ti. (Courtesy of Stephen Sass, Cornell University, Ithaca, N.Y.)

Examples of age precipitation hardenable alloys are given in Table 10.4.1. The age-hardened Mg–Al alloys are strong and very lightweight. The Cu–Be alloys which may contain *only* 1.7 wt. % Be can, when age-hardened, have a yield stress near 200,000 psi (compare with annealed copper in Table 1.3.5). This alloy is nonsparking when it strikes a steel pipe and hence is used for wrenches in the petroleum industry. Because it is strong and also a good electrical conductor, it is also used for commutator springs in motors.

**Table 10.4.1.** AGE PRECIPITATION HARDENABLE ALLOYS

| Solvent | Solute |
|---------|--------|
| Al      | Cu     |
| Al      | Mg     |
| Al      | Si     |
| Al      | Zn     |
| Ag      | Cu     |
| Au      | Cu     |
| Au      | Ni     |
| Cu      | Be     |
| Mg      | Al     |
| Ni      | Ti     |
| Pb      | Te     |

### Q. 10.4.1

To achieve age precipitation hardening we must have a system in which the solubility of solute in solvent (a)_____ as the temperature decreases. The phase diagram for the Cu–Al system could, on the aluminum-rich side, be considered as a phase diagram for the (b)_____ system. The three steps in the heat treatment of an Al–4% Cu alloy are (c)_____. When the precipitate particles first form they are (d)_____. However, when overaging occurs, they become (e)_____.

* * *

## ANSWERS TO QUESTIONS

**10.4.1** (a) decreases; (b) Al–CuAl$_2$; (c) solutionizing, quenching, aging; (d) coherent; (e) incoherent precipitates.

### 10.5 THE Fe–C SYSTEM

Figure 10.5.1 shows the Fe–C phase diagram. It also shows the Fe–Fe$_3$C phase diagram. Fe$_3$C is a crystalline compound containing 6.67 wt. % C. Fe$_3$C is metastable and under equilibrium conditions decomposes virtually completely to iron and graphite. However, except in the presence of a high carbon concentration and a catalyst such as silicon, Fe$_3$C (and not graphite) forms when a liquid iron-carbon alloy is cooled at typical rates. By a typical rate we mean by cooling in air or faster, i.e., at rates typically employed in industry. Thus under such circumstances it is the Fe–Fe$_3$C diagram which concerns us even though one of the components is metastable (however, its lifetime at room temperature in the absence of catalysts is of the order of millions of years).

The phase designated by $\alpha$ in Figure 10.5.1 has the bcc structure; it is called **ferrite**. The $\gamma$ phase has the fcc structure; it is called **austenite**. Carbon is dissolved interstitially in these structures as described in Section 6.4. The solubility of carbon in $\alpha$-iron is about 0.025%C at the eutectoid temperature of 723°C.

Let us suppose we cool down the $\gamma$ phase for an 0.8% alloy starting at 800°C. At about 770°C we *would under equilibrium conditions* (incredibly slow cooling) begin to find precipitated graphite; thus two phases would be present, $\gamma$ and $g$, where $g$ is graphite. This precipitation *would* continue until at 738°C the remaining $\gamma$ (which now would have a composition of 0.69%C) would transform to $\alpha + g$. After this with further cooling there would be only slight changes; additional slight amounts of graphite would precipitate out due to the decreasing solubility of carbon in the bcc structure. What *actually* happens when the alloy is cooled from 800°C at typical cooling rates is that the 0.8% alloy remains as a single $\gamma$ phase all the

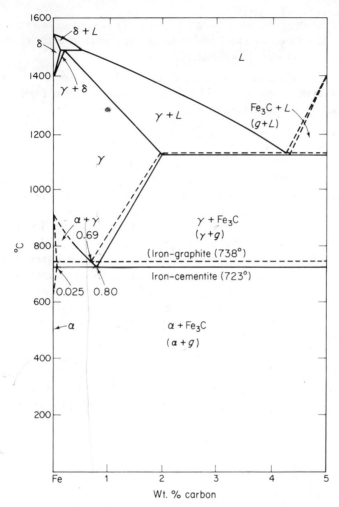

FIGURE 10.5.1. Fe–C and Fe–Fe₃C systems. [From *Metals Handbook*, American Society of Metals, Cleveland (1948).]

way down to 723°C at which time it then transforms to two phases, $\alpha$ + Fe$_3$C, by a eutectoid reaction. The Fe$_3$C is metastable (i.e., it is thermodynamically unstable relative to iron and graphite) but the decomposition is so sluggish that no observable changes occur. This can be attributed to the high interfacial energy between graphite and iron and hence to the low probability of nucleation of graphite [see Equation (9.12.2)]. Thus the fact that the metastable state is formed and persists is a result of the relative sluggishness of the reactions involving precipitation of graphite. We shall therefore assume, if the carbon content is below 1.98 wt. % C, that the

Fe–Fe$_3$C system is an equilibrium system. This system is extremely important in the study of **steel** which is an alloy of iron with less than 1.98 wt. % C. Fe$_3$C is called **cementite**.

PEARLITE. Let us consider the Fe–Fe$_3$C diagram further, in particular, the eutectoid transformation,

$$\gamma \to \alpha + Fe_3C. \tag{10.5.1}$$

For simplicity, we can ignore the slight amount of carbon in the $\alpha$ phase and consider it simply as bcc iron. An important nonequilibrium effect is associated with the eutectoid transformation because of the finite cooling rate. This is an analog of the effect which took place when the eutectic transformation occurred under nonequilibrium conditions. In other words, because carbon must diffuse in the transformation of (10.5.1), a lamellar structure of $\alpha$-iron and Fe$_3$C forms. This is a **eutectoid structure** which in steel is called **pearlite.** The lamellar spacing is determined by the temperature at which the eutectoid transformation occurs. Because of the slowness of diffusion in the solid state, large supercooling is possible. Thus it is possible to quench the 0.8 wt. % C alloy rapidly to 400°C without obtaining any transformation and then to allow the transformation to proceed at that temperature. As the temperature at which the transformation occurs is lowered (from below 723°C), the lamellar spacing decreases. This spacing profoundly affects the tensile yield stress, with smaller spacings corresponding to higher yield stresses. A micrograph of pearlite is shown in Figure 10.5.2.

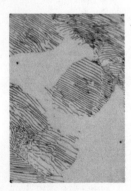

FIGURE 10.5.2. Photomicrograph of pearlite microstructure. ($\times 1000$.)

## Q. 10.5.1

In discussing the microstructure of steel, reference is made to the (a)_____ phase diagram even though (b)_____ is not an equilibrium structure. An alloy of 0.8 wt. % C at 900°C under equilibrium conditions has what crystal structure? (c)_____. When this alloy is cooled below 723°C it undergoes (d)_____, resulting in a (e)_____ structure called (f)_____. The eutectoid transformation is a (g)_____ -controlled process.

\* \* \*

## EXAMPLE 10.5.1

(a) A steel contains 0.4 wt. % C. Calculate the fraction of $\alpha$ and $\gamma$ at $723 + \epsilon°C$, where $\epsilon$ is arbitrarily small. (b) A steel contains 0.4 wt. % C. What do we see in the microscope if this is cooled to below 723°C after the $\gamma$ is transformed?

*Answer.* (a) We draw a tie line at $723 + \epsilon°C$ and apply the lever rule.

$$\% \alpha = \frac{0.80 - 0.40}{0.80 - 0.025} \times 100\% = 51.6\%$$

$$\% \gamma = 48.4\%.$$

*Note:* The $\gamma$ has the eutectoid composition. See Figure 10.5.3(b) for the microstructure. (b) Below 723°C, the 51.6% $\alpha$ which was already present at $723 + \epsilon°C$ is colder and otherwise unchanged while the $\gamma$ which has the eutectoid composition transforms to pearlite. Remember that pearlite contains both $\alpha$ and $Fe_3C$. See Figure 10.5.3(c) for the microstructure. If the austenite is supercooled to below approximately 400°C before it transforms, then **bainite,** a feathery arrangement of $Fe_3C$, is formed. Bainitic steels have a high toughness.

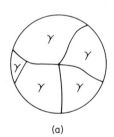
(a)

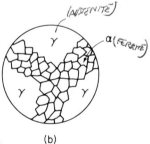
(b)

(c)

FIGURE 10.5.3. Sketch of the microstructure of 0.4 wt. % C steel at various temperatures. (a) Austenized at 900° C. (b) Slightly above the eutectoid temperature of 723°C. (c) Cooled to slightly below eutectoid temperature; $\gamma$ of (b) transforms to coarse pearlite.

MARTENSITE. Let us next consider austenite (the $\gamma$ phase of the $Fe-Fe_3C$ system) of eutectoid composition (0.8 wt. % C) at 800°C; let us place a sample of this in ice water. It is then cooled down so rapidly that the migration of carbon atoms is negligible so that separate particles of $\alpha + Fe_3C$ cannot form. At this temperature iron prefers the bcc structure but with a negligible amount of dissolved carbon; however, in the present case there is 0.8% carbon present.

**EXAMPLE 10.5.2**

When austenite is quenched to room temperature, what is obtained?

*Answer.* We might answer cold austenite by analogy with Example 10.4.1; this is wrong! See the following.

When the $\gamma$ phase is quenched below a certain temperature called $M_S$ (**martensite starts temperature**), **martensite** begins to form. This is a new crystalline form, a body-centered tetragonal crystal. (A tetragonal crystal has a unit cell with only 90-deg angles and with the magnitude of **a** equal to the magnitude of **b**; i.e., $a = b \neq c$.) A body-centered tetragonal crystal is like a bcc crystal which has been stretched along one of the [100] directions. The $c/a$ ratio of the martensite is proportional to its carbon content as illustrated in Figure 10.5.4. Of course, as the carbon content approaches zero, $c/a$ approaches 1; i.e., in the limit bcc iron is formed.

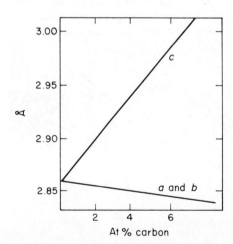

FIGURE 10.5.4. Lattice parameters of martensite versus carbon content.

Martensite can be considered to be a bcc crystal which is distorted along one axis as a result of the presence of the carbon atoms (which the bcc crystal itself does not dissolve to any appreciable extent). The formation of martensite does not require diffusion. The martensite transformation usually results in the formation of platelets which form by a shearing mechanism at velocities of the order of sound velocities in the material. This is called a **diffusionless transformation** (the eutectic, age precipitation, and eutectoid transformations considered previously all take place by diffusion).

### Q. 10.5.2

When $\gamma$-iron is cooled very rapidly there is no time for (a)_____ to form. When the temperature is cooled far below 723°C, the $\gamma$-iron is (b)_____ and transforms into (c)_____, which has a (d)_____ crystal structure. Martensite platelets are formed by (e)_____ at (f)_____ velocities. The martensite transformation is a (g)_____ transformation.

\* \* \*

Below a temperature called $M_F$ (**martensite finishes temperature**) no more austenite is transformed to martensite. There is usually some retained austenite. $M_S$ and $M_F$ vary with the carbon content in the manner shown in Figure 10.5.5. When steel is quenched from the $\gamma$ range to a tem-

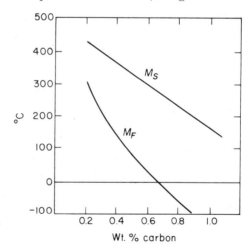

FIGURE 10.5.5. $M_S$ and $M_F$ temperatures for carbon steel.

perature between $M_S$ and $M_F$, the fraction of martensite formed (in most cases) depends only on the final temperature and not on the time at which the specimen is held at that temperature. Additions of other alloying elements can displace the curves shown in Figure 10.5.5. Martensite is *very* hard and strong. Dislocations move only with great difficulty through martensite because they interact strongly with the carbon. The martensite

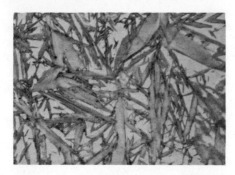

FIGURE 10.5.6. Micrograph of martensite with some retained austenite. (×1500.) (Courtesy U.S. Steel Corporation.)

crystals are often in the form of convex lenses or plates. A micrograph of martensite is shown in Figure 10.5.6.

The term **martensitic transformation** is usually reserved for transformations which occur by a displacement mechanism (such as shear) in which individual atoms execute well-defined and correlated movements. The shape change caused by this shear causes an initially smooth surface to become serrated as a result of the transformation. The nature of the surface relief suggests that the transformation is a homogeneous deformation (a homogeneous deformation is one in which straight lines remain straight lines, although their lengths may change and the angles between pairs of these lines may change). Also, in martensitic transformations there is a definite orientation relationship between the martensite crystal and the parent crystal. Finally, the plane separating the two phases (martensite and parent) is a particular crystallographic plane of the parent called the **habit plane**.

The martensitic transformation in steel is just one of many martensitic transformations. One of the simplest is the transformation in cobalt from hexagonal closest packed to face-centered cubic. In this case, the motion of partial dislocations (special dislocations whose Burgers vector is less than a repeat distance) on successive planes will change the packing of the closest packed layers from the $\cdots ABAB \cdots$ sequences of the hcp crystal to the $\cdots ABCABC \cdots$ sequence of the fcc crystal. The partial dislocation can move from plane to plane (normal to the $\cdots ABAB \cdots$ sequence) if a screw dislocation lies in the normal direction. This is called a **pole mechanism**. It is thought that the pole mechanism is the likely mechanism in martensitic transformations [see C. M. Wayman, *Introduction to the Crystallography of Martensitic Transformations*, The Macmillan Company, New York (1964)].

Martensitic transformations are not restricted to metals. Thus the transformations of RbCl, RbBr, and RbI at high pressure from the NaCl-type of the CsCl-type crystal structure shows the features characteristic of a martensitic transformation [see P. J. Reddy and A. L. Ruoff, in *Physics of Solids at High Pressure*, ed. by C. T. Tomizuka and R. M. Emrick, Academic Press, Inc., New York (1965), p. 510].

A partial list of materials which undergo martensitic transformations is shown in Table 10.5.1. Many of these are discussed in detail in the book by Wayman.

Table 10.5.1. SYSTEMS SHOWING MARTENSITIC TRANSFORMATIONS

Cu–10–15 at. % Al
Co
Ti
Carbon steel
Stainless steel
Au–47.5 at. % Cd
Rubidium halides

**EXAMPLE 10.5.3**

A single crystal of martensite is more stable than the $\gamma$-iron below $M_S$. Why does a sample of the steel only partially transform at a given temperature between $M_S$ and $M_F$ to martensite?

*Answer.* Martensite platelets originate at many points when the material is cooled to a given temperature below $M_S$. When numerous platelets are formed, two additional energy terms arise: interfacial energy and elastic strain energy. Both terms change the energetics so that additional lowering of the temperature is required before more martensite can be formed.

There is currently considerable research underway involving not only how martensitic transformations occur but why.

STEEL. Steel is an iron-carbon alloy which involves carbon compositions up to 1.98 wt. % C. The most common heat treatment of steel has as its purpose the production of martensite in order to increase the strength of the steel. This is the three-step process:

1. Austenitizing.
2. Quenching.
3. Tempering.

**Austenitizing** is the process of converting the steel to austenite by annealing it at a sufficiently high temperature (in the stable $\gamma$ region).

**Tempering** (heating to moderate temperatures after quenching) reduces the stresses induced by the martensitic transformation (martensite has a smaller volume than austenite) and converts some of the martensite to carbide. While the Egyptians knew earlier that quenching caused hardness in a carburized iron, the Greeks by 400 B.C. had learned that tempering relieved the material of some of its brittleness.

**EXAMPLE 10.5.4**

A long 12-in.-diameter solid cylinder of steel at 900°C is quenched in oil at room temperature. Does martensite form?

*Answer.* Three possibilities exist: (1) The cooling rate is so slow that all the $\gamma$ phase decomposes by diffusion into pearlite (and either $\alpha$ or $Fe_3C$) before $M_S$ is reached; (2) the cooling rate is so fast that none of the $\gamma$ phase has decomposed by diffusion when $M_S$ is reached and the $\gamma$ phase transforms to martensite; (3) the cooling rate is intermediate so that both pearlite and martensite are formed. Thus we need to be *quantitative*.

**EXAMPLE 10.5.5**

Alloying elements such as manganese, vanadium, nickel, etc., in steel slow down the transformation of $\gamma$ phase to pearlite. How does this decreased transformation rate affect the possibility of obtaining martensite during quenching?

*Answer.* It *greatly* enhances it even when the cooling rate is slow as along the axis of the cylinder of Example 10.5.4 since if pearlite is not formed, martensite can be. This will be discussed quantitatively in Section 12.8.

When nickel (or certain other elements) is added in sufficient quantity to iron the austenite phase (the $\gamma$ phase, i.e., the fcc phase) can be stabilized to room temperature and below; the resultant steel is called an **austenitic steel**. Ordinary **stainless steel** (18% Cr, 8% Ni) is an austenitic steel.

To summarize, the types of steel which we have encountered are pearlitic, bainitic, martensitic, and austenitic.

**Q. 10.5.3**

The most common heat treatment of steel has as its purpose the production of (a)_____ which results in (b)_____. After a steel is quenched, it is usually (c)_____ in order to (d)_____ _____. The addition of alloying elements slows down the rate of (e)_____ and enhances the possibility of obtaining martensite. If sufficient nickel is added to iron, the $\gamma \rightarrow \alpha$ transformation temperature can be (f)_____.

\* \* \*

CAST IRON. The Fe–Fe₃C system contains a eutectic point at 4.3 wt. % C. Systems in the composition range shown by the eutectic line in Figure 10.5.1 are known as **cast irons**. In **white cast iron** (lower part of the carbon range) the carbon is present in the form of $Fe_3C$; white cast iron is therefore very hard. In **gray cast iron** (approximately eutectic composition) the carbon is present in the form of graphite flakes (the addition of the

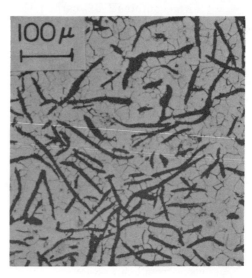

FIGURE 10.5.7. Gray cast iron.

catalyst silicon accelerates the formation of graphite, so that graphite rather then $Fe_3C$ forms). See Figure 10.5.7. It is the black graphite flakes which give gray cast iron its characteristic appearance. These large flat flakes act as stress concentrations in the iron, so that gray cast iron is rather weak in tension. If instead of the flat flakes of graphite in gray cast iron, the graphite particles were approximately spherical, the stress concentration factors

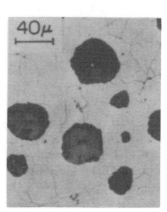

FIGURE 10.5.8. Nodular cast iron.

would be less. Such a structure can be obtained by adding magnesium to the melt; it is called **nodular cast iron** and has higher strength and ductility than gray cast iron. See Figure 10.5.8.

### Q. 10.5.4

If iron contains 2–6% carbon, the alloy is known as (a)_____. If the carbon is present in $Fe_3C$ only the material is (b)_____, while if it is present as graphite flakes it is called (c)_____, and as graphite spheres it is called (d)_____.

\* \* \*

## ANSWERS TO QUESTIONS

**10.5.1** (a) $Fe$–$Fe_3C$; (b) $Fe_3C$; (c) the fcc structure; i.e., it is $\gamma$-iron; (d) a eutectoid transformation; (e) lamellar; (f) pearlite; (g) diffusion.

**10.5.2** (a) Pearlite, (b) highly unstable, (c) martensite, (d) tetragonal, (e) a shearing mechanism, (f) very high, (g) diffusionless.

**10.5.3** (a) Martensite, (b) a large increase in strength, (c) tempered, (d) relieve the stresses and increase the ductility, (e) pearlite formation, (f) lowered to below room temperature.

**10.5.4** (a) A cast iron, (b) white cast iron, (c) gray cast iron, (e) nodular cast iron.

## 10.6 SEGREGATION IN BINARY ALLOYS DURING SOLIDIFICATION

This section provides a quantitative discussion of the nonequilibrium freezing of binary alloys in special cases.

THE DISTRIBUTION COEFFICIENT. Consider a portion of a binary phase diagram of the $A$-$B$ system as shown in Figure 10.6.1.
Define

$$m_s = \text{slope of solidus}$$

$$m_l = \text{slope of liquidus}.$$

Both of these slopes are assumed to be constants in the discussion which follows. Then for the solidus we have

$$T_m - T = -m_s C_s, \quad (10.6.1)$$

where $C_s$ is the concentration (in units of mass per unit volume) of $B$ in $\alpha$ at temperature $T$, and for the liquidus

$$T_m - T = -m_l C_l, \qquad (10.6.2)$$

where $C_l$ is the concentration of $B$ in $L$ at temperature $T$.
Combining these equations gives

$$K \equiv \frac{C_s}{C_l} = \frac{m_l}{m_s}. \qquad (10.6.3)$$

$K$ is called the **distribution coefficient** or the **segregation coefficient**.

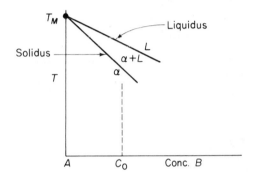

FIGURE 10.6.1. Portion of a binary phase diagram.

## EXAMPLE 10.6.1

An alloy of the $A$-$B$ system of Figure 10.6.1 of composition $C_0$ (the concentration of $B$) is cooled from the liquid phase. The distribution coefficient is $K$. What is the composition of the first tiny particle of solid which forms?

Answer. $\qquad\qquad C_s = KC_0 \qquad (10.6.4)$

since at the instant the first solid forms the composition of the liquid is still $C_0$.

SEGREGATION. Consider a liquid binary alloy, of the $A$-$B$ system whose phase diagram is shown in Figure 10.6.1, with composition $C_0$ which is dropped slowly into a cooled region as shown in Figure 10.6.2. The first drop to solidify at $x = 0$ has a concentration $C_s(0)$, i.e., it is the drop formed at $x = 0$. Since the initial concentration of $B$ in the liquid is $C_0$ we have

$$C_s(0) = KC_0. \qquad (10.6.5)$$

We assume that the liquid in Figure 10.6.2 is well stirred so that the concentration of the liquid is uniform; i.e., it does not vary with position. We also assume that very little change occurs within the solid because of the slowness of diffusion in the solid. The freezing takes place at the interface

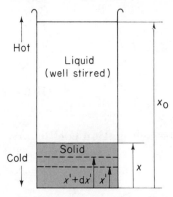

FIGURE 10.6.2. Solidification illustrating segregation.

at $x$. We use here the **concept of local equilibrium**; i.e., we assume that at the interface Equation (10.6.3) holds so that

$$C_s(x) = KC_l(x); \tag{10.6.6}$$

this is an equilibrium assumption; it is assumed to be true at the interface even though the process is a dynamic one.

We now use a material balance on the solute $B$ which enables us to complete our description of this problem (to simplify the mathematics we have ignored volume changes). We have (with reference to Figure 10.6.2)

$$C_l(x)\left(1 - \frac{x}{x_0}\right)x_0 + \int_0^x C_s(x')\,dx' = C_0 x_0. \tag{10.6.7}$$

**EXAMPLE 10.6.2**

State in words the material balance used to derive (10.6.7).

*Answer.* The amount of component $B$ in the liquid (which has uniform composition throughout at any instant) plus the amount of component $B$ in the solid (in which the composition varies with position in the solid) equals the total amount of component $B$ present in the original alloy.

Using (10.6.6), (10.6.7) becomes

$$\frac{C_s(x)}{K}\left(1 - \frac{x}{x_0}\right)x_0 + \int_0^x C_s(x')\,dx' = C_0 x_0. \tag{10.6.8}$$

This integral equation may be differentiated with respect to $x$ to give

$$\frac{1}{K}\frac{dC_s(x)}{dx}\left(1 - \frac{x}{x_0}\right) - \frac{C_s(x)}{Kx_0} + \frac{C_s(x)}{x_0} = 0 \qquad (10.6.9)$$

and is, as previously noted, subject to (10.6.5). It is a linear first-order differential equation with separable variables. The solution to the problem is

$$C_s(x) = KC_0\left(1 - \frac{x}{x_0}\right)^{K-1}, \qquad x < x_0. \qquad (10.6.10)$$

It is possible to have $K$ less than or greater than 1 in different systems. In the case of Figure 10.6.1, $K < 1$. Suppose $K = \frac{1}{10}$ and $C_0 = 0.1$ wt. % $B$ and we wish to obtain a sample of component $A$ which has much less than 0.1 wt. % $B$. Then by carrying out this process we can purify (by a factor up to 10) the first portion of this bar. After solidification we crop off the half near $x_0$ (see Figure 10.6.2) which contains most of the $B$ impurity and melt the remaining material and repeat the process. After repeated steps of this sort we achieve nearly pure $A$ (assuming no other impurities are present or become involved because of handling).

The segregation achieved in the previous discussion is a maximum. Had the liquid solution not been highly stirred, less segregation would have occurred. Had some diffusion taken place in the solid solution, less segregation would have occurred. Segregation such as this occurs at usual cooling rates in casting. An example is the variation of concentration across the individual crystals that grow from the melt, known as coring.

In the problem considered above it was assumed that heat conduction was very rapid so that the problem of carrying away the heat of solidification could be completely ignored. This simplifies the picture considerably. In actual casting practice the role that heat conduction plays is a vital one and must be properly accounted for.

### Q. 10.6.1

A liquid alloy contains 2% of an impurity $B$ whose segregation coefficient is 0.1. This liquid begins to freeze and the first drop of solid to form has a composition of (a)_____. If this liquid freezes uniaxially from $x = 0$ to $x = L$, the concentration of impurity will (b) (increase or decrease) as $x$ increases. The segregation will be a maximum if the remaining liquid is (c)_____ and if (d)_____ in the solid. Coring is an example of (e)_____.

\* \* \*

**ZONE REFINING AND LEVELING.** A technique of great importance in the semiconductor industry achieves the needed purity for semiconductor crystals, about $10^{-8}$ at. % impurity, by moving a melted zone along a bar of the fairly pure material. It is clear that (1) if the melt is richer in one component than the just solidified solid and (2) if this zone is moved from one end of a bar to the other by moving a narrow furnace or other source of heat, some of this component will be moved to the far end. By repeated passes in the same direction the concentration of the component may be made arbitrarily low in principle. This technique is called **zone refining.** A few solutes concentrate in the solid and therefore move the other way. Some impurities (unwanted components) have $K = 1$ and are not cooperative (no zone refining occurs). The phase diagrams for the systems give the essential information on distribution coefficients. The material is held in the narrow liquid zone against gravitational forces because of the surface energy. Hence the diameter of a material which may be zone-refined is limited by the presence of gravitational fields.

The process of zone refining can be described quantitatively. The

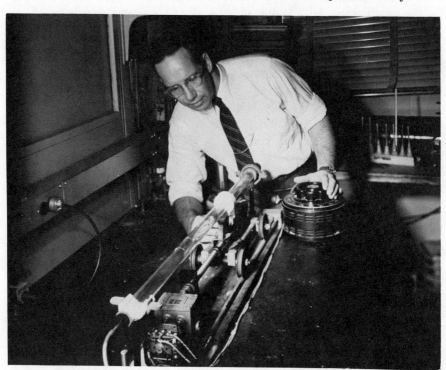

FIGURE 10.6.3. W. G. Pfann adjusting zone-refining apparatus. (Courtesy of Bell Telephone Laboratories, Murray Hill, N.J.)

analysis of this problem proceeds in the same fashion as the analysis of one-dimensional segregation.

If the passes are run back and forth, homogenization of a nonuniform alloy may be achieved. This is called **zone leveling.** Often homogeneity of an alloy is very important, as in doped silicon for microelectronic circuitry.

Zone refining was developed by W. G. Pfann at the Bell Telephone Laboratories and has played a major role in the rapid development of the semiconductor industry. Pfann is shown with a simple device for zone refining in Figure 10.6.3.

# ANSWERS TO QUESTIONS

**10.6.1** (a) 0.2% $B$, (b) increase, (c) stirred, (d) there is no diffusion, (e) segregation.

# REFERENCES

### Explanation of Phase Diagrams

Gordon, Paul, *Principles of Phase Diagrams in Materials Systems*, McGraw-Hill Book Company, Inc., New York (1968). This book and the next book are excellent sources for extending your knowledge of phase equilibria, including how phase diagrams are obtained.

Rhines, F. N., *Phase Diagrams in Metallurgy*, McGraw-Hill Book Company, Inc., New York (1965).

### Phase Diagrams

Levin, E. M., Robbins, C. R., and McMurdie, H. F., *Phase Diagrams for Ceramists*, American Ceramic Society, Columbus, Ohio (1964). Contains numerous phase diagrams of ceramic systems (inorganic systems).

Hansen, M., *Constitution of Binary Alloys*, McGraw-Hill Book Company, Inc., New York (1958). This book and the two supplements which follow contain much of what is known about binary systems involving the elements.

Elliott, R. P., *Constitution of Binary Alloys, First Supplement*, McGraw-Hill Book Company, Inc., New York (1965).

Shunk, F. A., *Constitution of Binary Alloys, Second Supplement*, McGraw-Hill Book Company, Inc., New York (1969).

*Metals Handbook*, American Society for Metals, Cleveland (1948). Contains a number of phase diagrams of particular importance in the metallurgical industry.

### Phase Transformations

Shewmon, P. G., *Transformations in Metals*, McGraw-Hill Book Company, Inc., New York (1970). An excellent introduction to phase transformations as well as other transformation processes.

### Zone Refining

Pfann, W. G., "Zone Refining," *Scientific American* (December 1967), p. 62. This is a popular exposition of zone refining.

Pfann, W. G., *Zone Melting*, John Wiley & Sons, Inc., New York (1966). This book describes in detail the experiments and theory of zone refining. It can be read without difficulty by the student who has progressed this far.

### Uniaxial Eutectic Solidification

Kraft, R. W., "Controlled Eutectics," *Scientific American* (February 1967), p. 86.

# PROBLEMS

**10.1** List five "why" questions concerned with phase equilibria.

**10.2** For an alloy of 30 wt. % Sn and 70 wt. % Pb under equilibrium conditions,
(a) What phases are present at 184°C?
(b) What is the composition of each phase?
(c) What is the weight fraction of each phase?
(d) What phases are present at room temperature (25°C)?
(e) What is the composition of each phase?
(f) What is the weight fraction of each phase?

**10.3** The ethylene glycol-water phase diagram has the general feature of the Pb–Sn diagram. Of what engineering importance is this?

**10.4** In the NaF–MgF$_2$ system of Figure 10.2.7 under equilibrium conditions, calculate the weight fraction of each phase present at 810°C for a 40 mol. % MgF$_2$ alloy.

**10.5** Give a simple explanation of why a eutectic alloy might yield a lamellar structure when cooled rapidly.

**10.6** A 30 wt. % Sn–70 wt. % Pb alloy is cooled rapidly (and hence under nonequilibrium conditions). Discuss as quantitatively as possible the microstructure which is obtained.

**10.7** An alloy of 10 wt. % Sn–90 wt. % Pb is cooled *very* slowly (so slowly that diffusion in the solid state takes place over the entire sample).

(a) What is the composition of the last drop of liquid which freezes?
(b) At what temperature does this occur?
(c) If the same alloy were cooled rapidly, what would be the composition of the last drop to freeze?
(d) At what temperature would this occur?

**10.8** An alloy contains 90 wt. % Ni–10 wt. % Cu.
(a) If cooled very slowly, at what temperature does it become entirely solid?
(b) What is the composition of the last drop to freeze?
(c) If cooled very rapidly, what is the composition of the last drop to freeze?
(d) At what temperature does this occur?
(e) How might the rapidly cooled structure appear in a micrograph?
(f) What tensile load would this rapidly cooled alloy carry at 1200°C?

**10.9** Name some of the necessary attributes of a system $A-B$ if it is to show age precipitation hardening characteristics.

**10.10** Ordinary pure gold is very weak. Describe a mechanism for making it stronger for use in jewelry, etc.

**10.11** The lead sheathing applied to the outside of electrical insulation on underground electrical cable has to have greater mechanical strength than pure lead. Suggest a way of achieving this.

**10.12** (a) What are the three steps in an age precipitation hardening heat treatment?
(b) Describe what happens on the atomic scale in each step.

**10.13** List five possible applications of phase diagrams.

**10.14** Suppose a crucible containing lead is heated to 400°C and then allowed to cool.
(a) Plot the temperature versus the time.
(b) *Note:* Lead exhibits a **thermal arrest** at the melting point. What does this mean?
(c) The $T$-$t$ curve is called a **cooling curve**. Sketch the cooling curve for the 30 wt. % Sn alloy.
(d) Explain how such cooling curves at intervals of 5% composition could be used to construct the phase diagram.

**10.15** Discuss how X-ray diffraction could be of help in finding the line separating the $\alpha$ and $\alpha + \beta$ regions of the Pb–Sn phase diagram.

**10.16** Explain how you would stir the liquid during zone refining.

**10.17** Steel contains a number of interesting phases which make it useful, which helps explain why hundreds of millions of tons of steel are produced throughout the world annually.
(a) Is $Fe_3C$ a stable phase?
(b) Is fine pearlite a stable two-phase configuration?
(c) Is bainite a stable two-phase configuration?
(d) Is martensite a stable phase?

**10.18** (a) Sketch a micrograph of a 1.5 wt. % C steel at 725°C.
(b) At room temperature, if air cooled.

**10.19** Explain why an age precipitation hardened Cu–Be alloy is a good electrical conductor.

**10.20** What is the principal difference between a binary ceramic system and a binary metal system.

**10.21** The $Na_2O$–$SiO_2$ system has a eutectic point. $SiO_2$ has a very high melting point, and pure $SiO_2$ glass has a high softening point. Explain why $Na_2O$ is added.

**10.22** Describe the nonequilibrium cooling of a peritectic alloy. Why is the structure called *surrounding?*

**10.23** Because of the presence of graphite flakes gray cast iron has a high damping capacity. Give another reason gray cast iron might be used instead of steel even though its strength and ductility are lower. *Hint:* See the phase diagram.

# MORE INVOLVED PROBLEMS

**10.24** Analyze carefully the advantages and disadvantages of carrying out zone refining and crystal growth in a satellite space station (where $g = 0$).

**10.25** Show that the expression in (10.6.10) is in fact a solution of (10.6.9).

# SOPHISTICATED PROBLEMS

**10.26** Assume a uniaxial eutectic transformation is taking place as shown in Figure 10.3.1 and below.

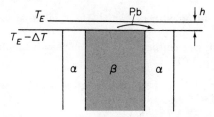

PROBLEM FIGURE 10.26.

The diffusion which causes the separation into $\alpha$ and $\beta$ takes place between $T_E$ and $T_E - \Delta T$ in a thin layer (thin with respect to the average diffusion distance). On this basis, assuming one-dimensional diffusion, derive Equation (10.3.1). Assume that the height, $h$, of the supercooled liquid is independent of the velocity.

# Prologue

The engineer is concerned with three electrochemical processes: generation of electricity by batteries and fuel cells using direct chemical conversion; electrolytic processes such as electrorefining, electroplating, and electrolysis; and corrosion. To understand these processes it is necessary to understand both the *equilibrium* and the *kinetic behavior* of the *electrochemical cell*. Electrochemical cells include *dissimilar electrode cells, differential concentration cells, differential temperature cells,* and others. Corrosion always involves an electrochemical cell. To prevent corrosion this cell must be made inoperative. Techniques for doing this are discussed; these techniques are based on an understanding of how the cell operates.

# 11

# ELECTROCHEMICAL PROPERTIES

## 11.1 INTRODUCTION

Take your choice of those that can best aid your action.—Shakespeare, *Coriolanus*.

There are three important areas where electrochemical processes are directly involved:
1. Direct chemical generation of electricity by batteries and fuel cells. Batteries have an important place in our lives in flashlights, automobiles, submarines, etc. The development of a battery or fuel cell which could produce sufficient power to propel our automobiles would help solve the air pollution problem.
2. Electroplating (which can include electrorefining) and electrolysis. Aluminum and many other metals are produced by electrorefining processes. Electroplating is an important commercial process. For example, memory drums for computers are produced by plating the base metal with copper and then with a nickel-cobalt layer which serves as the actual memory core. The reverse process, namely, electromachining, is also important commercially. In this case material is removed from a component. This is really controlled corrosion.
3. Corrosion. Corrosion is an important economic factor. It is estimated that it costs more than ten billion dollars in the United States annually.

To understand these various phenomena we must understand the electrolytic cell.

Figure 11.1.1 shows one example of an electrolytic cell. The electrodes are of different materials so we call this a **dissimilar electrode cell.** An **electrochemical cell** is always composed of four parts. These are two electrodes (one of copper, one of zinc in this case), an external electrical conductor (where the conduction is by electrons), and an electrolyte (where electrical conduction takes place by the motion of ions). Some cells also contain a semipermeable membrane which allows one or more ions to diffuse

480   *Electrochemical Properties*

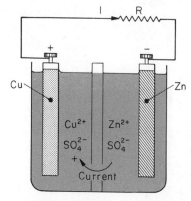

FIGURE 11.1.1. Dissimilar electrode cell with a porous barrier. The metal ions can diffuse through the barrier which, however, prevents direct mixing.

through, but does not permit the transfer of the solvent or a gas. In the present example the electrolyte is a liquid solution of $Cu^{2+}$, $Zn^{2+}$, and $SO_4^{2-}$ ions in water. However, the electrolyte could be a solid, in which the ions move rapidly; recall that solid NaCl is an ion conductor at high temperatures. There are certain ionic conductors through which the ions move rapidly at room temperature. Beta-alumina is such a compound. This is a layer-like structure with sodium ions on certain layers and the diffusion of sodium ions along these layers is very rapid.

In the cell shown in Figure 11.1.1 the Zn electrode dissolves, adding $Zn^{2+}$ ions to the solution, which then move toward the copper electrode. Meanwhile $Cu^{2+}$ ions move from the solution and plate out on the Cu electrode. Thus positive ions move from the Zn electrode to the Cu electrode through the electrolyte liquid. Hence the motion of *positive* charge within the cell is as shown, i.e., from the zinc electrode to the copper electrode. Electrical current, by convention, is defined as the flow of positive charge past a point per unit time. *Note:* In the external circuit, the current is actually due to electrons moving from the zinc electrode to the copper electrode but this is equivalent to positive charges moving from the copper electrode to the zinc electrode. The sign of the electrodes refers to the motion of positive charge in the external circuit; positive charges would, of course, move toward the negative electrode, in this case, the zinc electrode.

An **electrode** is an electrical conductor, at the surface of which conduction changes from conduction by electrons to conduction by ions.

### Q. 11.1.1

Every electrochemical cell has (a)____ parts. These are (b)_____
_____. Conduction in the
(c)_____ must be by ion motion. Some cells also contain (d)_____
_____.

\* \* \*

Sec. 11.1  Introduction  481

The positive electrode is called the cathode; the negative electrode is called the anode. A chemical reduction reaction always occurs at the **cathode**. In the case of the cell shown in Figure 11.1.1, the reduction reaction is

$$Cu^{2+} + 2e^- \rightarrow Cu \quad \text{(reduction)}. \quad (11.1.1)$$

A **reduction reaction** by definition involves the consumption of electrons. Reduction is the process whereby a positive ion such as $Ag^+$ is reduced to the metallic state. (As an example, when CuS ore is processed to produce copper metal we say the copper is reduced.)

An **oxidation reaction** involves the production of electrons and occurs at the **anode**. In the cell shown in Figure 11.1.1, the oxidation reaction is

$$Zn \rightarrow Zn^{2+} + 2e^- \quad \text{(oxidation)}. \quad (11.1.2)$$

In this particular chemical cell there is a flow of **cations** (positively charged ions) to the cathode, which is why it is called the cathode. There is no net flow of **anions** (negatively charged ions) in the present case. In other electrical cells, there may be a net flow of anions to the anode and no net cation flow. In still other cells, cations diffuse to the cathode, while anions diffuse to the anode.

## Q. 11.1.2

The reduction reaction takes place at the (a)_____. It involves the (b)_____ of electrons. If there is a net motion of cations in a cell the cations diffuse to the (c)_____.

\* \* \*

There is an electric potential difference between the electrodes of a cell. If there is a very high external resistance $R$, there will be a very small current flow. Under these conditions there will be a maximum electric potential difference $E°$ between the two electrodes, as shown in Figure 11.1.2. The chemical reaction taking place is

$$Zn + Cu^{2+} \rightarrow Zn^{2+} + Cu, \quad E° = 1.1 \text{ V}. \quad (11.1.3)$$

A cell is said to operate reversibly if all the energy available in the form of chemical energy performs useful work. A cell operates reversibly only if the current flow approaches zero. If the cell is operated irreversibly (high currents), much of the available chemical energy is dissipated (as heat) within the battery and hence this portion of the available energy does not do useful work. When a cell operates irreversibly, its potential difference is less, often much less, than $E°$; this is discussed further in Section 11.3 (e.g., see Figure 11.3.4). For the present we restrict our discussion to reversible processes.

The value of $E°$ given in (11.1.3) depends not only on the nature of the electrodes (pure copper and pure zinc) but on the concentrations of the

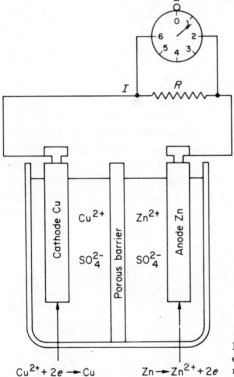

FIGURE 11.1.2. Cu–Zn cell operating at very low current has a maximum voltage of 1.1 volts.

ions in solution. More precisely, it depends on quantities called the activities of the ions. Activities are quantities which are functions of concentration and which, for very low concentrations (such as $10^{-4}$ m), are nearly equal to the concentrations. An understanding of activity requires a knowledge of thermodynamics. The detailed significance of the term activity is discussed in texts on electrochemistry. For our purposes it is sufficient to know that a specific activity for a particular ion corresponds to specific concentrations which could be looked up in a handbook. The value of $E°$ given in (11.1.3) refers to a situation when the activities of both the $Cu^{2+}$ ions and the $Zn^{2+}$ ions in solution are 1.

Suppose the cell shown in Figure 11.1.2 operates reversibly and in so doing eventually plates 1 mole of Cu onto the Cu electrode. Let us make $F$ (the **Faraday equivalent**) equal to the magnitude of 1 mole of charge. This equals Avogadro's number times the magnitude of electronic charge, i.e., $N_0 e = F = 96,500$ C/mole. When 1 mole of $Cu^{2+}$ is transferred, two Faraday equivalents are transferred because of the valence of 2. Let $n$ be the charge of the species being transferred ($n = 2$ for $Cu^{2+}$). Then $nF$ is the charge transferred when a mole of Cu is transferred. This charge is transferred across a potential difference $E°$. Hence it could produce an amount of work $nFE°$; i.e., the energy of the cell is reduced by $nFE°$ in the process. The change in energy of the chemical system under conditions of constant pressure and temperature is the change in the Gibbs free energy per mole:

$$\Delta G° = -nFE°. \qquad (11.1.4)$$

In chemical terms, the reaction takes place because it results in a lowering of the Gibbs free energy of the system. Note that the cell reaction takes place if the potential $E°$ of the reaction is positive [see Equation (11.1.3)] and if the free energy of the reaction is negative. The work done *by* the chemical system is $nFE°$. (If the reaction is not carried out reversibly, less useful work will be done when one mole of Cu is plated out.)

It should be noted that the cell potential represented by the reaction of (11.1.3) is 1.1 V only if the activity of each of the ions is 1.

## Q. 11.1.3

When a cell is operated reversibly the current flow is (a)_____ _____. The total charge of $6.02 \times 10^{23}$ charges, each of magnitude $e$, is called (b)_____. The change of the Gibbs free energy per mole is related how to the equilibrium cell potential? (c)_____ ____.

\* \* \*

There are various kinds of electrolytic cells:

1. **Dissimilar electrode cells.**
2. **Differential concentration cells.** Both electrodes are copper, say, but they are immersed in solutions of different concentrations, say 0.01 molal $CuSO_4$ on one side, 0.001 molal $CuSO_4$ on the other. (The number of moles $m$ added to 1 liter of solvent equals the **molality** $m$.)
3. **Differential temperature cells.** Here the temperature of one electrode is different than the temperature of the other.
4. **Differential pressure cells.**

**EXAMPLE 11.1.1**

Consider the concentration cell mentioned above. Which electrode would be the cathode and which the anode?

*Answer.* The cell potential would be exactly zero if the two concentrations were equal. Hence the reactions which occur will be in such a direction as to cause these two concentrations to be equal. Hence in the 0.01 molal $CuSO_4$, the reaction

$$Cu^{2+} + 2e^- \rightarrow Cu$$

will occur (and this decreases the copper ion concentration at that electrode) while in the 0.001 molal $CuSO_4$, the opposite reaction will occur (and this increases the copper ion concentration at that electrode). Hence the cathode is in the high-concentration region while the anode is in the low-concentration region.

Thermodynamics can be used to show that the potential difference for this cell is

$$E = \frac{RT}{nF} \ln \frac{a_{0.01}}{a_{0.001}}.$$

(This equation is not derived here.) Here $a_{0.01}$ is the activity of the $Cu^{2+}$ in the 0.01 molal solution. In dilute solutions such as these the activities can be roughly approximated by the concentrations. Hence the equilibrium potential of this concentration cell at 300°K is

$$E = \frac{RT}{nF} \ln 10 = \frac{8.31 \times 300 \times 2.3}{2 \times 96{,}500} = 0.03 \text{ V.}$$

## Q. 11.1.4

There are at least four ways of generating a potential difference in a cell. These are (a)_____
_____.

\* \* \*

## ANSWERS TO QUESTIONS

**11.1.1** (a) Four; (b) two electrodes, external conductor, and electrolyte; (c) electrolyte; (d) a semipermeable membrane.

**11.1.2** (a) Cathode, (b) consumption, (c) cathode.

**11.1.3** (a) Arbitrarily close to zero, (b) the Faraday equivalent, (c) $\Delta G^0 = -nFE^\circ$.

**11.1.4** (a) Dissimilar electrodes, differential concentration, differential temperature, and differential pressure.

## 11.2 HALF-CELL POTENTIALS

There are numerous electrode combinations possible. Suppose there are 100 different half-cells which could be combined with 100 other half-cells. A **half-cell** is an electrode plus the solution and atmosphere surrounding the electrode (the Cu and $CuSO_4$ solution of Figure 11.1.1 is a half-cell). We could then make a total of 10,000 different cells. Would we have to measure 10,000 cell potentials or would a smaller number of measurements suffice to define all 10,000 cell potentials? The answer is that 200 measured quantities would suffice. These are called **electrode potentials** or **half-cell potentials**. We would write each **electrode reaction** as an oxidation process;

$$Cu \rightarrow Cu^{2+} + 2e^-.$$

We would compare each half-cell with a standard reference half-cell (which we shall discuss later). The potential of the resultant cell is then arbitrarily defined to be the *electrode* potential of the particular electrode (relative to the standard reference electrode whose potential is arbitrarily set at zero). For the above half-cell we have, if the activity of the copper ions surrounding the copper is 1, by experiment

$$E° = -0.337 \text{ V}.$$

Likewise, if the activity of the zinc ions surrounding the zinc electrode is 1,

$$Zn \rightarrow Zn^{2+} + 2e^-$$

would give $E° = 0.763$ V against the standard reference half-cell. Consequently we have

$$\begin{array}{ll} Cu^{2+} + 2e^- \rightarrow Cu & 0.337 \text{ V} \\ \underline{Zn \rightarrow Zn^{2+} + 2e^-} & \underline{0.763 \text{ V}} \\ Cu^{2+} + Zn \rightarrow Cu + Zn^{2+} & 1.100 \text{ V} \end{array}$$

In other words by combining the oxidation electrode reactions in an appropriate fashion we can obtain the overall cell reaction and the resultant cell potential.

Table 11.2.1 gives a few **standard oxidation electrode potentials** which are defined for unit activities (or unit pressures (1 atm) if a gas is involved) and at a temperature of 25°C. In some textbooks and in some of the corrosion literature the electrode potentials are written as reduction potentials instead of oxidation potentials. This means that all the $E°$ signs are reversed. However, when the total cell reaction is written down, such as (11.1.3), the calculated cell voltage is the same.

Table 11.2.1. STANDARD OXIDATION HALF-CELL POTENTIALS*

| Electrode Reaction | $E°$(volt) | |
|---|---|---|
| $Li = Li^+ + e^-$ | 3.045 | Anodic |
| $Na = Na^+ + e^-$ | 2.714 | ↑ |
| $Mg = Mg^{2+} + 2e^-$ | 2.37 | |
| $Al = Al^{3+} + 3e^-$ | 1.66 | |
| $Zn = Zn^{2+} + 2e^-$ | 0.763 | |
| $Fe = Fe^{2+} + 2e^-$ | 0.440 | |
| $Cr^{2+} = Cr^{3+} + e^-$ | 0.41 | |
| $Cd = Cd^{2+} + 2e^-$ | 0.403 | |
| $Sn = Sn^{2+} + 2e^-$ | 0.136 | |
| $Pb = Pb^{2+} + 2e^-$ | 0.126 | |
| $Fe = Fe^{3+} + 3e^-$ | 0.036 | |
| $\frac{1}{2}H_2 = H^+ + e^-$ | 0 | Neutral |
| $Cu = Cu^{2+} + 2e^-$ | −0.337 | |
| $Ag = Ag^+ + e^-$ | −0.7991 | |
| $Hg = Hg^{2+} + 2e^-$ | −0.854 | |
| $Cl^- = \frac{1}{2}Cl_2 + e^-$ | −1.359 | |
| $Au = Au^{3+} + 3e^-$ | −1.50 | ↓ |
| $F^- = \frac{1}{2}F_2 + e^-$ | −2.65 | Cathodic |

\* W. M. Latimer, *Oxidation Potentials*, Prentice-Hall, Inc., Englewood Cliffs, N.J., (1952).

**EXAMPLE 11.2.1**

The half-cell potentials shown in Table 11.2.1 would enable one to calculate equilibrium electrochemical cell potentials for how many cells in all?

*Answer.* There are 18 electrode reactions in the table. For example, the lithium electrode could be combined with 17 others. Hence there are data here for 153 cells. This assumes that only standard half-cells are used. Otherwise there could be an infinite number.

**EXAMPLE 11.2.2**

What is the cell potential if the electrodes are a standard Ag electrode and a standard Zn electrode?

*Answer.* $2Ag^+ + Zn \rightarrow 2Ag + Zn^{2+}$ is obtained by adding

$$Zn \rightarrow Zn^{2+} + 2e^- \quad 0.763$$
$$Ag^+ + e^- \rightarrow Ag \quad \underline{0.799}$$
$$1.564$$

Hence the cell has a potential of 1.564 V and Ag is the positive electrode. *Note:* If the potential calculated is positive, the reaction proceeds in the assumed direction. Since reduction takes place at the Ag electrode, it is the cathode and hence the positive electrode.

### Q. 11.2.1

The standard oxidation potential for copper is defined for the case in which (a)_____. The advantage of defining standard oxidation potentials is (b)_____

_____. Oxidation is defined as a partial reaction in which electrons are (c)_____.

\* \* \*

STANDARD HYDROGEN ELECTRODE. What is the **standard reference half-cell?** This is a cell in which the reactive species is hydrogen. The electrode may be some (relatively) inert material such as platinum surrounded by a solution containing $H^+$ ions in HCl solution with an activity of 1 (a molality of 1.25 in this case) in the presence of hydrogen gas at a pressure of 1 atm at 25°C as shown in Figure 11.2.1. Thus the electrode

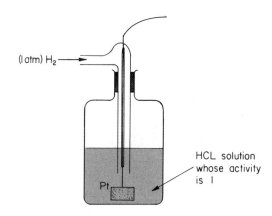

FIGURE 11.2.1. Standard hydrogen reference electrode. $T = 25°C$, $P_{H_2} = 1$ atm, $a_{H^+} = 1$.

reaction is

$$\tfrac{1}{2}H_2 \rightarrow H^+ + e^-, \qquad E° = 0 \text{ V};$$

the electrode potential is *arbitrarily* set equal to zero. This half-cell would be connected to another type of half-cell by a semipermeable barrier through which $H^+$ ions and possibly also $Cl^-$ ions or other ions in the other half-cell could pass but $H_2$ gas itself could not pass.

### EXAMPLE 11.2.3

A cell is made from a standard hydrogen electrode and a second hydrogen electrode in which the $H^+$ activity is 1 but the hydrogen pressure is 2 atm. The cells are connected by a semipermeable membrane through which $H^+$ only diffuses. Which is the anode and which is the cathode?

*Answer.* Clearly there will be no potential difference if the two pressures were the same. The process which occurs is of such a nature as to make the two pressures change in such a way that they do become equal. Thus the high-pressure electrode is the anode. Hydrogen ions will be produced here and will diffuse to the low-pressure electrode where they will combine with electrons to produce hydrogen gas. It is a simple problem in courses in thermodynamics to show that the equilibrium potential of this cell is

$$E = \frac{RT}{2F} \ln 2.$$

### Q. 11.2.2

The standard hydrogen electrode has three parts (a)_____

_____. The standard reference voltage for this cell operated at 25°C is (b)_____.

\* \* \*

The standard oxidation electrode potential for copper is $-0.337$ V. But what do we mean when we say copper? Is it a perfect single crystal? Is it a pure polycrystalline material? Is it a cold-rolled sheet of copper? Or does it not matter? The answer is that the electrode potential can be changed by

1. Cold working.
2. Impurities.

The potential of a grain boundary is slightly different than the potential of a grain center. These small differences can be very important in corrosion applications. For example, an ordinary iron nail is highly cold-worked at the tip and head. These parts of the nail will therefore have a different potential than the shank of the nail. If the nail is placed in the appropriate electrolyte, all four parts of an electrolytic cell are present and corrosion will occur. Similarly a piece of Monel may have a uniform composition or

**Table 11.2.2.** GALVANIC SERIES OF METALS AND ALLOYS IN SEA WATER*

| | |
|---|---|
| Anodic ↑ | Magnesium |
| | Zinc |
| | Alclad 3S |
| | Aluminum 3S |
| | Aluminum 61S |
| | Aluminum 63S |
| | Aluminum 52 |
| | Low steel |
| | Alloy steel |
| | Cast iron |
| | Type 410 stainless steel (active†) |
| | Type 430 stainless steel (active) |
| | Type 304 stainless steel (active) |
| | Type 316 stainless steel (active) |
| | Ni-resist |
| | Muntz metal |
| | Yellow brass |
| | Admiralty brass |
| | Aluminum brass |
| | Red brass |
| | Copper |
| | Aluminum bronze |
| | Composition G bronze |
| | 90:10 copper-nickel |
| | 70:30 copper-nickel—low iron |
| | 70:30 copper-nickel—high iron |
| | Nickel |
| | Inconel |
| | Silver |
| | Type 410 stainless steel (passive†) |
| Cathodic ↓ | Type 430 stainless steel (passive) |
| | Type 304 stainless steel (passive) |
| | Type 316 stainless steel (passive) |
| | Monel |
| | Hastelloy C |
| | Titanium |

\* From *Corrosion in Action*, International Nickel Co., New York.
† Stainless steel can be made passive by a treatment with oxidizing acids which combine with chromium in the steel and form chromate ions which are adsorbed on the surface and prevent further reaction.

**490** *Electrochemical Properties*

it may be cored as in Figure 10.3.6, in which case the electrode potential varies from position to position. Such differences are important in corrosion.

### EXAMPLE 11.2.4

A specimen of copper is cold-worked and the stored deformation energy is $5 \times 10^3$ J/mole. Such copper would tend to dissolve in an appropriate solvent and plate out on annealed copper. Why? What would be the cell potential if the cold-worked copper is the anode and annealed copper is the cathode of an electrochemical cell?

*Answer.* The free energy difference between the two phases is $5 \times 10^3$ J/mole. Hence by (11.1.4) the magnitude of the potential difference is

$$E = \frac{\Delta G}{nF} = \frac{5 \times 10^3 \text{ J/mole}}{2 \times 96{,}000 \text{ C/mole}} = 0.026 \text{ V}.$$

### Q. 11.2.3

If an iron nail is placed in an appropriate electrolyte, the shank of the nail will be the (a)_____ while the tip and the head will behave as (b)_____. If a pure sheet of Cu is one electrode in a $CuSO_4$ solution and a piece of bronze (90% Cu–10% Zn) is the other, there (c) (will or will not) be a potential difference.

\* \* \*

It is useful to have a **galvanic series** as shown in Table 11.2.2 in a real environment such as sea water. Note the position of magnesium, i.e., its tendency to undergo the oxidation reaction. It is not surprising that Mg wishes to return to the sea from where it came. In this table the metal near the top would be an anode in a cell with a lower metal.

## ANSWERS TO QUESTIONS

**11.2.1** (a) $Cu^{2+}$ ions having an activity of 1 are present around the copper and the temperature is 25°C; (b) that by combining several half-cells a huge number of cells can be obtained; the potential of these can be calculated from the standard oxidation potentials; (c) produced.

**11.2.2** (a) A platinum conductor (usually a wire gauze with a coating of finely divided platinum, called platinum black), a solution containing $H^+$ ions with an activity of 1 (these, of course, may be present as hydrated ions $H_3O^+$), with hydrogen gas at a pressure of 1 atm above the solution; (b) zero.

**11.2.3** (a) Cathode, (b) anodes, (c) will.

## 11.3 POLARIZATION AND OVERVOLTAGE

When electrochemical cells are operated at finite rates (rather than infinitesimal rates) they behave irreversibly and their cell potentials differ from the equilibrium (reversible) values, $E^0$. When the cell operates as a battery, the potential falls below the equilibrium value. If we are carrying out electrolysis at high current rates, the reverse is true; i.e., we must supply a voltage (opposite in sign) greater (in magnitude) than the equilibrium value (the value at negligible current rates).

There are three sources of these differences in potential:

1. Internal resistance of the cell (joule heating, i.e., $i^2R$ losses).
2. Concentration polarization.
3. Overvoltage.

All these are *kinetic* phenomena and are discussed next.

### Q. 11.3.1

The calculation of an equilibrium cell potential from the oxidation potentials of the half-cells applies only to the case where the current is (a)_____. The cell potential is always (b)_____ if there is a large current flow. This difference is due to (c)_____.

\* \* \*

CONCENTRATION POLARIZATION. Consider a cell of two identical electrodes of Cu dipped in the same $CuSO_4$ solution under an applied potential. When current flows

$$Cu \rightarrow Cu^{2+} + 2e^- \quad \text{(anode)}$$

while at the cathode

$$Cu^{2+} + 2e^- \rightarrow Cu \quad \text{(cathode)}.$$

Because transfer of $Cu^{2+}$ is diffusion-controlled the solution around the cathode will be depleted of $Cu^{2+}$. Hence we have under *dynamic conditions* a concentration cell. We have created two half-cells:

Cu with low $Cu^{2+}$ region    and    Cu with high $Cu^{2+}$ region,

and the reactions which would tend to occur because of the **concentration**

**polarization** cell *alone* are, respectively,

$$Cu \rightarrow Cu^{2+} + 2e^-, \qquad Cu^{2+} + 2e^- \rightarrow Cu.$$

That is, $Cu^{2+}$ would be produced in the low-$Cu^{2+}$ region and eliminated in the high-$Cu^{2+}$ region tending to form uniform concentration. The concentration cell potential *opposes* the applied potential. In electrolysis we would want to eliminate this **back voltage.** Vigorous stirring of the solution and an increase in temperature would succeed in doing this. However, in corrosion phenomena, a high back voltage may radically slow down the corrosion process; hence stirring would have an adverse effect. This is shown in Figure 11.3.1. Back voltage is due to the slowness of motion of the ions

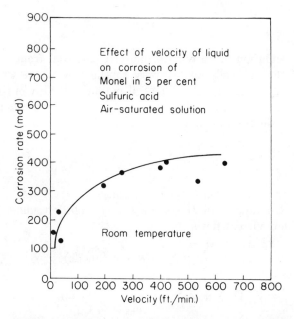

FIGURE 11.3.1. Effect of velocity of liquid on corrosion of Monel in sulfuric acid. The corrosion rate is milligrams per square decimeter (100 cm²) per day. (From *Corrosion*, The International Nickel Co., New York.)

in the electrolyte. If this motion is exceedingly fast then concentration polarization could not occur. In the absence of convection caused by agitation, the ions move by diffusion.

## Q. 11.3.2

When a cell operates at a finite rate, it is necessary that (a)_____
_____. Because this is a kinetic process (b)_____
_____ occurs. This results in (c)_____. Concentration polarization is ordinarily due to (d)_____.
It can be decreased by (e)_____
_____.

\* \* \*

OVERVOLTAGE. Overvoltage is due to the slowness of a chemical reaction at the electrode surface. Usually this is very small for metal deposition on metal. However, it is quite large for reactions involving liberation of gases. As an example, if we attempt electrolysis of an HCl solution

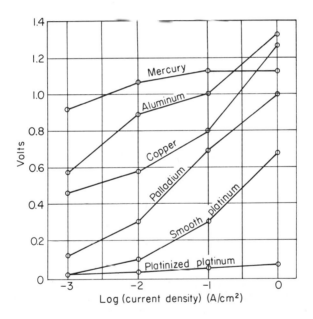

FIGURE 11.3.2. Hydrogen overvoltage on various electrode materials at various current densities.

of unit activity (using inert electrodes such as platinum), we would have to apply a voltage of 1.3595 V under equilibrium conditions (see Table 11.2.1) to produce hydrogen and chlorine gases:

$$Cl^- = \tfrac{1}{2}Cl_2 + e^-, \qquad E^0 = -1.3595 \text{ V}$$
$$\tfrac{1}{2}H_2 = H^+ + e^-, \qquad E^0 = 0 \text{ V}.$$

However, if we desire a finite current flow of, say, 1 A/cm² and both electrodes are ordinary platinum and the liquid is well stirred we would have to apply a voltage at least 0.7 V higher than this (the **overvoltage**); i.e., we must apply about 2.1 V. We could measure the overvoltage at an electrode where hydrogen is liberated by preparing a cell in which the overvoltage at the other electrode is negligible. Such results are shown in Figure 11.3.2. Platinized platinum has a very fine dispersion of platinum (called platinum black) on platinum sheet which acts as a catalyst for the hydrogen liberation process so the overvoltage is relatively small.

### Q. 11.3.3

Ordinarily, the overvoltage at an electrode where a metal dissolves or is deposited is (a)_____. However, the overvoltage is large when (b)_____. The overvoltage is due to (c)_____ _____. Hydrogen overvoltage on platinum can be decreased by (d)_____ _____.

\* \* \*

**EXAMPLE 11.3.1**

Can zinc be plated from an aqueous solution? For example, an ore solution contains $Zn^{2+}$ ions and we wish to plate out zinc (on zinc). The other electrode is inert.

*Answer.* Here the possible reactions and equilibrium voltages are

$$\tfrac{1}{2}H_2 \to H^+ + e^-, \qquad 0 \text{ V}$$
$$Zn \to Zn^{2+} + 2e^-, \qquad 0.763 \text{ V}.$$

Under equilibrium conditions, the answer is clearly "no" since zinc would tend to dissolve and hydrogen ions would tend to be converted to gas. However, because of the sizable overvoltage at high current densities, it is indeed possible to plate zinc from solution at rapid rates.

INTERNAL RESISTANCE. In an electrolytic cell there is a net flow of ions in an electrolyte as a result of the potential difference generated between the electrodes. The conductivity of an electrolyte in which only one ion, say $Cu^{2+}$, is responsible for charge transport is

$$\sigma = \frac{Dq^2}{k_B T},$$

where $D$ is the diffusion coefficient of the copper ions in the solution and $q$ is their charge. The resistivity $\rho$ of the solution is $\rho = 1/\sigma$. Knowing this and the geometry of the cell, the resistance of the electrolyte can be computed. This resistance is called the **internal resistance.** When the cell operates at a finite rate a certain fraction of the available energy $\Delta G^0$ will be dissipated as heat due to this internal resistance.

### Q. 11.3.4

There is a power loss in a cell operated at finite current due to the fact that (a)_____.

\* \* \*

FUEL CELLS. These kinetic phenomena which have just been discussed are of the utmost importance in the operation of batteries and fuel cells. A schematic of the hydrogen-oxygen fuel cell is shown in Figure 11.3.3. It has a reversible voltage of 1.22 V. However, when operated at finite power (finite current) the voltage output is considerably less as shown in Figure 11.3.4. The initial voltage drop is primarily due to overvoltage while the final rapid drop is due to concentration polarization. The reader should take note of the challenging materials problems associated with the

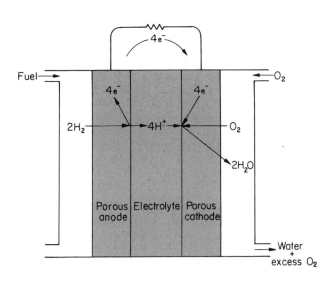

FIGURE 11.3.3. Schematic of the hydrogen-oxygen fuel cell.

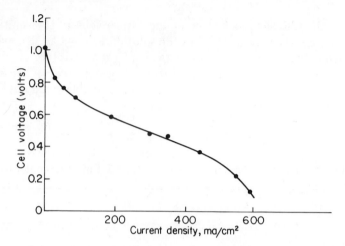

FIGURE 11.3.4. Cell voltage versus current density for a hydrogen-oxygen electrode with a 5 $N$ $H_2SO_4$ electrolyte.

porous electrodes and electrolyte, particularly since this fuel cell is usually operated at high temperatures rather than at room temperature.

## ANSWERS TO QUESTIONS

**11.3.1** (a) Arbitrarily close to zero, (b) less than the equilibrium (reversible) cell potential, (c) kinetic phenomena.

**11.3.2** (a) A net diffusion of ion(s) occurs, (b) concentration polarization, (c) a back voltage, (d) the slowness of diffusion, (e) rapid stirring of the electrolyte (if a liquid) or by raising the temperature.

**11.3.3** (a) Very small, (b) a gas is liberated, (c) a slow chemical reaction at the surface of the electrode, (d) coating the electrode with platinum black.

**11.3.4** (a) Energy is needed to cause a *net* flow of ions in the electrolyte.

### 11.4 CORROSION

DIFFERENTIAL OXYGEN CELL. An **oxygen concentration cell** is shown schematically in Figure 11.4.1. The oxygen concentration is high near the air-water interface where the reduction reaction

$$H_2O + \tfrac{1}{2}O_2 + 2e^- \to 2OH^- \qquad (11.4.1)$$

takes place.

Sec. 11.4  Corrosion  497

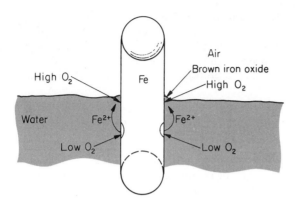

FIGURE 11.4.1. Corrosion of a steel (iron) pier or of a steel fence post. Oxidation concentration cell.

## EXAMPLE 11.4.1

The reaction represented by (11.4.1) might take place near the water surface (high-$O_2$ region) or below the water surface (low-$O_2$ region). Where does it preferentially occur?

*Answer.* We are considering a concentration cell. Thus Le Chatelier's principle (or the law of mass action) applies. The tendency for (11.4.1) to occur is greatest where the $O_2$ concentration is the greatest, hence near the surface.

Since there is a tendency for a reduction reaction to occur near the surface, there must be a tendency for an oxidation reaction to occur in the low-oxygen region; this is the reaction

$$Fe \rightarrow Fe^{2+} + 2e^-. \qquad (11.4.2)$$

Thus the iron would corrode away and form a pit, as shown in Figure 11.4.1, at the point where the concentration of oxidizing agent (oxygen) is the *least*. Thus **oxygen starvation** causes corrosion.

In general *corrosion* requires an electrolyte, a cathode, an anode, and an electrical connection. Note that all these are present in the cell of Figure 11.4.1. The corrosion of the metal below the chrome strip on an automobile is caused by oxygen starvation. Something as trivial as a speck of dirt on a metal bumper can be responsible for the formation of an **aeration cell** (a cell involving oxygen starvation) which results in a corrosion pit.

## EXAMPLE 11.4.2

Based on your knowledge of corrosion, describe the scientific procedures which you might use to prevent corrosion.

**498** *Electrochemical Properties*

*Answer.* Inasmuch as corrosion always involves an electrochemical cell, one logical answer is: Eliminate one or more of the four necessary components of the cell. Such methods will be discussed in the next section. Also, since corrosion takes place at the anode, another logical answer is: Make the material which you wish to protect the cathode. How might you do this?

## Q. 11.4.1

A steel fence post will rust (a) (just below or just above) the soil. In an oxygen cell, corrosion occurs (metal is removed) at the point where the oxygen concentration is the (b)_____. The pitting below a drop of tar on an automobile bumper is due to (c)_____. In general the process of corrosion involves (d)_____.

\* \* \*

DISSIMILAR METAL ELECTRODES. Suppose a ship has a bronze propeller and a steel hull. Clearly, an electrolytic cell is present [the oxidation potential of the bronze is about $-0.25$ V and of the steel (approximately iron) is about 0.44 V]. The iron hull would tend to dissolve, i.e., to behave as the anode.

**Intergranular corrosion** is another example of a dissimilar electrode cell. The atoms in the grain boundary have a higher free energy and hence tend to dissolve by the oxidation process, i.e., to behave as the anode. There may also be segregation at the grain boundaries so the composition is somewhat different from the interior. In addition grain boundaries often are the site of heterogeneous nucleation and hence of preferential precipitation. As an example, chromium steels if not cooled rapidly precipitate chromium carbide preferentially at the grain boundaries. This depletes the steel of chromium in the neighborhood of the boundaries and hence the steel surfaces are not protected from corrosion in these regions. These second-phase particles may also act as anodes and intergranular corrosion may be particularly bad.

DEZINCIFICATION. In certain brasses (Cu–Zn alloys) which are high in Zn content (called Muntz metal) the $\beta$ solid solution (bcc solution) is subject to **dezincification.** Here the zinc atoms are preferentially leached away from the $\beta$ phase and the porous copper is left behind. This reaction takes place in water with high chlorine content.

The preferential removal of one of the constituents of an alloy can take place in many systems. Figure 11.4.2 shows an example of the removal of iron from gray cast iron ($\alpha$-iron plus graphite flakes—alloy contains

FIGURE 11.4.2. Porous graphite residue of a corroded cast iron pipe elbow.

4.5 wt. % carbon) in which only the graphite is left behind. In the oxidation of steel, carbon from regions adjacent to the surface will rapidly diffuse to the surface and form CO gas. Hence the steel near the surface becomes **decarburized,** which results in a softening of the steel in these regions.

EFFECT OF STRESS ON CORROSION. The presence of stress enhances corrosion. Two types of processes are considered:

1. **Corrosion fatigue.**
2. **Stress corrosion cracking.**

When steel is subjected to cyclic loading in the absence of a corrosive environment it exhibits an endurance limit. However, if the same steel is subjected simultaneously to a corrosive environment, the stress required for

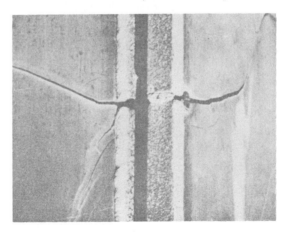

FIGURE 11.4.3. Stress corrosion cracking.

failure to occur decreases with an increasing number of cycles, i.e., the endurance limit ceases to exist.

The strength of a material may be drastically weakened by a crack formed by a combination of stress and corrosion. An example of such a crack is shown in Figure 11.4.3. Stress increases the free energy per unit volume due to the presence of strain energy. This increase can be particularly profound at the tip of a crack which therefore acts as the anode in a corrosive electrolytic cell.

**EXAMPLE 11.4.3**

Why is the crack tip the anode?

*Answer.* Because the highly strained metal, M, at the crack tip will tend to dissolve, $M \rightarrow M^{n+} + ne^-$, and redeposit, $M^{n+} + ne^- \rightarrow M$, as unstrained metal on the material which is nearly free from strain. The reaction $M \rightarrow M^{n+} + ne^-$ is an oxidation reaction which occurs at the anode.

Removal of material from the crack tip deepens the crack and increases the chance of fracture. In brasses such corrosion is called **season cracking** while in steels and irons it is called **caustic embrittlement.**

UNIFORM CORROSION. Uniform corrosion includes ordinary tarnishing, high-temperature oxidation (but this is not always uniform), and rusting. Uniform corrosion is usually measured in inches per year or in milligrams per square decimeter per day.

**Q. 11.4.2**

Because the atoms in grain boundaries in pure metals are not bonded as well as the atoms in the bulk, cells are set up in the metal with the grain boundary as the (a)_____. The endurance limit of a steel can vanish in the presence of (b)_____ during cyclic loading. Corrosion of a part loaded in tension is likely to lead to crack formation and fracture because (c)_____.

\* \* \*

# ANSWERS TO QUESTIONS

**11.4.1** (a) Just below, (b) least, (c) an oxygen or aeration cell, (d) an electrochemical cell.

11.4.2 (a) Anode, (b) a corrosive environment, (c) the highly stressed tip of a crack acts as the anode.

## 11.5 PROTECTING AGAINST CORROSION

There are several ways of avoiding corrosion difficulties:

1. Change materials.
2. Use special heat treatments.
3. Coat the material.
4. Use sacrificial anodes.
5. Use an impressed potential.
6. Add inhibitors to the "electrolyte."

HEAT TREATMENTS. We have already noted that the presence of a second phase or a composition variation such as in a cored Cu–Ni alloy gives rise to dissimilar electrode cells. In the case of a cored alloy a simple homogenization treatment removes the difficulty. In the case of second-phase precipitates, one may form larger precipitates so that the interfacial area is greatly decreased. For example, in a tempered martensitic steel the maximum rate of corrosion occurs for the finest form of $\epsilon$-carbide (a form of iron carbide which is produced when martensite decomposes during tempering).

NOBLE COATINGS. If the component is completely coated with another more noble metal (such as gold), a nonporous paint, etc., then the metal is isolated from the environment and cannot corrode. However, there is one serious drawback as illustrated in Figure 11.5.1. As is illustrated there, the material below a scratch through the paint on an automobile fender will corrode very rapidly because this material is an anode in the cell in which the noble coating is the cathode. Another important mechanism which usually operates in this situation is oxygen starvation; that is, an aeration cell is present.

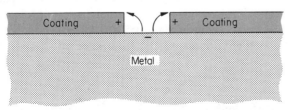

FIGURE 11.5.1. Pitting at a scratch in a noble coating.

### EXAMPLE 11.5.1

One difficulty with using porcelain coatings on steel is the stress buildup in the porcelain as a result of the difference in thermal expansion. Using the data of Table 4.1.1, compute the compressive stress in the porcelain when it is cooled to 25°C from 425°C. Assume that it is stress free at 425°C and that both components behave elastically. Assume that porcelain has $E = 750$ kbars and $\nu = 0.25$.

*Answer.* When the steel cools there is a strain $\epsilon_{st} = \alpha_{st} \Delta T = 11 \times 10^{-6} \times (-400)$. When the porcelain cools there is a strain $\epsilon_p = \alpha_p \Delta T = 4 \times 10^{-6} \times (-400)$. Hence the difference in strain is $\epsilon = \epsilon_{st} - \epsilon_p = -2.8 \times 10^{-3}$. The stress which must be applied to the porcelain is compressive in all directions of all the sheet. This is just the stress which would produce the same strain $\epsilon$. Therefore

$$\epsilon = \frac{\sigma}{E} - \nu \frac{\sigma}{E}.$$

Hence

$$\sigma = \frac{E}{1-\nu} \epsilon = \frac{750}{0.75} (-2.8 \times 10^{-3}) = -2.8 \text{ kbars}.$$

This is the reason that a hard bump causes a porcelain coating to spall.

OXIDE COATINGS. One look at the oxidation potential of aluminum tells us that this is a material which is extremely reactive to an oxidation environment. Aluminum readily forms an oxide layer. In fact aluminum is often systematically oxidized in such a way that an **adherent oxide film** forms on the surface. This is known as **anodizing**. Further oxidation can proceed only by diffusion of the aluminum (as an ion) through the ionic aluminum oxide layer (the electrolyte). The motion through this solid electrolyte occurs by solid state diffusion and is, of course, slow. The oxidation of metals which form adherent films follows the parabolic growth law discussed in Chapter 9 for the case of silicon. A solution which dissolves the oxide layer would wreak havoc with this mechanism of protection.

Materials such as magnesium, in which a large volume reduction occurs on oxidation, form a porous oxide scale rather than an adherent film (see Problem 11.11). This porous scale does not provide protection.

In stainless steels (containing say 18% Cr and 8% Ni) which are placed in an oxidizing environment, e.g., sulfuric or nitric acid, the reaction

$$Cr + 2O_2 + 2e^- \rightarrow (CrO_4)^{2-}$$

occurs. The chromate ions are adsorbed on the anode surface and prevent further reaction. We say then that steels are **passive** in the presence of oxidizing acids. However, they are **active** in the presence of other acids such as hydrochloric acid, since such acids do not form complexes which isolate the surface.

SACRIFICIAL COATING. Iron is often coated with zinc by dipping the iron in molten zinc (**galvanizing**). Zinc is less noble than iron (Table 11.2.1). Hence a scratch through the zinc will not have the same consequence as a scratch through a coating more noble than iron such as chromium. In fact in galvanized iron the zinc acts as the anode and therefore dissolves. This is known as a **sacrificial coating**.

### Q. 11.5.1

Coating a material with paint or with a noble metal, as in gold plating, prevents corrosion because (a)_____. A major drawback of this technique is that (b)_____
_____
_____. A cell in sea water has a zinc and an iron electrode; the zinc is the (c)_____. The steel in a galvanized steel roof does not rust even though there are obvious pinholes in the coating because (d)_____. In this case the zinc coating is called (e)_____.

\* \* \*

SACRIFICIAL ANODES. Sir Humphrey Davy was the first to use **sacrificial anodes** to control corrosion of ships hulls in 1824. [See I. A. Denison, *Corrosion* **3**, 295 (1947).]

Here chunks of zinc were connected to the ship's hull by electrical conductors and, as with the galvanized coating, the zinc rather than the iron of the ship's hull dissolved. The zinc anodes could be easily and periodically replaced.

IMPRESSED POTENTIALS. A steel pipeline buried in the ground will ordinarily act as an anode. We can apply an *external* voltage (**impressed potential**) so that the steel pipe becomes the cathode (see Figure 11.5.2) in any electrochemical cells which may exist.

RUST INHIBITORS. Your friendly service station attendant will be glad to furnish you with a can of rust inhibitor for your car cooling system

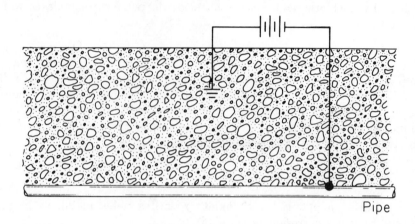

FIGURE 11.5.2. Impressed potential protection.

which probably contains chromates, phosphates, or tungstates (all of these being high oxide-containing complexes) which are adsorbed on the metal surface and isolate it from the electrolyte. This behavior is identical to the passivization of stainless steel. The most effective inhibitor of this sort is $TcO^{4-}$ (the oxide complex of the rare earth metal technicium). Why this is so is not known.

**EXAMPLE 11.5.2**

Suggest methods for preventing corrosion of an offshore oil platform. The structure is made of steel and there are three critical areas with corrosion rates in inches per year:

1. The immersed zone, 0.025 in./year.
2. The splash zone, 0.055 in./year.
3. The atmospheric zone, 0.005–0.010 in./year.

*Answer.* Because an electrolyte is always present in the immersed zone cathodic protection can be provided by using an impressed potential. Since this requires little labor for maintenance, it is a cheap and highly reliable method.

In the splash zone oxygen starvation is critical and the electrolyte is moving. Hence we would use isolation techniques or sacrificial coatings but could not use cathodic protection with an impressed potential (because of the absence of electrolyte).

In the atmospheric zone, paint could be used as an isolation technique.

## Q. 11.5.2

When large zinc rods are buried in the ground next to a steel bridge, they act as (a)_____. When an external voltage is applied to a pipe to protect it from corrosion the pipe is made the (b)_____ relative to the ground. This is known as (c)_____ protection. Rust inhibitors for an automotive cooling system work on the principle of (d)_____.

\* \* \*

## ANSWERS TO QUESTIONS

11.5.1 (a) It isolates the material from the electrolyte; (b) if the coating is scratched and water or some other electrolyte is present, the inert coating will act as the cathode and the metal below the scratch will act as the anode; (c) anode; (d) the zinc coating acts as the anode; (e) a sacrificial coating.

11.5.2 (a) Sacrificial anodes (the bridge is said to be cathodic), (b) cathode, (c) cathodic, (d) isolation.

## REFERENCES

### Corrosion

*Corrosion in Action*, International Nickel Company, New York. An excellent qualitative, short (47 pages) discussion of corrosion. It also contains a description of about a dozen experiments and a good list of references divided into several areas of subject matter.

*Corrosion*, International Nickel Company, New York (1949).

Swann, P. R., "Stress Corrosion Failure," *Scientific American* (February 1966), p. 72.

Uhlig, H. H., *Corrosion and Corrosion Control: An Introduction to Corrosion Science and Engineering*, John Wiley & Sons, Inc., New York (1971).

Bosich, J. F., *Corrosion Prevention for Practicing Engineers*, Barnes & Noble, Inc., New York (1970).

### Batteries and Fuel Cells

Grubb, W. T., and Niedrach, L. W., "Fuels Cells," in *Direct Energy Conversion*, ed. by G. W. Sutton, McGraw-Hill Book Company, Inc., New York (1966).

Liebhafsky, H. A., and Cairns, E. J., *Fuel Cells and Fuel Batteries*, John Wiley & Sons, Inc., New York (1968).

**Electrodeposition**

West, J. M., *Electrodeposition and Corrosion Processes*, Van Nostrand Reinhold Company, New York (1965).

Lowenheim, F. A., *Modern Electroplating*, John Wiley & Sons, Inc., New York (1963).

# PROBLEMS

**11.1** List five "why" questions in the area of electrochemical cells and corrosion.

**11.2** Phenolphthalein is an indicator which turns pink in the presence of excess hydroxyl ions. Potassium ferricyanide is an indicator which turns blue in the presence of ferrous irons. Design a simple laboratory experiment to show that in Figure 11.4.1 the high oxygen concentration region is the positive electrode, i.e., the cathode or the electrode where reduction occurs.

**11.3** A cell contains copper electrodes in $CuSO_4$ solution. In one cell the concentration is $10^{-4}$ m. In the other cell it is $10^{-3}$ m. What is the potential difference? Which electrode is positive?

**11.4** A car battery is in a car in Alaska on a cold night with no external heating. One student claims that when the starter is pressed, the battery will cool down and the electrolyte might even freeze. Another claims this is nonsense, and in fact the electrolyte will be heated. Who is correct and why?

**11.5** A sheet of copper metal is stressed elastically by a tension stress $\sigma$. What is the cell potential between copper in this stressed state and unstrained copper? *Hint:* The stored elastic energy can be equated to the change in Gibbs free energy.

**11.6** What is the cell potential between a large-angle grain boundary and a grain interior in copper? Assume that the copper is ultrapure so there is no possibility of segregation of impurities at the grain boundaries. Assume the grain boundary is $5 \times 10^{-8}$ cm thick. What would the anode be?

**11.7** If in Problem 11.6 there were impurity atoms segregated at the grain boundary, is it possible that the sign of the cell potential could be reversed? Assuming that ultrapure copper can be made and that impurities can be systematically added, could the cell potential between the grain boundary and grain interior be used to measure segregation?

**11.8** A pearlitic steel corrodes rapidly.
 (a) Why?
 (b) How may this be reduced?

**11.9** Explain why the head and tip of a nail would behave as anodes relative to the shank of the nail.

**11.10** There are three reasons the potential of an operating battery is less than the equilibrium potential. Discuss them.

**11.11** Magnesium has a density of 1.74 g/cm³ and MgO has a density of 3.65 g/cm³.
(a) If 1 cm³ of Mg is oxidized, find the oxide volume.
(b) Does Mg form an adherent MgO film?

**11.12** Austenitic stainless steel (18% Ni, 8% Cr plus small amounts of carbon which it is difficult to remove during manufacture) shows a rapidly decreasing solubility of carbon below 1300°F [see the book edited by E. E. Thum, *The Book of Stainless Steels*, American Society for Metals, Cleveland (1935)]. If this material is cooled slowly, chromium carbide is precipitated out. This precipitation process is most rapid at grain boundaries. The result is that the material is highly sensitive to grain boundary corrosion.
(a) Why?
(b) How would you eliminate this difficulty?

**11.13** Unused railroad tracks in a humid environment appear to corrode more rapidly than regularly used tracks. Why?

**11.14** Season cracking is an example of stress corrosion in brasses. It was found that brass cartridges stored near cavalry stables showed such behavior. Give some possible reasons.

**11.15** Tin plating is often used to protect iron. In fact, it is regularly used in making tin cans. A scratch in the tin is usually not a cause for concern, although a look at the oxidation potentials suggests otherwise. Explain.

**11.16** A replaceable Mg rod is sometimes inserted into hot water heaters. Describe its purpose.

**11.17** Pb should react with $H_2SO_4$ in our typical car batteries and produce $H_2$ and $PbSO_4$; i.e., the lead electrode should completely dissolve according to equilibrium considerations. It does not. Why?

**11.18** Two automobiles of the same type and year are owned by neighbors who regularly drive them to work over the same roads in the Northeast. One car is left out at night while the other is in a heated garage. Which shows more corrosion?

# MORE INVOLVED PROBLEMS

**11.19** Evaluate Sir Humphrey Davy's contribution to electrochemistry and corrosion.

**11.20** Discuss how the cathodic charging of steel introduces hydrogen and causes crack nucleation. See A. S. Tetelman and A. J. McEvily, Jr., *Fracture of Structural Materials*, John Wiley & Sons, Inc., New York (1967).

**508** *Electrochemical Properties*

**11.21** Dr. Joseph Kummer of the Ford Motor Co. has developed a Na–S battery in which the electrolyte is a solid called beta-alumina. Diffusion of $Na^+$ ions occurs extremely rapidly through this electrolyte. Write a paper on this battery. Include a discussion of the various scientific knowledge associated with the operation of this battery. Also discuss the engineering problems one might encounter in mass production of these fuel cells for automobile propulsion.

**11.22** Discuss in detail the various materials problems associated with the hydrogen-oxygen fuel cell.

**11.23** Describe at least one specific case in which electrochemical machining is used industrially for the rapid removal of material.

**11.24** Describe how chemical milling can be used for rapid material removal.

# SOPHISTICATED PROBLEM

**11.25** Gold is obtained from ores by leaching with cyanide solution (this was a revolutionary development in gold mining which made many low-concentration mines operational). The gold dissolves in such a solution because it forms a complex chemical ion with several $CN^-$ units attached. Describe how you would set up a transference cell to obtain the formula of the complex gold ion. This requires outside reading on the subject of transference cells.

# Prologue

The ultimate strength of materials, i.e., how strong materials might be, is discussed first. Expressions for the maximum possible fracture stress and the maximum possible shear yield stress are derived. Next, the reasons bulk materials are so weak are discussed. *Griffith theory* is used to explain low fracture stresses while the high mobility of dislocations explains low yield stresses. Then a study is made of the methods of obtaining stronger materials. These may involve making defect-free materials (fine fibers and whiskers) or using various strengthening mechanisms. Materials can be strengthened by adding solute atoms, plastic deformation, decreasing the grain size, having second-phase particles present (as a result of age precipitation or as the result of a eutectic or eutectoid transformation), forming martensite in steels, and forming composites. Why these strengthening mechanisms fail at high temperature is discussed. How *dispersion hardening* solves part of the high-temperature strength problem is studied. Strengthening mechanisms in polymeric materials are briefly reviewed.

# 12

# STRENGTHENING MECHANISMS

## 12.1 HOW STRONG CAN MATERIALS BE?

There's a way to do it better...find it. —Thomas A. Edison.

TENSILE STRENGTH. The detailed calculation of the forces needed to break chemical bonds is very involved. However, there is one case in which it is relatively simple to compute this. Consider the ionic NaCl *molecule* (such as exists in the vapor phase) which has a potential energy curve similar to that shown in Figure 5.5.1. This curve is based on the interaction potential energy

$$U = -\frac{e^2}{4\pi\epsilon_0 r} + \frac{b}{r^n}. \qquad (12.1.1)$$

Here $b$ and $n$ are empirical constants. Recall that the first term represents the coulombic attraction between the two ions and the second term represents the overlap repulsion. Also recall that $n \approx 10$.

The minimum in the potential curve corresponds to $dU/dr = 0$ and occurs at a value of $r$ which we call $r_0$. These facts can be used [by taking the derivative of $U$ given by (12.1.1)] to eliminate the parameter $b$; i.e., one gets

$$b = \frac{e^2}{4\pi\epsilon_0} \frac{r_0^{n-1}}{n}. \qquad (12.1.2)$$

The value of $r_0$ can be measured by electron diffraction. The minimum potential energy is, from (12.1.1) and (12.1.2),

$$U_0 = U(r_0) = -\frac{e^2}{4\pi\epsilon_0 r_0}\left(1 - \frac{1}{n}\right). \qquad (12.1.3)$$

Suppose now that a force $F$ is applied to the ions to pull them to a separation $r$ greater than $r_0$. What will be the force needed? This force is simply

$$F = \frac{dU}{dr} = \frac{e^2}{4\pi\epsilon_0 r^2} - \frac{nb}{r^{n+1}},$$

which by (12.1.2) becomes

$$F = \frac{e^2}{4\pi\epsilon_0} \left[ \frac{1}{r^2} - \frac{r_0^{n-1}}{r^{n+1}} \right]. \qquad (12.1.4)$$

We now ask two questions:

1. How is the force related to the displacement $x = r - r_0$ for very small displacements, such that $x \ll r_0$?
2. What is the maximum possible force which must be applied to completely separate the ions, i.e., to fracture the bond?

To answer the first question, let $r = r_0 [1 + (x/r_0)]$. Then (12.1.4) becomes

$$F = \frac{e^2}{4\pi\epsilon_0 r_0^2} \left\{ \frac{1}{[1 + (x/r_0)]^2} - \frac{1}{[1 + (x/r_0)]^{n+1}} \right\}.$$

Now by the binomial theorem

$$(1 + z)^p \doteq 1 + pz, \quad \text{for } zp \ll 1.$$

Likewise

$$\frac{1}{1 + y} \doteq 1 - y, \quad \text{for } y \ll 1.$$

By consecutive use of these expansions we have

$$F \doteq \frac{e^2}{4\pi\epsilon_0 r_0^2} \left\{ \left[ 1 - 2\frac{x}{r_0} \right] - \left[ 1 - (n+1)\frac{x}{r_0} \right] \right\} = kx$$

so that

$$k = \frac{e^2}{4\pi\epsilon_0 r_0^3} (n - 1). \qquad (12.1.5)$$

In other words for small displacements the molecule behaves like a linear spring with spring constant $k$; i.e., the applied force $F$ is directly proportional to the displacement $x$. If it is assumed that the effective cross-sectional area of the molecule normal to the bond is $r_0^2$, then the tensile stress $\sigma$ is related to the strain $\epsilon$ by

$$\sigma = \frac{F}{r_0^2} = \frac{k}{r_0}\frac{x}{r_0} = \frac{k}{r_0}\epsilon = E\epsilon. \qquad (12.1.6)$$

Hence $k/r_0 = E$, where $E$ is Young's modulus. This, of course, is the physical origin of Hooke's law for an ionic solid which is only complicated by the fact that it is necessary to sum over all the interactions.

For small vibrations the molecule behaves like a harmonic oscillator (see Section 9.7). The frequency $\nu$ of a harmonic oscillator is

$$\nu = \frac{1}{2\pi}\sqrt{\frac{k}{M}}.$$

The allowed energies of the oscillator are, as we have seen in Section 9.7,

$$u = (i + \tfrac{1}{2})h\nu, \qquad i = 0, 1, 2, \ldots,$$

so that by measuring the frequency of absorbed radiation which has an energy $h\nu$ the frequency of the oscillator can be measured. Hence $k$ is obtained and from this value of $k$, a value of $n$ can be calculated using (12.1.5). A value of $n \doteq 9$ is obtained in this way for the NaCl molecule.

Let us now return to the second question, namely, what is the maximum force that the ionic bond can withstand? This corresponds to finding the value of $F$ in (12.1.4) for which $dF/dr = 0$; this occurs for a value of $r$ which we call $r_m$. It follows that

$$\frac{r_0}{r_m} = \left(\frac{2}{n+1}\right)^{1/(n-1)}, \qquad (12.1.7)$$

and hence that

$$F_{max} = F(r_m) = \frac{e^2}{4\pi\epsilon_0 r_0^2}\left(\frac{r_0}{r_m}\right)^2 \frac{n-1}{n+1}. \qquad (12.1.8)$$

Combining (12.1.8) and (12.1.5), we obtain

$$F_{max} = \frac{kr_0}{n+1}\left(\frac{r_0}{r_m}\right)^2. \qquad (12.1.9)$$

The maximum stress is

$$\sigma_{max} = \frac{F_{max}}{r_0^2} = \frac{E}{n+1}\left(\frac{r_0}{r_m}\right)^2, \qquad (12.1.10)$$

where we have used $E = k/r_0$. When $n = 9$, this becomes

$$\sigma_{max} \doteq \frac{E}{15}.$$

Detailed calculations for perfect crystalline solids usually yield the result that the maximum allowable tensile stress is about

$$\sigma_{max} \approx \frac{E}{p}, \qquad (12.1.11)$$

where $5 < p < 15$.

### EXAMPLE 12.1.1

How much was the bond stretched when the force reached the maximum?

*Answer.* From (12.1.7), for $n = 9$,

$$\frac{r_m}{r_0} = (5^{1/8}) \doteq 1.22.$$

Hence it was stretched by 22%. In the solid state, bonds would be stretched by 5–15% before fracture of a perfect crystal.

### EXAMPLE 12.1.2

Estimate $\sigma_{max}$ for glass using the data of Table 1.3.3. This should be compared with the actual value for bulk glass in Table 1.3.5.

*Answer.* The value for $\sigma_{max}$ from (12.1.11) is, for $p = 5$,

$$\sigma_{max} = 138 \text{ kbars}.$$

The actual value is 0.7 kbars or only $\frac{1}{200}$ as large. However, freshly drawn glass fibers do have the estimated strengths.

### EXAMPLE 12.1.3

Estimate the value of $\sigma_{max}$ for the metal iron and compare with the value for gray cast iron and for an iron whisker. Use $p = 15$.

*Answer.*

$$\sigma_{max} = 133 \text{ kbars} \qquad \text{for } p = 15.$$

The value for gray cast iron is 1.4 kbars. The value for the strongest iron whisker is 131 kbars.

In the bulk form, materials are often only $\frac{1}{100}$ as strong as they might be.

### Q. 12.1.1

In an ionic molecule with an inverse power-repulsive potential with $n = 10$, the fraction of the binding energy due to coulombic attraction is (a)_____. The repulsive term in the potential plays a (b) (much larger or smaller) role in determining the stiffness constant. The calculated strength of a perfect ionic crystal is estimated to be $\sigma_{max} =$ (c)_____. Real materials in the *bulk* are (d) (never or usually) this strong.

\* \* \*

SHEAR STRENGTH. Not only do most materials fracture at strengths far below their ultimate potential strengths, but they also plastically deform at stresses far below that expected for perfect crystals. Plastic deformation is caused by shear stresses. In 1926 Frenkel proposed a theoretical model for computing the strength of perfect crystals [I. Frenkel, *Zeitschrift Physik* **37**, 572 (1926)].

We assume that one half of the crystal slides as a rigid body over the other half. Now we need a force law. We know that for small displacements Hooke's law is obeyed across the glide plane. We also know that the

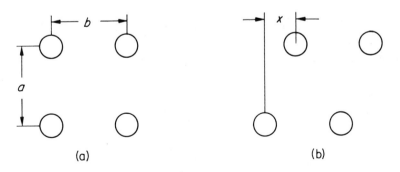

FIGURE 12.1.1. Atoms in a crystal. (a) Undeformed crystal. (b) Deformed crystal.

force law is periodic in the sense that if the atoms move one slip distance, the atomic arrangement is just as before. Now consider Figure 12.1.1 where the atoms in a simple cubic array are shown.

Let us now sketch the potential energy and the force as the top half of the crystal of Figure 12.1.1 slips over the lower half. Clearly, the potential is a minimum at the equilibrium positions, namely, $x = 0, \pm b, \pm 2b, \cdots$. Also, the midway positions ($x = \pm(b/2), \pm(3b/2), \cdots$) must correspond to a metastable equilibrium or a maximum potential. Thus the potential energy curve must have some general periodic form as shown in Figure 12.1.2(a).

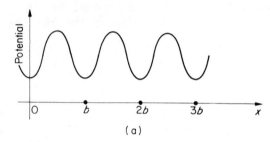

(a)

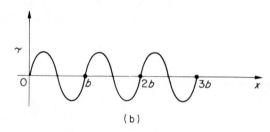

(b)

FIGURE 12.1.2. (a) Periodic potential energy curve for simultaneous slip. (b) Corresponding stress required for simultaneous slip.

The shear stress $\tau$ which causes this displacement is the derivative of the potential energy and hence has the form shown in Figure 12.1.2(b). Every student who has progressed to this point knows that the simplest periodic function that satisfies these conditions is

$$\tau = c \sin\left(\frac{2\pi x}{b}\right). \quad (12.1.12)$$

(You do not have to be a great scientist such as Frenkel to see that!) This may not be the best representation of the actual periodic function in this case, but because of its simplicity, Frenkel used it in his theory.

When the displacements are very small we must have linear behavior; i.e.,

$$\tau = G\gamma = \frac{Gx}{a},$$

where $G$ is the shear modulus, $x$ and $a$ are as shown in Figure 12.1.1, and $\gamma$ is the shear strain defined by (1.3.16). But by (12.1.12) we have for small $x$ (since $\sin y \doteq y$ for $y \ll 1$)

$$\tau = c\frac{2\pi x}{b} = \left(c\frac{2\pi a}{b}\right)\gamma$$

so that we conclude that $c = Gb/2\pi a$ and that

$$\tau = \frac{Gb}{2\pi a}\sin\left(\frac{2\pi x}{b}\right) = \tau_{max}\sin\left(\frac{2\pi x}{b}\right),$$

which when $a = b$ (as in the case of a simple cubic crystal) gives simply

$$\tau_{max} = \frac{G}{2\pi}. \quad (12.1.13)$$

Thus according to **Frenkel's model** there will be flow if the applied shear stress equals $G/2\pi$. This is an estimate of the shear strength of *perfect* crystals.

**EXAMPLE 12.1.4**

Using the data of Table 1.3.4, compute the maximum shear stress possible for copper. Compare this with the values given in Table 1.3.5. Recall that the maximum shear stress in a tensile specimen occurs on a plane inclined at 45 deg to the tensile axis and has a value of $\sigma_0/2$, where $\sigma_0$ is the yield strength [see Figure 1.3.3 and Equation (1.3.15)].

*Answer.* $\tau_{max} = 65$ kbars, which corresponds to a value of 130 kbars for the tensile yield stress. The strongest copper whisker is only half this strong. Annealed OFHC copper has a value of $\tau_{max} = 0.35$ kbars and a very carefully grown single crystal has a shear strength of only $10^{-5}$ kbars. Certainly bulk copper is incredibly weak compared to Frenkel's estimated strength for perfect crystals.

**Q. 12.1.2**

Frenkel proposed that when a perfect crystal deforms plastically due to the presence of (a)_____ stresses, the top half moves as a block over the bottom half. The stress needed to cause this deformation would vary (b)_____ with distance in the deformation direction. The form of the shear stress variation with distance was assumed to be (c)_____. The constant $k$ can be evaluated in terms of (d)_____. The maximum tolerable shear stress according to Frenkel's model is (e)_____. Bulk materials are (f) (usually or never) this strong.

\* \* \*

## ANSWERS TO QUESTIONS

**12.1.1** (a) 90% [see (12.1.3)]; (b) much larger; (c) $E/p$, where $5 < p < 15$; (d) never.

**12.1.2** (a) Shear, (b) periodically, (c) $\tau = k \sin(2\pi x/b)$, (d) the shear modulus $G$, (e) $G/2\pi$, (f) never.

### 12.2 WHY ARE BULK MATERIALS SO WEAK?

Glasses (such as ordinary window glass) are weak because they have tiny sharp cracks on the surface and these cracks have large stress concentrations at the tip. Other materials also have defects, such as pores and cracks within them, which act as stress concentrations.

Bulk crystalline materials are weak because of the presence of dislocations. They have such low yield stresses because flow by dislocation motion takes a stress very much lower than $G/2\pi$. They have such low fracture stresses because pileups of dislocations can cause large stress concentrations.

#### Q. 12.2.1

Glasses usually have strengths very much less than the theoretical strength $E/p$, where the number $p$ is about (a)_____, because of the presence of (b)_____. Crystalline materials usually have yield strengths less than $G/s$, where $s$ is about (c)_____, because of the presence of (d)_____.

\* \* \*

BRITTLE FRACTURE OF GLASS, ETC. Let us first consider another way of estimating the cleavage strength of materials. Suppose glass fractures across a plane normal to the tensile axis. A new surface area $2A$ is created, where $A$ is the cross-sectional area. The surface energy (energy per unit area) is $\gamma_S$ so the work necessary to create these surfaces is $W = 2\gamma_S A$. Now suppose that in causing the fracture, an average force $F$ acts over a distance $\delta$. The work done is $W = F\delta$. Hence

$$\sigma_{\text{fract}} = \frac{2\gamma_S}{\delta}. \tag{12.2.1}$$

For glass, $\gamma_S = 300$ ergs/cm² and assuming that $\delta \approx 0.3$ Å, we have $\sigma_{\text{fract}} = 200$ kbars. This is enormous compared to the actual fracture stress of bulk glass, 0.7 kbars.

Glass has tiny cracks at the surface about 1 $\mu$ deep. [The density of fused silica glass is less than that of the crystalline silica phase. Hence if at the surface there is a small quantity of material which transforms from the glassy state to the crystalline state (**devitrification**), a crack might appear to compensate for the volume change.]

Let us now briefly discuss the **Griffith theory** [A. A. Griffith, *Philosophical Transactions of the Royal Society of London* **A221**, 163 (1920)]. Suppose that there is formed at the surface of a plane body of *unit* thickness a very sharp crack of depth $c$ as shown in Figure 12.2.1. The surface energy of the crack is

$$U_s = 2c\gamma_S, \tag{12.2.2}$$

since new surface area $2c$ is created. The difference in elastic energy between this body with a crack and the same body without the crack when a tensile stress $\sigma$ is applied normal to the crack is

$$U_e = -\frac{\pi c^2 \sigma^2}{2E}, \tag{12.2.3}$$

a result first given by Inglis [C. E. Inglis, *Transactions of the Institute of Naval Architects* **55** *Part 1*, 219 (1913)].

An heuristic derivation of the last equation follows; it should be emphasized that it involves drastic approximations, although it leads to approximately the correct result. The strain energy density in a tensile specimen is $\sigma\epsilon/2 = \sigma^2/2E$, where $E$ is Young's modulus [see (1.3.11)]. Because of the presence of the crack, the material within the circle of approximate radius $c$ is essentially free of stress, as shown in Figure 12.2.1. This region has a volume of $\pi c^2/2$ (recall that the plate has unit thickness) and when multiplied by the strain energy density which would otherwise be present gives, except for the factor of $\frac{1}{2}$, Equation (12.2.3).

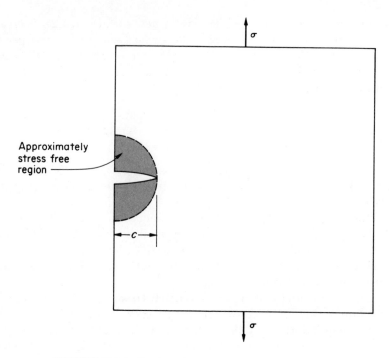

FIGURE 12.2.1. Crack at the surface of stressed glass plate.

The *increase* in energy when a crack is formed is

$$\Delta U = 2c\gamma_S - \frac{\pi c^2 \sigma^2}{2E}. \tag{12.2.4}$$

It would be helpful to the reader to sketch $\Delta U(c)$ assuming $\sigma$ is a constant (you need not know $\gamma_S$, $\sigma$, and $E$ to do this). The crack will grow spontaneously if $\partial \Delta U/\partial c < 0$ so the critical stress for a given crack can be calculated by setting $\partial \Delta U/\partial c = 0$. We have

$$\sigma_{\text{crit}} = \left(\frac{2E\gamma_S}{\pi c}\right)^{1/2}. \tag{12.2.5}$$

Once the stress reaches the critical value for the weakest crack present that crack will begin to propagate. It can be shown by the theory of elasticity that its maximum velocity is that of a sound wave. For comparison with experiment, see D. K. Roberts and A. A. Wells, *Engineering* **178**, 820 (1954).

## Q. 12.2.2

A successful model for the strength of glass was proposed by (a)_____. In his model there is a critical balance between (b)_____ energy, which increases when a crack grows, and (c)_____ energy, which decreases when the crack grows. When the rate at which these two energies change with crack depth is equal, the (d)_____ _____. If we write $\sigma_{cr} = K\sigma_{max}$, what is the value of the stress concentration factor $K$ assuming $\sigma_{max} = E/10$? (e)_____.

* * *

It has not been possible to test the Griffith equation precisely because of the smallness of the cracks on the surface. It is thought that for glasses (and other materials for which slip does not occur) Equation (12.2.5) is essentially correct.

### EXAMPLE 12.2.1

If a glass has $E = 7 \times 10^{11}$ dynes/cm² ($10^7$ psi) and $\gamma_S = 300$ ergs/cm² and $\sigma_{crit} = 7 \times 10^8$ dynes/cm², calculate the depth of the crack causing fracture.

*Answer.* From (12.2.5) we find $c \approx 3 \times 10^{-4}$ cm = 3 $\mu$.

If a piece of glass has a 3 $\mu$ crack on the surface, it will fracture at a tensile stress of only 0.7 kbars. Yet freshly drawn glass fibers (drawn in a vacuum) have tensile strengths of 3000 kbars. The surface processes which produce cracks take place rapidly in the presence of water vapor but only slowly in a vacuum. Cracks can also readily be produced when two fibers touch. The stress concentration factor at the tip of the 3 $\mu$ crack is $3000/0.7 \approx 4000$.

Materials which fracture without plastic deformation are called brittle materials and the process is called brittle fracture. Silica glasses and polymeric glasses are good examples. However, many materials are quasi-brittle (very nearly brittle) materials. Examples are sintered tungsten carbide and white cast iron.

By extending the Griffith fracture theory it can be shown theoretically that the compressive fracture stress is eight times larger than the tensile fracture stress. [See J. C. Jaeger, *Elasticity, Fracture and Flow*, John Wiley & Sons, Inc., New York (1962), p. 167.] This is experimentally found to be the case for glass and is also approximately true for sintered carbides. Such materials can be considered truly brittle materials. Gray cast iron which is partially ductile (so that the above theory does not apply) has a compressive fracture stress which is about three times as large as the tensile fracture strength. The Griffith theory can be modified when dealing with materials in which plastic deformation occurs in the highly stressed region. In this case, work is done not only in creating surface area but in plastic deformation. The Griffith theory can be extended by replacing $\gamma$ by $\gamma + p$, where $p$ is the plastic energy.

How do cracks begin in crystalline materials which fracture after plastic deformation? There are at least two possibilities. First, vacancies can be produced by the deformation process and these can diffuse together to form voids. Two nearby voids can have a large stress concentration between them, so that these voids join, etc.

Second, a large number $n$ of edge dislocations of the same sign and on the same slip plane can be pushed by the shear stress $\tau$ against a barrier such as a grain boundary, as shown in Figure 12.2.2. This is called a **pileup**.

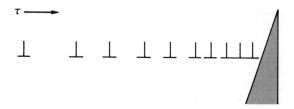

FIGURE 12.2.2. Pileup of edge dislocations.

The total shear stress acting on the barrier can be shown to be $n\tau$. The number of dislocations in such a pileup may be 1000 so there is a very large stress concentration here. Moreover, there is also a very large tensile stress present at the barrier. Suppose that the $n$ dislocations cover a length $c$ (normal to the dislocation lines and in the glide direction). The maximum *tensile* stress at the barrier due to these $n$ dislocations is exactly the same as that due to a crack of length $c$ in Griffith's theory (12.2.5). This stress at the barrier may be as large as the fracture stress of a perfect crystal; in that case a crack would be nucleated.

### Q. 12.2.3

In the case of materials which exhibit considerable ductility prior to fracture, fracture may be due to either (a)_____ _____ or (b)_____.
The tensile stress at the tip of a pileup of length $c$ is the same as (c)_____ _____.

\* \* \*

PLASTIC YIELDING IN CRYSTALLINE MATERIALS. Plastic yielding takes place at low stresses because of the movement of dislocations. Such movement produces plastic flow as was illustrated in Section 6.6.

**Slip** (or **glide**) in single crystals is illustrated in Figures 12.2.3 and 12.2.4. It can be compared to the shear distortion of a pack of cards. The individual segments shown in these figures are called **glide packets.** Large

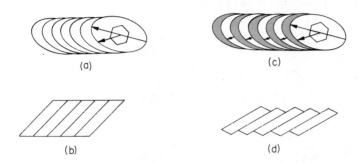

FIGURE 12.2.3. Mechanism of slip in a cylindrical crystal. (a) and (b), initial state (top and side views). (c) and (d), after slip. Slip directions are designated by arrows. [From E. Schmid and W. Boas, *Plasticity of Crystals*, F. A. Hughes, London (1950).]

numbers of dislocations move on the planes between the packets. The displacement of the glide packets takes place on crystallographic planes called **slip planes** or **glide planes** and in a crystallographic direction called the **slip direction.** Table 12.2.1 gives examples.

The glide packets might have thicknesses of $10^{-3}$–$10^{-4}$ cm. Often the glide packets themselves are made up of many smaller lamellae. We have restricted the examples in Table 12.2.1 to the simpler structures. The most prominent characteristic of slip is that the slip direction is universally the direction along which the atoms are closest packed while the slip plane is a plane in which the atoms are closely packed (but not necessarily the most closely packed plane in that crystal). *This means that the slip distance is as short as possible while the slip planes tend to be as far apart as possible.*

The combination of a slip plane and a slip direction is known as a **slip system.** Face-centered cubic metals have 12 slip systems; there are 4 slip planes $\{111\}$ with 3 slip directions $\langle 110 \rangle$ on each plane. Hexagonal close-packed zinc will have only 3 slip systems. The possibility of gliding confers plasticity to a crystal. A polycrystalline material containing crystals with low symmetry and hence few glide systems will tend to have low plasticity and be more brittle.

FIGURE 12.2.4. Appearance of glide lines on metal crystals. [From F. von Göler and G. Sachs, *Zeitschrift Physik* **55**, 581 (1929).]

Table 12.2.1. SLIP PLANES AND SLIP DIRECTIONS

| Structure | Type of Material | Material | Plane | Direction |
|---|---|---|---|---|
| Diamond cubic | Homopolar | Si | (111) | [1$\bar{1}$0] |
| Wurzite | Homopolar | InSb | (111) | [1$\bar{1}$0] |
| NaCl | Ionic | NaCl | (110) | [1$\bar{1}$0] |
| NaCl | Ionic | KCl | (110) | [1$\bar{1}$0] |
| fcc | Metallic | Al | (111) | [10$\bar{1}$] |
| fcc | Metallic | Cu | (111) | [10$\bar{1}$] |
| fcc | Metallic | Ni | (111) | [10$\bar{1}$] |
| bcc | Metallic | α-Fe | (101) | [11$\bar{1}$] |
| bcc | Metallic | α-Fe | (112) | [11$\bar{1}$] |
| bcc | Metallic | α-Fe | (123) | [11$\bar{1}$] |
| bcc | Metallic | Mo | (112) | [11$\bar{1}$] |
| bcc | Metallic | Na | (112) | [11$\bar{1}$] |
| hcp | Metallic | Zn | (0001) | [2$\bar{1}\bar{1}$0] |
| hcp | Metallic | Mg | (0001) | [2$\bar{1}\bar{1}$0] |
| Rhombohedral | Metallic | Bi | (111) | [10$\bar{1}$] |

At a certain critical stress a crystal begins to deform rapidly. When differently oriented crystals of the same material with the same history are loaded, slip will appear under the conditions given by **Schmid's law:** *Slip begins when the stress resolved on the slip plane in the slip direction reaches a certain value called the* **critical resolved shear stress,** $\tau_c$. In Figure 12.2.5

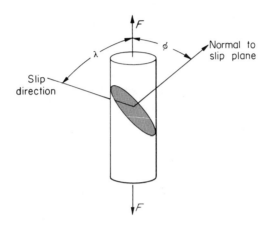

FIGURE 12.2.5. Single crystal in tension.

we show a crystal of uniform cross section $A$ loaded in tension. The shear stress resolved in the slip direction is the force in the slip direction divided by the area of the slip plane; i.e.,

$$\tau = \frac{F}{A} \cos \phi \cos \lambda. \tag{12.2.6}$$

To get a large range of $\lambda$ and $\phi$ when making such studies, crystals having few slip systems are chosen. Here zinc, which has the hcp structure, is a good example since slip occurs on the single basal plane in three directions. Figure 12.2.6 illustrates how well Schmid's law applies.

The critical resolved shear stress is extremely sensitive to impurities, surface conditions, etc. It is a structure-sensitive property. We shall return to these points later. In any case we note that $\tau_c$, the critical resolved shear stress, is small. The value predicted by Frenkel's equation for simultaneous slip in a perfect zinc crystal is $5 \times 10^4$ bars while the experimental value for zinc in Figure 12.2.6 is 1.8 bars.

The shear stress needed to cause deformation is low because the movement of dislocations is involved. A dislocation can move through

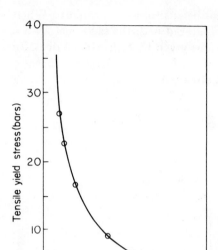

FIGURE 12.2.6. Variation of yield stress with orientation. The single crystals of 99.999 % zinc were tested at 25.0°C. The solid line is computed from Schmid's law with $\tau_c = 1.81$ bars $= 25.7$ psi. (1 bar $= 10^6$ dynes/cm$^2 = 14.5$ psi.) [From E. Schmid and W. Boas, *Plasticity of Crystals*, F. A. Hughes, London (1950).]

otherwise perfect pure metallic and ionic crystals at extremely low stresses. Peierls and Nabarro calculated that this stress would be

$$\tau_{\text{PN}} = \frac{2G}{1-\nu} e^{-2\pi a/(1-\nu)b}.$$

Here $G$ is the shear modulus, $\nu$ is Poisson's ratio, $a$ is the distance between planes, and $b$ is the slip distance. With $\nu = \frac{1}{3}$ and $a = b$ this predicts that $\tau_{\text{PN}} = 3.6 \times 10^{-4} G$. More recent calculations give an even smaller value. The stress needed to move a dislocation through an otherwise perfectly periodic lattice is called the **Peierls-Nabarro stress.**

Fred Young has found experimentally that the shear stress needed to move a dislocation in copper (when its motion is not impeded by the presence of other dislocations) is at least as small as 0.1 bars [F. W. Young, Jr., *Journal of Applied Physics* **33**, 963 (1962)]. This is less than $10^{-6}G$.

### Q. 12.2.4

A combination of slip plane and slip direction is called (a)_____. There are altogether (b)____ {110} ⟨111⟩ slip systems in bcc iron. Schmid's law states that a (c)_____ shear stress exists. The actual stress needed to move a dislocation in a metal such as copper (in the absence of impurities and other dislocations) is called the (d)_____. It is less than (e)\_\_\_\_ bars in Cu.

\* \* \*

## ANSWERS TO QUESTIONS

**12.2.1** (a) 5 to 15, (b) cracks, (c) 6 or $2\pi$, (d) dislocations.

**12.2.2** (a) Griffith; (b) surface; (c) stored elastic energy; (d) critical crack length has been reached; (e) $(200\gamma_s/\pi cE)^{1/2}$; in other words $\sigma_{\text{crit}} = (200\gamma_s/\pi cE)^{1/2}$ $(E/10)$ follows from (12.2.5).

**12.2.3** (a) Joining of voids which were formed from vacancies, (b) cracks caused by dislocation pileups, (c) that at the tip of a crack of length $c$ in Griffith's theory.

**12.2.4** (a) A slip system, (b) 12, (c) critical resolved, (d) Peierls-Nabarro stress, (e) 0.1.

## 12.3 GENERAL STRENGTHENING CONCEPTS

REDUCE THE EFFECT OF FLAWS. Cracks on the surface of glass, weak graphite flakes in cast iron, or pores in aluminum oxide all act as sites of stress concentration. Hence to strengthen these materials, we should eliminate these flaws. In the case of bulk silica glass this is presently impossible to do. One substitute procedure is to cool the glass in such a way that large compressive stresses will be present in the surface only (although the cracks are still present, they have been rendered less effective). This is called tempered glass. Another procedure is to bombard the glass with high energy potassium ions which will go into the glass near the surface and expand it. This also results in compressive stresses near the surface.

Fatigue cracks due to cyclic loading almost inevitably begin at the external surface. Many of the methods for reducing fatigue depend on improving the surface. Highly polished specimens fatigue less rapidly than ordinary machined specimens. If the surface is in compression, the material fatigues less rapidly. Nitriding of steel alloyed with aluminum produces aluminum nitride which expands the steel. Since nitriding introduces nitrogen only near the surface (a diffusion process analogous to carburizing) the surface layer is in compression. Another method of hardening the surface and introducing compressive stresses is shot peening.

Figure 10.5.7 is a micrograph of gray cast iron. The graphite flakes act as sharp cracks and greatly reduce the strength. Figure 10.5.8 is a micrograph of nodular cast iron in which the graphite is present as spheres. The nodular cast iron has a strength of more than twice that of gray cast iron.

ELIMINATE FLAWS. The most significant feature of Table 1.3.5 is that the strongest materials are fibers, filaments, whiskers, and wires. It is in general not possible to duplicate these strengths in the bulk; i.e., there is a strong size effect on strength when the particle becomes very small. This was illustrated in Figure 8.5.3 for the case of whiskers. The strongest whiskers are completely dislocation free. A similar size effect exists with glass fibers and with tungsten wires. Commercial grade glass fibers have strength of about 30 kbars (450,000 psi), although much stronger glass fibers have been grown. Very strong boron and carbon filaments have also been grown. The boron filaments are made by passing $BCl_3$ gas over a hot tungsten wire. The carbon filaments are made by heating rayon fibers and simultaneously pulling them at a very high temperature.

DISRUPT THE CONTINUUM. Because materials scientists have been unable to produce perfect crystals in bulk form, another approach usually is used for strengthening materials. Dislocations can move easily through an otherwise perfect crystal. However, this is not the case if there are various irregularities in the path of the dislocation. These may be point defects, solute atoms, other dislocations, grain boundaries, and second-phase particles. When tungsten wires are placed in alumina ceramic, the brittleness of the ceramic is decreased; thus if a crack is formed in the ceramic, it will stop at the wire. Filamentary composites owe their strengthening to the disruption of the continuum. As another example, the tiny precipitate particles in age-hardened aluminum act as barriers to the motion of dislocations and hence strengthen the alloy, again, by disrupting the continuum. Strengthening by this mechanism is discussed in the succeeding sections.

## Q. 12.3.1

There are three general procedures for obtaining strong materials. These are (a)_____ _____. The effect of a crack on the surface of glass can be reduced by introducing (b)_____ _____. Filaments, fibers, and whiskers of very small diameter are often extremely strong because (c)_____. Five packing imperfections which impede the motion of dislocations are (d)_____ _____.

\* \* \*

## ANSWERS TO QUESTIONS

**12.3.1** (a) Reduce the effect of flaws, eliminate flaws, and disrupt the continuum; (b) compressive stresses parallel to the surface at the surface; (c) of the absence of flaws; (d) point defects, solute atoms, other dislocations, grain boundaries, and second-phase particles.

### 12.4 SOLUTE STRENGTHENING

SUBSTITUTIONAL SOLUTES. Figure 12.4.1 shows a strong impurity effect on the critical resolved shear stress $\tau_c$ (at $-60°C$) in mercury monocrystals containing silver [by K. M. Greenland, *Proceedings of the Royal Society (London)* **A163**, 28 (1937)]. An impurity content of 1 part in $10^4$ raises $\tau_c$ by a *factor* of 5. (Compare this effect with the change in elastic modulus which would be only a fraction of a percent.) The result of adding substitutional impurities at higher concentration is usually not so strong.

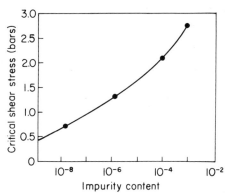

FIGURE 12.4.1. Impurity effects on $\tau_c$ (1 bar = $10^6$ dynes/cm$^2$ = 14.5 psi) for Hg crystals.

Consider the edge dislocation of Figure 6.6.1. Just above the dislocation line the atoms are crowded in and the stress is compressive while just below the dislocation line the atom-to-atom distances are stretched and the material is in tension. Suppose that the atoms are spheres and that solute atoms which are 10% larger are present. If present in the bulk of the crystal a big solute atom would cause considerable strain energy in the crystal. However, if this large solute atom is present in the tensile region below the dislocation line, it could easily be accommodated without causing so much strain energy. Hence the energy of the system is lower if the solute atom is at the dislocation. Therefore a force will be needed to separate the dislocation from the solute atom; this is called the **pinning force**. Mathematical models can be used to calculate approximately the force needed to tear dislocations from the solute atoms which can be considered as fixed at low temperatures since they can only move by diffusion. Note that solute atoms will tend to segregate at dislocations because this lowers the energy of the system. The extra solute atoms near the dislocation are called **impurity atmospheres**.

INTERSTITIAL SOLUTES. Interstitial atoms interact strongly with the stress field of a dislocation and hence tend to prevent its motion. It is this interaction which is in part responsible for the high strength of martensite in steel. Because the carbon distorts the bcc lattice into a tetragonal lattice in martensite, this interaction is often referred to as a **tetragonal distortion**. This results in strong pinning. A similar pinning mechanism exists for carbon in bcc iron.

### Q. 12.4.1

Because of the manner in which yield strength is affected by impurities, yield strength is called a structure (a) (sensitive or insensitive) property. One reason solute atoms are attracted to dislocations is that (b)_____. The force needed to pull the dislocation away from a solute atom is called the (c)_____.

\* \* \*

## ANSWERS TO QUESTIONS

12.4.1 (a) Sensitive, (b) the elastic strain energy of the system decreases, (c) pinning force.

### 12.5 STRAIN HARDENING

A rereading of the portion of Section 6.6 concerned with dislocations would be helpful at this time. It was noted in Chapter 2 that when materials are plastically deformed they are strengthened; i.e., their yield stress is increased. The effect is even more pronounced for high-purity single crystals, as shown in Figure 12.5.1. Here the critical resolved shear stress is plotted versus the amount of plastic deformation. Consider the results for magnesium. The crystal tested had a critical resolved shear stress of 8 bars (1 bar = $10^6$ dynes/cm$^2$ = 14.5 psi). However, after plastic deformation of

200%, the resolved shear stress required to cause further deformation was over 90 bars. This strengthening is called **strain hardening.** Note that while it is large in the hcp metals (Mg and Zn), it is *enormous* in the fcc metals (Ni, Cu, Ag, Au, and Al). Slip in both of these crystal structures takes place on the closest packed layers (the one basal plane or (001) plane of the hcp crystal and one or more of the four octahedral planes or {111} planes of the fcc crystal). Strain hardening is much greater in the fcc crystals because slip can take place on intersecting slip systems (**multiple slip**).

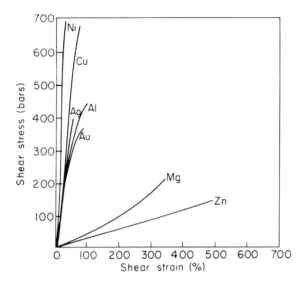

FIGURE 12.5.1. Effects of cold working on the critical resolved shear stress of single crystals. Mg and Zn are hcp crystals while the others are fcc crystals.

As slip occurs in crystals, dislocations move. The shear strain for slip on a single plane is, as we show next, given by

$$\gamma = \rho b s, \qquad (12.5.1)$$

where $\rho$ is the dislocation density (centimeters of length per cubic centimeter), $b$ is the Burgers vector (centimeters), and $s$ is the average distance a dislocation moves (centimeters). Thus if one edge dislocation moves entirely across a unit cube (and hence a unit distance) it causes a shear strain $b$ (see Figure 12.5.2 and recall that $\gamma = \tan \psi$). If $\rho$ dislocations each move across the unit cube (with many parallel slip planes involved), they cause a shear strain $\rho b$, and if they each move a distance $s$ instead of unit distance, they cause the strain $\rho bs$.

Moreover, as slip occurs dislocations multiply. A newly grown crystal may have only $10^6$ centimeters of dislocation length per cubic centimeter while one that has been highly cold-worked (plastically deformed at tem-

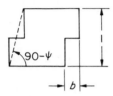

FIGURE 12.5.2. Side view of unit cube after dislocation has moved through from left to right.

peratures below about $T_m/4$) might have $10^{11}$ cm/cm$^3$. Two questions now arise:

1. Why and how does dislocation multiplication occur?
2. Why does an increased dislocation content increase the yield stress, i.e., cause strain hardening?

DISLOCATION MULTIPLICATION. In the first discussion of dislocations in Section 6.6, the dislocations were presented as straight lines (Figures 6.6.1 and 6.6.4). However, the possibility that they were curved lines was later considered in Figure 6.6.6(c) and the possibility that they were jogged was considered in Section 9.10 (see Figure 9.10.1). Consider now a screw dislocation. Shear stress pushes this dislocation through a crystal. At some point, a portion of the dislocation meets some barrier, perhaps a precipitate particle. This portion of the screw dislocation bows out on a second glide plane (inclined at some angle to the first) and moves past the particle. This is **cross slip**. As an example, a screw dislocation on a fcc crystal might glide on the (111) plane and part of it then cross slips to a (11$\bar{1}$) plane as in Figure 12.5.3. Other portions might cross slip to some other plane. Portions which have cross-slipped and are moving on a second plane may later cross-slip to another plane, perhaps parallel to the first. Eventually, the dislocation line is a complex tangle many times longer than the initial dislocation line.

Another process of dislocation multiplication involves large jogs on screw dislocations. A jog segment on a screw dislocation is an edge dislocation (since the Burgers vector of a dislocation line cannot change and since the jog segment is normal to the screw dislocation). The jog (whose Burgers vector is parallel to the screw dislocation) can glide along the dis-

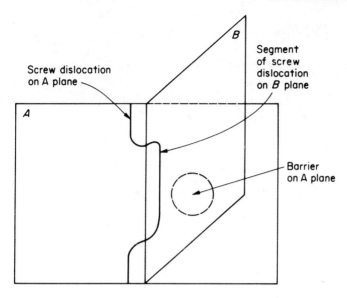

FIGURE 12.5.3. Schematic of cross slip.

location line but not normal to it. Hence when the screw dislocation moves it leaves a pair of parallel dislocations (called a **dislocation dipole**) in its wake, as illustrated in Figure 12.5.4. If the jog is short (between adjacent planes), then either a string of vacancies or interstitial atoms is produced instead of a dislocation dipole. A transmission electron micrograph illustrating dipole formation is shown in Figure 12.5.5.

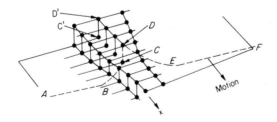

FIGURE 12.5.4. Jog $CD$ on a screw dislocation line. A pair of parallel edge dislocation lines $CC'$ and $DD'$ are created. The segments $AB$ and $EF$ are screw dislocations.

Finally, another process of dislocation multiplication proposed by Frank and Read in 1950 is illustrated in Figure 12.5.6. The portions of the line normal to the paper are pinned (unable to move). The remaining portion bows out under the applied stress as shown. Eventually a new loop is generated in addition to the original configuration. The process can be re-

534  *Strengthening Mechanisms*

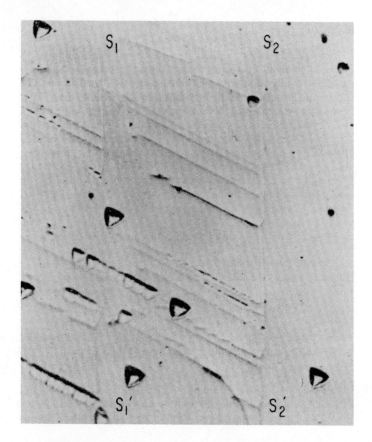

FIGURE 12.5.5. Debris left by moving dislocations $SS'$ in silicon ($\times 1000$.). (Courtesy of W. C. Dash, General Electric Corporation, Schenectady, N.Y.)

peated over and over, generating many loops in the process. This is now called a **Frank-Read source.** It is often true that the dislocation density $\rho$ varies with stress according to

$$\rho = a(\tau - d)^2, \qquad (12.5.2)$$

where $a$ and $d$ are constants characteristic of the material.

### Q. 12.5.1

The increase of yield stress as a result of plastic deformation is called (a)_____. It is much larger in single crystals of fcc metals as a rule than in hcp crystals because of the presence of (b)_____ in the

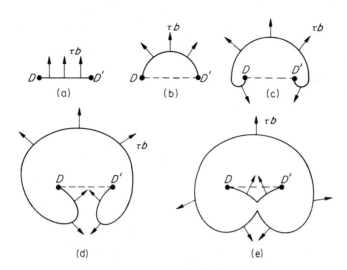

FIGURE 12.5.6. Frank-Read source. The plane of the figure is the slip plane of dislocation $DD'$; the dislocation leaves the plane of the figure at the fixed points $D$ and $D'$. An applied stress produces a glide force $\tau b$ on the dislocation and makes it bulge. The initially straight dislocation (a) acquires a curvature proportional to $\tau$. If $\tau$ is increased beyond a critical value corresponding to position (b), where the curvature is a maximum, the dislocation becomes unstable and expands indefinitely. The expanding loop doubles back on itself, (c) and (d). Unit slip occurs in the area swept out by the bulging loop. In (e) the two parts of the slipped area have joined; now there is a closed loop of dislocation and the section $DD'$ is ready to bulge again and give off another loop. [From W. T. Read, *Dislocation in Crystals*, McGraw-Hill Book Company, Inc., New York (1953).]

fcc crystals. Three mechanisms of dislocation multiplication are (c)_____
_____.

\* \* \*

WHY STRAIN HARDENING? Energy is required to create new dislocations. An estimate of this energy per unit length of dislocation line is

$$U_l \approx Gb^2, \qquad (12.5.3)$$

where $G$ is the shear modulus and $b$ is the Burgers vector. This energy can

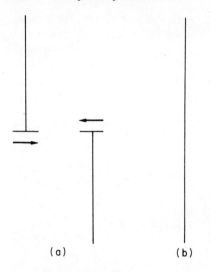

FIGURE 12.5.7. Dislocation annihilation. Two half planes in (a) end at dislocation lines. These combine in (b), resulting in a plane which extends through the crystal and hence no dislocation.

be estimated by calculating the stress field around a dislocation and then calculating the strain energy associated with this stress. (See the previously cited Ruoff, *Materials Science*, Sections 22.2 and 22.3.)

Of the work done in plastically deforming a crystal, only 5–7% is associated with the creation of the dislocations which remain in the crystal after the plastic deformation ceases. The remainder is dissipated as heat. There are many dissipative processes. Probably the most important two are dislocation annihilation and dislocation rearrangement. For instance,

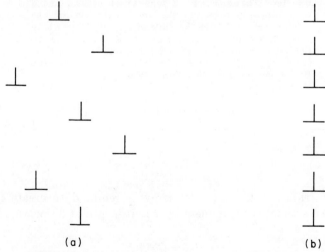

FIGURE 12.5.8. Dislocation rearrangement. The dislocations wall in (b) has a lower energy than the arrangement in (a).

Sec. 12.5  Strain Hardening  537

two dislocations of opposite sign on the same glide plane attract each other and glide together as in Figure 12.5.7; they annihilate each other and in the process dissipate the energy which was required to create them. Dislocations also interact with each other and rearrange themselves in ways which reduce the energy of the system. One example is shown in Figure 12.5.8. The wall has lower potential energy than the distributed dislocations. Thus,

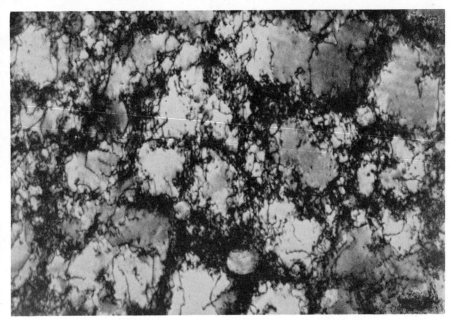

FIGURE 12.5.9. Dislocations forming a tangled cell structure. Polycrystalline copper with 2% $SiO_2$ after 0.34 strain at 77°K. (×40,000.) [After A. Kelly, *Strong Solids*, Oxford University Press, New York (1966).]

when a dislocation moves into a wall, the energy of the system is lowered and heat is given off. During deformation, cells bounded by dislocation walls are formed [the cell walls formed at low temperature are not as well defined as those in Figure 12.5.8(b)]; these are called **subcells**. An example is shown in Figure 12.5.9. Each subcell may vary in orientation by a degree or so from a neighboring subcell. They are really subgrains within grains.

### Q. 12.5.2

Two widely separated edge dislocations of opposite sign on the same glide plane (a) <u>(attract or repel)</u> each other. When they come together, the combination can be written as a vectorial reaction $\mathbf{b}_1 + \mathbf{b}_2 \rightarrow \mathbf{b}_3$. In the case of two edge dislocations of opposite sign on the same glide plane,

$b_2$ equals (b)_____ and $b_3$ equals (c)_____. There is an energy (d) (increase or decrease) if the reaction occurs in the present case. If there is a decrease in the energy, the reaction *does* take place. The energy of the system would (e) (increase or decrease) when two dislocations of the *same* sign on the slip plane come together.

\* \* \*

As deformation proceeds, the dislocation content in the crystal increases, and it becomes more difficult for a dislocation to move through the crystal. For example, as a dislocation slices through a forest of dislocations which are perpendicular to it, jogs are created. This takes energy. Hence the more dislocations in the forest, the larger is the stress required to push a dislocation through the forest. If the jogs created are on a screw dislocation, they greatly impede its motion since dipole formation is now required (see Figure 12.5.4). Hence a large stress must be applied to push the dislocations through the crystal. To summarize this phenomenon, the more dislocations which are intersected, the more jogs which will be created and the more stress which will be needed to move the dislocations. There are several other mechanisms of strain hardening which are not discussed here.

## ANSWERS TO QUESTIONS

**12.5.1** (a) Strain hardening; (b) multiple slip; (c) repeated cross slip, dipole formation, and the Frank-Read source.

**12.5.2** (a) Attract; (b) $-b_1$; (c) 0; (d) decrease [inasmuch as the energy of the reactants is $Gb_1^2 + Gb_2^2$ and since $b_1 = -b_2$, the energy of the product is zero because the Burgers vector of the product is zero]; (e) increase [since the energy of the product is $G(2b)^2$ compared to the energy of the reactants $2Gb^2$, this reaction would not occur spontaneously; a force would therefore be needed to push the dislocations together].

### 12.6 STRENGTHENING BY GRAIN BOUNDARIES

Grain size strongly influences yield stress, fracture stress, and ductility, as shown in Figure 12.6.1. These data illustrate a rather general point, that decreasing grain size is accompanied by increasing strength and by increasing ductility. This may be accounted for in a general way on the basis that grain boundaries resist the passage of dislocations; in a fine-grained material there are more boundaries and hence more barriers, and therefore the material is stronger. A similar strength-grain size relationship exists for nearly every polycrystalline material that has been investigated.

Why do grain boundaries resist the passage of dislocations? One reason is shown in Figure 12.6.2; the explanation is given in the legend.

With respect to fracture, the grain size diameter can be considered as a potential crack length. Note that dislocations generated on a specific

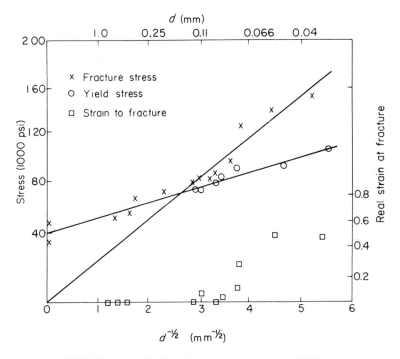

FIGURE 12.6.1. Yield and fracture stresses at $-195°C$ as a function of grain size, $d$, for a low carbon steel. Single crystal cleavage stresses plotted at $d^{-\frac{1}{2}} = 0$. [After J. R. Low in *Relation of Properties to Microstructure*, American Society for Metals, Cleveland (1954), p. 167.]

slip plane could form a pileup no longer than the diameter of the grain. Recall from Section 12.2 that at the tip of such a pileup there is a stress concentration equivalent to a crack of the same length. Therefore, according to the Griffith theory, the fracture stress could be expected to depend on the inverse square root of the grain size as is often found (see Figure 12.6.1).

It is extremely difficult to produce bulk materials with a grain size of less than about 10 $\mu$, although in the past several years considerable

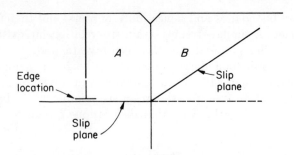

FIGURE 12.6.2. Edge dislocation approaching a grain boundary. If the edge dislocation could continue on the plane in crystal $B$ shown by the dashed line, it would pass through the boundary. However the slip plane in $B$ is shown by the solid line. Note that the Burgers vector, $\mathbf{b}_B$, of the edge dislocation moving on the slip plane in $B$ is different than the Burgers vector, $\mathbf{b}_A$, of the original edge dislocation moving in $A$. Therefore $\mathbf{b}_A \to \mathbf{b}_B + \mathbf{b}_{GB}$ where $\mathbf{b}_{GB}$ is the Burgers vector of a dislocation located at the boundary. Creation of this extra dislocation $\mathbf{b}_{GB}$ requires work.

progress has been made in the production of microduplex structures (very fine-grained two-phase materials). This is an active area of research.

### Q. 12.6.1

Grain boundaries act as barriers for dislocations because there is a change in orientation of (a)_____. Thus for an edge dislocation to move into the second grain, which is tilted with respect to the first, (b)_____ as well as glide would have to occur. Since a pileup can be no longer than a grain, the grain diameter is an effective maximum (c)_____.

* * *

## ANSWERS TO QUESTIONS

**12.6.1** (a) Glide planes, (b) climb, (c) crack length (prior to propagation).

### 12.7 SECOND-PHASE STRENGTHENING

MOTT-NABARRO FORCE. Consider the motion of the dislocation of Burgers vector $b$ through a unit cube of material as shown in Figure 12.5.2 as a result of the applied shear stress $\tau$. The work done in causing the deformation is the applied force times the displacement $b$. The applied force is simply $\tau$ times the area of one, i.e., $\tau$. Hence, $W = \tau b$. Let us *suppose* instead that there is a force which we call $f$ which pushes the unit length of dislocation across the unit distance. The work is then the force acting on the dislocation times the distance which the dislocation moves (which is one unit here), so $W = f$. It follows that the force per unit length of dislocation is given by

$$f = \tau b. \tag{12.7.1}$$

This is the **Mott-Nabarro** force. (This force is defined only because it is useful to think in terms of a force acting on a dislocation line.)

PARTICLE STRENGTHENING. We consider a dislocation line of Burgers vector $b$ which moves in a crystal and is pushed against precipitate particles by an applied shear stress $\tau$ as shown in Figure 12.7.1. As the dislocation line

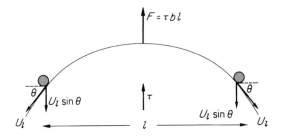

FIGURE 12.7.1. Shear stress causes bowing of dislocation.

is pushed by the force $fl = \tau bl$, it becomes curved and its length increases. There is an increase in energy per unit length of line created equal to $U_l$. The work done in creating new dislocation line of length $x$ is $W = U_l x$. Hence $U_l$ is a force acting *along* the dislocation line; this is called a **line tension**. By equating the appropriate force components shown in Figure 12.7.1, we have

$$\tau bl = 2U_l \sin \theta.$$

We noted earlier that $U_l \approx Gb^2$. Hence the maximum stress needed to push the dislocations past the barrier (leaving a loop around the precipitate particle as shown in Figure 12.7.2) is

$$\tau = \frac{2Gb}{l}.$$

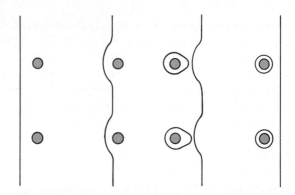

FIGURE 12.7.2. Progress of a dislocation along a slip plane containing second-phase particles. Successive stages shown in four views, from left to right.

The tensile yield stress $\sigma_0$ is twice as large since the maximum shear stress in a tensile specimen is half the tensile stress. Therefore

$$\sigma_0 = \frac{4Gb}{l}. \qquad (12.7.2)$$

**EXAMPLE 12.7.1**

Estimate in units of $b$ the value of $l$ needed to obtain $\sigma_0 = 80,000$ psi in aluminum at room temperature.

*Answer.* Since $G = 4 \times 10^6$ psi (see Table 1.3.4), $l = 200b$. The slip direction for aluminum is the $[10\bar{1}]$ direction (see Table 12.2.1) and the repeat distance is $a/\sqrt{2}$, i.e., half the face diagonal of the cube; since $a = 4.05$ Å (see Table 6.2.1), $b = 2.86$ Å.

An aluminum alloy contains 1.7 at. % copper. See Figure 10.4.1. When this alloy is quenched from the $\kappa$ range, a penny-shaped precipitate called a GP II precipitate is formed with a thickness of 50 Å and a diameter of 500 Å. This precipitate is a Guinier-Preston zone of the second type; it is a copper-rich region and is apparently an ordered structure which is coherent to the lattice. The analysis leading to (12.7.2) assumes that the particles cannot be cut by the dislocation. Some particles, for example, coherent precipitate particles, can be cut if the stress on the dislocation is sufficiently high.

**EXAMPLE 12.7.2**

Estimate the distance $l$ between precipitate particles when GP zones as just described are formed in an aluminum alloy containing 1.7 at. % copper. The composition of the precipitate can be taken to be roughly $CuAl_2$ (although it is not the compound) and it can be assumed that all the copper in the alloy is present in these precipitates.

*Answer.* As an approximation we shall assume that the volume fraction of precipitate equals three times the atom fraction of copper. Then the volume fraction of precipitate is 0.051. Let $N$ be the number of precipitate particles per unit volume and $v$ the volume of one particle. Then $Nv = 0.051$, and since we can calculate $v$ from the given dimensions, we have $N = 5 \times 10^{15}/cm^3$. Then $l \approx 1/N^{1/3} \approx 6 \times 10^{-6}$ cm = 600 Å $\approx 210b$. Hence, the strength should be about 76,000 psi (over 5 kbars). Compare with the data in Table 1.3.5.

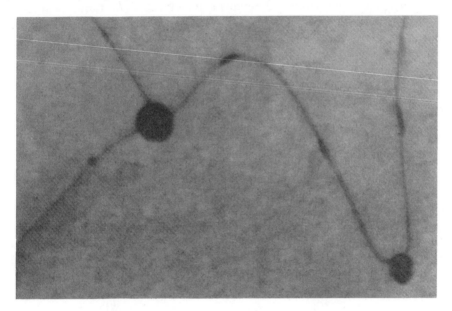

FIGURE 12.7.3. Electron micrograph illustrating age precipitation hardening of MgO. (Courtesy of G. Thomas and J. Washburn, University of California, Berkeley, Cal.)

To obtain sizable increases in strength the precipitate particles must have a spacing of less than 1000 Å. Figure 12.7.3 shows precipitate particles, in the ionic material MgO, acting as barriers to dislocations. Data on various commercial age-precipitation-hardened alloys based on aluminum, magnesium, copper, titanium, etc., matrices are given in E. R. Parker, *Materials Data Book*, McGraw-Hill Book Company, Inc., New York (1967).

## Q. 12.7.1

When a shear stress $\tau$ acts in the glide plane of a dislocation (with the direction of the stress parallel to the Burgers vector) it exerts a force on unit length of the dislocation line equal to (a)____. The work required to create a unit length of dislocation line is about (b)____. If $U_l x$ represents the work required to create a line of length $x$, then $F = \partial(U_l x)/\partial x = U_l$ represents the (c)_____. By balancing the force due to line tension with that due to the Mott-Nabarro force, it is possible to calculate the (d)_____.

* * *

LAMELLAR STRUCTURE. When a eutectoid steel (see Figure 10.5.1) is cooled below the eutectoid temperature at 723°C, it forms pearlite, a structure consisting of alternate layers of hard $Fe_3C$ and relatively soft $\alpha$-iron or ferrite. We would expect the dislocation movement to take place

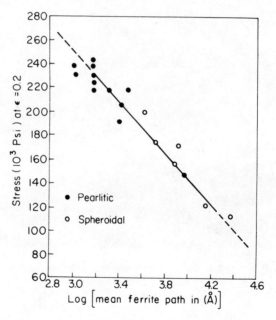

FIGURE 12.7.4. Relation between pearlite lamellar spacing and strength. [From J. H. Holloman and L. D. Jaffee, *Ferrous Metallurgical Design*, John Wiley & Sons, Inc., New York (1947).]

primarily in the ferrite. Hence, decreasing the lamellar spacing decreases the mean free path of the dislocations; i.e., they can effectively move no further than the ferrite thickness. Hence we expect the strength to increase as the lamellar spacing decreases. This is shown to be the case by the data in Figure 12.7.4. The pearlite spacing decreases as the isothermal transformation temperature decreases.

Patented steel wire (piano wire) is an alloy of about 0.9 wt. % C, 0.4% Mn, and 0.2% Si which is transformed to pearlite at 500°C (from austenite at 1000°C). This results in a very fine pearlite which is then drawn to a real strain $\epsilon = 4$ (note the enormous ductility of fine pearlite steels). The resultant material has a strength of about 600,000 psi (40 kbars). Here strengthening depends on a combination of strengthening by second-phase particles and strengthening by strain hardening.

COMPOSITES. Composites were previously mentioned in Section 8.5. It was noted there that the tremendous strengths of whiskers, filaments, and wires (see Table 12.7.1) could be utilized by placing them in a ductile matrix.

Table 12.7.1. STRENGTHS OF FILAMENTS, WIRES, AND WHISKERS

| | | | |
|---|---|---|---|
| Asbestos | 750,000 psi | 50 kbars | Short fibers |
| Glass | 500,000 psi | 33 kbars | Filaments |
| Boron | 400,000 psi | 27 kbars | Filaments |
| Graphite | 300,000 psi | 20 kbars | Filaments |
| Beryllium | 180,000 psi | 14 kbars | Wire |
| Patented steel | 600,000 psi | 40 kbars | Wire |
| Tungsten | 550,000 psi | 37 kbars | Wire |
| Iron | 1,900,000 psi* | 140 kbars | Whiskers |
| Graphite | 2,800,000 psi* | 200 kbars | Whiskers |
| Sapphire | 7,500,000 psi* | 60–500 kbars | Whiskers |

* Maximum values are quoted. The stronger sapphire whiskers *usually* have strengths of about 1,000,000 psi.

Matrices are commonly ductile metals or plastics. However, graphite filaments have also been embedded in bulk graphite (by first embedding in rayon and then heating to decompose the rayon).

The role of the matrix is to distribute the stress from one fiber to another and in so doing it must support shear stresses. We shall not delve into the mechanism of stress distribution in a composite at this time. There are a few simple rules which composites often obey. Figure 12.7.5 illustrates the **volume fraction rule:** The tensile strength of the composite equals the tensile strength of the first phase times the volume fraction of that phase plus the tensile strength of the second phase times the volume fraction of that phase. A similar rule applies to the elastic moduli.

The strength of composites based on fibers will often vary drastically with direction. For instance, aligned silica fibers in a "pure" aluminum matrix result in a composite which has a tensile strength of 110,000 psi along

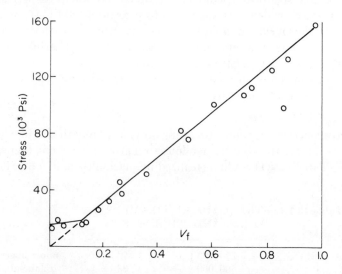

FIGURE 12.7.5. Measured ultimate tensile strength of copper reinforced with continuous brittle tungsten wires of 0.5-mm diameter, as a function of volume fraction of tungsten. [After A. Kelly, *Strong Solids*, Oxford University Press, Inc., New York (1966).]

the fibers but has a tensile strength of only 14,000 psi at angles ranging from 45 to 90 deg to the fiber axis. There are several mechanisms of failure:

1. Fracture of the fibers.
2. Shear failure of the matrix along the fibers.
3. Fracture of the matrix in tension normal to the fibers or failure of the fiber-matrix interface.

Another type of composite is shown in Figure 12.7.6. Here, very hard particles of tungsten carbide are mixed with several percent cobalt, pressed together, and sintered above the melting point of cobalt. The cobalt forms a continuous matrix between the carbide particles. Such very hard materials are used as cutting tools for high-strength steels, as rollers for rolling sheet of high-strength alloys, etc.

### Q. 12.7.2

The motion of a crack in the matrix of a composite consisting of a brittle matrix and ductile wires would be stopped by the (a)_____.

In a GRP composite, a broken fiber represents a crack. This crack does not propagate because the stresses are redistributed by the (b)_____.
The volume rule for the tensile strength $\sigma$ of a composite having a volume fraction $f$ of glass fibers of strength $\sigma_f$ and a fraction $1 - f$ of matrix of strength $\sigma_m$ is (c)_____.

* * *

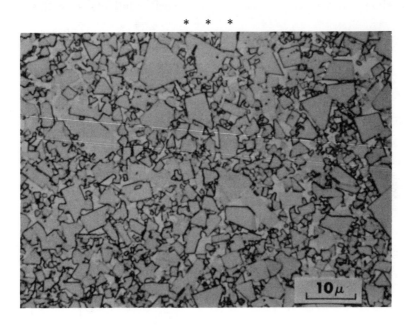

FIGURE 12.7.6. Sintered tungsten carbide. (Courtesy of General Electric Company. Schenectady, N.Y.)

## ANSWERS TO QUESTIONS

**12.7.1** (a) $\tau b$; (b) $Gb^2$; (c) line tension; (d) the yield stress, or the stress needed for dislocations to move past barriers.

**12.7.2** (a) Ductile wire, (b) ductile matrix, (c) $\sigma = f\sigma_f + (1 - f)\sigma_m$.

## 12.8 STRENGTHENING BY MARTENSITIC TRANSFORMATION

When steel is quenched rapidly from high temperature, there is no time for the $\gamma$-iron (also called austenite) to transform to the lamellar structure called pearlite. Pearlite consists of alternate layers of $\alpha$-iron (also called ferrite) and $Fe_3C$ (also called cementite). Instead martensite forms as was discussed in Section 10.5. Martensite forms by a shearing transformation involving the motion of dislocations, and dislocations are profusely produced during the transformation. Hence martensite is a highly dislocated structure and the dislocations are pinned by the tetragonal distortions of the carbon. Moreover, the size of the martensite platelets is small, so there

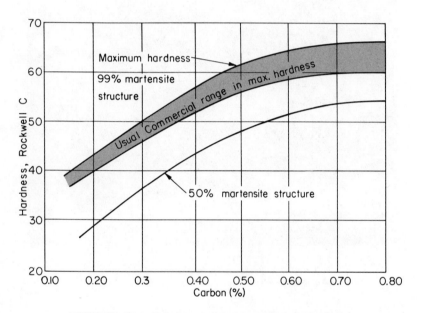

FIGURE 12.8.1. Relation of carbon content to the hardness of plain carbon steel. Data apply approximately to low and medium alloy steels. [From *Metals Handbook*, American Society for Metals, Cleveland (1948), p. 497.]

is grain boundary hardening. Recall that the $c/a$ lattice parameter ratio for the tetragonal martensite increases as the carbon content increases (see Figure 10.5.4). Hence we expect the strength and hardness to increase as the carbon content increases. This is indeed the case, as shown in Figure 12.8.1.

As is clear from Figure 12.8.1, if martensite is not formed, the hardness of the steel (and hence the strength) is lower. Therefore the *kinetics* of the transformation from austenite must be considered in detail.

Figure 12.8.2 illustrates the transformations which take place when a eutectoid steel is cooled very rapidly from the austenite ($\gamma$-iron) range to a fixed temperature and is then held at that temperature. The resultant curve is called a **time-temperature-transformation curve (TTT curve)**. Suppose the steel is cooled rapidly from 800 to 550°C. After 3 sec at 550°C some pearlite has formed. Some arbitrary definition, such as the presence of

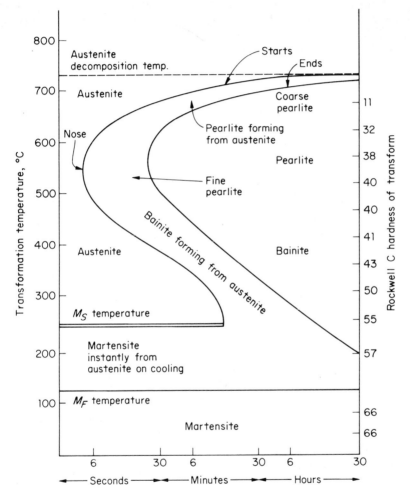

FIGURE 12.8.2. TTT curve for austenite decomposition in a eutectoid steel.

5% pearlite, is used to define the transformation-starts line. To carry out this analysis, the experimentalist would rapidly quench the specimen in water when $t = 3$ sec, and would subsequently polish and etch the tiny specimen and examine it microscopically for the presence of pearlite (no additional pearlite will form during the rapid quench or subsequently at room temperature). If a specimen is held for 30 sec at 550°C, it nearly completely transforms to pearlite. The presence of 95% pearlite can be arbitrarily set to define the transformation-ends line.

If the specimen had been cooled from 800 to 200°C in less than 1 sec, so that no pearlite was formed, then some martensite would be formed, nearly instantaneously, when the $M_S$, or martensite starts temperature, is reached. The fraction of martensite formed depends on the temperature only and not the time. If cooled to below $M_F$, or the martensite finishes temperature, essentially all of the austenite is transformed to martensite; there may be some retained austenite.

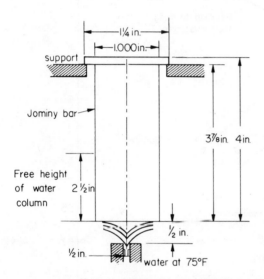

FIGURE 12.8.3. Jominy end-quench hardenability test.

The times to begin the pearlite transformation at any given temperature and to complete it at that given temperature can be increased considerably by the addition of alloying elements such as nickel and manganese to the steel. Hence even large sections of alloy steel, which take a long time to cool, can be transformed into martensite. The ability of a steel to form martensite during a quench is called its **hardenability.**

A convenient hardenability test is the **Jominy test.** Here a specimen (see Figure 12.8.3) is taken from the furnace at the austenitizing temperature, placed in a holder, and quenched at one end by a jet of water. Heat losses to the air are negligible. (The specimen initially has two long flat parallel ground surfaces on opposite sides of the cylinder parallel to the axis of the cylinder.) When the specimen is cooled to room temperature, Rockwell-C hardness measurements are made at $\frac{1}{16}$-in. intervals from the quenched end. The results are shown for two steels, each of which contain 0.4 wt. % C, in

Figure 12.8.4. We note that the maximum hardness in *each* case is about $R_c$-58. The *maximum* hardness is obtained when 100% martensite is produced and is determined only by the carbon content, as noted earlier. The 1040 steel is a plain carbon steel with 0.40% C. It has a low hardenability. The 4140 steel is a low alloy steel (0.80–1.10% Cr, 0.18–0.25% Mo, and 0.40% C).

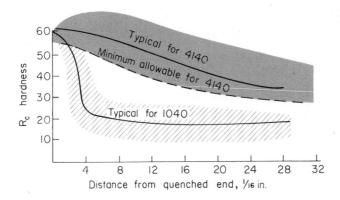

FIGURE 12.8.4. Jominy curves for two 0.40 wt. % C steels. The last two digits in the alloy number give the carbon content.

The hardenability of 4140 steel is considerably higher than that of 1040 steel. Figure 12.8.5 shows the hardness versus cross section of quenched rods of the same material.

The American Iron and Steel Institute (AISI) and the Society of Automotive Engineers (SAE) have established standard four-numeral designations such as 4140 for carbon and alloy steels. These designations and the properties of the materials may be found in the latest edition of the Metals Handbook or in E. R. Parker, *Materials Data Book*, McGraw-Hill Book Company, Inc., New York (1967).

High-alloy martensitic steels may have yield strengths approaching 300,000 psi (20 kbars).

One standard engineering practice involves the carburization of a low carbon steel (diffusion of carbon into the surface) followed by an appropriate heat treatment to produce a layer of hard martensite at the outside surface of the steel. The surface is said to be **case hardened.** The hard surface surrounds a tough, highly ductile core.

Such hard martensite layers at the surface can be produced in other ways. For instance, a steel may be cooled to yield pearlite. Then only the

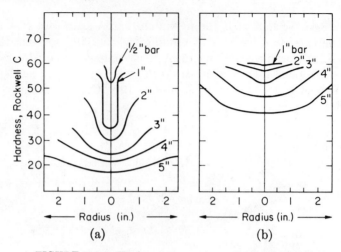

FIGURE 12.8.5. Hardness versus cross section for hardened bars of various diameters. (a) SAE 1040 quenched in water. (b) SAE 4140 quenched in water.

surface is heated to a high temperature (either by a flame or by induction heating) at which austenite is produced. Quenching then produces martensite.

### Q. 12.8.1

Martensite hardening is due to (a)_____
_____
_____. If a steel is cooled slowly, it will transform entirely to (b)_____, while if cooled sufficiently rapidly it will form (c)_____. The maximum hardness of a steel is determined only by the (d)_____. The hardenability, which is a measure of (e)_____, is increased by (f)_____. A convenient measure of the hardenability of a steel is provided by (g)_____.

\* \* \*

## ANSWERS TO QUESTIONS

12.8.1 (a) the large number of dislocations produced by the transformation, the fact that these are strongly pinned, and the fact that the martensite plates are small so that grain boundary strengthening can occur; (b) pearlite;

(c) martensite; (d) carbon content; (e) the ability of a steel to form martensite during a quench; (f) the addition of alloying elements; (g) the Jominy test.

## 12.9 STRENGTHENING AT HIGH TEMPERATURES

Materials which have been strengthened by grain boundaries, strain hardening, precipitation hardening, and martensite formation lose their strength at high temperatures. Dislocation climb involving the motion of vacancies can occur at high temperature and extensive annihilation and rearrangement of dislocations take place. Grain growth occurs at elevated temperature. Tiny precipitate particles grow into large precipitate particles as some particles grow while others shrink and disappear (**overaging**). Carbides are precipitated from martensite. All the strengthening mechanisms discussed thus far were based on *metastable structures*. It is useful to discuss some of the atomic mechanisms associated with the loss of strength at high temperature, for if high-temperature strength is to be obtained, the operation of these mechanisms must be prevented.

### Q. 12.9.1

Materials which owe their strength to strain hardening become weak at high temperatures because (a)_____
_____. Materials which owe their strength to precipitation hardening lose their strength at high temperatures because (b)_____. Steel which has been strengthened by martensite formation loses its strength at high temperatures because (c)_____
_____. Materials which are strong because of their small grain sizes are weak at high temperatures because (d)_____.

\* \* \*

DISLOCATION REARRANGEMENT AT HIGH TEMPERATURES. Figure 12.9.1 shows micrographs of silicon iron after cold working and various amounts of annealing. The points at which the dislocations intersect the surface are made visible by etching. The pits are called etch pits. The top photograph shows the specimen after a 1-hr anneal at 700°C. The original slip planes are still visible but there has been considerable dislocation rearrangement. The middle photograph shows that the dislocations are lining up perpendicular to the original glide planes (after a 1-hr anneal at 875°C) and the bottom photograph shows that the alignment is complete (after a 1-hr anneal at 1060°C). We noted in Figure 12.5.8(b) that dislocation align-

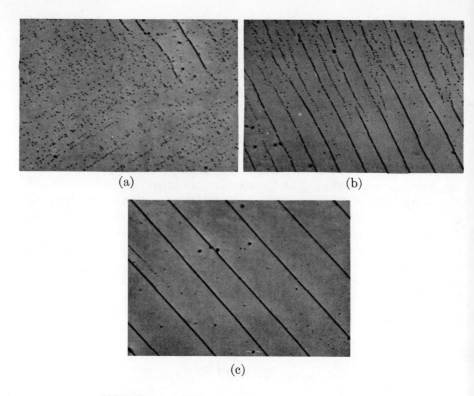

FIGURE 12.9.1. Dislocation rearrangement during recovery of cold-worked silicon iron. Slip was on the $(01\bar{1})$ plane only. Photographs taken after annealing 1 hr at (a) 700°C, (b) 875°C, and (c) 1060°C. ($\times$ 500.) [From W. R. Hibbard, Jr., and C. G. Dunn, in *Creep and Recovery*, American Society for Metals, Cleveland (1957), p. 52.]

ment could sometimes occur by glide and results in reduced energy. However, suppose the dislocations are aligned on a slip plane as in Figure 12.9.2; this is a very high energy configuration compared to that in Figure 12.5.8(b). Clearly it is not possible to go from one of these configurations to the other by glide alone. Climb is also required. Climb involves the motion of vacancies, as has already been discussed in Section 9.10. When climb occurs the edge dislocations move normal to the glide plane. Climb combined with glide can convert the structure of Figure 12.9.2 to that of 12.5.8(b) and the latter has a much lower energy. The process by which the dislocations line up to form cell walls and well-defined cells at high temperature is called **polygonization.**

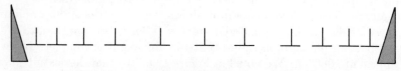

FIGURE 12.9.2. Pileup of dislocations on a glide plane.

RECRYSTALLIZATION AND GRAIN GROWTH. When a polycrystalline aggregate which was highly deformed is annealed, **recrystallization** occurs; i.e., the old grain structure is destroyed and replaced by a new grain structure. This involves a major reduction in the strain hardening. When a polycrystalline aggregate is heated to a high temperature, **grain growth** can occur. This involves the motion of grain boundaries, which involves the jumping of atoms, next to and in the grain boundary, from one crystal to another.

Consider grains in a sheet structure. The ideal case is the hexagonal network of Figure 8.3.2. Note the 120-deg angles which are highly stable (recall the argument used with Figure 8.3.2). Hence, grains with more or

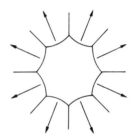

FIGURE 12.9.3. Eight-sided grain with 120-deg junctions.

less than six sides will tend to have 120-deg angles also at the junctions, so that their sides are curved as in Figures 12.9.3 and 12.9.4. If the eight-sided grain grows *outward*, there is a decrease in the surface area (and hence total surface energy). If the four-sided grain grows inward, there is a decrease in the surface area (and hence total surface energy) and such a grain will shrink as shown in the sequence (a)–(d).

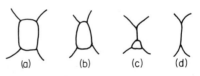

FIGURE 12.9.4. Elimination of four- and three-sided grains during growth of neighboring grains.

An heuristic derivation of an expression for the kinetics of grain growth follows: It is assumed that the grain growth rate is proportional to the total grain boundary energy per unit grain volume. This is reasonable since all the grain boundary area represents energy in excess of that in a perfect single crystal. Then

$$\frac{dD}{dt} \propto \frac{\gamma D^2}{D^3},$$

$\gamma$ = grain boundary energy (energy per area)
$D$ = grain diameter
$t$ = time

and upon integrating, we obtain

$$D^2 - D_0^2 = kt. \tag{12.9.1}$$

The constant $k$ contains a host of parameters, including temperature, geometry, grain boundary energy, grain boundary diffusion coefficient, etc.

All the annealing processes which involve a return toward the single crystal, including dislocation rearrangement, polygonization, recrystallization, and grain growth, are together called **recovery processes.** These processes normally occur at a moderate to rapid rate at $T_m/4$ to $T_m/2$. They can be slowed down some by the addition of impurities, especially if those impurities form quite stable precipitates.

OVERAGING

**EXAMPLE 12.9.1**

The fine precipitates of Example 12.7.2 are not in equilibrium. Why not?

*Answer.* The $5 \times 10^{15}$ particles/cm$^3$ have a total interfacial area of about $5 \times 10^5$ cm$^2$, while if all the "CuAl$_2$" were present as one particle, its surface area would be less than 1 cm$^2$. Because the interfacial energy is about 1000 ergs/cm$^2$, the total interfacial energy is thus about $10^9$ ergs/cm$^3$ in the one case and negligible in the other. The fine precipitates have a higher free energy. The free energy of the system can decrease by particle growth (larger ones grow at the expense of smaller ones) thus reducing the total interfacial energy.

When one precipitate particle grows at the expense of another, diffusion is involved.

**EXAMPLE 12.9.2**

If the distance between particles in Example 12.7.2 is $l = 600$ Å and the time for diffusion is $10^8$ sec (about 3 years), estimate the diffusion coefficient needed if appreciable diffusion is to occur.

*Answer.* A good estimate is that $l^2 \approx 4Dt$. This yields $D \approx 10^{-19}$ cm$^2$/sec. The self-diffusion coefficient of aluminum at room temperature is estimated from the data of Table 9.10.3 to be $D \approx 10^{-22}$ cm$^2$/sec. In the present case both Cu and Al must diffuse in opposite directions and the process would be controlled by the diffusion of the slowest-moving atom. If the value of $10^{-22}$ cm$^2$/sec is a good estimate for this, overaging should not occur at room temperature.

However, since the diffusion coefficient varies strongly with temperature according to Equation (9.10.6), $D$ will increase rapidly as the temperature is increased.

**EXAMPLE 12.9.3**

Estimate the aging time if $l = 600$ Å at 327°C, assuming that the $D$ for Al in Table 9.10.3 correctly describes the diffusion of both copper and aluminum in the alloy.

*Answer.*

$$D = 0.35 \times 10^{-(28,750/1.987 \times 2.3 \times 600)} \approx 10^{-11} \text{ cm}^2/\text{sec}.$$

Then using $l^2 = 4Dt$, we have $t \approx 1$ sec.

Clearly, age-precipitation-hardened alloys overage very rapidly at high temperature. It is of interest to point out that cyclic stressing can also cause overaging since excess point defects are produced during plastic deformation (dragging of small jogs on dislocations). These excess defects not only accelerate the diffusion which causes overaging but accelerate climb as well. An important method of particle strengthening which does not suffer from overaging is discussed next.

DISPERSION-HARDENED ALLOYS. Aluminum powder very rapidly forms a thin oxide coating of $Al_2O_3$. Such powder when sintered forms **sintered aluminum powder,** or **SAP**. The strengthening mechanism in the aluminum by the oxide particles is the same as with the age-precipitation-hardening alloys. There is one very important difference. The $Al_2O_3$ particles are extremely stable thermally (compared to the GP zones in the Cu–Al system) and the rate at which one $Al_2O_3$ particle grows at the expense of another is very slow. $Al_2O_3$ has a melting point of 2050°C and oxygen has a very low solubility in aluminum. (If the oxygen does not dissolve in the aluminum, one oxide particle cannot grow at the expense of another since this must involve oxygen diffusion through the aluminum.) An illustration of the strengthening of Al–$Al_2O_3$ alloys is shown in Figure 12.9.5 where a progressive increase in strength with decreasing interparticle spacing is established. In this system the dispersed phase ($Al_2O_3$) is an incoherent dispersion; the same qualitative effect is obtained with coherent precipitates, but the theoretical considerations are somewhat different.

Nickel hardened by a dispersion of $ThO_2$ also has good strength at high temperatures. It can be produced by starting with a mixture of nickel oxide and thorium oxide (thoria); the nickel is reduced to metal but the more stable $ThO_2$ remains. It is called **TD nickel.**

The second-phase particles impede but do not stop dislocation motion at high temperatures because the dislocations can climb over the particles and then glide to the next barrier. Thus at sufficiently high temperatures creep will still occur. The rate of creep is controlled by the rate at which

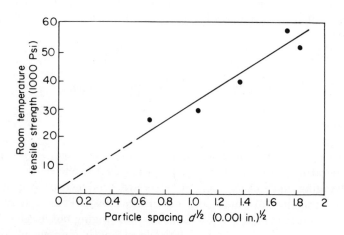

FIGURE 12.9.5. Variation of tensile strength of Al–$Al_2O_3$ alloys with particle spacing. [E. Gregory and N. J. Grant, *Trans. AIME* **200**, 247 (1954).]

vacancies diffuse to or from dislocations and allow climb to occur. The rate of creep is therefore controlled by the following product: number of vacancies times how fast they diffuse; i.e., it depends on the self-diffusion coefficient [see Equations (9.10.5) and (9.10.6)]. Therefore the creep rate is expected to depend on temperature according to

$$\dot{\epsilon} \propto e^{-Q_D/k_B T}.$$

This should be compared with Equation (2.2.2).

### Q. 12.9.2

Tiny $Al_2O_3$ particles distributed in aluminum give a material called (a)_____. The advantage of this material is (b)_____.

\* \* \*

## ANSWERS TO QUESTIONS

**12.9.1** (a) Extensive dislocation annihilation and rearrangement occurs, (b) overaging occurs, (c) martensite decomposes and forms $\alpha$-iron and carbide, (d) grain growth occurs.

**12.9.2** (a) A dispersion-hardened alloy, (b) that particle growth is extremely slow because it involves the diffusion of oxygen through aluminum.

## 12.10 STRENGTHENING MECHANISMS IN POLYMERS

Because polymers differ considerably in structure from other materials, strengthening of them is considered separately. Several hundred million tons of iron and steel are produced throughout the world every year. However, the *volume* of plastics produced annually exceeds the volume of all metals produced. The applications of plastics are diverse and certainly in many cases we are concerned with high strength (nylon rope, for example).

Polymeric materials are usually used because of their mechanical properties, although their use as electrical insulators (dielectrics) also is important. Often plastics are used because of their ease of production by casting, molding, extruding, blowing, etc. [See A. X. Schmidt and C. A. Marlies, *Principles of High Polymer Theory and Practice*, McGraw-Hill Book Company, Inc., New York (1948).]

Herman Mark suggests three principles of polymer strengthening, shown in Figure 12.10.1. These three principles have been discussed in specific cases in Chapter 7 and are reviewed here.

Nylon is an example where good packing plus numerous strong secondary bonds leads to strong crystals. If the crystals are oriented in one

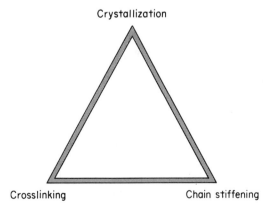

FIGURE 12.10.1. [Polymer strengthening. After H. Mark, *American Scientist* **55**, 265 (1967).]

560  Strengthening Mechanisms

direction (as in a drawn fiber), the resultant material will be particularly strong. Polyethylene without side chains is another example.

Phenolics are an example of a highly crosslinked structure (a network structure). They owe their strength to the large number of primary bond crosslinks. The epoxies owe their strength to a similar structure. Diamond could be considered as the extreme limit of a three-dimensional carbon network.

Polystyrene is an example of a linear molecule with a large side group (see Table 7.4.1). The large side group prevents free rotation of the chains and hence results in chain stiffening. Consequently, polystyrene is a hard plastic at room temperatures. Polymethylmethacrylate (lucite or plexiglas) is a similar example and there are many more.

### Q. 12.10.1

The three methods of increasing the strength of polymers are (a)_____
_____.

\* \* \*

## ANSWERS TO QUESTIONS

**12.10.1** (a) Crystallization, network formation, and chain stiffening.

### 12.11 STRENGTHENING OF VISCOUS MATRICES

In many applications the matrix of a composite may be a highly viscous material such as asphalt which is reinforced with gravel and sand. Likewise when fiber composites are used at high temperatures, the matrix material creeps readily and thus also behaves in a viscous manner, that is, as a highly viscous fluid.

Figure 12.11.1 shows the variations in viscosity of a fluid to which spheres are added. This could be considered to be a model for sand or gravel in asphalt and hence is important in highway behavior. It could also be considered to be a model for wet concrete. Usually we add much more water to the concrete mix than is needed for hydration because the viscosity of the mix has to be reduced to a point where the mix can readily be handled. In so doing we obtain a weaker dry concrete which is an elastic matrix composite similar (except for particle size) to grinding wheels which are also important composites.

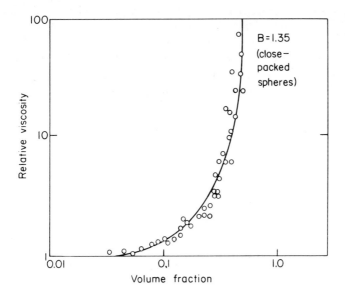

FIGURE 12.11.1. Variation of relative viscosity with volume fraction. [Experimental data after J. G. Brodnyan, *Transactions of the Society of Rheology* **3**, 61 (1959).]

The solid line which closely fits the experimental points in Figure 12.11.1 follows **Mooney's equation** for the viscosity ratio

$$\frac{\eta}{\eta_0} = \exp\left(\frac{2.5 V_r}{1 - \beta V_r}\right). \quad (12.11.1)$$

Here $\eta_0$ is the viscosity of the pure liquid, $V_r$ is the volume fraction of spheres (which can be as large as 0.741 for closest packing of spheres), and $\beta = 1.35$ for the closest packing case. It should be noted that there is a very rapid increase in viscosity as the available solid volume space is nearly occupied; i.e., the composite behavior approaches that of a solid. The viscosity coefficient, however, does not go to infinity as (12.11.1) predicts (besides the material would fracture at high stresses anyhow).

It is clear, however, that adding a strong second phase to a viscous matrix will enormously increase its resistance to creep flow.

# REFERENCES

### Dislocations

Hull, D., *Introduction to Dislocations*, Pergamon Press, Inc., Elmsford, N.Y. (1965). A good introduction to dislocation theory.

## Strengthening Mechanisms

Felbeck, D. K., *Introduction to Strengthening Mechanisms*, Prentice-Hall, Inc., Englewood Cliffs, N.J. (1968). An elementary discussion.

Kelly, A., *Strong Solids*, Oxford University Press, Inc., New York (1966). This book gives an excellent integrated view of strengthening.

A. Kelly and R. B. Nicholson, Eds., *Strengthening Methods in Crystals*, Elsevier Publishing Co., Ltd., New York (1971).

## Special Materials

Kingery, W. D., *Introduction to Ceramics*, John Wiley & Sons, Inc., New York (1960).

Hove, J. E., and Riley, W. C., *Modern Ceramics*, John Wiley & Sons, Inc., New York (1965).

Rauch, H. W., Sr., Sutton, W. H., and McCreight, L. R., *Ceramic Fibers and Fibrous Composite Materials*, Academic Press, Inc., New York (1968). Excellent discussion of high-temperature composites.

Holliday, L., Ed., *Composite Materials*, American Elsevier Publishing Company, Inc., New York (1966). A wide range of materials are described.

Frazer, A. H., "High Temperature Plastics," *Scientific American* (July 1969), p. 96.

Frazer, A. H., *High Temperature Resistant Polymers*, John Wiley & Sons, Inc., New York (1968).

Winters, R. F., Ed., *Newer Engineering Materials*, The Macmillan Company, New York (1969).

## Strength Data

Parker, E. R., *Materials Data Book for Engineers and Scientists*, McGraw-Hill Book Company, Inc., New York (1967).

# PROBLEMS

**12.1** List five "why" questions concerned with strengthening mechanisms.

**12.2** (a) Outline the procedure used to obtain the parameters $b$ and $n$ in the potential energy expression (12.1.1) for the NaCl molecule.

(b) Assuming that $b$ and $n$ in (12.1.1) have been determined, that $b$ is given by (12.1.2), and that $r_0$ is known and $n = 9$, outline the procedure for computing the force needed to pull the ions permanently apart.

**12.3** (a) According to models presented in the text, what is the ultimate fracture strength of a material?
(b) The ultimate shear strength?

**12.4** Why are bulk materials so weak compared to the ultimate tensile strength which they could have?

**12.5** Assuming that the ultimate tensile strength is $E/10$, derive the stress concentration factor due to a crack of depth $c$ using the Griffith theory.

**12.6** A window glass has a tensile fracture stress of 10,000 psi. According to Griffith's theory the compressive strength should be what?

**12.7** A sintered tungsten carbide fractures at a tensile stress of 125,000 psi.
(a) Estimate the compressive fracture stress.
(b) Suggest methods for increasing the strength.

**12.8** How is the Griffith theory modified if plastic flow precedes the crack front?

**12.9** Discuss the origin of fracture in materials which exhibit some ductility.

**12.10** (a) What is meant by a pileup of edge dislocations?
(b) Why are pileups important?

**12.11** (a) An ordinary annealed crystal might have what dislocation density?
(b) A heavily cold-worked crystal?

**12.12** How are the unit tangent vector and the Burgers vector related for the screw and edge dislocations?

**12.13** (a) How is the direction of glide motion of an edge dislocation related to the unit tangent vector of the dislocation line?
(b) A screw dislocation?

**12.14** Explain on the basis of the nature of bonding why the Peierls-Nabarro stress is low for copper and high for germanium.

**12.15** What is meant by the Mott-Nabarro force on a dislocation line?

**12.16** Show that it is not possible for a dislocation to stop within a crystal (excluding the formation of a dislocation loop).

**12.17** One of the important equations of dislocation theory relates the shear strain to the number of dislocations and the distance they move. Derive this relationship.

**12.18** Show that a pileup causes a shear stress concentration factor of $n$, where $n$ is the number of dislocations in the pileup. *Hint:* Assume a stress $\tau$ has pushed the dislocations against the barrier. Now allow the barrier to move a distance $\delta x$.

**12.19** (a) Describe the process of dislocation climb.
(b) Why is climb important?

**12.20** If a screw dislocation line becomes jogged by intersection, why must further motion of the jog be nonconservative, assuming that the direction of motion of the screw dislocation remains unchanged?

**12.21** (a) Describe the operation of the Frank-Read source.
(b) What other mechanisms of dislocation multiplication exist?

**12.22** (a) Discuss the binding of impurity atoms to dislocation lines.
(b) Why does this make the motion of dislocations more difficult?

**12.23** Why does strain hardening occur?

**12.24** What grain size would be needed in order that the yield stress of the low carbon steel of Figure 12.6.1 be increased to 2,000,000 psi? Criticize your extrapolation assumptions.

**12.25** About what minimum distance between centers of precipitate particles is necessary for achieving sizable strengthening?

**12.26** Explain why precipitation-hardened alloys are not useful for
(a) High-temperature applications.
(b) Unlimited cyclic loading applications.

**12.27** Assuming it was possible to make very fine pearlite, what mean ferrite path would be needed to obtain a strength of $10^6$ psi in the steel of Figure 12.7.4.

**12.28** Give several reasons why medium to high carbon martensite is so strong.

**12.29** (a) Define hardness and hardenability.
(b) Why is the hardenability of a steel such an important commercial property?

**12.30** It is desired to make a shaft of 4-in. diameter having 0.40% carbon and at least 50% martensite throughout. Will SAE 1040 suffice or must we use the expensive 4140 steel?

**12.31** Give some reasons for the size effect on tensile strength shown by whiskers, metal wires, and glass filaments.

**12.32** Discuss some of the problems one might have in making composites of
(a) Continuous filaments.
(b) Short fibers.

**12.33** Make a list of ten different composites of which at least two are natural and at least two are synthetics known to the Romans.

**12.34** (a) What is dislocation annihilation?
(b) Describe how it can occur by glide.
(c) By climb.

**12.35** Describe the process of polygonization.

**12.36** What is the driving force for recrystallization of a cold-drawn wire?

**12.37** (a) What is the driving force for grain growth?
(b) Describe the overall process.

**12.38** Why does high-temperature creep have the same activation energy as self-diffusion?

**12.39** Give three mechanisms of strengthening of polymeric materials.

## MORE INVOLVED PROBLEMS

**12.40** Give a quantitative example of why yield strength in fcc crystals falls off rapidly at about $0.5 T_m$.

**12.41** Derive the stress field of a screw dislocation in an elastically isotropic media. *Hint:* Begin by showing that the shear strain parallel to the dislocation line is $b/2\pi r$.

**12.42** An imaginary one-dimensional NaCl crystal of $N$ sodium ions and $N$ chlorine ions has the potential energy

$$U = -\frac{NAe^2}{4\pi\epsilon_0 r} + \frac{B}{r^n}$$

with $A = 1.38$.
(a) Find the expression for $U_0$.
(b) Find the expression for the stiffness constant $k$.
(c) Do both $U_0$ and $k$ (or $E$) increase by the same ratio over the same quantities for the NaCl molecule?
(d) Find the expression for $\sigma_{max}$. Does it increase by the same ratio as $E$ over the same quantity for the molecule?

## SOPHISTICATED PROBLEMS

**12.43** Discuss the structure changes which occur during forming of patented steel wire. See J. D. Embury and R. M. Fisher, *Acta Metallurgica* **14**, 147 (1966).

**12.44** OFHC 99.95% annealed copper has a yield strength of 10,000 psi. A properly heat-treated alloy of copper with 1.7 wt. % Be can have a yield strength of 200,000 psi. The phase diagram can be found in the *Metals Handbook*. Explain what the heat treatment does.

**12.45** A strong steel rod is sawed off. The ends are carefully ground and then lapped together. The rod is then joined with a *very* thin layer of indium. Indium has a melting point of 156°C and can readily be indented by a thumbnail at room temperature. The new rod is tested in tension and the indium joint does not break until a stress of 150,000 psi is reached. This is about what is expected of a perfect indium crystal. Explain.

**12.46** (a) Discuss the formation of a crack by slip in MgO. See Y. T. Chou and R. W. Whitmore, *Journal of Applied Physics* **32**, 1920 (1961).
 (b) Discuss the rapid growth of a crack in MgO.

# Prologue

The nature of the valence electrons in crystals is discussed. A simple model for a metal involves the assumption that the valence electrons form a valence gas which behaves in a classical fashion. This model has certain weaknesses, many of which are removed by the *quantized electron gas* (whose behavior is based on a quantum model discussed in Chapter 5, namely, the particle in a box model). However, in order to understand many of the properties of solids (why is copper a metal, germanium a semiconductor, and diamond an insulator) it is necessary to take into account the fact that electrons move in a *periodic potential*. This leads to the concept of *forbidden energy bands* and *allowed energy bands*. The manner in which the allowed levels are occupied determines whether a solid is a metal, semiconductor or, insulator. The terms *valence band* and *conduction band* are introduced. The *Fermi surface* is introduced. The origin of resistivity in metals is described. *Intrinsic* and *extrinsic* semiconductors are studied, as are recombination processes and minority carrier diffusion lengths. The *p-n junction* is then described and we explain how it acts as a *rectifier*. We briefly study the *Zener diode*, the *solar cell*, and *thermoelectricity*. The basis of the *n-p-n junction transistor* (which acts as an amplifier) is then explained. There is a brief discussion of *inverted populations* and *lasers*.

# 13

# ELECTRONS IN CONDENSED PHASES

## 13.1 THE ELECTRON GAS

Chance favors the well prepared mind—Louis Pasteur

In 1905, Drude proposed that a metal contains free electrons whose movement was similar to the motion of atoms in an ideal gas. This is called the **classical free electron gas model** or **Drude's model**. We recall that the ionization potential of a metal atom is quite small (see Table 5.4.3). Thus in a metal such as sodium (atomic number 11) one electron per atom might be ionized (sodium is monovalent chemically) and this electron would be free to move throughout the metal which can be considered as a box. In Drude's model, the ion cores consisting of the nucleus and ten electrons remain at the lattice sites and the valence electrons behave as ideal gas particles.

The root-mean-squared velocity of an "ideal gas" particle is, from Section 9.1,

$$v_{\text{rms}} = \sqrt{\overline{v^2}} = \sqrt{\frac{3k_B T}{m}} \tag{13.1.1}$$

(from $\frac{1}{2}m\overline{v^2} = \frac{3}{2}k_B T$). Ordinarily the gas particles (electrons) move about randomly with a mean free path $\lambda$ between collisions (Drude assumed that the electrons collided with the ion cores and that $\lambda$ was the spacing between ions). Their average velocity (not speed) is zero. We are concerned here with the effects of the presence of an electric field **E** acting in the $x$ direction on an electron which has just undergone a collision at the point $x = 0$. On the average the $x$ component of velocity in the absence of the field will be zero, so $\dot{x} = 0$ (we use the notation $\dot{x} = dx/dt$). The electric field exerts a force $-e\mathbf{E}$ on the electron in the positive $x$ direction. It is accelerated under this applied field and moves a distance $\lambda$ during a time $\tau$ before it undergoes a collision. The time $\tau$ is called the **relaxation time.** The process then begins anew. Newton's equation of motion during this interval is

$$m\ddot{x} = -eE$$

so that

$$\dot{x} = \frac{-eE}{m}t$$

and the average velocity $\bar{\dot{x}}$ during the time interval $\tau$ is

$$\bar{\dot{x}} = \frac{-eE\tau}{2m}.$$

This average velocity is called the **drift velocity.** It is very small relative to $v_{\text{rms}}$ for typical fields. The magnitude of the drift velocity per unit field, i.e., $\bar{\dot{x}}/E$, is often called the **charge mobility** $\mu$. Hence

$$\mu = \frac{e\tau}{2m}. \tag{13.1.2}$$

Now the net electric charge which crosses unit area per unit time (called the current density $J$) is given by the product of the number $N$ of charge carriers per unit volume times the charge on each carrier times the drift velocity,

$$J = N(-e)\bar{\dot{x}} = \frac{Ne^2\tau}{2m}E.$$

This is Ohm's law, $J = \sigma E$, with the conductivity given by

$$\sigma = \frac{Ne^2\tau}{2m} = Ne\mu, \tag{13.1.3}$$

where we have used (13.1.2) to obtain the latter equality.

The relaxation time $\tau$ can be expressed as

$$\tau = \frac{\lambda}{v_{\text{rms}}}$$

where $v_{\text{rms}}$ is given by (13.1.1) in Drude's model. Hence we have

$$\sigma = \frac{Ne^2\lambda}{2mv_{\text{rms}}}. \tag{13.1.4}$$

Drude suggested that λ was the distance between the ions. However, using the experimental data for the conductivity of metals at 20°C, a value of λ of 10–100 Å is obtained. Clearly one of the shortcomings of the Drude model is that λ cannot be predicted. It is, however, to the credit of the model that it does lead to the correct form of Ohm's law.

### Q. 13.1.1

The electrical conductivity can be written as the product of the three terms (a)_____. The charge mobility is defined as the (b)_____.
The charge mobility in terms of the time between collisions is (c)_____.

\* \* \*

The specific heat of crystalline sodium due to the ion vibrations is approximately $C_v^L = 3R$ at high temperatures. According to Drude's model the electrons would have the same specific heat as an ideal monatomic gas; i.e., $C_v^E = 3R/2$. Hence the total specific heat for copper should be

$$C_v = C_v^L + C_v^E = 3R + \tfrac{3}{2}R.$$

In fact, the total specific heat is found experimentally to be very close to $C_v^L$ so in the Drude model the electronic contribution has been vastly overestimated (by a factor of about 100 at room temperature).

The Hall effect was described in Section 3.2. There it was shown that in a material for which one charge carrier dominates, the predicted Hall coefficient for free charge carriers is given by

$$R_H = \frac{1}{Nq}.$$

Thus a Drude metal would have a negative Hall coefficient, since $q = -e$. The ratio of the measured value of Hall coefficients to the predicted value is given in Table 13.1.1. If Drude's model is correct, the ratio should be 1. For the monovalent metals this is approximately true, but for the divalent metals, the prediction fails miserably (even in sign).

DIFFICULTIES WITH DRUDE'S THEORY. There are several obvious shortcomings to the classical free electron theory of Drude:

1. There is no way of knowing a priori when the valence electrons are free. Why are they not free in diamond but free in copper?

2. $\lambda$ is an ill-defined quantity which cannot be calculated a priori.
3. There is a **specific heat paradox**.
4. Many metals show a **Hall coefficient anomaly** which suggests that the charge carriers are positive. Quantum mechanics is required to resolve these difficulties. Drude's concept of an electron gas is correct, but the electrons must be described by quantum mechanics rather than classical mechanics.

Table 13.1.1. RATIO OF EXPERIMENTAL HALL COEFFICIENT TO CLASSICAL THEORETICAL VALUE

| Metal | $R_H$(measured)$/[1/-Ne]$ |
|---|---|
| Li | 1.3 |
| Na | 0.9 |
| K  | 0.9 |
| Rb | 1.0 |
| Cs | 1.1 |
| Cu | 0.8 |
| Ag | 0.8 |
| Au | 0.7 |
| Be | $-5.0$ |
| Cd | $-0.5$ |

## Q. 13.1.2

According to Drude's model the electronic specific heat per mole of copper should be (a)_____. According to Drude's model the Hall coefficient of a metal should be (b)_____. According to Drude's model the mean free path between electron collisions is the (c)_____.

\* \* \*

**EXAMPLE 13.1.1**

Compare the values of the drift velocity for a field of 100 V/m with $v_{\text{rms}}$ at room temperature.

*Answer.* Let us use experimental data to compute $\bar{\dot{x}}$. The electrical resistivity is the reciprocal of conductivity, $\rho = 1/\sigma$. Using the data for copper in Table 3.2.1 and Equation (13.1.3) and estimating that there are $8 \times 10^{28}$ charge carriers/m³, we have for the charge mobility

$$\mu = \frac{1}{\rho Ne} \approx \frac{1}{1.67 \times 10^{-8} \times 8 \times 10^{28} \times 1.6 \times 10^{-19}} \approx 10^{-2} \text{ m}^2/\text{V-sec}.$$

Hence $\bar{\dot{x}} \approx 10^{-2} \times 100 = 1$ m/sec.
However,

$$v_{rms} = \sqrt{\frac{3 \times 1.38 \times 10^{-23} \times 300}{9 \times 10^{-31}}} \approx 10^5 \text{ m/sec}.$$

Although we shall see later that it is necessary to use quantum mechanics rather than classical mechanics to describe the behavior of the electron, this result is of the correct order of magnitude. Note that $\bar{\dot{x}} \ll v_{rms}$.

## ANSWERS TO QUESTIONS

**13.1.1** (a) $Ne\mu$; (b) average magnitude of the net velocity attained in a unit electrical field, i.e., $\bar{\dot{x}}/E$; (c) $e\tau/2m$.

**13.1.2** (a) $3R/2$, (b) $-1/Ne$, (c) iron core spacing.

## 13.2 THE QUANTIZED ELECTRON GAS

In classical theory the total energy $E$ of a particle is given by the sum of the kinetic and potential energy,

$$\frac{p^2}{2m} + U = E,$$

where $p^2 = p_1^2 + p_2^2 + p_3^2$ is the magnitude of the momentum squared. In quantum mechanics these various energy terms behave as operators on the wave function $\psi$ (previously discussed in Chapter 5) and the operator for $p_1$ is $(h/2\pi i)(\partial/\partial x_1)$, etc., where $i = \sqrt{-1}$. The one-dimensional wave equation (which we consider for simplicity) is

$$\frac{h^2}{8\pi^2 m} \frac{d^2\psi}{dx_1^2} + (E - U)\psi = 0.$$

We assume that the particle is restricted to the region $0 < x_1 < L_1$, that $U = 0$ on this domain, and that $U \to \infty$ as $x_1 \to 0$ and $U \to \infty$ as $x_1 \to L_1$. This is called the one-dimensional case of a **particle in a box with infinite walls.** When the particle is in the box we assume that there are no potential energy interactions (with other electrons or with the ion cores). This is identical to the assumption made in the theoretical discussion of the ideal gas law in Section 9.1; hence $U = 0$. However, when the particle collides with a wall, the latter behaves as if it is infinitely rigid; hence there is no chance whatsoever of the particle escaping from the box.

Since the particle cannot exist in the region where $U = \infty$, $\psi$ must be zero there; i.e., the probability, $\psi^2 dx_1$, of finding the particle in the element of length $dx_1$ (where $U = \infty$) must be zero. Consequently, since $\psi$ must be a continuous function of $x_1$,

$$\psi = 0 \quad \text{at } x_1 = 0 \quad \text{and} \quad \psi = 0 \quad \text{at } x_1 = L_1.$$

A nontrivial solution for $\psi$ exists if and only if the total energy $E$ takes on one of the discrete values

$$E = \frac{h^2 n_1^2}{8mL_1^2}, \quad n_1 = 1, 2, 3, \ldots. \tag{13.2.1}$$

(Recall the discussion of eigenvalue problems in Section 5.2.) The possible solutions are standing waves

$$\psi = \sqrt{\frac{2}{L_1}} \sin \frac{n_1 \pi}{L_1} x_1. \tag{13.2.2}$$

**EXAMPLE 13.2.1**

Starting with $E = p_1^2/2m$ and de Broglie's relation $p_1 = h/\lambda_1$, where $h$ is Planck's constant and $\lambda_1$ is the wavelength ($\lambda$ itself is used in this chapter for mean free path but there should be no confusion), give an alternative derivation of (13.2.1) based on the assumption that the solutions are standing waves.

*Answer.* We have $E = h^2/2m\lambda_1^2$. What are the allowed $\lambda_1$'s? The standing waves must always have nodes at $x_1 = 0$ and $x_1 = L_1$. The longest wave which can be used is $\lambda_1 = 2L_1$. In general $n_1 \lambda_1 = 2L_1$. Hence $\lambda_1 = 2L_1/n_1$, which when substituted into the equation for $E$ gives

$$E = \frac{h^2 n_1^2}{8mL_1^2}.$$

The student should plot several of the wave functions in (13.2.2) at this time.

The two-dimensional problem has solutions if and only if

$$E = \frac{h^2}{8m} \left[ \frac{n_1^2}{L_1^2} + \frac{n_2^2}{L_2^2} \right]. \tag{13.2.3}$$

Similarly a particle in a three-dimensional box with $U = 0$ within the box of

**Sec. 13.2**               *The Quantized Electron Gas*    573

dimensions $L_1$, $L_2$, and $L_3$ has possible energy states

$$E = \frac{h^2}{8m}\left[\frac{n_1^2}{L_1^2} + \frac{n_2^2}{L_2^2} + \frac{n_3^2}{L_3^2}\right], \tag{13.2.4}$$

where the $n$'s are integers.

In addition to the quantum numbers $n_1$, $n_2$, and $n_3$, the electron has a fourth quantum number associated with electron spin. There are two spin states. According to the Pauli exclusion principle, each electron must have a different set of quantum numbers. There are therefore two allowed states per energy level given by (13.2.4). It is quite clear that at the absolute zero of temperature, none of the electrons have zero energy and many have sizable energies.

You are given that a cubic meter of metal contains $6 \times 10^{28}$ free electrons. Assuming that the energy levels described by (13.2.4) are filled, from the lowest possible up to some energy $E_F$, what is $E_F$? To answer this question we consider for purposes of demonstration the two-dimensional case first. As a mathematical convenience, we assume that $L_1 = L_2 = L$. Figure 13.2.1 shows a representation of the quantities $(n_1^2 + n_2^2)$ plotted in $n_1 n_2$ space. Note that each circle represents a different set of two quantum numbers (since there are spins). There is one circle per each unit area in $n_1 n_2$ space. Note that for sufficiently large values of $n_1$ and $n_2$ a constant

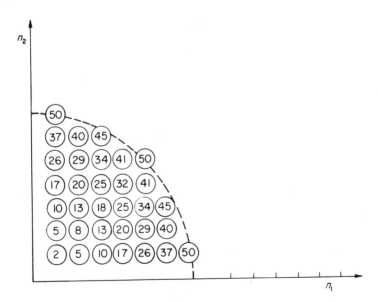

FIGURE 13.2.1. The quantity $(n_1{}^2 + n_2{}^2)$ is shown within the small circles. The dashed line which is the arc of a circle is effectively a constant energy profile.

value of $n_1^2 + n_2^2$ [and hence, by (13.2.3), a constant value of energy] is represented by the arc of a circle. Therefore the area of the quadrant of this circle represents the number of possible energy levels, i.e., half the number of wave functions. (In Figure 13.2.1, the area of the quadrant is 38.5 while there are 33 energy levels as shown by the small circles.) Let us define a radius in $n_1 n_2$ space so that

$$r^2 = n_1^2 + n_2^2 = \frac{8mL^2 E}{h^2},$$

where we have used (13.2.3) with $L_1 = L_2 = L$. Then if $\mathfrak{N}$ is the total number of allowed states, each having a distinct wave function, with energy less than or equal to $E$,

$$\frac{\mathfrak{N}}{2} = \frac{\pi r^2}{4}$$

or

$$\mathfrak{N} = \frac{\pi r^2}{2} = \frac{4\pi m L^2 E}{h^2}.$$

The total area $A$ of the two-dimensional box is $L^2$ so $E = (h^2/4\pi m)(\mathfrak{N}/A)$. As an illustration, suppose that we know the number of free electrons per unit area in a layer of graphite. Then at the absolute zero of temperature these free electrons have energies ranging from nearly zero up to the value of $E = (h^2/4\pi m)(\mathfrak{N}/A)$. In Drude's model (the classical model) *all* the electrons have zero energy at absolute zero.

In three dimensions

$$r^2 = n_1^2 + n_2^2 + n_3^2 = \frac{8mL^2}{h^2} E \qquad (13.2.5)$$

and the volume in one octant of $n_1 n_2 n_3$ space which contains $\mathfrak{N}$ allowable states is

$$\frac{\mathfrak{N}}{2} = \frac{1}{8}\frac{4}{3}\pi r^3 = \frac{\pi}{6}\left(\frac{8mL^2}{h^2} E\right)^{3/2}.$$

An allowable state in three dimensions involves $n_1$, $n_2$, $n_3$, and spin. In this case $V = L^3$ and

$$N = \frac{\mathfrak{N}}{V} = \frac{\pi}{3}\left(\frac{8mE}{h^2}\right)^{3/2}. \qquad (13.2.6)$$

Sec. 13.2    The Quantized Electron Gas    575

The quantity $N$ is the number of allowable states (four quantum number states) per unit volume.

If the total number of electrons per unit volume is $N_F$, then the highest filled allowable state (when energy levels are filled consecutively starting with the lowest) will have an energy $E_F$ called the **Fermi energy at absolute zero of temperature.** From (13.2.6) we have

$$E_F = \frac{h^2}{8m}\left(\frac{3N_F}{\pi}\right)^{2/3}. \tag{13.2.7}$$

The constant energy surface in $n_1 n_2 n_3$ space which has $E = E_F$ is called the **Fermi surface.** For the present model it is a spherical surface [see (13.2.5)].

**EXAMPLE 13.2.2**

Compute the Fermi energy for copper on the basis of the above model.

*Answer.* Note first of all that $E_F$ is *independent of the size of the crystal.* $N_F$ can be computed using the measured lattice parameter of copper and the fact that there are four atoms per unit cell. Copper atoms are monovalent and contribute one electron per atom to the electron gas. Hence $N_F = 8.5 \times 10^{28}$ meters$^{-3}$. Since Planck's constant $h = 6.63 \times 10^{-34}$ J-sec and $m = 9.11 \times 10^{-31}$ Kg we have from (13.2.7)

$$E_F = 11.2 \times 10^{-19} \text{ J}$$

or

$$E_F = 7.0 \text{ eV}.$$

Note that this is very large compared to thermal energy $k_B T$ at room temperature, which is only $\frac{1}{40}$ eV.

[If you were to compute the maximum energy of an argon atom if a mole of such atoms occupied 22.4 liters and each energy state up to the maximum were filled, you would find it to be only $10^{-7}$ times as large as the Fermi energy of copper (because the mass of an argon atom is about $40 \times 1800$ times larger than the electron mass and the number of argon atoms per unit volume is only $5 \times 10^{-4}$ times as large). Note that the maximum energy of such a system of argon atoms is exceedingly small compared to $k_B T$ at room temperature. It is for this reason that classical mechanics can be used in dealing with gas atoms at room temperature.]

It is common to define a **Fermi temperature,** $T_F$, by

$$k_B T_F = E_F.$$

Note that the Fermi temperature is many times the melting temperature of the solid. See Table 13.2.1. Also a **Fermi velocity** is defined by $E_F = \frac{1}{2}mv_F^2$.

Table 13.2.1. CALCULATED FERMI SURFACE PARAMETERS FOR FREE ELECTRONS

|    | Electron Concentration, $N/V$ (per m³) | Fermi Velocity, $v_F$ (m/sec) | Fermi Energy, $E_F$ (eV) | Fermi Temperature $T_F = E_F/k_B$ (°K) |
|----|---|---|---|---|
| Li | $4.6 \times 10^{28}$ | $1.3 \times 10^6$ | 4.7 | $5.5 \times 10^4$ |
| Na | 2.5  | 1.1  | 3.1 | 3.7 |
| K  | 1.34 | 0.85 | 2.1 | 2.4 |
| Rb | 1.08 | 0.79 | 1.8 | 2.1 |
| Cs | 0.86 | 0.73 | 1.5 | 1.8 |
| Cu | 8.50 | 1.56 | 7.0 | 8.2 |
| Ag | 5.76 | 1.38 | 5.5 | 6.4 |
| Au | 5.90 | 1.39 | 5.5 | 6.4 |

Note that the Fermi velocity is still small (but not completely negligible) relative to the speed of light.

It can be shown that the average kinetic energy of the electrons is $\frac{3}{5}E_F$ at absolute zero (perpetual motion!).

### Q. 13.2.1

The valence electrons in a metal cannot have any arbitrary energy and are restricted to (a)_____. The possible solutions for the wave functions are (b)_____. Only (c)_____ valence electrons in a metal can occupy the same energy state. At the absolute zero of temperature the valence electrons occupy energy states extending from (d)_____ _____. The Fermi energy of metals is about (e)____ eV, while thermal energy at room temperature is about (f)____ eV.

\* \* \*

THERMAL EFFECTS. Figure 13.2.2 shows the probability $F(E)$ of an energy state being filled at $T = 0$ and at $T > 0$. Of course, at $T = 0$, every allowable state is occupied up to $E = E_F$. Above $E_F$, none of the states are occupied. At finite temperatures, some of the electrons near $E_F$ are thermally excited to higher energy states. Since the thermal energy is $k_BT$, only electrons within an energy range of about $k_BT$ below the Fermi surface can be excited. One could say that all the other electrons (those for which $E_F - E > k_BT$) are unaware of the temperature $T$; such electrons therefore do not contribute to the specific heat; they remain in their unexcited state. Only those electrons within $k_BT$ of $E_F$ have thermal energy (approximately equal to that of an ideal gas) which we approximate by $3k_BT/2$. The fraction

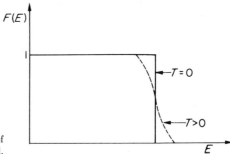

FIGURE 13.2.2. Probability of an electron state being occupied.

of the latter is approximately $k_BT/E_F$. Hence the specific heat per mole of electrons is estimated to be

$$C_v^E \approx \left(\frac{k_BT}{E_F}\right)\frac{3}{2}k_BN_0 = \frac{T}{T_F}\frac{3}{2}R.$$

But $T/T_F \approx \frac{1}{100}$ at room temperature. Hence the electronic specific heat is only about $\frac{1}{100}$ of the classical value as first shown by A. Sommerfeld in 1928. [An exact derivation, which makes use of the exact form of $F(E)$, called the **Fermi-Dirac distribution,** is too long to be given here; the results are essentially the same; for a derivation of this, see J. E. Mayer and M. G. Mayer, *Statistical Mechanics,* John Wiley & Sons, Inc., New York (1940). In general the **Fermi energy** (at any temperature) is the value of the energy for which $F(E) = \frac{1}{2}$. In the case of metals the Fermi energy is essentially independent of temperature. This is not the case for semiconductors.]

Note that the electronic specific heat is directly proportional to the temperature. At a few degrees Kelvin, where the lattice specific heat is proportional to the cube of the temperature, the electronic specific heat exceeds the lattice specific heat. An example of this is shown in Figure 13.2.3.

The Sommerfeld model explains the specific heat paradox but it does not explain the other shortcomings of Drude's model. In both models, it was assumed that the electron moved in a zero potential field. This is incorrect, because the electrons move in the periodic potential of the ion cores. When this is taken into account, all the difficulties encountered with Drude's model are eliminated. The following sections deal with this problem.

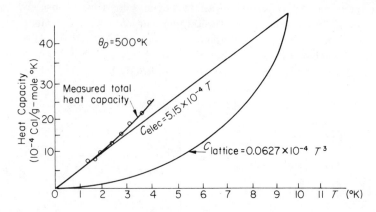

FIGURE 13.2.3. Specific heat data for a 20–80 V–Cr alloy. [From C. H. Cheng, C. T. Wei, and P. A. Beck, *The Physical Review* **120**, 426 (1960).]

Another aspect of the particle in a box model which is incorrect is the assumption that the potential jumps to infinity outside the box, which, of course, leads to the conclusion that the probability of finding the electron outside the metal is zero. This, of course, is incorrect. Actually the energy which an electron at the Fermi surface must gain to escape from a metal (an energy called the **work function,** $\phi$) is about 5 eV. The probability of escape is proportional to $e^{-\phi/k_B T}$. Hence electron emission from a hot filament depends very strongly on temperature.

## Q. 13.2.2

(a) Sketch $F(E)$ at $T = 0$ and at $T > 0$ for a metal such as sodium:

_____.

The Fermi energy for $T > 0$ for metals appears to be (b)_____.
The electronic specific heat is roughly what fraction of $3R/2$? (c)_____.
The energy for which the probability of that energy state being occupied is one half is called (d)_____.

\* \* \*

## ANSWERS TO QUESTIONS

**13.2.1** (a) Discrete energies, (b) standing waves, (c) two, (d) just above zero to the Fermi energy, (e) 5, (f) $\frac{1}{40}$.

**13.2.2** (a) See Figure 13.2.2, (b) the same as at $T = 0$, (c) $T/T_F$ or about $\frac{1}{100}$ at room temperature, (d) the Fermi energy.

## 13.3 ELECTRONS IN A PERIODIC POTENTIAL

A completely free electron has energy given by

$$E = \frac{p^2}{2m} = \frac{h^2}{2m\lambda^2} = \frac{h^2 k^2}{8\pi^2 m} \tag{13.3.1}$$

where we have used de Broglie's relation ($p = h/\lambda$) and where we have defined the **wave vector**

$$\mathbf{k} = \frac{2\pi}{\lambda} \mathbf{s} \tag{13.3.2}$$

where **s** is the unit vector in the direction of propagation of the plane wave whose wavelength is $\lambda$ (not to be confused with the mean free path for which $\lambda$ has also been used).

For mathematical simplicity we shall now assume that the wave propagates in the $x_1$ direction only:

$$E = \frac{p_1^2}{2m} = \frac{h^2}{2m\lambda_1^2} = \frac{h^2 k_1^2}{8\pi^2 m}. \tag{13.3.3}$$

A plot of this function is shown in Figure 13.3.1. Note that it is a parabola. This represents a continuum; i.e., all values of $k_1$, and $E$ are possible.

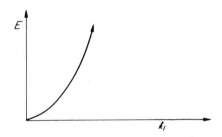

FIGURE 13.3.1. Energy versus wave vector for free electron.

The wave function which represents a free electron moving in the $x_1$ direction is

$$\psi \propto e^{ik_1 x_1} \tag{13.3.4}$$

as can be shown by substitution into the one-dimensional Schrödinger equation with $U = 0$. In the general case the solution is

$$\psi \propto e^{i\mathbf{k}\cdot\mathbf{r}}. \tag{13.3.5}$$

For the quantized electron in a one-dimensional box (discussed in the previous section) only certain values of $\lambda_1$, or $k_1$, are possible:

$$n\lambda_1 = 2L_1 \tag{13.3.6}$$

$$k_1 = \frac{n\pi}{L_1}, \quad n = 1, 2, 3, \ldots. \tag{13.3.7}$$

We can use the same curve in Figure 13.3.1 to represent the $E$ vs. $k_1$ relation but we must remember that now it is not a continuous curve but rather a representation of *very* closely spaced levels.

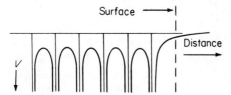

FIGURE 13.3.2. Potential energy in a crystal in a given direction along a given path.

An electron in a crystal moves through a periodic potential such as that shown in Figure 13.3.2 rather than in a potential $U = 0$ as assumed in the previous section. A periodic potential of period $a$ has $U(x \pm na) = U(x)$, where $n$ is any integer.

This periodicity of $U(x)$ has a resultant profound effect on the energy vs. wave vector curve as shown in Figure 13.3.3. The effect is to

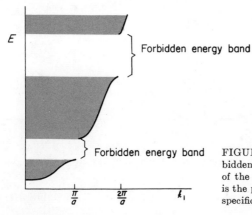

FIGURE 13.3.3. Origin of forbidden energy bands as a result of the periodic potential. Here $a$ is the period of the potential in a specific direction.

introduce **forbidden energy bands** (**energy gaps**) as first shown by Kronig and Penney. The actual proof that the $E(k)$ curves have the shape shown in Figure 13.3.3 is too involved to consider here (see the previously cited Ruoff, *Materials Science*, Problems 27.20–27.30). A result also shown there is that *the $E(k)$ curve for a band is parabolic near the bottom of a band and is also parabolic near the top of the band.* [The quantized electron in a box ($U = 0$) had a single allowed energy band which includes all energies from zero to infinity, although only very closely spaced discrete energies within the band are allowed.] However, when a periodic potential is present, there are *large* intervals of energy for which no possible energy levels exist separated by intervals in which there are discrete levels available (**allowed bands**). Note that the discontinuities in $E$ are periodic in $k$ with period $\pi/a$ (we would expect a different $E$ vs. $k$ curve in another direction). Hence, when we include waves with negative $k$, discontinuities in $E$ exist in the one-dimensional case when

$$k = \pm \frac{n\pi}{a} \quad \text{where } n = 1, 2, 3, \ldots. \quad (13.3.8)$$

In the present example we started with a free electron and described how the introduction of a periodic potential led to energy bands. One could also start with the electrons tightly bound to the isolated atoms and bring the atoms together and show that the energy levels spread out into bands as shown in Figure 13.3.4. This tight binding approximation is discussed

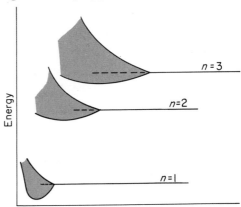

FIGURE 13.3.4. Formation of energy bands in the tight binding approximation.

with unusual clarity in R. L. Sproull, *Modern Physics*, John Wiley & Sons, Inc., New York (1963) and will not be repeated here. The energy levels for the two extreme cases, namely, free electrons and tight binding, are shown along with the actual intermediate case in crystals in Figure 13.3.5. The band in the solid which corresponds to the outermost or valence electrons

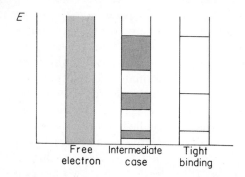

FIGURE 13.3.5. Allowed energy levels for different binding strengths.

of the atom is called the **valence band**. The next higher allowed band is given the name **conduction band**.

### Q. 13.3.1

The motion of electrons in (a)_____ gives rise to energy bands. Near the bottom of an allowed energy band the $E$ vs. $k$ curve is (b)_____. Near the top of an allowed band the $E$ vs. $k$ curve is (c)_____ _____. The bands in which no discrete energy levels exist are called (d)_____.

\* \* \*

## ANSWERS TO QUESTIONS

**13.3.1** (a) A periodic potential; (b) parabolic, as is the case for the free electron; (c) also a parabola but one which is inverted; (d) forbidden bands.

### 13.4 BRILLOUIN ZONES

We have noted that the discontinuities in $E$ occur when

$$k = \pm \frac{n\pi}{a}, \quad \text{where } n = 1, 2, 3, \ldots \quad (13.4.1)$$

in the one-dimensional case. The negative $k$'s involve waves moving in the negative direction. The first region of $k$ values where there is no discontinuity extends over $-\pi/a < k < \pi/a$. It is called the **first Brillouin zone**. The zones for the one-dimensional case are shown in Figure 13.4.1. It can be

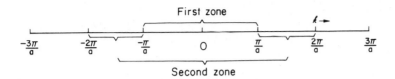

FIGURE 13.4.1. Brillouin zones in one-dimensional lattice.

shown that Equation (13.4.1) is the equation for Bragg reflection from a one-dimensional lattice. It is in fact true that the periodicity of the potential introduces discontinuities in the $E(k)$ curves exactly at the values of $k$ corresponding to the Bragg condition.

The condition for Bragg reflection [of a wave whose wave vector is $k = (k_1, k_2)$] from a two-dimensional square lattice can be written

$$k_1 n_1 + k_2 n_2 = \frac{\pi}{a}(n_1^2 + n_2^2), \qquad (13.4.2)$$

where $n_1$ or $n_2$ may be positive, negative, or zero except that both may not be zero. Examples of two-dimensional Brillouin zones for a simple square lattice are shown in Figure 13.4.2. The energy shows a discontinuity as the $k$ values cross a zone boundary. We note that the total area of each zone is the same.

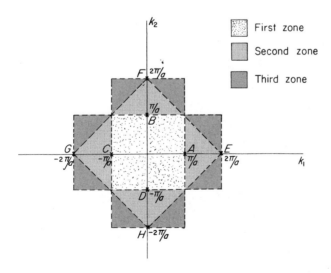

FIGURE 13.4.2. First three Brillouin zones for a two-dimensional square lattice.

The condition for Bragg reflections from a cubic lattice is

$$k_1 n_1 + k_2 n_2 + k_3 n_3 = \frac{\pi}{a}(n_1^2 + n_2^2 + n_3^2). \tag{13.4.3}$$

Here the $n_1$, $n_2$, and $n_3$ take on the same values as do $H$, $K$, and $L$ in diffraction theory (Section 6.8). [For a derivation of this expression, see Charles Kittel, *Introduction to Solid State Physics*, John Wiley & Sons, Inc., New York (1968), p. 52. It is necessary to read about the reciprocal lattice and the Ewald construction.] Inasmuch as the discussion here is introductory, it will be limited to one- and two-dimensional Brillouin zones since these exhibit the qualitative features needed.

NUMBER OF STATES IN AN ENERGY BAND. How many different wave functions are possible in an energy band?

Let us consider for mathematical simplicity a one-dimensional lattice. Let there be $M$ identical atoms of spacing $a$ so that the total length of the array is

$$d = Ma. \tag{13.4.4}$$

The allowed wavelength $\lambda$ for standing waves are, in the terms of the integers, $n = 1, 2, \ldots, n_{\max}$,

$$\frac{\lambda}{2} = \frac{d}{n} = \frac{Ma}{n}. \tag{13.4.5}$$

Now for the first zone the maximum magnitude of the wave vector is $\pi/a$ as is clear from Figure 13.4.1 and

$$k_{\max} = \left(\frac{2\pi}{\lambda}\right)_{\max} = \frac{\pi}{a} \tag{13.4.6}$$

so that by (13.4.5) and (13.4.4) we have

$$\left(\frac{2\pi}{\lambda}\right)_{\max} = \frac{\pi n_{\max}}{d} = \frac{\pi n_{\max}}{aM},$$

which combined with (13.4.6) gives $n_{\max} = M$. If we also include electron spin states, then the number of allowed states in the first band is $2M$, i.e., twice the number of atoms.

The above result applies for simple cubic, body-centered cubic, face-centered cubic, and hexagonal closest packed crystals. For diamond cubic crystals the number of allowed states per allowed energy band is $4M$. A metal such as sodium which is monovalent contributes one electron per atom. Since it has the bcc crystal structure, the valence band is half full. However, in the diamond form of carbon, which has a valence of 4, there are four electrons per atom in the valence band and so the valence band is full. These are significant points to which we shall return later.

### Q. 13.4.1

The extent of $k$ space over which there is no discontinuity in $E$ is called (a)_____. The discontinuities in $E$ occur at values of $k$ which correspond to (b)_____. It can be shown that the number of states available in an allowed energy band in a one-dimensional simple crystal is (c)_____. The number of states available in an allowed energy band per atom in a bcc crystal is (d)____ and in a diamond cubic crystal (e)____.

\* \* \*

## ANSWERS TO QUESTIONS

**13.4.1** (a) A Brillouin zone, (b) Bragg reflections, (c) two per atom, (d) two, (e) four.

## 13.5 CONDUCTIVITY

The $E$ vs. $k$ curve of Figure 13.3.3 is decidedly not that of a classical free electron of mass $m$ and charge $-e$ and having by (13.3.1) kinetic energy

$$E = \frac{h^2}{8\pi^2 m} k^2. \tag{13.5.1}$$

For such a classical free electron

$$\frac{dE}{dk} = \frac{h^2}{4\pi^2 m} k$$

and

$$\frac{d^2 E}{dk^2} = \frac{h^2}{4\pi^2 m}$$

and

$$m = \frac{h^2}{4\pi^2 (d^2 E/dk^2)}. \tag{13.5.2}$$

We could take the approach that we shall attribute classical free electron behavior to the electron moving in a periodic potential (the electron does not, in fact, exhibit such behavior); if a particle of mass $m^*$ and charge $-e$ behaves in a classical fashion in the presence of a field **E** in the $x$ direction, it obeys the equation $m^*\ddot{x} = -eE$. If classical free electron behavior is ascribed to the electron, then the **effective mass** $m^*$ of the electron must be given by Equation (13.5.2). Near the bottom of the valence band the $E$ vs. $k$ curve does indeed have the parabolic form $E \propto k^2$ and the electron does indeed behave like a classical free electron.

If the energy at the top of the valence band is called $E_0$, then near the top of the band, the $E$ vs. $k$ curve has according to the Kronig-Penney model the form of an inverted parabola, $E - E_0 \propto -k^2$. Here the electron behaves as a free classical particle whose effective mass $m^*$ is $-m$. In other words if we wish to ascribe classical free particle behavior to an electron moving in a periodical potential (a fiction), we have to introduce the idea that the mass varies according to (13.5.2) (a compensating fiction). A detailed analysis reveals that for $k = 0$, $m^* = m$, that $m^*$ increases as $k$ increases and that $m^* \to \infty$ as the inflection point of the $E$ vs. $k$ curve is approached. For $k = \pi/a$, $m^* = -m$, and $m^*$ decreases as $k$ decreases so that $m^* \to -\infty$ as the inflection point is approached. Moreover, in a full band the distribution of electrons with negative mass is exactly the same as that with positive mass; i.e., for every electron with a specific mass $m_1^*$, there is a corresponding electron with mass $-m_1^*$. There are several important implications to be drawn from this:

1. The application of an electric field to a completely filled band results in no conductivity (the particles having negative $m^*$ will move in a direction opposite to those with positive $m^*$). Hence materials in which the valence band is filled (and the conduction band is empty) are insulators.
2. If a band is only half full, it will conduct, and conduction will be due to the motion of electrons with $m^* > 0$.
3. If a valence band is nearly full (only a tiny fraction of electrons missing) it will conduct and the charge carriers will behave as positive charge carriers.

The last statement needs further discussion. Consider a full band. We take one electron whose charge is $-e$ and whose effective mass is $-m$ from the band, so the band now takes on a net charge $e$ and mass $m$ relative to the filled band. We call the missing electron a **hole**. Rather than considering the conduction to be due to all the electrons except one in the band, we could attribute the conduction to the hole and attribute to the hole the net charge $e$ and mass $m$ of the band with one electron missing. This introduces a symmetry into conductivity in that in a nearly empty band (a band containing a few electrons) the conductivity will be due to electrons of charge $-e$ and mass $m$ while in a nearly full band (a band containing a few holes) the conductivity will be due to holes of charge $e$ and mass $m$.

Sec. 13.5                                                     Conductivity

## Q. 13.5.1

For a completely free electron the $E(k)$ relation is (a)_____.
If an electron in a periodic potential is to be considered as a free particle, then its mass must vary according to (b)_____. This means that near the top of a band a (c)_____ mass must be attributed to an electron. In a completely filled band, the application of an electric field results in acceleration of equal numbers of electrons in opposite directions and hence (d)_____. A band with one electron missing from near the top of the band behaves as if it has a (e)_____. Conduction in such a band is, for convenience, attributed to the motion of (f)_____.

\* \* \*

CONDUCTIVITY IN MONOVALENT METALS. Sodium metal has a bcc structure. It is a monovalent metal. The outer electron of sodium is the 3s electron, which leaves the atom when the positive sodium ion is formed. This is the valence electron. In sodium metal this electron forms part of the electron gas or electron sea. The first band of states available for these valence electrons is called the valence band. In a bcc crystal, the valence band can have up to two electrons per atom, so that in sodium the valence band is only half full. All the electrons in the valence band are nearly free electrons, so that the Fermi surface is approximately a sphere; i.e., the electrons are not affected much by the periodic potential.

When an electrical field is applied to sodium the net result is to move the Fermi "sphere," as shown in Figure 13.5.1. (For simplicity we show the Fermi surface as a circle and the Brillouin zone as a square; i.e., we discuss a two-dimensional case.)

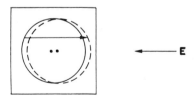

FIGURE 13.5.1. Square represents view of first Brillouin zone in a simple square metal. Solid circle represents Fermi surface (plotted in $k_1 k_2$ space) in absence of field. Dashed sphere represents Fermi surface in the presence of the field **E**.

### EXAMPLE 13.5.1

Estimate the energy gained by an electron between collisions due to the presence of the applied field. Assume that there are 100 V across 1 m length and that the mean free path $\lambda = 200$ Å.

*Answer.* A collision scatters the electron in random directions. Then the applied field **E** acts on it while it moves the distance λ. The force acting has magnitude $eE$ so the work done is the force times the distance the particle moves or $eE\lambda = e\ 100\ \text{V/m} \times 200 \times 10^{-10}\ \text{m} = 2 \times 10^{-6}\ \text{eV}$.

Thus note that the perturbation due to the electrical field is extremely small (and hence vastly exaggerated in Figure 13.5.1) since the Fermi energy is several electron volts.

Recall that there are numerous available states within the Brillouin zone in Figure 13.5.1. The vast majority of electrons are in states common to each circle in Figure 13.5.1. A tiny fraction changes states as shown by the arrow. This tiny fraction always involves electrons *very* near (at) the Fermi surface. Hence the effective mass $m^*$ associated with conductivity [see Equation (13.1.3)] is the effective mass evaluated at the Fermi surface. Likewise the velocity of the conducting electrons is the velocity evaluated at the Fermi surface. To calculate the conductivity of the metal it is necessary to know the mobility and hence the relaxation time $\tau$.

Bloch has proved quite generally that in a perfect static crystal (no crystalline imperfections and no atom vibrations) the electron would not change states, except in the presence of changing external fields. Hence there would be no resistivity in such a metal. Note how this differs from Drude's original idea in which it was assumed that the ions were static and that the electron mean free path λ was equal to the interatomic spacing. Real metals show finite resistivity due to a finite path length λ or time $\tau$ between "collisions" as a result of scattering of the electrons. This scattering is a consequence of the departure of the lattice from perfect periodicity. These imperfections include

1. Lattice vibrations or phonons.
2. Point, line, and planar defects. Point defects may include vacancies, impurities, etc.

If the crystal does not have the latter stacking imperfections (defects) it would ordinarily be called perfect. In this case the scattering and hence the resistivity would be entirely due to phonons. As a result of this the resistivity at high temperatures is directly proportional to the temperature. At very low temperatures the resistivity due to scattering by phonons is very small ($\rho \propto T^5$) and the resistivity due to scattering by stacking imperfections may be larger. At the absolute zero temperature there is a finite resistivity (**residual resistivity**) due to these imperfections. This is illustrated in Figure 13.5.2 where the mean free path variation with temperature is shown (the resistivity is inversely proportional to the mean free path).

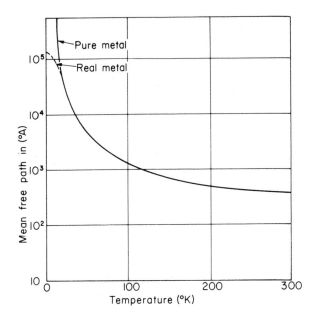

FIGURE 13.5.2. The electron mean free path in copper.

## EXAMPLE 13.5.2

Calculate the relaxation time $\tau$ and the mean free path $\lambda$ for electrons in copper at 20°C.

*Answer.* We obtain $\rho$ from Table 3.2.1. Using the fact that $\sigma = 1/\rho$ and that $\sigma = Ne^2\tau/m$ [in more exact scattering theory the factor $\frac{1}{2}$ is absent from (13.1.3)] and data from Table 13.2.1, we have in the MKS system

$$\tau = \frac{(9.11 \times 10^{-31} \text{ Kg})}{(8.5 \times 10^{28} \text{ m}^{-3})(1.67 \times 10^{-8} \text{ }\Omega\text{-m})(1.6 \times 10^{-19} \text{ C})^2}$$

$$= 2.5 \times 10^{-14} \text{ sec.}$$

Hence

$$\lambda = v_F\tau \approx 3.9 \times 10^{-8} \text{ m} \approx 390 \text{ Å}.$$

In very thin films at low temperatures, electrons are scattered predominantly from the surface and hence the film thickness is effectively $\lambda$. Hence, below a certain thickness, the resistivity would become size-dependent; i.e., it would depend on the sample thickness. A value of $\lambda = 450$ Å is obtained at 20°C for pure copper in this way. [See F. W. Reynolds and G. R. Stillwell, *The Physical Review* **88**, 418 (1952).]

**EXAMPLE 13.5.3**

"Pure" copper has a resistivity ratio (defined in Section 3.3) of 50,000. What is the mean free path at 4°K?

*Answer.* $\lambda = 450 \text{ Å} \times 50{,}000 = 0.2$ cm.

## Q. 13.5.2

If there were no atomic vibrations and no lattice imperfections, the electrical conductivity of copper would be (a)_____. In a nearly pure metal, the mean free path at high temperatures, and hence the mobility, is determined by (b)_____, while at temperatures very near to absolute zero, $\lambda$ is determined by (c)_____.

\* \* \*

DIAMOND AS AN INSULATOR. The valence band in the diamond cubic crystal contains four allowed states per atom and the tetravalent carbon atom contributes four electrons. The valence band is filled. The conduction band (the allowed band immediately above the valence band) is empty. Thus at room temperature and below, pure diamond is an insulator (see Table 3.2.1). However, at quite high temperature pure diamond is a semiconductor.

SEMICONDUCTORS. The energy bands for pure semiconductors (or insulators) are shown in Figure 13.5.3. At low temperature in the pure material the valence band is filled and the conduction band is empty. The material behaves as an insulator. However, depending on the size of the energy gap and the temperature, the material may behave as a semiconductor. Thus if the energy gap is small, electrons can be thermally excited

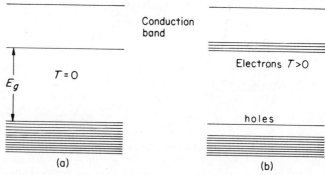

FIGURE 13.5.3. Energy bands in pure semiconductors. (a) At absolute zero. (b) At room temperature.

at moderate temperatures from levels near the top of the valence band to levels near the bottom of the conduction band. Hence a certain number of electrons exists in the conduction band and an equal number of holes exists in the valence band. Let $N_n$ be the number of negative charge carriers per unit volume and $N_p$ be the number of positive charge carriers. Then, if we consider the formation of these carriers to be a chemical reaction,

full valence band + empty conduction band $\rightleftharpoons$ electrons + holes,

an equilibrium constant can be written for the reaction in the form

$$N_n N_p = K_{eq} = A e^{-E_g/k_B T} \tag{13.5.3}$$

where $E_g$ is the energy gap. Since $N_n = N_p$ for the pure material,

$$N_n = A^{1/2} e^{-E_g/2k_B T}.$$

Rigorous analysis (see the previously cited Ruoff, *Materials Science*, Section 28.2) can be used to show that

$$N_n = 2 \left( \frac{2\pi m^* k_B T}{h^2} \right)^{3/2} e^{-E_g/2k_B T}. \tag{13.5.4}$$

The conductivity [see (13.1.3)] is the result of the mobility $\mu$ of both charge carriers:

$$\sigma = N_n e \mu_n + N_p e \mu_p. \tag{13.5.5}$$

It can be shown that at high temperatures the mobility varies as $1/T^{3/2}$. Hence the conductivity varies as

$$\sigma \propto e^{-E_g/2k_B T}$$

and the resistivity varies as

$$\rho \propto e^{E_g/2k_B T},$$

as has previously been noted [see Equation (3.3.1)].

Because the concentration of charge carriers in a semiconductor is small compared to a metal, their conductivity is correspondingly less (see Table 3.2.1). Semiconductors will be studied further in Section 13.6.

### Q. 13.5.3

At temperatures approaching absolute zero, a pure semiconductor would be (a)_____. However, at temperatures for which $k_B T \approx \frac{1}{20} E_g$, these materials are (b)_____, containing an equal number of (c)_____. An equilibrium constant which represents the concentration of charge carriers is (d)_____.

\* \* \*

DIVALENT METALS. The number of allowable states per band per atom in an hcp crystal is two. Hence cadmium, which has two valence electrons per atom, could be expected to have a full valence band and to be an insulator. This is not the case. The reason for this is illustrated in Figure 13.5.4, where the $E$ vs. $k$ curves in two different directions are

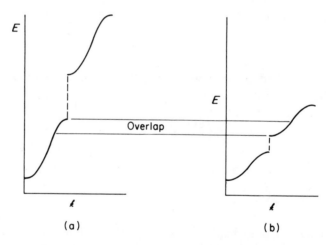

FIGURE 13.5.4. Energy versus wave vector ($E$ vs. $k$) curve in two directions in Cd.

shown. [Note that since the periodicity and the nature of the potential seen by the electron in two different directions is, as a rule, different, it is reasonable to expect different $E$ vs. $k$ curves in different directions.] The energy band for these two cases are shown in Figure 13.5.5. The result is that the conduction band overlaps with the valence band. Consequently the valence band is not quite full; i.e., it has some holes, and there are some electrons in the conduction band. Conduction will therefore be due to both electrons and holes as given by Equation (13.5.5). Suppose that the mobility of the electrons in the conduction band is negligible compared to the mobility of holes in the valence band. Then the conductivity is due to the motion of holes and the material will exhibit a positive Hall coefficient as is the case for Be and Cd (see Table 13.1.1).

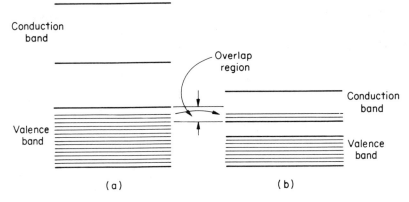

FIGURE 13.5.5. Energy bands in Cd corresponding to the $E$ vs. $k$ curves of Figure 13.5.4.

## ANSWERS TO QUESTIONS

**13.5.1** (a) $E = (h^2/8\pi^2 m)k^2$, (b) $m = h^2/[4\pi^2(d^2E/dk^2)]$, (c) negative, (d) a zero drift velocity and no conductivity, (e) positive mass and positive charge, (f) holes.

**13.5.2** (a) Infinite; (b) atom vibrations; (c) impurities, dislocations, etc.

**13.5.3** (a) An insulator, (b) semiconductors, (c) holes and electrons, (d) Equation (13.5.3).

## 13.6 INTRINSIC SEMICONDUCTORS

ENERGY GAP. An **intrinsic semiconductor** is a semiconductor in which the concentration of charge carriers is independent of impurities. This would be the case for a perfectly pure material. In such a case the concentrations of holes and electrons are equal and from Equation (13.5.4) are given by

$$N_n(\text{electrons/m}^3) = 4.83 \times 10^{21} \left(\frac{m^*}{m}\right)^{3/2} T^{3/2} e^{-E_g/2k_B T}. \quad (13.6.1)$$

The energy states are shown in Figure 13.6.1. Here $m^*/m$ is the ratio of the effective mass to the free electron mass. Actually $m^*$ is the geometrical mean of the effective mass of a hole near the top of the valence band and the mass of an electron at the bottom of the conduction band. For the present it is sufficient to assume that $m^*/m = 1$. Then only $E_g$ is needed to

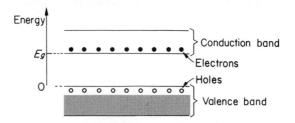

FIGURE 13.6.1. Energy diagram for intrinsic semiconductors.

compute the charge carrier concentration. Examples of $E_g$ are given in Table 13.6.1.

Table 13.6.1. ENERGY GAPS IN SEMICONDUCTORS

| Element | $E_g$ (eV) | Compound | $E_g$ (eV) |
|---|---|---|---|
| Diamond | 5.3 | AlP | 3.0 |
| Silicon | 1.1 | GaP | 2.25 |
| Germanium | 0.72 | AlSb | 1.52 |
| Gray tin* | 0.08 | GaAs | 1.34 |
|  |  | GaSb | 0.70 |
|  |  | InAs | 0.33 |
|  |  | InSb | 0.18 |

* Gray tin is the diamond cubic polymorph of tin. Tin is usually found in the metallic phase which has a tetragonal crystal structure.

**EXAMPLE 13.6.1**

(a) Compute the number of electrons per unit volume in the conduction band in germanium at 300°K. (b) Compute the number $N_a$ of germanium atoms per unit volume at 300°K.

*Answer.* (a) $N_n = 1.4 \times 10^{19}/\text{m}^3$ using $E_g = 0.72$ eV and Equation (13.6.1). (b) Germanium is diamond cubic and hence has eight atoms/unit cell. The lattice parameter is (Table 6.2.1) 5.66 Å. The number of atoms per cubic meter is therefore

$$N_a = \frac{8}{(5.66)^3 \times 10^{-30}} = 4.4 \times 10^{28} \text{ atoms/m}^3.$$

Hence $N_n/N_a = 3.1 \times 10^{-10}$ at 300°K in pure germanium.

Sec. 13.6                                                    *Intrinsic Semiconductors*    **595**

In an intrinsic semiconductor the Fermi energy lies midway in the energy gap.

MOBILITIES. The conductivity of a semiconductor is a result of the motion of both charge carriers so that

$$\sigma = N_n e \mu_n + N_p e \mu_p, \qquad (13.6.2)$$

where $\mu_n$ is the (charge) mobility of the negative carrier (electrons) and $\mu_p$ is the mobility of the holes.

For the case where two charge carriers are present the Hall coefficient can be shown to be

$$R_H = -\frac{1}{e} \frac{N_n \alpha^2 - N_p}{(N_n \alpha + N_p)^2}, \qquad (13.6.3)$$

where

$$\alpha = \frac{\mu_n}{\mu_p}.$$

For an intrinsic semiconductor, inasmuch as $N_n = N_p$,

$$R_H = -\frac{1}{N_n e} \frac{(\alpha - 1)}{(\alpha + 1)}. \qquad (13.6.4)$$

Some values of mobilities are given in Table 13.6.2.

**Table 13.6.2.** MOBILITIES IN UNITS OF m²/V-sec at 25°C

| Element | $\mu_n$ | $\mu_p$ | Compound | $\mu_n$ | $\mu_p$ |
|---|---|---|---|---|---|
| Diamond | 0.18 | 0.12 | GaP | 0.045 | 0.002 |
| Silicon | 0.14 | 0.048 | GaAs | 0.85 | 0.45 |
| Germanium | 0.39 | 0.19 | GaSb | 0.50 | 0.085 |
| Gray tin | 0.20 | 0.10 | InSb | 8.00 | 0.070 |

**EXAMPLE 13.6.2**

Is pure germanium an *n*-type or *p*-type semiconductor?

*Answer.* Although $N_n = N_p$, $\mu_n > \mu_p$. Hence $R_H$ is negative. It is *n*-type, i.e., conduction by electrons predominates.

At high temperatures, the resistivity of a semiconductor is due to "collisions" between the conduction electrons and the vibrating atom. This is called an electron-phonon interaction. It can be shown that the mobility, $\mu_L$, based on these lattice interactions is given by

$$\mu_L = BT^{-3/2},$$

where $B$ is a parameter depending on the material. The calculation of $B$ is a very complex problem in quantum mechanics.

At room temperature the mobilities of carriers in semiconductors may be $10^2$ or $10^3$ times those in metals. There are a number of scattering mechanisms which become important at low temperatures. For semiconductors the term *low temperatures* may mean in the vicinity of room temperature rather than at several degrees Kelvin as in metals. In semiconductors charge carriers are scattered by (1) ionized impurities, (2) neutral impurities, (3) grain boundaries, and (4) dislocations. These mechanisms are discussed in the article by K. Lark-Horovitz and V. A. Johnson in *The Science of Engineering Materials* (J. E. Goldman, ed.), John Wiley & Sons, Inc., New York (1957), p. 336.

### Q. 13.6.1

The general expression for conductivity when both holes and electrons are involved is $\sigma =$ (a)_____. Intrinsic silicon is predominantly (b)____ -type and shows a (c)_____ Hall coefficient.

\* \* \*

## ANSWERS TO QUESTIONS

**13.6.1** (a) $N_n e \mu_n + N_p e \mu_p$, (b) $n$, (c) negative.

## 13.7 EXTRINSIC SEMICONDUCTORS

An **extrinsic semiconductor** is a semiconductor in which the charge carriers are due to the presence of impurities. The semiconducting elements listed in Table 13.6.1 all have the diamond cubic structure illustrated in Figure 6.2.3. Thus in silicon, each silicon atom is tetrahedrally bonded to four other silicon atoms. In the pure crystal the outer electrons are all in the localized pair bonds and hence do not contribute to conduction. From the free electron approach the valence band is filled and the conduction band is empty and there is no conductivity. The III-V compounds listed in Table 13.6.1 have the zinc blende structure (see Figure 6.4.10). Thus in the compound GaAs each gallium atom is bonded tetrahedrally to four arsenic atoms and each arsenic atom is bonded tetrahedrally to four gallium atoms; as with the semiconducting elements such as silicon there are four valence electrons associated with each atom.

Let us consider what happens when a pentavalent impurity atom such as phosphorus is purposely added to a semiconducting material such as germanium (this is called **doping**) as a substitutional impurity. After the four covalent bonds are taken care of there is still an excess of one positive charge on the phosphorus ion and an excess electron. This electron is very weakly bound to the ion as we shall now show. A simple model assumes that the electron and ion form a hydrogen-like atom in a dielectric medium whose

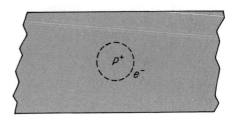

FIGURE 13.7.1. Hydrogen-like atom in $\kappa \gg 1$ medium.

dielectric constant is $\kappa$, as shown in Figure 13.7.1. This atom would be described by Schrödinger's equation (5.3.1) but with the potential given by [in MKS units]

$$U = -\frac{e^2}{4\pi\epsilon_0 \kappa r}. \qquad (13.7.1)$$

For germanium $\kappa = 16$, while for a vacuum $\kappa = 1$. The solution for $\kappa = 1$ is given by Equations (5.4.1) and (5.4.2) and it obviously follows for the present case that with $Z = 1$,

$$E = -\frac{13.6}{\kappa^2 n^2} \text{ (eV)}, \qquad n = 1, 2, \ldots. \qquad (13.7.2)$$

Thus for germanium the ground state energy ($n = 1$) is $E = -0.053$ eV. This means that an energy of 0.053 eV is needed to excite the electron from the bound state on phosphorus to the fully ionized state ($n = \infty$) in which the electron is now at the bottom of the conduction band. Thermal energy, $k_B T$, equals about 0.025 eV at room temperature so that thermal energy is more or less sufficient to strip the fifth electron from phosphorus and to place it in the conduction band of germanium. Impurity elements which can easily contribute an electron to the conduction band in this way are called **donors**. Donors are usually pentavalent substitutional impurities, but they may also involve interstitial impurities with a different valence such as lithium.

If a trivalent atom such as boron is added to germanium, it enters a substitutional site. It is tetrahedrally coordinated and to complete the four paired bonds, the boron must extract an electron from the valence band which leaves a hole there. Such an impurity is called an **acceptor**. The formation of holes in the valence band and electrons in the conduction band

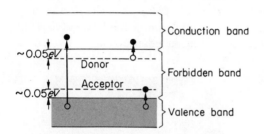

FIGURE 13.7.2. Formation of electrons in the conduction band or holes in the valence band.

is illustrated in Figure 13.7.2. The ionization energies of some solutes in germanium and silicon are shown in Table 13.7.1. Note that the values are of the same order of magnitude as predicted.

Table 13.7.1. IONIZATION ENERGIES OF SOLUTES (eV)

|  | Site | In Silicon | In Germanium |
|---|---|---|---|
| Donors |  |  |  |
| $Li^+$ | Interstitial | 0.033 | 0.0093 |
| $P^+$ | Substitutional | 0.045 | 0.012 |
| $As^+$ | Substitutional | 0.049 | 0.0127 |
| $Sb^+$ | Substitutional | 0.039 | 0.0096 |
| Acceptors |  |  |  |
| $B^-$ | Substitutional | 0.045 | 0.0104 |
| $Al^-$ | Substitutional | 0.057 | 0.0102 |
| $In^-$ | Substitutional | 0.16 | 0.0112 |
| $Cu^-$ | Interstitial | 0.49 | 0.04 |

Lark-Horowitz and his associates at Purdue in 1942 showed that trivalent impurities in tetravalent elements (germanium, silicon, and tin) act as acceptors while pentavalent impurities act as donors. The lithium atom or ion is in an interstitial position. For small ionization energies it can be assumed that at room temperature each donor atom donates one electron to the conduction band and that each acceptor atom generates one hole in the valence band. Materials in which the charge carriers are primarily due to donors or acceptors are called extrinsic semiconductors.

## Q. 13.7.1

A bonding characteristic of the common semiconductor elements and compounds is (a)_____. All these pure materials have an average of (b)_____ valence electrons per atom. Phosphorus has (c)_____ valence electrons. When silicon is doped with phosphorus, there is (d)_____ than is needed for tetrahedral bonding. The ionization energy of the fifth electron can be approximately calculated on the basis of a model (e)_____.

\* \* \*

It can be shown that the product $N_n N_p$ is equal to an equilibrium constant $K_{eq}$ which is the same as for the intrinsic material (where $N_n = N_p = N_i$; here $N_i$ is the electron or hole carrier concentration in the intrinsic material) at the same temperature. Hence

$$N_p N_n = K_{eq} = N_i^2. \tag{13.7.3}$$

This is a very useful expression inasmuch as $N_i$ is given by Equation (13.6.1) and

$$N_p N_n \propto e^{-E_g/k_B T} \tag{13.7.4}$$

regardless of whether the material is intrinsic or extrinsic. Thus if "pure" silicon is heavily doped with a donor (phosphorus, say) such that the donor concentration $N_D \gg N_i$, then $N_n \doteq N_D$ and $N_p$ can be calculated from (13.7.3); $N_p$ would be very small, i.e., $N_p \ll N_i$.

## EXAMPLE 13.7.1

How much phosphorus added to germanium would cause it to be extrinsic rather than intrinsic at 300°K?

*Answer.* We can consider the crossover point to be one where the concentration of electrons in the conduction band due to donors equals the concentration due to excitation from the valence band in the intrinsic material. The latter was given in Example 13.6.1(b). Hence the mole fraction of phosphorus needed is only $2.5 \times 10^{-6}$.

Thus a sample of germanium doped with an atom fraction of phosphorus to $10^{-4}$ would definitely be extrinsic in the neighborhood of room temperature. Consequently the electron concentration in the valence band would not vary much with temperature in a small temperature range near room temperature. Since the mobility of the electron in germanium varies slowly with temperature, the conductivity is nearly independent of temperature in this range.

*In an extrinsic p-type semiconductor the Fermi energy lies just above the top of the valence band. In an extrinsic n-type semiconductor the Fermi energy lies just below the bottom of the conduction band.*

## Q. 13.7.2

A semiconductor may be intrinsic or extrinsic depending on the (a)_____. The Fermi energy for phosphorus-doped silicon lies (b) _____.

\* \* \*

RECOMBINATION OF EXCESS CARRIERS. To properly discuss the *p-n* junction and in particular the *n-p-n* transistor, it is necessary to consider the minority carrier diffusion distance.

Consider an extrinsic semiconductor which has an equilibrium concentration of carriers at a given temperature. If this is irradiated with light of a sufficiently high frequency such that

$$h\nu > E_g,$$

an electron and a hole will be created for each photon absorbed. These excess carriers continuously combine at impurity centers (called recombination centers) and disappear. This process is called **recombination**.

Suppose that we irradiate an extrinsic *n*-type material ($N_n \gg N_p$). The radiation would only cause negligible change in concentration of electrons in the conduction band but would cause a major change in concentration of holes [recall that because Equation (13.7.3) holds, $N_p$ would be exceedingly small and hence negligible prior to irradiation]. In this case the electrons are the **majority carriers** and the holes are the **minority carriers.**

The concentration of minority carriers, holes in the present case, decreases exponentially with time when radiation ceases and can be characterized by a **minority carrier lifetime** $\tau$. The minority carrier concentration is proportional to

$$e^{-t/\tau}$$

when irradiation ceases. The quantity $\tau$ can be greatly decreased by the presence of energy states lying deep in the energy band which are caused by various impurity atoms and other defects in the crystal structure. Hence the materials scientist can increase $\tau$ by increasing the purity and perfection of the crystal.

It is often useful to use the **diffusion length** (the distance the carrier moves before recombination), rather than the lifetime $\tau$. The semiconductor literature uses the relationship

$$L = \sqrt{D\tau} \qquad (13.7.5)$$

for the diffusion depth. A carrier diffusion coefficient is related to the charge mobility $\mu$ and the charge $q$ by

$$D = \frac{k_B T \mu}{q}, \qquad (13.7.6)$$

a relation which is called the **Nernst relation** (and sometimes the Einstein relation).

Typical values of minority carrier diffusion lengths are $10^{-1}$–$10^{-3}$ cm. This is an important parameter in the design of semiconductor devices such as the $n$-$p$-$n$ transistor, which is discussed later.

## Q. 13.7.3

Light of sufficient energy per photon, incident on a semiconductor, creates (a)_____. If the semiconductor is $n$-type, then the holes are called (b)_____. The holes and electrons combine and disappear in a time called the (c)_____. During its lifetime, a minority carrier moves on the average a distance called the (d)_____. The diffusion coefficient of the minority carrier is directly proportional to (e)_____. The minority carrier lifetime can be greatly increased by (f)_____.

\* \* \*

## ANSWERS TO QUESTIONS

**13.7.1** (a) Tetrahedral bonds; (b) four; (c) five; (d) one more electron; (e) which is analogous to the hydrogen atom, except that the electron moves in a medium of dielectric constant $\kappa$.

**13.7.2** (a) Temperature and the nature of the doping, (b) just below the conduction band.

**13.7.3** (a) Pairs of holes and electrons; (b) minority carriers; (c) minority carrier lifetime $\tau$; (d) diffusion length, $L = \sqrt{D\tau}$; (e) the mobility of that carrier; (f) making the crystal purer and more nearly perfect.

## 13.8 THE $p$-$n$ JUNCTION

Consider a single crystal of silicon. To the right of a certain interface it is doped with a donor (phosphorus) and is hence $n$-type and to the left of that interface it is doped with an acceptor (boron) and is hence $p$-type. The interface region is called a $p$-$n$ junction. In this case, it is an idealized $p$-$n$ **junction** because of the sharp transition. In a real crystal, there would be a gradual change from $p$-type to $n$-type. The crystal can be considered as a box in which the electrons are moving. *The Fermi energy is everywhere the same in the crystal.* Hence an energy diagram for the $p$-$n$ junction appears as shown in Figure 13.8.1. Recall that in $p$-type extrinsic

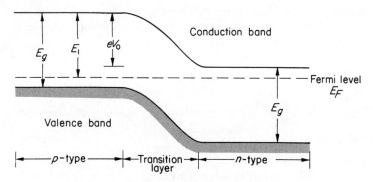

FIGURE 13.8.1. Energy diagram for *electrons* at a $p$-$n$ junction in equilibrium with no applied voltage.

material the Fermi energy is just above the valence band while in $n$-type extrinsic material the Fermi energy is just below the conduction band. The difference in energy between the bottom of the conduction band in the two cases is $eV_0$, where $V_0$ is called the **contact potential**.

The concentration of conduction electrons on the right side ($n$-type) is high and on the left side ($p$-type) is low. The ratio of electron concentration on the left to electron concentration on the right is given by the Boltzmann factor $e^{-eV_0/k_BT}$.

If the probability of a conduction electron moving from the $p$ region to the $n$ region is 1, the probability of the reverse process, i.e., the electron moving up the potential barrier of height $eV_0$, is only $e^{-eV_0/k_BT}$; i.e.,

$$p \propto e^{-eV_0/k_BT}.$$

(This is similar to the process of activated jumps discussed in Section 9.8.) Let $j_n$ be the flux of electrons to the left in Figure 13.8.1 and $j_f$ be the flux of electrons to the right (the forward direction). Under no external applied voltage, there is no net current, so

$$j_f = j_n.$$

Suppose we now apply an external voltage, $V$, such that the $p$ region is made more positive relative to the $n$-type region. This is called **forward bias**. This is shown in Figure 13.8.2.

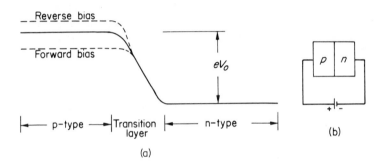

FIGURE 13.8.2. (a) Electron energy diagrams for a $p$-$n$ junction with applied voltages. For forward bias the $p$-type is made positive relative to the $n$-type. The curve for no applied voltage is the top curve of Figure 13.8.1. (b) Forward-biased $p$-$n$ junction.

The probability of an electron moving to the right is unchanged so the current flow to the right is still $j_f$. The probability of the electron moving up the potential barrier to the left (which now has height $eV_0 - eV$) is now proportional to $e^{-(eV_0-eV)/k_BT}$ and hence the electron flow to the left is now $j_n e^{eV/k_BT}$. Consequently, the net electron flux through the junction to the left is

$$j_n^{\text{net}} = j_n e^{eV/k_BT} - j_f = j_n(e^{eV/k_BT} - 1) \qquad (13.8.1)$$

and the net electrical current density to the *right*, $J_n$, due to electrons is

$$J_n = e j_n^{\text{net}}.$$

See Figure 13.8.3. It can be shown that an expression of the same type holds for the flow of holes so

$$J = J_n + J_p = J_0(e^{eV/k_BT} - 1), \qquad (13.8.2)$$

where $J_0$ is a constant. Hence, as soon as $eV \gg k_BT$, $J$ increases exponentially as $V$ increases.

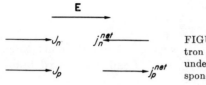

FIGURE 13.8.3. Fluxes of electron and holes, $j_n^{net}$ and $j_p^{net}$, under an applied field **E** and corresponding *charge* flow.

For a **reverse bias**, $V$ is negative. As $V$ increases negatively $J$ approaches $-J_0$, called the **saturation current density.** The current versus voltage for a specific $p$-$n$ junction is shown in Figure 13.8.4.

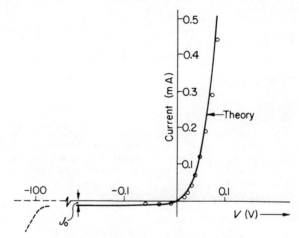

FIGURE 13.8.4. Current versus voltage for a $p$-$n$ junction. Dashed line is experimental. Solid line is theoretical. [After W. Shockley, *Electrons and Holes in Semiconductors*, Van Nostrand Reinhold Company, New York (1950).]

### Q. 13.8.1

The Fermi energy everywhere in a solid is the (a)_____. Hence, in a $p$-$n$ junction the top of the valence band of the $p$-type material is an energy $eV_0$ (b)_____ than the top of the valence band of the $n$-type material. Therefore the probability of an electron going from the $n$-type to the $p$-type is (c)_____ times the reverse process. If a forward bias, $V$, is applied, the probability in (c) is now (d)_____. Hence the net electron flow to the left is (e)_____. The total charge flux, or current density, as it is called, in a $p$-$n$ junction is (f)_____.

\* \* \*

The saturation current density is given by

$$J_0 = J_n^0 + J_p^0,$$

where $J_p^0$ can be shown to be

$$J_p^0 = \alpha(1+\alpha)^{-2} \frac{\sigma_i^2}{\sigma_n L_p} \frac{k_B T}{e}.$$

$J_n^0$ is given by a similar expression. Here

$\alpha = \mu_n/\mu_p$

$\sigma_i = $ conductivity of intrinsic material

$\sigma_n = $ conductivity of $n$-type material

$L_p = $ diffusion length of holes in $n$-type material.

A derivation of this expression is given in W. Shockley, *Electrons and Holes in Semiconductors*, Van Nostrand Reinhold Company, New York (1950). One important parameter which effects $J_p^0$ is $L_p$, the minority carrier diffusion distance. As noted earlier this can be increased by increasing the minority carrier lifetime, which can be achieved by improving the purity and the perfection of the crystal.

RECTIFICATION. As is clear from Figure 13.8.4, if a voltage $V = 10 \cos \omega t$ is applied across the $p$-$n$ junction, current will readily flow during the positive portion of the cycle and will essentially not flow during the negative portion of the cycle. The input has been **rectified** so that the output is a dc voltage with a ripple. A typical low-cost battery charger has such a ripple. Other electrical devices can be used to smooth out the ripple.

In addition to the $p$-$n$ junction rectifier, there are other important rectifiers. Metal-semiconductor rectifiers, of which the most common are the copper oxide and the selenium rectifier, are important devices in low-frequency power engineering (high amperages). They also depend on junction effects for rectification.

BREAKDOWN. At high reverse voltages a **breakdown** in the $p$-$n$ junction rectifier occurs and large currents begin to flow. Current increases of several orders of magnitude then occur at nearly constant voltage. Recall that carriers in semiconductors have a large mobility and hence a large mean free path. Under an electrical field **E** such carriers can achieve between collisions an energy $eE\lambda$, which is correspondingly large, i.e., of the order of 1 eV. Such high energy carriers are capable of producing by collision electron-hole pairs which in turn are accelerated and produce additional carriers. This is known as the **avalanche effect** (an avalanche on a mountainside is often started by one region of snow falling and knocking other regions loose). When the semiconductor is flooded with carriers, high currents can pass. The current flow then rapidly increases by several orders of magnitude with only small voltage changes (see Figure 13.8.4). This suggests a constant voltage control device which is called the Zener diode.

SOLAR CELLS. Figure 13.8.5 illustrates a solar cell. In this device, solar energy is used to create electron-hole pairs. The additional *minority*

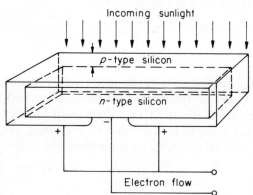

FIGURE 13.8.5. Schematic of a solar cell.

carriers on each side of the junction diffuse across it in opposite directions. Since these are also oppositely charged the electrical current is additive. There is a critical length (thickness of the $p$ region) involved in the design of the solar batteries. Only those minority carriers created by radiation within a diffusion length of the junction will contribute to the current; the others will recombine and hence decrease the efficiency of the device. This is the reason for the thinness of the $p$-type layer. Using the absorption spectrum of the silicon, we can show that most of the solar radiation will be absorbed in a distance of about $10^{-4}$ cm from the surface. Note that the current in the $p$ region is parallel to the junction, and hence if this region is made too thin, its resistance and hence joule heating losses will increase and the efficiency of the device will decrease. For a discussion of solar cells see the article by J. F. Elliott, "Photovoltaic Energy Conversion," in the book edited by George W. Sutton, *Direct Energy Conversion*, McGraw-Hill Book Company, Inc., New York (1966).

The theoretical efficiency of a silicon solar cell is 20% and operating cells with efficiencies of 15% have already been made. A cell such as shown in Figure 13.8.5 might have an operating voltage of 0.4 V and a power output of 12 mW/cm². The reader should calculate the number of square miles of solar cells in sunny Arizona which would be needed to produce the annual United States electrical energy output of 1 trillion kWh.

THERMOELECTRIC EFFECTS. Thermoelectric heating and cooling can be explained in terms of the energy diagram in Figure 13.8.1. The electrons which move to the left over the barrier are high kinetic energy electrons (hot electrons). In the $p$-type material, these electrons have high potential energy but low kinetic energy (cold electrons). Thus hot electrons have been removed from the $n$-type material and cold electrons introduced into the $p$-type material: The temperature in both regions near the junction and hence at the junction is lowered. If electrons were to flow in the opposite direction, the junction temperature would be raised. (The flow of holes would result in temperature changes of the same sign as the electron flow.) Thus a device as shown in Figure 4.6.2 would be heated at one junction and cooled at the other. The Peltier coefficient in semiconductors is about 100 times larger than in metals.

## Q. 13.8.2

One of the most important device features of a $p$-$n$ junction is that it can (a)_____. At high reverse biases, a $p$-$n$ junction undergoes (b)_____. This is the basis for a voltage control device known as a (c)_____. The current generated in a solar cell is the result of (d)_____. Most of the solar energy is absorbed in a layer within a distance of (e)____ cm from the surface.

\* \* \*

## ANSWERS TO QUESTIONS

13.8.1 (a) Same, (b) higher, (c) $e^{-eV_0/k_BT}$, (d) $e^{-(eV_0-eV)/k_BT}$, (e) $j_n(e^{eV/k_BT} - 1)$, (f) $J_0(e^{eV/k_BT} - 1)$.

**13.8.2** (a) Rectify alternating currents, (b) breakdown, (c) Zener diode, (d) the production of minority carriers by radiation on both sides of the junction, (e) $10^{-4}$.

## 13.9 THE JUNCTION TRANSISTOR

In 1947 Bardeen and Brattain invented the point contact transistor. This was rapidly followed by the invention of the $n$-$p$-$n$ junction transistor by Shockley in 1948. This was an invention based on fairly sophisticated scientific knowledge. Shockley had synthesized from the knowledge base built up by Bloch, Kronig and Penney, Wilson, Frenkel, Schottky, Lark-Horowitz, and Bardeen and Brattain and in the process had created a device for amplifying power which was to cause a *revolution* in the communications and information transfer and processing industry. This revolution will profoundly affect all of us. It has just begun. Between 1948 and 1970 the increase in density of electronic circuits was a factor of 100,000. This involves the use of **large-scale integration,** LSI, i.e., building into and on a single crystal disc of silicon various $p$-$n$ junctions, $n$-$p$-$n$ junctions, electrical conductors, etc., with up to $10^5$ components in.$^2$ Such systems were discussed in Section 1.1.

An $n$-$p$-$n$ junction transistor is shown schematically in Figure 13.9.1. In the absence of the applied potentials (due to two power supplies and the

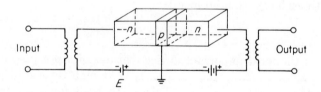

FIGURE 13.9.1. $n$-$p$-$n$ junction transistor.

input voltage) the energy levels appear as in Figure 13.9.2(a). This is exactly what we would expect from placing two $p$-$n$ junctions back to back. We now connect the power supplies and an input voltage. In Figure 13.9.2(b), the potential curve is shown for the forward-biased left junction and reverse-biased right junctions of Figure 13.9.1. Since the left junction is now forward-biased a large number of electrons are injected from the left into the $p$ region (**electron injection**). The $n$ region at the left is called the **emitter**. The $p$ region is very thin so that electrons are not destroyed by recombination [this means the thickness is less than the diffusion distance for the minority carriers (the electrons) in the $p$ region].

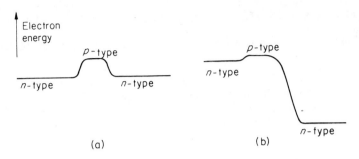

FIGURE 13.9.2. Electron energy versus distance in $n$-$p$-$n$ junction. (a) Unbiased. (b) Biased as an amplifier as in Figure 13.9.1.

The $p$ region, called the **base**, is therefore flooded with electrons (compared to the usual electron concentration there which is quite small). This flow of electrons into the $p$ region from the $n$ region on the left depends very strongly (exponentially) on the voltage of the emitter battery, $V_E$, plus the oscillating input voltage, $V_i$,

$$j \propto e^{e(V_E+V_i)/k_BT} = e^{eV_E/k_BT} e^{eV_i/k_BT}. \tag{13.9.1}$$

Because of the thinness of the $p$ region, the injected electrons do not recombine there but can pass into the $n$ region on the right, which is called the **collector** (think of these electrons as coasting down the potential barrier). The donor concentration (in the $n$ region) is made much higher than the acceptor concentration in the base. Therefore, almost all the current is due to electrons, and the holes make very little contribution.

The current flow in the collector loop is nearly the same as in the emitter loop. The input impedance $Z_i$ (impedance is the effective resistance in a circuit containing not only resistors but capacitors and inductors) is high compared to the impedance in the emitter. Hence

$$V_i = Z_i i_i. \tag{13.9.2}$$

Similarly, for the output,

$$V_0 = Z_0 i_0. \tag{13.9.3}$$

Since

$$i_i \doteq i_0 \tag{13.9.4}$$

(there is virtually no current flow through the lead connected to the base) the voltage gain is

$$\frac{V_0}{V_i} \doteq \frac{Z_0}{Z_i}. \tag{13.9.5}$$

Gains of the order of $10^5$ can be achieved.

610  *Electrons in Condensed Phases*

**EXAMPLE 13.9.1**

What sweeps the excess electrons from the $p$ region through the collector and the output impedance?

*Answer.* The voltage of the collector battery, $V_C$. The power output relative to the power input is, of course, similarly amplified. The energy for this amplification comes from the collector battery.


**EXAMPLE 13.9.2**

Suppose the base of the $n$-$p$-$n$ junctions were 0.1 in. thick instead of very thin. What would happen?

*Answer.* The electrons injected into the $p$ region from the emitter would flow around the emitter circuit only. The collector circuit would not "know about" the existence of this current.

The **injection efficiency** is the ratio of $i_0/i_i$ and can be of the order of 0.99 for a good design or nearly zero for a poor design as in Example 13.9.2.

## Q. 13.9.1

For an $n$-$p$-$n$ junction to act as an amplifier, it is necessary to have, in addition to the input potential, (a)_____. As a result of the forward bias on the $n$-$p$ junction on the left in Figure 13.9.1, electrons (b)_____. If the $p$ region were very thick, the electrons which are (c)_____ there would either recombine in the $p$ region or (d)_____.
If the $p$ region is very thin, the electrons diffuse into the $n$ region where they are swept through the collector and output impedance by the (e)_____.

\* \* \*

# ANSWERS TO QUESTIONS

**13.9.1** (a) Two voltage supplies, (b) flow into the $p$ region, (c) minority carriers, (d) flow around the circuit on the left in Figure 13.9.1, (e) collector power supply.

## 13.10 LASERS

The emission of radiation from an excited atom as discussed in Chapter 5 is an example of **spontaneous emission,** i.e., independent of the incident radiation. There is also the possibility of stimulated emission. Absorption is clearly stimulated. The probability of such absorption taking place in unit time is $\rho_\nu B_{12}$, where $B_{12}$ is the probability of transition from a state of energy $E_1$ to a state of energy $E_2$, where $E_2 > E_1$ and $\rho_\nu$ is the energy density of the incident radiation of frequency $\nu$, where $h\nu = E_2 - E_1$.

There is also the reverse stimulated process $\rho_\nu B_{21}$ called **stimulated emission.** Here, the emitted radiation is *phase coherent* with the incident radiation. Einstein first showed that $B_{12} = B_{21}$. This says that the probability of external radiation inducing an upward transition from an atom in state 1 to 2 is exactly equal to the probability of the same radiation inducing a downward transition from an atom in state 2 to 1. The calculation of $B_{12}$ is under certain conditions a straightforward, although a tedious, quantum mechanical problem [H. Eyring, J. Walter, and G. E. Kimball, *Quantum Chemistry,* John Wiley & Sons, Inc., New York (1949), p. 108].

Einstein showed that the probability coefficient for *spontaneous* emission from a system in state 2 to 1 could be obtained from a thermodynamical argument as follows. If the number of systems in the state with energy $E_2$ is $N_2$ and the number in the state with energy $E_1$ is $N_1$, then, according to the Boltzmann distribution, for equilibrium

$$\frac{N_2}{N_1} = \frac{e^{-E_2/k_B T}}{e^{-E_1/k_B T}} = e^{-(E_2 - E_1)/k_B T}. \tag{13.10.1}$$

Similarly for equilibrium, the number of upward transitions must equal the number of downward transitions or

$$N_1 \rho_\nu B_{12} = N_2 \rho_\nu B_{21} + N_2 A_{21}, \tag{13.10.2}$$

where $A_{21}$ represents the probability of spontaneous emission occurring in unit time. $A_{21}$ can be obtained from these two equations.

### EXAMPLE 13.10.1

Would photoluminescent radiation be phase coherent or incoherent?

*Answer.* This represents spontaneous emission from excited atoms. The emission from one atom is independent from another so the emission from different atoms takes place at different times. The total radiation is phase incoherent.

INVERTED POPULATIONS

**EXAMPLE 13.10.2**

Consider the case where $E_2 > E_1$ and $E_2 - E_1 = 2.3$ eV. The ground state of the atom (or ion) is $E_1$. There are no other nearby energy levels except for $E_2$. What is the fraction of atoms in the excited state at room temperature?

*Answer.* Recall that $k_B T \approx \frac{1}{40}$ eV at room temperature. From (13.10.1) we have

$$\frac{N_2}{N_1} = e^{-2.3 \times 40} = 10^{-40}.$$

From the previous problem it is clear that the number of excited states in thermal equilibrium at room temperature in a system capable of emitting visible radiation is very small. To get any net emission it is necessary to increase the number of electrons in excited states above the equilibrium concentration. To get high-intensity emission it is necessary to have a large number of electrons in the excited state (leaving few in the ground state). Such a condition is called a **population inversion.**

LASERS. As a result of work by J. Weber (1952) and C. H. Townes (1955), the **laser** was developed. Laser is an acronym for *light amplification by stimulated emission of radiation.* The radiation which *lasers* emit is phase coherent, it has high intensity, and it has an extremely sharp emission line (small frequency spread) and narrow beam divergence. (Radio waves emitted from an antenna are also phase coherent but with a much different wavelength.) These four unique characteristics of the laser in the range of visible and near-visible radiation lead to a number of interesting applications. Because of the phase coherence of the radiation they emit, lasers can be used to carry signals by amplitude modulation; the possibilities for communications are enormous. A laser source is an ideal source for interferometry measurements, spectroscopy, holography, high-speed photography, etc. Because of the narrow beam, it can provide localized heating for surgical purposes. Because of the high intensity (particularly of the $CO_2$ laser) and the high focusing, lasers can be used in such materials processes as drilling of holes in diamonds to be used for wire drawing, cutting of inch thick titanium plate, welding of different metals, etc.

There are four major types of lasers: the chemical laser, the gas-discharge laser, the semiconductor diode laser, and the homogeneous optically pumped liquid or solid state laser. We shall briefly consider only one of these, the ruby laser. A crystal of corundum, $Al_2O_3$, is doped with 0.05%

Cr. The chromium is present as a trivalent substitutional ion, $Cr^{3+}$, for $Al^{3+}$. Pure corundum is an insulator which is transparent and colorless. The $Cr^{3+}$ ion is a color center which gives the material its characteristic ruby color (by providing an absorption band). Figure 13.10.1 is a simplified picture of the energy level diagram.

The ion can be excited by radiation of about 4100 Å wavelength (blue-violet). Such light can readily be provided by a fluorescent xenon tube. The electron then falls from the excited state of the absorption band to a metastable state by a nonradiative process; instead of a photon being

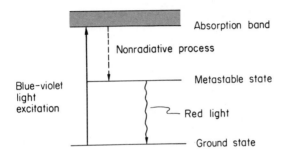

FIGURE 13.10.1. Simplified picture of energy states of the ruby laser.

created, a phonon is created, i.e., a lattice vibration is excited to a higher energy (increasing the temperature of the ruby rod). This nonradiative process is extremely rapid. The metastable state has, relatively speaking, a very long lifetime (for spontaneous decay) of about $5 \times 10^{-3}$ sec.

**EXAMPLE 13.10.3**

How far can light travel in a medium of refractive index 1.75 in $5 \times 10^{-3}$ sec? The speed is $c/1.75$.

*Answer.* The speed of light in a vacuum is $3 \times 10^8$ m/sec and in the above medium, which could be $Al_2O_3$, $3 \times 10^8/1.75 = 1.71 \times 10^8$ m/sec.

Hence the distance is $8.55 \times 10^5$ m. If light is traveling back and forth in a crystal 8.55 cm long, it could make $5 \times 10^6$ round trips through the crystal after a specific electron was excited before that electron would spontaneously decay to the ground state.

Because of the long lifetime of the metastable state, the ions can essentially all be "pumped" into the metastable state quickly by a xenon flashtube. Hence an inverted population is created. If now a single emission occurs and this in turn induces (stimulates) a second emission, the latter will be phase coherent with the former. If these in turn stimulate other emissions, a situation can rapidly be reached wherein we have a high-density coherent radiation supply in the crystal. A schematic of the system is shown in Figure 13.10.2.

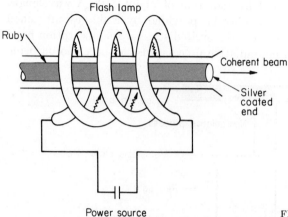

FIGURE 13.10.2. Ruby laser.

The coherent radiation is reflected back and forth from the silver mirrors. Finally there is essentially a population explosion of emission and a high-intensity pulse of short duration then is emitted from the lightly silvered face.

## Q. 13.10.1

Under equilibrium conditions, the concentration of electrons in an excited state 2 divided by the concentration in the ground state 1 is (a)_____. The operation of the laser depends on first attaining (b)_____. After an inverted population is formed with a long lifetime for spontaneous decay, a spontaneous emission occurs which can then cause (c)_____. Stimulated emission is (d)_____ with the stimulating radiation. This radiation can in turn stimulate additional emission which is phase coherent.

\* \* \*

## ANSWERS TO QUESTIONS

**13.10.1** (a) $N_2/N_1 = e^{-(E_2-E_1)/k_BT}$, (b) an inverted population, (c) stimulated emission, (d) phase coherent.

# REFERENCES

Heath, F. G., "Large Scale Integration in Electronics," *Scientific American* (February 1970), p. 22. This article discusses the great strides in miniaturization made in one decade.

Henisch, H. K., "Amorphous Semiconductor Switching," *Scientific American* (November 1969), p. 30. This article discusses a type of switching based on what appears to be very rapid phase transformations.

Sproull, R. L., *Modern Physics*, John Wiley & Sons, Inc., New York (1963). This book has an excellent presentation of the subject of electrons in solids.

Shockley, W., *Electrons and Holes in Semiconductors*, Van Nostrand Reinhold Company, New York (1950). This is a classic on the subject published only two years after the invention of the $n$-$p$-$n$ junction transistor.

Holmes, R. A., *Physical Principles of Solid State Devices*, Holt, Rinehart and Winston, Inc., New York (1970). An upper-class book concerned entirely with electrons in solids.

Kittel, C., *Introduction to Solid State Physics*, John Wiley & Sons, Inc., New York (1968), Chapters 7–9. This has an excellent discussion of electrons in solids for the instructor.

Weber, S., Ed., *Optoelectronic Devices and Circuits*, McGraw-Hill Book Company, Inc., New York (1964). This is a superb collection of articles by different authors which bridges the gap between pure theory and application.

# PROBLEMS

**13.1** List at least five "why" questions concerned with electrons in solids.

**13.2** (a) Discuss the essential features of Drude's theory of conductivity of metals.
(b) What are the shortcomings of Drude's theory?

**13.3** Discuss in an approximate way how Sommerfeld explained the electronic specific heat.

**13.4** What is the origin of energy gaps?

**13.5** (a) How many possible energy states per atom are there for the valence band of a simple one-dimensional crystal?
(b) A bcc crystal?
(c) An hcp crystal?
(d) A diamond cubic crystal?

**13.6** Give typical values for the electrical conductivity coefficient of metals, semiconductors, and insulators.

**13.7** Explain the origin of the essential differences among the conductivities of copper, diamond, and doped silicon at 20°C.

**13.8** Where does the Fermi energy lie for
(a) An intrinsic semiconductor?
(b) A doped $p$-type semiconductor?
(c) A doped $n$-type semiconductor?

**13.9** Of what use are Hall effect studies in the science of semiconductors?

**13.10** If the silicon described in Table 3.2.1 contains only one pentavalent impurity, what is the atom fraction of this impurity?

**13.11** From the viewpoint of building a temperature-sensing device, why would a thermistor be preferred to a metal resistor?

**13.12** Sketch $\ln N_n$ (or $\ln N_p$) vs. $1/T$ from the melting point to very low $T$ for a germanium crystal which becomes intrinsic at 300°K.

**13.13** Could the energy gap of a semiconductor be measured by absorption of radiation? If so, describe in detail how you would do this. *Hint:* If radiation can excite an electron from an occupied state to an allowed level, it will be absorbed. If not, it will not be absorbed.

**13.14** Metals are not transparent to visible radiation. Explain why. Photons in the visible range have an energy of 1.7–3.2 eV.

**13.15** A commercial microscope is designed for transmission of infrared radiation of energy 1.0 eV through silicon. Explain why silicon does not absorb this radiation.

**13.16** The "radius" of the ground state electron in hydrogen is given by (5.1.14). What is the radius for the ground state electron bound to the phosphorus ion, $P^+$, in germanium?

**13.17** A semiconductor of unknown purity shows a Hall coefficient of zero at room temperature. How can this be?

**13.18** (a) Discuss why a solar cell should have a large energy gap.
(b) Why should this energy gap be restricted, however, to the range 1–3 eV?

**13.19** Estimate the power output of a square mile of solar cells on a bright (typical) Arizona day. How many square miles would be needed to produce the annual power output in the United States of 1 trillion kWh?

**13.20** Discuss the critical design features of a $n$-$p$-$n$ junction transistor.

**13.21** If you were told to learn three fairly simple equations in semiconductor physics which you could use to form the basis for further discussion, list the equations which you would choose and why.

# MORE INVOLVED PROBLEMS

**13.22** A CdS photodetector receives radiation of 4000 Å wavelength over an area of $2 \times 10^{-6}$ m² and with an intensity of 40 W/m². The energy gap is 2.4 eV.
(a) Calculate the number of electron-hole pairs generated per second if each quantum generates a pair.
(b) Calculate the increase in conductivity if the electron lifetime is $10^{-3}$ sec and $\mu_n = 10^{-2}$ m²/V-sec.

**13.23** In an *n-p-n* junction, the electrons injected into the base move across by diffusion. If an alternating high-frequency signal is applied, many of these electrons will not have time to reach the collector. Show that there is a cutoff frequency which varies inversely with the square of the thickness of the base.

**13.24** Discuss quantitatively the analogy between extrinsic semiconductors and extrinsic conduction in ionic crystals such as NaCl doped with $CdCl_2$.

**13.25** For some devices it is desirable to have the semiconductor intrinsic, while for other devices this would be disastrous. Discuss.

**13.26** The junction transistor is not a thermodynamically stable device. Its behavior deteriorates reversibly at moderate operating temperatures while at still higher temperatures the deterioration is irreversible.
(a) Explain each of these phenomena.
(b) Give an energy which is associated with each.

**13.27** Impurities in silicon such as Fe, Co, and Mn have large ionization potentials (roughly $E_g/2$ in silicon). Of what possible use are such impurities?

**13.28** Selenium is a high **dark-resistance material** with a resistivity greater than $10^{12}$ Ω-m (see Table 3.2.1 for comparison). Illumination profoundly lowers the resistivity: This is the basis of **xerography**. Discuss xerography. See, e.g., R. M. Schaffert and C. D. Oughton, *Journal of the Optical Society of America* **38**, 991 (1948).

**13.29** Check a chemistry book such as A. F. Wells, *Structural Inorganic Chemistry*, Oxford University Press, Inc., New York (1962) and study the types of compounds formed by silicon. Based on your studies invent a way to obtain ultrapure silicon.

# SOPHISTICATED PROBLEMS

**13.30** Discuss the manufacture of the selenium rectifier. Use outside references. A starter is T. S. Hutchison and David C. Baird, *The Physics of Engineering Solids*, John Wiley & Sons, Inc., New York (1963), p. 276.

**13.31** Discuss the basis of operation of the field effect transistor.

**13.32** Charge carriers in a semiconductor in a magnetic field tend to move in paths having circular projection in the plane perpendicular to the magnetic field. Show that the angular frequency $\omega$ (**cyclotron frequency**) is given by

$$\omega = \frac{eB}{m^*}.$$

**13.33** Describe how recombination lifetimes are measured by the light probe method. See W. C. Dunlap, Jr., *An Introduction to Semiconductors*, John Wiley & Sons, Inc., New York (1957).

**13.34** Derive the general expression for the Hall coefficient.

**13.35** Show how the Hall effect can be used as the basis for designing a wattmeter.

**13.36** The **Esaki** or **tunnel diode** involves tunneling in heavily doped semiconductors. Describe its operation. See, e.g., C. A. Wert and R. M. Thomson, *Physics of Solids*, McGraw-Hill Book Company, Inc., New York (1964), p. 263.

**13.37** Using experimentally available values of the absorption coefficient versus wavelength for GaAs [see G. W. Sutton, *Direct Energy Conversion*, McGraw-Hill Book Company, Inc., New York (1966), p. 11], explain why a photon from solar radiation is either absorbed in a layer of $10^{-4}$–$10^{-6}$ cm from the surface or not absorbed at all. The **absorption coefficient,** $\alpha$, is defined by

$$\frac{dI}{dL} = -\alpha I,$$

where $I$ is the intensity and $L$ is the distance of penetration.

**13.38** (a) Show that the thermal conductivity of a gas of density $\rho$ is

$$k_T = \tfrac{1}{3}\rho C_v \bar{v} \lambda,$$

where $C_v$ is the specific heat per atom. *Hint:* The procedure to use is similar to that used in Section 9.2 to derive the viscosity coefficient, except that energy transfer is considered here.

(b) Show using the classical expressions for conductivity by an electron gas that the ratio

$$\frac{k_T}{\sigma T} = L \quad \text{(the Lorentz constant)},$$

where $\sigma$ is the electrical conductivity. Use the quantum mechanical expression for the specific heat.

(c) Show that the Lorentz number is

$$L = \frac{\pi^2}{3}\left(\frac{k_B}{e}\right)^2.$$

Compare with the values for metals in Tables 3.2.1 and 4.2.1. We have assumed that the mean free path for the thermal and electrical behavior are the same; i.e., all types of collisions affect the two equally. Advanced analysis shows that this is nearly true at high temperatures but not at low temperatures.

# Prologue

All materials have a *diamagnetic* contribution to their magnetic behavior, the origin of which is discussed briefly. Many materials exhibit paramagnetism which is due to the presence of *permanent magnetic moments* associated with atoms. In some materials (primarily in the transition and rare earth metals) there is a strong *cooperative* interaction between the magnetic dipoles. This leads to *spontaneous alignment* and the formation of *domains* (regions in which the dipoles are similarly aligned). These materials are *ferromagnetic*. In *ferrimagnetic* materials, a fixed fraction of dipoles is aligned in one direction; the remainder are antiparallel, but there is a net magnetic dipole moment in one direction. A number of oxides, in particular the *ferrites*, exhibit this behavior. Domains are also present in these materials. They are strongly magnetic, but have high electrical resistance. Hence they exhibit small eddy losses. There is an upper limit on the size of a domain due to energetic considerations. Tiny elongated particles about 1000 Å in diameter exist as *elongated single domains*. These can be used to make hard (permanent) magnets. However, a large block of a single-phase material, such as iron, consists of many domains. The reasons for this are studied in this chapter. A ferromagnetic material is soft if the domain walls can move easily and is hard if the domain walls are pinned. Mechanisms for pinning domains are discussed. Methods for producing very hard magnets (oriented elongated single domains), by both mechanical and metallurgical means, are discussed.

# 14

# MAGNETISM

## 14.1 DIAMAGNETISM

Magnetic behavior was described in Sections 3.6–3.9 and the origin of magnetic dipoles was described in Section 5.9. It is assumed in this chapter that the reader is familiar with that material; a quick review by the reader may be needed. Let us now consider the origin of diamagnetism (induced magnetization).

In Section 5.9 we noted that the magnetic dipole moment $\mathbf{p}_m$ of the atom is, in terms of the angular momentum of the electrons, $\mathbf{L}$,

$$\mathbf{p}_m = -\frac{e}{2m}\mathbf{L}. \qquad (14.1.1)$$

An inert gas atom such as helium, in the ground state, will have zero orbital momentum as well as a zero spin momentum (since the two electrons will have opposite spin). Such an atom therefore has no permanent magnetic dipole moment. However, it can be shown that when such atoms are placed in a magnetic field, they will on the average have a net angular momentum which is directly proportional to the applied field. The exact proof of this involves a rather sophisticated application of quantum mechanics and is not given here. Instead a simple classical mechanics model first considered by Langevin in 1905 is described.

Let us now consider the motion of an electron around a positive ion core in the presence of a field of magnetic induction $\mathbf{B}$ (whose direction is normal to the plane of the Bohr orbit). The inward force on the electron for which the magnitude of the velocity is $v$ is $evB$ due to the magnetic field and $e^2/4\pi\epsilon_0 r^2$ due to Coulomb attraction,

$$F_{\text{inward}} = evB + \frac{e^2}{4\pi\epsilon_0 r^2}.$$

The inward acceleration of a particle moving in a circular orbit with constant angular velocity $\omega$ is $v^2/r$ (where $v = \omega r$). Hence by Newton's second law,

$$evB + \frac{e^2}{4\pi\epsilon_0 r^2} = \frac{mv^2}{r}.$$

We want to calculate the *small change* in the angular momentum, $mvr$, due to the presence of the field. As $B$ is increased slowly the velocity $v$ increases, but the orbit stays the same. We next calculate $v$ from the previous equation, obtaining

$$v = \frac{eBr/m + (e/\sqrt{\pi\epsilon_0 rm})\sqrt{1 + (eBr/m)^2(\pi\epsilon_0 rm/e^2)}}{2}.$$

Note that for $B = 0$, we have

$$v_0 = \frac{e}{\sqrt{4\pi\epsilon_0 rm}}.$$

We are interested in $\Delta v = v - v_0$. For static values of $B$ ranging from zero to the highest values currently obtainable in the laboratory, the expression for $\Delta v$ can be approximated to a high degree of accuracy (a tiny fraction of 1%) by

$$\Delta v \doteq \frac{erB}{2m}.$$

(To prove this, expand $v$ in a Taylor series about $x = 0$ and consider the effect of dropping quadratic and higher order terms in $B$.) Therefore, the change in angular momentum induced by the field is

$$m\,\Delta vr = \frac{er^2 B}{2}.$$

Hence, the induced magnetic dipole moment is, by (14.1.1),

$$p_m = -\frac{e^2 r^2}{4m} B \doteq -\frac{e^2 r^2}{4m} \mu_0 H,$$

where we have assumed that in the expression $B = \mu_0(H + M)$ [see (3.6.11)], $M \ll H$.

The magnetization, **M**, of a material is the *net* dipole moment (the vectorial sum) per unit volume. If there are $N$ atoms per unit volume, each with $Z$ electrons,

$$M = Np_m = -\frac{NZe^2 \overline{r^2}}{4m} \mu_0 H = \chi_m H, \qquad (14.1.2)$$

where $\overline{r^2}$ is the mean value of $r^2$ and $\chi_m$ is the magnetic susceptibility (previously defined in Chapter 3). Note that $\chi_m$ is negative for induced magnetism (diamagnetism); this means that the flux lines are slightly repelled by the material.

While the classical model leads to a reasonable estimate of $\chi_m$ for diamagnetic solids, it should be emphasized that a direct calculation of $\chi_m$ is a quantum mechanical problem. For a spherical electron distribution in an atom or ion with $Z$ electrons and for $N$ atoms per unit volume, the result is

$$\chi_m = \frac{-NZe^2}{6m} \mu_0 \overline{r^2}, \qquad (14.1.3)$$

where $\overline{r^2}$ is the mean squared distance of the electrons from the nucleus. This can be evaluated from a knowledge of the wave function $\psi$ of the electrons. In fact

$$\overline{r^2} = \int_0^\infty \psi r^2 \psi 4\pi r^2 \, dr.$$

Note that the diamagnetic susceptibility is essentially *independent* of temperature. Table 14.1.1 shows some experimental values of the magnetic susceptibilities.

Table 14.1.1. MAGNETIC SUSCEPTIBILITIES

| Material | $\chi_m$ (dimensionless) |
|---|---|
| Cu | $-0.7 \times 10^{-6}$ |
| Au | $-2.8 \times 10^{-6}$ |
| He (STP)* | $-0.8 \times 10^{-10}$ |
| Ne (STP) | $-3.0 \times 10^{-10}$ |
| Ar (STP) | $-9.0 \times 10^{-10}$ |
| Kr (STP) | $-13.0 \times 10^{-10}$ |
| Xe (STP) | $-19.6 \times 10^{-10}$ |

* STP means at 0°C and 1 atm.

All materials have a diamagnetic component of susceptibility. However, $\chi_m$ will be positive for many of these materials because the paramagnetic or ferromagnetic component is much larger, as we shall see in the following sections.

### Q. 14.1.1

All materials have (a)_____ component of susceptibility. The magnetic susceptibility of this component for solids is about (b)_____.

The quantum mechanical expression for the diamagnetic susceptibility of a gas of $N$ atoms per unit volume and whose atoms have an atomic number $Z$ is equal to (c)_____.

\* \* \*

## ANSWERS TO QUESTIONS

**14.1.1** (a) A diamagnetic, (b) $-10^{-6}$, (c) $-(NZe^2\overline{r^2}\mu_0/6m)$.

## 14.2 PARAMAGNETISM

As noted in the previous section, the phenomenon of diamagnetism is a result of magnetic dipole formation due to the presence of an external magnetic field. However, paramagnetism is a result of the orientation of (already existing) permanent magnetic dipoles due to the presence of an external magnetic field. We shall study paramagnetism because of the insight it provides into ferromagnetism and ferrimagnetism.

It was noted in Section 5.9 that the permanent magnetic dipole of an atom or ion may be due to the orbital angular momentum, the spin angular momentum, or a combination of these. In this section we consider atoms or ions with permanent dipole moments. In the absence of a magnetic field these dipoles are randomly oriented because of thermal energy so the net magnetic dipole moment (the vector sum of all the dipoles) per unit volume, i.e., the magnetization, is zero.

A field **B** exerts a torque on a dipole whose moment is $p_m$ given by

$$\mathbf{T} = \mathbf{p}_m \times \mathbf{B}$$

or

$$T = p_m B \sin \theta,$$

where $\theta$ is the angle between the dipole and the field.

The potential energy of the dipole in this field is

$$\text{P.E.} = -p_m B \cos \theta. \tag{14.2.1}$$

The potential energy is least if the dipole is parallel to the field, $\theta = 0$, and the highest if they are antiparallel.

Because of thermal agitation, the dipoles in the presence of a field are not all at minimum potential energy, i.e., aligned with $\theta = 0$. Rather, at the largest static fields generated in the laboratory and at room temperature or higher the distribution is nearly (not quite) random. However, there is a slight bias in favor of the dipoles being parallel to the field. Hence there is a net magnetization in the direction of the field. This is paramagnetism.

**EXAMPLE 14.2.1**

Approximately how large must the magnetic induction be for the orientation energy to be comparable to the thermal energy at room temperature?

*Answer.* The maximum change in energy due to orientation is, from (14.2.1), $p_m B$. We must have

$$p_m B \approx k_B T.$$

Assuming $p_m = 5\beta$, where $\beta$ is the Bohr magneton (discussed previously in Section 5.9), we have (using MKS units)

$$B = \frac{k_B T}{5\beta} = \frac{1.38 \times 10^{-23} \times 300}{5 \times 9.27 \times 10^{-24}} \approx 10^2 \text{ Wb/m}^2.$$

This is an enormous field. The largest static magnetic fields available in the laboratory are only about 20 Wb/m², so that $p_m B \ll k_B T$. Thus the typical magnetic field only slightly perturbs the random distribution.

**Q. 14.2.1**

Paramagnetism is due to the (a)_____
_____. The potential energy of a dipole $p_m$ in a field **B** is (b)_____. For the magnetic fields ordinarily obtainable in research laboratories, the quantity $p_m B/k_B T$ is (c)_____.

\* \* \*

In the absence of a field the number of dipoles in a unit volume of given material between $\theta$ and $\theta + d\theta$ (see Figure 14.2.1) is

$$dN = c\, dA = c(2\pi \sin \theta)\, d\theta, \qquad (14.2.2)$$

where the proportionality constant $c$ can be evaluated by noting that integration over all $\theta$ gives the total number of dipoles, $N$, in a unit volume of the material. Figure 14.2.1 shows the octant of a sphere. Here one fourth of the area element $dA$ (on the surface of the sphere) is shown by the

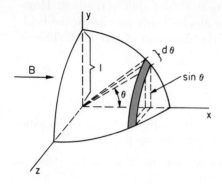

FIGURE 14.2.1. Octant of a unit (radius) sphere.

hatched element. In the presence of the magnetic field the fraction $dN$ is weighted by the Boltzmann factor so that

$$dN = c\, 2\pi \sin\theta e^{-\text{P.E.}(\theta)/k_B T}\, d\theta, \tag{14.2.3}$$

where P.E.$(\theta)$ is given by (14.2.1). Again integration over all $\theta$ gives the total number of dipoles.

**EXAMPLE 14.2.2**

Write down the expression for the average value of dipole moment (averaged over all orientations) in the direction of **B**.

*Answer.* The component of the dipole moment of a given dipole in the direction of **B** is $p_m \cos\theta$. Hence the average value of the dipole moment in the direction of the field **B** is

$$\bar{p}_m = \frac{\int (p_m \cos\theta)\, dN(\theta)}{\int dN(\theta)}. \tag{14.2.4}$$

Once we have $\bar{p}_m$, we have the magnetization since

$$M = N\bar{p}_m. \tag{14.2.5}$$

Combining (14.2.4) with (14.2.3) leads to the rather imposing relation

$$\bar{p}_m = \frac{\int_0^\pi p_m \cos\theta \sin\theta e^{p_m B \cos\theta / k_B T}\, d\theta}{\int_0^\pi \sin\theta e^{p_m B \cos\theta / k_B T}\, d\theta}, \tag{14.2.6}$$

first obtained by Langevin in 1905 who integrated it and obtained

$$\bar{p}_m = p_m \left[ \operatorname{ctnh} x - \frac{1}{x} \right] = p_m L(x), \qquad (14.2.7)$$

where, for convenience, we have defined

$$x \equiv \frac{p_m B}{k_B T}. \qquad (14.2.8)$$

The function in brackets in (14.2.7) is now called the **Langevin function**, $L(x)$. Students who are concerned with the details of performing the integration leading to (14.2.7) should see Problem 14.18. Physically the Langevin function is a fraction which equals (on the average) the component of a unit dipole moment in the direction of the applied field. It is expected that this fraction would increase as the field increases and would decrease as the temperature increases.

$L(x)$ has an initial slope of $\frac{1}{3}$ so that

$$L(x) \doteq \frac{x}{3}, \qquad x \ll 1, \qquad (14.2.9)$$

as shown in Figure 14.2.2. Recall from Example 14.2.1 that for practical

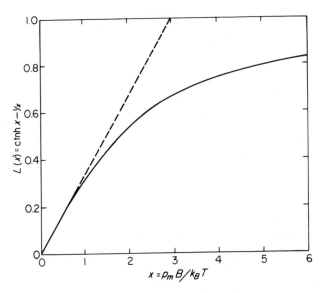

FIGURE 14.2.2. Langevin function.

experimental considerations $x \ll 1$ except at low temperatures. Hence for these conditions, it follows from (14.2.5) and (14.2.7)–(14.2.9) that

$$M = Np_m L(x) \doteq Np_m \frac{x}{3} = \frac{Np_m^2 B}{3k_B T}.$$

If we assume that $M \ll H$, then $B \doteq \mu_0 H$ and

$$M \doteq \frac{Np_m^2 \mu_0}{3k_B T} H = \chi_m H.$$

Hence the magnetic susceptibility of a paramagnetic substance is

$$\chi_m = \frac{Np_m^2 \mu_0}{3k_B T}. \tag{14.2.10}$$

Materials which exhibit this behavior are said to obey the **Curie law.** Note that $\chi_m$ is inversely proportional to the temperature. As $T \to 0$, $x \to \infty$ and $L(x) \to 1$, so $\bar{p}_m \to p_m$ or $M \to M_S$, where $M_S$ is the saturation magnetization. Then only a tiny magnetic field such as the earth's magnetic field would suffice to cause complete alignment, i.e., magnetization saturation.

### Q. 14.2.2

Let $\theta$ be the angle between a given dipole moment $\mathbf{p}_m$ and the field $\mathbf{B}$. Then the component of $\mathbf{p}_m$ in the direction of $\mathbf{B}$ is (a)_____. If $dN(\theta)$ represents the number of dipoles lying between $\theta$ and $\theta + d\theta$ (relative to the direction of $\mathbf{B}$), then the average value $\bar{p}_m$ in the direction of $\mathbf{B}$ is (b)_____. The magnetization $M$ is related to the number of dipoles per unit volume $N$ and $\bar{p}_m$ by (c)_____. When $x = p_m B/k_B T \ll 1$, $\bar{p}_m =$ (d)_____. Langevin's theory predicts that the paramagnetic susceptibility is related how to the temperature? (e)_____.

\* \* \*

The induced susceptibility (diamagnetism) is always present in all materials while the effect due to permanent dipoles (paramagnetism) is only sometimes present. Usually when paramagnetism is present it is an order of magnitude larger than the diamagnetism of the material and so the material has a positive magnetic susceptibility.

When one studies the experimental data of solids carefully as Weiss did in 1912, one finds approximately the **Curie-Weiss relation**

$$\chi_m = \frac{C}{T - T_C}, \qquad (14.2.11)$$

where $C$ is an experimental parameter and $T_C$ is an experimental parameter. This means that the magnetic dipoles will all be aligned when $T - T_C \to 0$ (rather than when $T \to 0$) upon application of a tiny magnetic field. Langevin's theory does not account for $T_C$. Weiss (1912) assumed that this deviation was due to the interaction of the dipoles. There is a magnetic induction field generated by a magnetic dipole of moment $\mathbf{p}_m$ whose magnitude along the direction of the magnetic dipole is

$$B = \frac{\mu_0 p_m}{2\pi r^3}. \qquad (14.2.12)$$

[See H. E. Duckworth, *Electricity and Magnetism*, Holt, Rinehart and Winston, Inc., New York (1961), p. 259.]

### EXAMPLE 14.2.3

Estimate the field at one dipole of magnitude $4\beta$ (the Bohr magneton) due to a similar adjacent parallel dipole in a solid.

*Answer.* We assume that $r \approx 2$ Å $= 2 \times 10^{-10}$ m. Then from (14.2.12)

$$B \approx \frac{4\pi \times 10^{-7} \times 4 \times 10^{-23}}{2\pi \times 10^{-29}} \approx 1 \text{ Wb/m}^2.$$

The interaction energy is given by (14.2.1) which for the case of Example 14.2.3 gives an interaction energy between two dipoles of about $4 \times 10^{-23}$ J or about $3 \times 10^{-4}$ eV. When all the interactions in the crystal are summed, the interaction energy per dipole (attractive binding energy) would still be of this order, say $10^{-3}$ eV.

### EXAMPLE 14.2.4

To what thermal energy $k_B T_C$ does $10^{-3}$ eV correspond?

*Answer.*

$$T_C = \frac{10^{-3}}{8.62 \times 10^{-5}} \approx 10°K.$$

The answer to Example 14.2.4 implies that at about 10°K or below the attractive energy between all the dipoles is such that they would align in the absence of a field, i.e., spontaneously. Note that if this attractive energy were very small, the dipoles would not become aligned except very near $T = 0$; $T = 0$ is, of course, unattainable. It is the presence of this binding energy which causes the appearance of the $T_C$ in (14.2.11). The quantity $C$ in this theory is $Np_m^2\mu_0/3k_B$; if $T_C \to 0$, the result is the same as in the Langevin theory.

OCCURRENCE OF PARAMAGNETISM. Paramagnetism is exhibited by

1. Atoms and molecules which have an odd number of electrons. *Example:* Alkali metal atoms in the gas state.
2. Atoms or ions with only partially filled shells. *Example:* $Mn^{2+}$, which has the configuration $Ar(3d)^5$. The shells of argon are full.
3. Certain compounds with an even number of electrons. *Example:* Oxygen molecule.
4. Metals.

PARAMAGNETISM DUE TO CONDUCTION ELECTRONS. The free electron gas in a metal such as sodium might be expected to have a susceptibility of the Curie form of (14.2.10), where $p_m = \beta$ (the Bohr magneton) and $N$ is the number of free electrons per unit volume. Recall, however, that this electron gas obeys Fermi-Dirac statistics (see Figure 13.2.2). Hence only those *unpaired* electrons near the top of the distribution can exhibit paramagnetism. [The situation is analogous to the electronic specific heat (see Section 13.2).] Roughly, only a fraction of about $T/T_F$, where $T_F$ is the Fermi temperature, are unpaired and can contribute to paramagnetism. Thus $N$ should be replaced by $NT/T_F$, resulting in a paramagnetic susceptibility *independent* of temperature.

## Q. 14.2.3

The paramagnetic susceptibility of solids is of the order of (a)_____ _____. According to Weiss's modification of the Curie law, the magnetic susceptibility varies how with temperature (b)_____. According to Weiss, $T_C$ owes its origin to the presence of (c)_____ _____. The interaction energy of one dipole with all the other dipoles in a crystal is about (d)_____ eV. The paramagnetic susceptibility of a free electron gas varies how with temperature? (e)_____ _____.

\* \* \*

A DIVERSION TO A DIELECTRIC ANALOGY. It should be noted that the diamagnetic susceptibility is independent of temperature while the paramagnetic susceptibility is, according to the Langevin theory, inversely proportional to the temperature. An analogy with electric dipoles is worthy of mention at this time. The dielectric susceptibility due to induced effects is independent of temperature (but always positive) while the dielectric susceptibility due to orientation of permanent electric dipoles is inversely proportional to the temperature. An approximate treatment of the induced dielectric susceptibility in molecules such as argon (which have no permanent electric dipoles) leads to

$$\chi_{el} = N 4\pi\epsilon_0 R^3, \quad (14.2.13)$$

where $R$ is the atom radius (see the previously cited Ruoff, *Materials Science*, Section 30.4). A treatment completely analogous to that discussed in this section leads to the expression for the orientation susceptibility:

$$\chi_{el} = \frac{N p_{el}^2}{3 k_B T} \quad (14.2.14)$$

(see Ruoff, *Materials Science*, Section 30.4). Recall from Chapter 3 that $\chi_{el}$ is related to the dielectric constant.

A measurement of the temperature dependence of the dielectric constant of a gas, such as HCl, would enable us to calculate the electric dipole of the molecule. It is from measurements of this type and from Equation (14.2.14) that we can deduce that the bond in HCl is partially ionic, i.e., that the hydrogen has a net charge of $e/6$ and the chlorine of $-e/6$, as was discussed in Section 5.8.

Values of dipole moments are shown in Table 14.2.1. The **debye** is a commonly used unit for dipole moment; it equals $3.33 \times 10^{-30}$ C-m.

Table 14.2.1. PERMANENT ELECTRIC DIPOLE MOMENTS OF SIMPLE MOLECULES

| Molecules | $10^{-30}$ C-m | debyes |
|---|---|---|
| HCl | 3.5 | 1.05 |
| CsCl | 35.0 | 10.5 |
| $H_2O$ | 6.2 | 1.87 |
| $NH_3$ | 4.9 | 1.47 |
| $CO_2$ | 0.0 | 0.0 |

## Q. 14.2.4

The dielectric susceptibility due to induced dipoles in atoms which have no permanent dipole is related how to the atomic volume? (a)_____. The dielectric susceptibility due to orientation of permanent dipoles is related how to the temperature? (b)_____. Since we would expect the oxygen atom to have a net negative charge and since $p_{el} = 0$ for the $CO_2$ molecule, we can conclude that (c)_____.

* * *

## ANSWERS TO QUESTIONS

14.2.1 (a) Orientation of permanent magnetic dipoles by a magnetic field, (b) $-\mathbf{p}_m \cdot \mathbf{B}$ or $-p_m B \cos \theta$, (c) much less than 1.

14.2.2 (a) $p_m \cos \theta$, (b) $\int p_m \cos \theta \, dN(\theta)/\int dN(\theta)$, (c) $M = N\bar{p}_m$, (d) $p_m x/3$, (e) it is inversely proportional to the temperature.

14.2.3 (a) $10^{-4}$ to $10^{-5}$ as may be computed from (14.2.10), (b) $\chi_m \propto 1/(T - T_C)$, (c) interactions between the dipoles, (d) $10^{-3}$, (e) it is independent of temperature.

14.2.4 (a) It is directly proportional to it, (b) it is inversely proportional to it, (c) the $CO_2$ molecule is a linear molecule.

## 14.3 FERROMAGNETISM

Ferromagnetic materials approximately obey the Curie-Weiss relation; the value of $T_C$ is very high; in the case of iron it is 1043°K. This means that above 1043°K, the susceptibility obeys (14.2.11). However, below 1043°K there is some spontaneous alignment of dipoles (with $B = 0$). This is called **spontaneous magnetization**. It is due to the interactions which take place between the dipoles. These interactions tend to overcome the thermal energy if the temperature is not too high. As the temperature is lowered further and further below $T_C$, these interactions become more and more important and the thermal energy becomes less and less important. Therefore, the spontaneous magnetization increases as the temperature decreases, as is shown in Figure 3.6.6.

There is a magnetic induction field associated with a magnetic dipole as noted in Section 14.2. If there is a net alignment of dipoles, then there is a net field $B_i$ due to all these dipoles. This is called an **internal field**. Weiss postulated in 1912 that this internal field is directly proportional to the magnetization,

$$B_i = aM.$$

Then the total effective field $B'$ acting on a dipole is

$$B' = B + aM.$$

This leads directly to the Curie-Weiss relation and to a magnetization versus temperature curve of the form shown in Figure 3.6.6. (See Ruoff, *Materials Science*, Section 32.4.)

Weiss in 1912 did not understand the origin of the very large interactive energy which causes ferromagnetism (note that the $T_C$ in iron is about 100 times as large as that calculated from ordinary dipole interactions as in Example 14.2.4).

Iron can be found either magnetized or demagnetized. To explain this, Weiss in 1912 postulated the presence of **domains** which are regions which are small compared to the ordinary sample size and within which the material is magnetized in a given direction (see Figure 1.2.9). The random orientation of domains leads to no net macroscopic magnetization. The reason for the formation of domain structures is described in Section 14.5. Their existence was demonstrated by Francis Bitter in 1931.

The magnetization behavior is related to the ease or difficulty of domain boundary motion and domain rotation. This is studied in Section 14.6.

## Q. 14.3.1

Spontaneous magnetization is due to (a)_____
_____. In a ferromagnetic solid there is a strong (b)_____ associated with the spontaneously aligned dipoles. Weiss suggested that the reason iron could be found as either magnetized or demagnetized iron in the ferromagnetic state was that (c)_____
_____.

\* \* \*

THE ORIGIN OF FERROMAGNETISM. The interaction energy between the magnetic dipoles in a solid calculated by classical theory is only about $\frac{1}{100}$ of that needed to explain ferromagnetism ($10^{-3}$ eV instead of 0.1 eV per dipole or $10^7$ J/m$^3$ instead of $10^9$ J/m$^3$). The situation is somewhat analogous to the bonding of two hydrogen atoms to form a hydrogen molecule. Classical interaction forces could not explain the magnitude of this interaction either. However quantum mechanical concepts could. A most elementary model for the hydrogen bond is given in Section 5.6. The essence of this model is that each electron has become delocalized and may be found in any part of the molecule. The electron from one atom exchanges positions (continuously) with the electron from the other atom. The electrons have opposite spin. This **exchange interaction** is similar to the interaction which leads to ferromagnetism. However, in the latter case the energy is lowered when the spins are aligned or parallel.

The $d$-band electrons of iron can be considered as quasi-free. They therefore obey Fermi-Dirac statistics; i.e., an available energy level can be occupied by no more than two electrons. We would therefore expect the spins to be paired (the electron configuration of iron is $Ar3d^64s^2$). If the $d$ electrons were unpaired, the Fermi energy of the $d$ band would be increased since then there would be only one electron per allowed level. However, there is an energy decrease due to the exchange interaction which occurs when the unpaired spins are aligned.

It is convenient to define a quantity called the **density of states** $\rho(E)$ such that $\rho(E)\, dE$ represents the number of energy states (each of which can be occupied by two electrons) lying between $E$ and $E + dE$. In those elements in which the $d$ band is narrow and only partially filled there is a high density of states at the Fermi level, called $\rho_F$. See Figure 14.3.1. The number of additional states occupied if the electrons are unpaired is approximately $\rho_F\, \Delta E$ if $\rho$ is approximately constant. Hence the larger $\rho_F$ is, the smaller is the energy increase $\Delta E$ due to unpairing. In such elements it is possible for the energy decrease due to the exchange interaction to be larger than the energy increase due to unpairing. This situation (in which there is a net decrease in the energy of the system if the spins are unpaired) exists for only a few transition elements ($3d$ band) and rare earth ($4f$ band) elements. Table 14.3.1 shows the principal ferromagnetic elements.

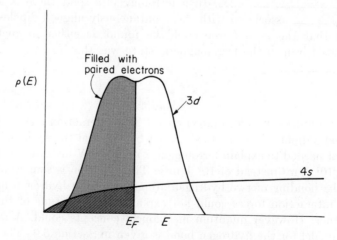

FIGURE 14.3.1. Sketch of density of states versus energy of typical $3d$ band and $4s$ band for typical ferromagnetic material. Note that when the electrons are unpaired the energy increases for the $3d$ band is not nearly as large as for $s$ bands.

## Table 14.3.1. PRINCIPAL FERROMAGNETIC ELEMENTS

| Element | Electron Configuration | Atomic Radius Å | $M_S$(at 0°K) ($10^6$ A/m) | Number of Bohr Magnetons per Atom | $T_C$ (°K) | $T_m$ (°K) |
|---|---|---|---|---|---|---|
| Fe | $3d^6 4s^2$ | 1.24 | 1.69 | 2.2 | 1043 | 1808 |
| Co | $3d^7 4s^2$ | 1.25 | 1.36 | 1.7 | 1404 | 1753 |
| Ni | $3d^8 4s^2$ | 1.25 | 0.47 | 0.6 | 631 | 1728 |
| Gd | $4f^7 5d^1 6s^2$ | 1.78 | 5.66 | 7.12 | 289 | 1585 |

**Hund's rule** states that the electron spins are arranged in such a way that they make a maximum contribution to angular momentum.

**EXAMPLE 14.3.1**

What is the maximum spin magnetic moment of a $d$ shell? An $f$ shell?

*Answer.* There are ten possible electrons in a $d$ shell. Five of these can have a plus spin, and five can have minus spins. A shell with *only* five $d$ electrons would according to Hund's rule have five unpaired electrons, i.e., either $5+$ or $5-$. A $d$ shell with more or less than five electrons would have a net spin magnetic moment less than 5 Bohr magnetons. For the $f$ shell, the maximum spin moment would be 7 Bohr magnetons.

Note that in the isolated iron atom we would expect according to Hund's rule five spins up and one down or a net electronic moment of four magnetons (for reasons which we shall not give here the orbital moment of the $d$ electrons is zero). In the condensed metal the number of Bohr magnetons per atom is experimentally found to be 2.2 instead of 4, which would be expected on the basis of Hund's rule (there is no simple explanation for this). As we shall see in the next section, Hund's rule is very successful in predicting the dipole moment in ferrimagnetic compounds, which are insulators and not metals.

**Q. 14.3.2**

The large quantum mechanical attraction between magnetic dipoles is due to (a)_____. Ferromagnetism occurs only in a few metals with (b)_____. According to Hund's rule the isolated $Fe^{3+}$ ion would have (c)____ unpaired spins.

\* \* \*

# ANSWERS TO QUESTIONS

**14.3.1** (a) Unusually strong interactions between dipoles, (b) internal field, (c) ferromagnetic domains exist.

**14.3.2** (a) Exchange interactions, (b) partially filled narrow $d$ bands or partially filled narrow $f$ bands, (c) five.

## 14.4 ANTIFERROMAGNETISM AND FERRIMAGNETISM

ANTIFERROMAGNETISM. In certain compounds of the transition metals the exchange interaction is such that the energy is minimized when neighboring dipoles are antiparallel. In such materials, called antiferromagnetic, below a certain critical temperature called the Néel temperature, spin pairing begins to occur spontaneously. The extent of spin pairing increases as the temperature is lowered. Thus above $T_C$ these materials are paramagnetic and their susceptibility increases as the temperature decreases, while below $T_C$ their susceptibility decreases as the temperature decreases. An example is MnO, which from the chemical viewpoint has the NaCl-type structure. However, from the magnetic viewpoint the unit cell has twice the lattice parameter and thus has eight times the volume since the dipoles of nearest neighbor $Mn^{2+}$ ions are antiparallel. See Figure 14.4.1.

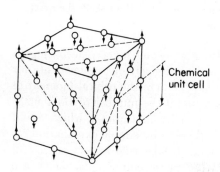

FIGURE 14.4.1. Structure of antiferromagnetic MnO as found by neutron diffraction. Only the $Mn^{2+}$ ions are shown. They alternate with $O^{2-}$ ions along the $\langle 100 \rangle$ directions. Note that the "magnetic unit cell" has twice the length of the "chemical unit cell." [From C. G. Shull, E. O. Wollan, and W. A. Strauser, *The Physical Review* **81**, 483 (1951).]

Neutron diffraction can be used to study this behavior, since the neutron has a permanent magnetic moment which interacts with the permanent magnetic moment of the $Mn^{2+}$ ions.

FERRIMAGNETISM. The ferrites are a class of ionic crystals of composition $MeFe_2O_4$, where Me is a metal ion; these crystals have the inverted spinel structure. The compound $Fe_3O_4$, i.e., $Fe_2^{3+}O_3^{2-}$-$Fe^{2+}O^{2-}$, is an example. So is $NiFe_2O_4$. The cubic unit cell of $Fe_3O_4$ contains 56 ions: $32O^{2-}$, $8Fe^{2+}$, and $16Fe^{3+}$. One fourth of the unit cell is shown in Figure 14.4.2.

The iron ions occupy two types of sites, eight which are tetrahedrally coordinated and sixteen which are octahedrally coordinated. Eight of the $Fe^{3+}$ ions occupy the tetrahedral sites and eight occupy eight of the octahedral sites. The eight $Fe^{2+}$ ions occupy the remaining eight octahedral sites. The spin moments of the two groups of $Fe^{3+}$ ions are antiparallel so

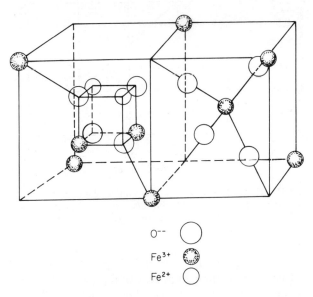

FIGURE 14.4.2. One fourth of the unit cell of $Fe_3O_4$, lodestone.

the $Fe^{3+}$ ions contribute no net moment. The eight $Fe^{2+}$ ions have their moments aligned so the solid has a net moment. Solids which have a resultant moment because of incomplete cancellation are called **ferrimagnetic materials**.

### EXAMPLE 14.4.1

If orbital momentum does not contribute, what dipole moment is expected for $Fe^{2+}$ ions according to Hund's rule?

*Answer.* The electron configuration is $Ar(3d)^6$. From Hund's rule we would expect a spin configuration for the $d$ electrons of 5 up and 1 down or a net spin of $4\beta$ (Bohr magnetons). This leads to a saturation magnetization at low temperature (all the dipoles aligned) of $0.5 \times 10^6$ A/m for $Fe_3O_4$, as calculated in Example 5.9.3. This is in close agreement with the experimental value of $0.48 \times 10^6$ A/m.

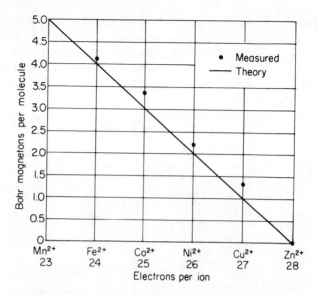

FIGURE 14.4.3. Magnetic moments of divalent ions in some ferrites. [After J. E. Goldman, *The Science of Engineering Materials*, John Wiley & Sons, Inc., New York (1957).]

The fact that we can approximately predict the magnetization saturation of ferrites in general is shown in Figure 14.4.3.

Ferrites, because of their high resistivity (a factor of $10^6$–$10^9$ higher than common magnetic metal alloys), show much smaller losses in high frequency applications because the eddy current losses are less. The ferrites can also be used for information storage.

### Q. 14.4.1

When spontaneous alignment occurs in MnO, half the dipoles are aligned (a)_____. When spontaneous alignment occurs in the compound $Fe_3O_4$, the dipoles associated with (b)_____ ions are aligned. Compounds with the formula $MeFe_2O_4$ are called (c)_____. These have important commercial applications as magnets because of their high (d)_____.

\* \* \*

## ANSWERS TO QUESTIONS

**14.4.1** (a) Antiparallel to the other half, (b) the $Fe^{2+}$, (c) ferrites, (d) electrical resistivity.

## 14.5 DOMAINS

It was previously pointed out that Weiss proposed the existence of domains in crystals. It is the purpose of this section to discuss the origin of domains. We have noted earlier that exchange energy is responsible for spontaneous magnetization and that the magnitude of this exchange energy is on the order of $10^9$ J/m$^3$. Hence we might at first expect a bar magnet to be magnetized as in Figure 14.5.1(a). Let us assume that this is a single crystal

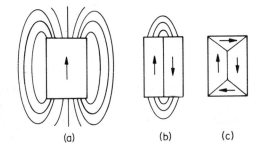

FIGURE 14.5.1. (a) Single domain established by exchange energy. (b) Magnetostatic energy, roughly proportional to the spatial extension of the field, about halved. (c) Magnetostatic energy reduced virtually to zero by closure domains.

of iron with the $\langle 100 \rangle$ directions parallel to the edges of the bar (a rectangular parallelepiped). It is a fact, which we shall discuss in more detail later, that iron prefers to be magnetized in one of the $\langle 100 \rangle$ directions rather than, say, a $\langle 111 \rangle$ direction. In other words the energy of the system is lower if it is magnetized in the [100] direction rather than in the [111] direction. For the present, we assume that iron is spontaneously magnetized in the $\langle 100 \rangle$ direction only.

Note the flux lines in Figure 14.5.1(a). There is a large energy, the stored magnetic energy discussed in Section 3.6, associated with this magnetic field. This is called **magnetostatic energy.** This energy can be greatly reduced if domains are formed as in Figure 14.5.1(b). Note how extensively the spatial extension of the field has been decreased. The magnetostatic energy is roughly proportional to the spatial extension of the field. In Figure 14.5.1(c), additional domains are formed, which essentially completely eliminate the macroscopic dipole and the magnetostatic energy.

Is there a limit to the number of domains which will be formed? When a domain structure is formed domain boundaries are formed. There is an energy per unit area associated with such boundaries analogous to grain boundary energy. However, there is an important difference. Grain boundaries are relatively narrow (a few atoms), as has been clearly shown by use of the field ion microscope. Domain boundaries are wide (300 atoms); the reason for this will be studied later. They have energies (in iron) of $2 \times 10^{-3}$ J/m$^2$ (for comparison purposes grain boundary energies are of the order of 1 J/m²). Thus the size and number of domains formed is limited by the domain wall boundary energy; magnetostrictive effects also play a role in determining domain size and shape, as is discussed next.

## Q. 14.5.1

Ferromagnetic (or ferrimagnetic) domains form in order to lower the (a)_____ associated with the (b)_____ dipole. When domains form in a crystal, (c)_____ also form. These limit the smallness of the domains because of (d)_____.

\* \* \*

MAGNETOSTRICTION. When a crystal becomes magnetized it changes dimensions. This is called **magnetostriction.** The longitudinal strain versus field for three crystallographic directions in iron is shown in Figure 14.5.2. There is also a tiny linear fractional increase in volume of about

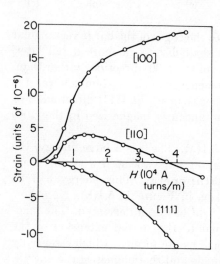

FIGURE 14.5.2. Longitudinal magnetostrictive strain versus field in iron. [After A. Webster, Proceedings of The Royal Society of London, **A109**, 570 (1925).]

$4 \times 10^{-11}$ per unit field for fields up to $2.5 \times 10^4$ A/m. Thus a crystal magnetized in the [100] direction will stretch in the [100] direction but will shrink in the [010] or [001] direction (since the volume is nearly a constant).

**EXAMPLE 14.5.1**

A field of $2.5 \times 10^4$ A/m is applied in the [100] direction of an iron crystal. What is the strain in the [010] direction?

*Answer.* The strain in the [100] direction is, from Figure 14.5.2, $\epsilon = 15 \times 10^{-6}$. For small strains the fractional change in volume $\Delta V/V = \epsilon_{[100]} + \epsilon_{[010]} + \epsilon_{[001]}$. In the present case $\epsilon_{[010]} = \epsilon_{[001]}$ and

$$\frac{\Delta V}{V} = \epsilon_{[100]} + 2\epsilon_{[010]} = 15 \times 10^{-6} + 2\epsilon_{[010]}.$$

But the volume magnetostriction is

$$\frac{\Delta V}{V} = 4 \times 10^{-11} \times 2.5 \times 10^4 = 10^{-6}.$$

Hence

$$\epsilon_{[010]} = -7 \times 10^{-6}.$$

Consider now Figure 14.5.1(b). Both slabs will expand along the [100] direction and *contract* normal to it in the [010] and [001] directions. There are no constraints and hence no stresses are introduced so that there is no elastic strain energy. Next consider 14.5.1(c). Here domains are present at the top and bottom which close the magnetic field loops within the material. These are called **closure domains**. Note that the closure domain would tend to *expand* in the [010] direction. It is constrained from doing so by the vertical domains, so that stresses are generated. Hence there is elastic strain energy present because of magnetostriction. The strain energy can be decreased by reducing the volume of the closure domain, as shown in Figure 14.5.3. This results in an increase in domain boundary area

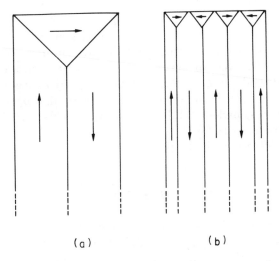

FIGURE 14.5.3. Elastic strain energy in (a) can be decreased by decreasing the volume of closure domain as in (b).

and hence domain boundary energy. The total balance between these various energy effects (along with the presence of externally induced strains, imperfections, etc.) determines the domain structure of a crystal.

It should be noted that not only ferromagnetic but also ferrimagnetic and antiferromagnetic materials exhibit domain structures.

MAGNETOCRYSTALLINE ANISOTROPY. When a pure single crystal of a ferromagnetic or ferrimagnetic material with no net moment is placed in a magnetic field it magnetizes at a relatively small field. The magnitude of the field needed to achieve full magnetization depends on crystallographic direction, as shown in Figure 14.5.4. The quantity $\mu_0 \int H \, dM$ represents the

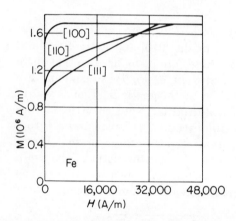

FIGURE 14.5.4. Magnetization curves for single crystals of iron.

change in the Gibbs free energy (as is shown in texts in thermodynamics). This quantity is nearly zero for the [100] direction in iron. We say that the [100] direction is the **easy direction** of magnetization. However, it is quite large for the [111] direction, which is called the **hard direction**. Hence a sizable energy is needed to convert an iron crystal from complete [100] magnetization to complete [111] magnetization. This is called the **magnetocrystalline energy** for the [111] direction. The effect is referred to as **magnetocrystalline anisotropy**. Without it there would be no hard magnets.

### Q. 14.5.2

The magnetization of a crystal induces (a)_____; this is called magnetostriction. Because of magnetostriction there is (b)_____ _____ associated with closure domains. By reducing the volume of the closure domains, the (c)_____ can be reduced. The integral $\mu_0 \int H\, dM$ for a given direction minus the integral $\mu_0 \int H\, dM$ for the easy direction of magnetization is called the (d)_____ _____. In iron the easy direction is the (e) _____ direction.

\* \* \*

For nickel the easy direction is [111] while [100] is the hard direction. For cobalt (which has the hcp structure) the easy direction is [0001] and the hard direction is [10$\bar{1}$0].

THE NATURE OF DOMAIN BOUNDARIES. A domain wall or **Bloch wall** is shown in Figure 14.5.5. In iron it has a thickness of about 1000 Å. Let

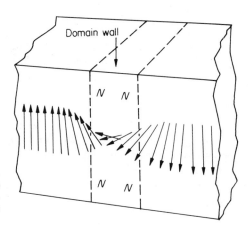

FIGURE 14.5.5. Domain wall. All magnetic moments lie in the plane of the wall. The $N$'s represent poles formed on the surface of the material.

us examine the reason for this. It is shown in advanced solid state physics texts that the exchange interaction between neighboring spins (dipoles) of magnitude $S_1$ and $S_2$ is

$$U = -2JS_1S_2 \cos \theta,$$

where $\theta$ is the angle between $\mathbf{S}_1$ and $\mathbf{S}_2$ and $J$, called the exchange integral, is a constant of proportionality. For small angles, since

$$\cos \theta = 1 - \frac{\theta^2}{2!} + \frac{\theta^4}{4!} - \cdots,$$

$$U \doteq -2JS^2 \left(1 - \frac{\theta^2}{2}\right)$$

so the increase in the exchange energy of neighboring spins due to rotation is $JS^2\theta^2$. Clearly the spins must rotate as we change from one domain to another. From the viewpoint of the exchange energy it is desirable to have the rotation between neighbors as small as possible and hence to have a very wide boundary.

**EXAMPLE 14.5.2**

Explain why wide walls are favored by exchange interactions.

*Answer.* The total rotation is 180 deg ($\pi$ rad). This takes place by having $n$ successive rotations of $\theta$ radians each. Hence, $n\theta = \pi$. Thus the energy for all the $n$ rotations in a boundary (per cross-sectional area of one dipole) is

$$nJS^2\left(\frac{\pi}{n}\right)^2 = JS^2\frac{\pi^2}{n}.$$

Thus this extra energy decreases as $n$ increases, i.e., as the boundary gets wider.

However, in passing from one [100] domain to a [$\bar{1}$00] domain in iron the direction of spin rotates relative to the crystal, passing through various harder magnetization directions, such as the [111] direction (see Figure 14.5.4). Thus from the viewpoint of magnetocrystalline energy a narrow boundary is preferred. The balance of these two energy terms determines the width of the boundary.

Note in Figure 14.5.5 the poles at the surface where the domain wall intersects the surface. Since lines of magnetic induction emerge from the surface here, tiny particles of a ferromagnetic material would be attracted here. This is the basis of the Bitter patterns which are shown in Figure 1.2.9 and which were discussed in Section 8.2. Domains can also be made visible by using the reflection of polarized light from a surface; such light is rotated by a magnetized surface in direct proportion to the magnetization.

**Q. 14.5.3**

Exchange energy favors (a)_____ domain wall while magnetocrystalline anisotropic energy favors (b)_____ wall. At the point where domain walls emerge from a surface, the **B** field is oriented how with respect to the surface? (c)_____. Hence iron filings in a colloidal suspension will be attracted here because (d)_____
_____.

\* \* \*

SUMMARY OF DOMAIN WALL FORMATION. Spontaneous magnetization is caused by the exchange energy. The direction of the spontaneous magnetization is determined by the magnetocrystalline anisotropy. Domains form in order to decrease the magnetostatic energy associated with a macroscopic dipole (a permanent bar magnet). The size and shape of the domains in a solid are determined by domain boundary energy and also the magnetostriction strain energy associated with closure domains. The width of the domain wall itself is determined by a balance between exchange energy and magnetocrystalline anisotropic energy.

## ANSWERS TO QUESTIONS

**14.5.1** (a) Magnetostatic energy, (b) macroscopic, (c) domain walls, (d) domain wall energy.

**14.5.2** (a) A strain, (b) stress or elastic strain energy, (c) elastic strain energy, (d) magnetocrystalline energy (sometimes the anisotropic energy), (e) [100].

**14.5.3** (a) A wide, (b) a narrow, (c) it is perpendicular to it, (d) the ferromagnetic filings attract flux (and vice versa).

### 14.6 MAGNETIZATION PROCESSES ACCORDING TO DOMAIN THEORY

Ferromagnetic materials are soft or hard depending on the ease or difficulty of moving domain boundaries and of rotating domains. In iron, it is much easier to magnetize the material by domain wall motion than by domain rotation. This is illustrated in Figure 14.6.1. The difficulty of domain rotation is determined by the magnetocrystalline anisotropic energy. If this is large, then the material may be soft or hard depending on the ease of domain wall motion. At low fields domain wall motion may take place reversibly, but at higher fields it takes place irreversibly with energy being dissipated as heat. Domain wall motion is a highly structure-sensitive property.

HARD MAGNETS. To make really hard magnets, we should eliminate the possibility of domain boundary movement. The best way to do this is to eliminate the domain boundary altogether. If a ferromagnetic material is divided into very fine elongated particles, these exist as single domains, as first suggested by Louis Néel in 1950. Note that if the width of these tiny elongated particles were about equal to the domain wall thickness, it would be expected that domains would not form.

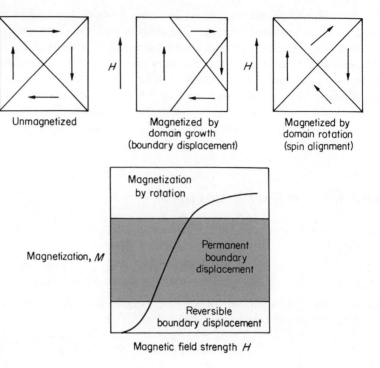

FIGURE 14.6.1. Relation of magnetization behavior to applied field.

**EXAMPLE 14.6.1**

Discuss how such **elongated single domains,** ESD, could be used to make a permanent magnet.

*Answer.* These particles are placed in a liquid matrix, such as Pb. or epoxy resin prior to curing. This mixture is placed in a magnetic field so that the particle domains are all oriented in the same direction. The matrix would then be solidified. The resultant permanent magnet is known as an **elongated single domain permanent magnet.** Theory suggests that the energy product could be about five times as large as for Alnico V, i.e., about $2 \times 10^5$ J/m³.

The General Electric Company produces an ESD magnet material called Lodex which may be stamped, cold-pressed, etc. This is an alloy of 65Fe–35Co and is electrodeposited in mercury as ESD. The particles are removed from the liquid and coated with antimony. They are then placed in a lead matrix. The resultant material is ground to powder, each grain of which is a single magnet. Because lead is a soft malleable material, this material can be used to make magnets of complicated configuration.

Sec. 14.6    Magnetization Processes    647

The Westinghouse Electric Corporation makes a product named Westro by sintering ESD ferrite. The ferrite used is strontium ferrite, which has a very large magnetocrystalline anisotropy. Recall that sintering is the powder metallurgy process which depends on the elimination of surface energy as the driving force and on diffusion to accomplish the mass transport.

**EXAMPLE 14.6.2**

A refrigerator door needs a rubber gasket for stopping air flow and a magnetic latch. Develop a technique for "killing both birds with one stone."

*Answer.* Strontium ferrite particles are dispersed in the rubber prior to curing. The domains are aligned in a magnetic field and the rubber is cured. This composite material is both the seal and latch.

Alnico V is one of the hardest permanent magnet materials readily available. It is a cast alloy of iron, cobalt, nickel, aluminum, and copper. The first Alnico alloy was roughly $Fe_2NiAl$ and was invented by T. Mishima in Japan. This was improved by F. W. Tetley of England, who added cobalt and copper. When properly heat-treated this material is actually a fine powder magnet since it consists of two phases, one of which is a rod-like precipitate as was shown by transmission electron microscopy by Nesbitt and Heidenreich. Figure 14.6.2 illustrates this structure. This structure is

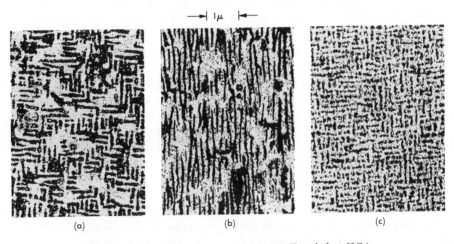

(a)    (b)    (c)

FIGURE 14.6.2. Microstructure of Alnico V cooled at 2°C/sec from 1300°K. (a) (100) plane, no field. (b) (010) plane, with field along [100] during cooling. (c) (100) plane, with field along [100] during cooling. [Courtesy of R. D. Heidenreich and E. A. Nesbitt. For further information, see *Journal of Applied Physics* **23**, 366(1952).]

obtained by pouring the melt into a cylindrical mold which has hot walls and a cold bottom. It therefore cools uniaxially in the presence of a magnetic field, separating in the process into a magnetic component which is rich in iron and cobalt and a nonmagnetic component rich in nickel and aluminum. In this respect it is somewhat analogous to the uniaxial formation of eutectic composites discussed in Section 10.3.

There is an excellent elementary article on the production of cobalt-samarium magnets, the hardest of all magnets, by J. J. Becker, "Permanent Magnets," *Scientific American* (December 1970), p. 92.

## Q. 14.6.1

If elongated single crystals of a ferromagnetic material have a diameter of only 1000 Å, they exist as (a)_____. Such single domain magnets can lead to very hard bulk magnets because (b)_____
_____.
Alnico V is really an elongated single domain magnet material obtained by cooling an appropriate melt (c)_____
_____.

\* \* \*

SOFT MAGNETS. The various mechanisms of pinning domain boundaries are very important in determining the behavior of soft magnetic materials. In this particular application we are interested in eliminating such pinning so that boundary displacement occurs readily and reversibly. It is therefore necessary to understand the origin of pinning if we are to eliminate it and produce soft magnets.

Becker noted that inasmuch as most materials show magnetostriction (which introduces strains, and if the domain is constrained, stresses), domain configuration must be related to local strains due to foreign atoms, grain boundaries, or other metallurgical inhomogeneities. These interactions are eliminated if the magnetostriction is zero. This is illustrated by the very high permeability of permalloy materials at compositions at which the magnetostriction is zero. Nickel and iron exhibit opposite magnetostriction so it is not too surprising that an alloy can be made which has zero magnetostriction.

Another pinning mechanism was suggested by Kersten. A large nonmagnetic inclusion in a crystal would, if located at a domain boundary, decrease the total boundary energy since it decreases the domain boundary area as shown in Figure 14.6.3. When a field is applied a certain critical energy has to be supplied by the field to move the boundary because of the need to create the additional domain boundary. The coercive force is not large for this mechanism. For example, cementite particles of the optimum size in a 1.5 wt. % C steel cause a coercive force of only 300 A/m.

Néel suggested another pinning mechanism at large nonmagnetic inclusions. There would be a pole density on the surface of such an inclusion within a domain. Magnetostatic energy is associated with such poles. Néel

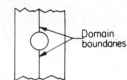

FIGURE 14.6.3. Nonmagnetic precipitate at domain boundary.

noted that this energy could be decreased by the creation of V-shaped spike-like domains of reverse or perpendicular magnetization near the inclusion. As a domain wall moves past, it is attracted to the spike; a coercive force must be applied to tear it away.

Another important factor, in addition to eliminating imperfections, in obtaining soft magnetic materials is eliminating magnetocrystalline anisotropy (because this would permit free domain rotation). The addition of molybdenum to permalloy results in nearly zero anisotropy and an increased softness.

## Q. 14.6.2

To produce soft magnetic materials, eliminate at least three causes of pinning. These are (a)_____
_____. In addition to making domain wall motion easy, another way of producing a soft magnet is (b)_____
_____.

\* \* \*

The field needed to magnetize iron to $M = 1.5 \times 10^6$ A/m in the [111] direction is about 20,000 A/m, while in the [100] direction only about 20 A/m are required. This suggests that transformer sheet should be manufactured with the plane of the sheet parallel to the cube faces of the crystal. Goss (1935) has developed techniques for rolling and heat-treating polycrystalline iron in such a way that most of the crystals have a direction of easy magnetization [100] within several degress of the rolling direction.

The hysteresis losses in such grain-oriented sheet is about half that of nonoriented sheet. An example of such a **cube-textured material** is shown in Figure 14.6.4.

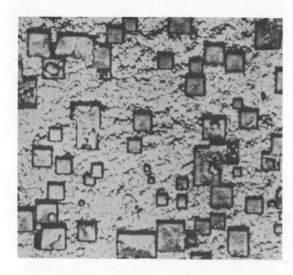

FIGURE 14.6.4. Cube-textured steel. (×500.) (Courtesy of Karl Foster, Westinghouse Electric Corporation, Pittsburgh, Pa.)

Q. 14.6.3

Because iron is very soft in the [100] direction (compared to the [111] direction) transformer sheet, ideally, should be made of (a)_____.

Goss has developed a technique for making (b)_____ in which the crystals exhibit a specific type of preferred orientation in the polycrystalline iron.

\* \* \*

## ANSWERS TO QUESTIONS

**14.6.1** (a) Single domains, (b) there is no domain wall motion which could lead to demagnetization, (c) uniaxially in the presence of a magnetic field.

**14.6.2** (a) Crystalline imperfections, nonmagnetic inclusions, and magnetostriction; (b) to choose a material with zero magnetocrystalline anisotropy.

**14.6.3** (a) Single crystals with two of the ⟨100⟩ directions in the plane of the sheet with one of these directions parallel to the flux lines, (b) cube-textured sheet.

## 14.7 MAGNETIC BUBBLES

Certain orthoferrite materials (materials with the formula $RFeO_3$, where R is a rare earth or yttrium), garnets, and hexagonal ferrites exhibit high magnetocrystalline anisotropy. It is possible to grow thin plates of such materials in which the direction of easy magnetization is normal to the plates while the hard direction lies in the planes of the plates. In the absence of a magnetic field two types of domains are simultaneously present. The magnetization in each is normal to the plate and extends *through* the plate. The magnetization of the two types of domains is, of course, opposite; the total volume (or area) of the two types is the same. Figure 14.7.1 illus-

FIGURE 14.7.1. Magnetic domain structure in an orthoferrite ($Sb_{0.55}Tb_{0.45}FeO_3$) plate. This same domain structure extends downward through the platelet which is 35 mm thick. The c-axis of the orthorhombic crystal is normal to the plate. The domains which are observed through crossed polarizers are made visible by the Faraday effect. (Courtesy of Raymond Wolfe, Bell Laboratories, Murray Hill, N.J.)

trates such a domain structure. It is of interest to note that this structure can be altered by "writing" on the surface with a thin magnetized wire.

When a magnetic field is applied normal to the plate of solid material at room temperature, the domain structure changes, i.e., one type of domain dominates at the expense of the other, as shown in Figure 14.7.2. Note that with a still larger field only tiny cylinders of one domain remain. These are called **magnetic bubbles.** Depending on the material, these bubbles may have diameters of 0.05 to 500 $\mu$. Increasing the field still further would collapse the bubbles and the material would then be entirely magnetized in one direction.

(a)

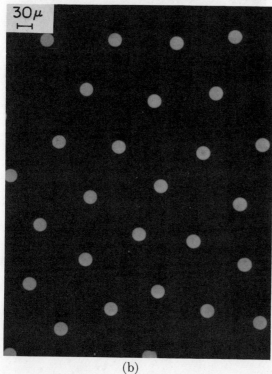

(b)

FIGURE 14.7.2. Effect of magnetic field on domain structure in plate. (a) A small bias field (800 amps/m) is applied normal to the platelet after the long strip domains have been cut into many closed domains by repeatedly passing the tip of a magnetized wire over the plate. (b) The bias field is increased to 3600 amps/m. Each of the closed domains has shrunk to a bubble of diameter 30 $\mu$. (Courtesy of Raymond Wolfe, Bell Laboratories, Murray Hill, New Jersey.)

A. H. Bobeck and other scientists at the Bell Laboratories have developed methods for systematically moving these cylindrical domains (bubbles) throughout the plate. In one such device a thin film permalloy (a soft magnetic material) is deposited on the plate as shown in Figure 14.7.3. Then a rotating magnetic field in the plane of the plate moves the bubble from T to T.

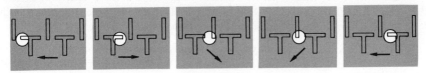

FIGURE 14.7.3. T-bar bubble mover. The arrow represents a small magnetic field applied in the plane of the wafer. This field rotates as shown. As it rotates the direction of magnetization in the permalloy changes, attracting the bubble as shown.

Bubble densities of greater than $10^6/in.^2$ have been obtained, with propagation rates in excess of $10^6$ steps/sec. Moreover, the energy required to "step" a bubble is less than $10^{-2}$ of the energy needed to switch a small transistor. Hence this device provides for the possibility of a high-density information storage system.

For an excellent discussion of magnetic bubbles, see A. H. Bobeck and H. E. D. Scovil, "Magnetic Bubbles," *Scientific American* (June 1971), p. 78.

## Q. 14.7.1

Magnetic bubbles occur in (a)_____ of materials such as orthoferrites, garnets, and hexagonal ferrites which have a high (b)_____ _____. The direction of easy magnetization is (c)_____ _____ to the plates, while the hard direction (d)_____.

\* \* \*

## ANSWERS TO QUESTIONS

14.7.1 (a) Thin plates, (b) magnetocrystalline anisotropy, (c) perpendicular, (d) lies in the plates.

## REFERENCES

G. E. *Permanent Magnet Manual*, General Electric Co., Edmore, Michigan. Contains design data on various hard magnetic materials plus some worked out design problems.

Keffer, F., "The Magnetic Properties of Materials," *Scientific American* (September 1967), p. 222. Clear, very elementary discussion.

Becker, J. J., "Permanent Magnets," *Scientific American* (December 1970), p. 92. This provides an excellent historical perspective of the discoveries which led to the hard cobalt-samarium magnets.

Bobeck, A. H., and H. E. D. Scovill, "Magnetic Bubbles," *Scientific American* (June 1971), p. 224. This is an excellent article which discusses both the origin of bubbles and some of the devices using bubbles.

Bozorth, R. M., "The Physics of Magnetic Materials," in *The Science of Engineering Materials* (J. E. Goldman, ed.), John Wiley & Sons, Inc., New York (1957). Excellent nonmathematical discussion.

Jacobs, I. S., "Role of Magnetism in Technology," *Journal of Applied Physics* **40**, 917 (1969). A thorough discussion of the applications of magnetic phenomena. Includes magnetic information storage.

Bozorth, R. M., *Ferromagnetism*, Van Nostrand Reinhold Company, New York (1951). A classic book concerned with the structure sensitive nature of magnetism.

# PROBLEMS

**14.1** List at least five "why" questions concerned with the phenomena of magnetism.

**14.2** Discuss why you would expect the $Na^+$ ion to be diamagnetic and the Na atom to be paramagnetic.

**14.3** Would the magnitude of $\chi_m$ for $Na^+$ ions be greater or smaller than for neon atoms (assuming equal concentrations)?

**14.4** Discuss why the paramagnetic susceptibility of conduction electrons is independent of temperature.

**14.5** Calculate the saturation magnetization of $NiO-Fe_2O_3$ ferrite assuming there is no orbital moment. Compare with experiment. The lattice parameter is 8.34 Å.

**14.6** What is the origin of ferromagnetism?

**14.7** (a) Why does a ferromagnetic crystal form domains?
 (b) Closure domains?
 (c) Why does it prefer many closure domains to a few?
 (d) What limits the size of domains?

**14.8** Describe the processes which occur during various portions of an $M$-$H$ curve in ferromagnetic materials.

**14.9** Name two techniques used to form very hard magnets.

**14.10** List all the structural factors which must be controlled to obtain a soft magnetic material.

**14.11** List the criteria for formation of magnetic bubbles.

## MORE INVOLVED PROBLEMS

**14.12** Show that the expression $\mathbf{T} = \mathbf{p}_m \times \mathbf{B}$ is consistent with

$$\text{P.E.} = -\mathbf{p}_m \cdot \mathbf{B}.$$

**14.13** Give at least two techniques for measuring $M$-$H$ curves.

**14.14** Write an essay on the current status of magnetic memory devices (except for the magnetic bubble).

**14.15** Write an essay on the magnetic bubble.

**14.16** A paramagnetic solid has ions with permanent magnetic dipoles. It is placed in a high magnetic field and cooled to 4°K. The field is then turned off. Discuss what happens to the system.

**14.17** Discuss the historical development of the hard samarium-cobalt magnet material.

**14.18** In evaluating the denominator of (14.2.6) note that

$$\sin\theta e^{x\cos\theta}\,d\theta = -\frac{1}{x}d(e^{x\cos\theta})$$

when $x$ can be treated as a constant. Use this to obtain (14.2.7). You may also wish to substitute $y = \cos\theta$.

## SOPHISTICATED PROBLEMS

**14.19** One of the important techniques of measuring magnetic fields is by studying nuclear magnetic resonance. Describe. For an elementary discussion, see C. A. Wert and R. M. Thomson, *Physics of Solids*, McGraw-Hill Book Company, Inc., New York (1964), Chapter 21.

**14.20** Calculate the diamagnetic susceptibility of hydrogen atoms if there are $N$ atoms per cubic meter.

**14.21** Discuss the economic feasibility of levitating a train on permanent magnet tracks.

# Prologue

The superconducting state is a new state of matter. The thermodynamics of the phase transformation from the normal to the superconducting state is discussed. The origin of the Meissner effect (flux exclusion) is studied and is shown to lead to the concept of bound electron pairs (*Cooper pairs*). The *energy gap* within the valence band at the Fermi surface is described. An expression for the *coherence length* (the distance over which the electrons in a bound pair interact) is derived. It is noted that superconductivity owes its origin to the *macroscopic de Broglie wave* associated with the motion of the assembly of Cooper pairs. The fact that there is *flux penetration* near the surface of a bulk Type I superconductor and that this has associated with it *supercurrents* is discussed. The *flux penetration depth* is introduced. It is pointed out that flux penetration (and associated supercurrents) occur only in quantum jumps, called *fluxoids*. The difference between Type I and Type II superconductors is based on the ratio of coherence length to flux penetration depth. The methods of *pinning* fluxoids in Type II superconductors, which is necessary if they are to carry high currents, are discussed.

# 15

# SUPERCONDUCTIVITY

## 15.1 THE SUPERCONDUCTING STATE

The discussion of this chapter assumes a knowledge of the phenomenological aspects of superconductivity discussed in Section 3.10. A quick review of that material by the reader would be appropriate.

A superconductor is not just a perfect ordinary conductor. It is a new state of matter.

The superconducting state is more stable than the normal state when its Gibbs free energy is lower. Consider the two states at a cryogenic temperature $T$ in the absence of a magnetic field; let $G_n^0$ be the Gibbs free energy per unit volume of the normal state and let $G_s^0$ be the Gibbs free energy per unit volume of the superconducting state. We then place both samples in identical magnetic fields. Then, since the Gibbs magnetic energy density is given by $-\int B \, dH$, the respective Gibbs free energies in the presence of a field $H$ are

$$G_n \doteq G_n^0 - \frac{\mu_0 H^2}{2}$$

since in the normal state $B = \mu H \doteq \mu_0 H$, and

$$G_s = G_s^0$$

since in the superconducting state $B = 0$. The two states are in equilibrium with each other when $H = H_c$, the critical field at the given temperature. Under equilibrium conditions $G_n = G_s$. Hence it follows that

$$G_n^0 - G_s^0 = \frac{\mu_0 H_c^2}{2}. \qquad (15.1.1)$$

A plot of the free energies is shown in Figure 15.1.1. See Figure 3.10.1 for the function $H_c(T)$. Hence by measuring $H_c$ at a given temperature, the difference between the two Gibbs free energies at that temperature (in the absence of a field) is measured.

The energy difference at absolute zero temperature can be approximately calculated. Only a small fraction of electrons lying close to the Fermi surface, a fraction of about $T_c/T_F$, where $T_F$ is the Fermi temperature, contribute to this difference (this means that the probability of a given electron contributing is about $T_c/T_F$). Since the energy change is about $k_B T_c$ per electron, the difference in the Gibbs free energy per electron at

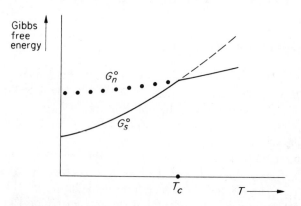

FIGURE 15.1.1. Gibbs free energy versus temperature for the normal phase, $G_n^0$, and for the superconducting phase, $G_s^0$. A solid line designates the region where a given phase is stable.

absolute zero is approximately $k_B T_c^2/T_F$. If this is multiplied by the number of valence electrons per unit volume, the difference in the Gibbs free energy per unit volume is obtained [see (15.1.1)]; i.e., $H_c(0)$ is predicted.

## Q. 15.1.1

Below $T_c$, the superconducting phase is stable; its free energy (in the absence of a field) is (a)_____ than the free energy of the normal state. The quantity $G_n^0 - G_s^0$ (difference in Gibbs free energy per unit volume) equals (b)_____. This difference at absolute zero temperature can be approximately calculated on theoretical grounds; the result is (c)_____.

\* \* \*

It is a simple thermodynamics problem (for an expert in thermodynamics) to show starting with (15.1.1) that the entropy difference (per unit volume) is

$$S_n - S_s = -\mu_0 H_c \frac{dH_c}{dT},$$

the enthalpy difference (per unit volume) is

$$h_n - h_s = -\mu_0 T H_c \frac{dH_c}{dT},$$

and the specific heat difference (per unit volume) is

$$c_n - c_s = -\mu_0 H_c \frac{dH_c}{dT} - \mu_0 T \left(\frac{dH_c}{dT}\right)^2 - \mu_0 T H_c \frac{d^2 H_c}{dT^2}.$$

Note (see Figure 3.10.1) that when $H_c = 0$ at $T = T_c$ both the entropy and enthalpy difference are zero, although the specific heat difference is not. Such a phase transition is called a **second-order transition**. (Ordinarily, as in melting, there is a finite entropy and enthalpy difference at the phase transition point. Such transformations are called **first-order transitions**.)

## ANSWERS TO QUESTIONS

**15.1.1** (a) Less; (b) $\mu_0 H_c^2/2$; (c) $N k_B T_c^2/T_F$, where $N$ is the number of valence electrons per unit volume.

## 15.2 FUNDAMENTAL CONCEPTS

COOPER PAIRS. The Type I superconductor exhibits the Meissner effect; i.e., while in the superconducting state, $B = 0$ within the bulk material; i.e., it excludes flux lines. Since $B = \mu_0(H + M)$ this is equivalent to saying that $M = -H$ (as shown in Figure 3.10.2). Such a material, for which the magnetic susceptibility $\chi_m = -1$, is called a perfect diamagnetic material. Recall from Section 14.1 that typical values of diamagnetic susceptibilities for solids are $-10^{-6}$. You should also recall that the computed values of diamagnetic susceptibilities of typical solid materials are consistent with a mean squared electron radius equal to $\overline{r^2} \approx 1$ Å$^2$ [see Equation (14.1.2)].

In deriving Equation (14.1.2) the assumption was made that $M \ll H$. If this assumption is not made, the expression for $\chi_m$ is for $Z = 1$

$$\chi_m = -\frac{\dfrac{N e^2 \overline{r^2} \mu_0}{4m}}{1 + \dfrac{N e^2 \overline{r^2} \mu_0}{4m}}$$

When $r \approx 1$ Å, $\chi_m \approx -10^{-6}$. However, if $r \approx 10^4$ Å, $\chi_m \doteq -1$. Thus if at low temperatures the valence electrons were very loosely coupled to the ion with a large effective radius rather than completely free as with the free electron gas, the material would approach perfect diamagnetic behavior.

In the typical normal metal the valence electrons are free to move throughout the metal (the electron gas) and therefore make no diamagnetic contribution. Let us assume, however, that at a certain low temperature pairs of electrons attract each other. Imagine, e.g., that a sort of electron pair molecule exists, with the distance between the electrons being of the order of 10,000 Å. If all the valence electrons were present as such bound pairs, this would lead to the appropriate large diamagnetic susceptibility (according to our approximation).

It is a complex problem in quantum mechanics to show that at absolute zero temperature the energy of the overall system is decreased if the electrons form bound pairs, called **Cooper pairs.** These bound electrons can be decoupled by thermal vibration of the atoms and hence depend on their vibrational frequency (and consequently on the Debye temperature of the solid). There is direct evidence of this since in many cases $T_c$'s for different isotopes of the same elements are inversely proportional to the square root of the atomic masses (the **isotope effect**).

Slightly above absolute zero temperatures, entropy effects enter, and some of the pairs dissociate in order to decrease the Gibbs free energy of the system. Finally, above a temperature $T_c$ all the pairs are dissociated; high temperature favors the greater disorder of the unbound or normal electrons.

Note that a bound pair is localized; i.e., it has zero net velocity. The distance over which the electrons in the bound pair attract each other is called the **coherence length.**

### Q. 15.2.1

It is a complex problem in quantum mechanics to show that there can be a net attraction between (a)_____ in certain solids at low temperature. This attraction leads to the formation of (b)_____. The bound pair of electrons has a net velocity of (c)_____. The distance of attraction of the bound pairs is called the (d)_____.

* * *

ENERGY GAP. In the superconducting state there is an energy band gap $E_g$ at the top of the Fermi surface within the valence band as shown in

Figure 15.2.1. This has been shown by several experiments, one of which, the specific heat, is discussed briefly now. (There are certain superconductors which have no apparent energy gap; these are not considered further here.)

Recall that the energy levels in a metal within an allowed band are extremely close together (perhaps $10^{-10}$ eV; $k_B T \approx 10^{-3}$ eV when $T = 12°$K). The specific heat of the superconducting electrons is *zero* (since there is no nearby energy level to which they can be excited). The specific heat of the superconducting state, however, is not zero; this is due to the presence of a small fraction of normal electrons (electrons not bound into pairs) which have been formed as a result of pairs being excited across the energy gap. Above the energy gap there are numerous states lying close together.

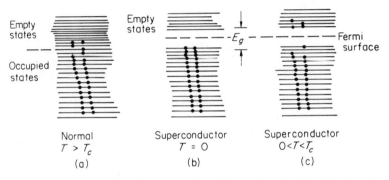

FIGURE 15.2.1. Electron levels in conductors. (a) Normal state. (b) Superconductor at $T = 0$. (c) Superconductor at $0 < T \ll T_c$.

The situation of forming some normal electrons in the superconducting state is somewhat analogous to the formation of an electron plus a hole in an intrinsic semiconductor. In that case also the Fermi level lies midway in the energy gap. Recall that the concentration of electrons (or holes) varies as $\exp[-E_g/2k_BT]$. A similar situation applies in the present case. The electronic specific heat of a normal conductor is directly proportional to the temperature

$$C_n^E = \gamma T, \qquad (15.2.1)$$

as was shown in Section 13.2. Thus the electronic specific heat of the superconductor is

$$C_s^E = \gamma T_c e^{-E_g/2k_BT}. \qquad (15.2.2)$$

The differences in specific heat between the normal and superconducting state are illustrated in Figure 15.2.2. Since gallium has a low value of $H_c(0)$,

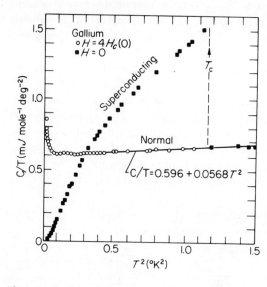

FIGURE 15.2.2. $C/T$ vs. $T^2$ curves for gallium in the normal and superconducting states. [From N. E. Phillips, *The Physical Review* **134A**, 385 (1964).]

the application of a small field makes possible the retention of the normal state at all temperatures.

## Q. 15.2.2

In many but not all superconductors there is (a)_____ at the Fermi surface of magnitude equal to about (b)_____. The specific heat of superconducting electrons (the actual bound pairs) is (c)_____. However, the superconducting state exhibits a specific heat because (d)_____.

* * *

THE COHERENCE LENGTH. An approximate quantum mechanical derivation of the coherence length of a Cooper pair is based on the uncertainty principle

$$\Delta p \, \Delta x = \frac{h}{2\pi}.$$

Physically this means that a wave packet which has a momentum uncertainty of $\Delta p$ has a spatial extent of $\Delta x$. The variation of kinetic energy $\Delta(p^2/2m) = p \, \Delta p/m$ approximately equals the energy gap $E_g$. Hence $\Delta p \approx E_g m/p = E_g/v_F$, where $m$ is the electron mass and $v_F$ is the Fermi velocity (defined in Section 13.2). Therefore the coherence length is approximately

$$\Delta x = \frac{h v_F}{2\pi E_g}.$$

The quantity

$$\xi_0 = \frac{h v_F}{2\pi^2 E_g} \qquad (15.2.3)$$

is called the **coherence length** (the additional $\pi$ arises from more exact analysis). Usually $v_F$ is large for high-conductivity metals and smaller for low-conductivity metals. $E_g$ is roughly proportional to $T_c$. $\xi_0 \approx 5000$ Å for Type I superconductors and $\xi_0 \approx 50$ Å for Type II superconductors.

### Q. 15.2.3

The coherence length increases as the (a)_____ increases and decreases as the (b)_____ increases. The coherence length of Type II superconductors is (c)_____ than the coherence length of Type I superconductors.

\* \* \*

## ANSWERS TO QUESTIONS

**15.2.1** (a) Two electrons, (b) Cooper pairs, (c) zero, (d) coherence length.

**15.2.2** (a) An energy gap; (b) $10^{-3}$ eV; (c) exactly zero; (d) a certain tiny fraction of bound pairs are excited across the energy gap, causing dissociation into normal electrons.

**15.2.3** (a) Fermi velocity, (b) energy gap, (c) smaller.

### 15.3 COLLECTIVE DE BROGLIE WAVE

Given that Cooper pairs form, why does this lead to superconductivity?

It is helpful to review the origin of electric resistance in normal metals. In this case the electrons have a high kinetic energy; they are more or less described by the particle in the box model. Their maximum kinetic energy at absolute zero is the Fermi energy $E_F$ and their average kinetic energy is $3E_F/5$. Considered as classical particles their Fermi velocity is about $10^6$ m/sec (see Table 13.2.1). Their Fermi momentum is about $10^{-24}$ kg-m/sec; hence from the de Broglie relation they have wavelengths of about 5 Å. This wavelength is of the magnitude of the spacing between atoms. As was mentioned in Chapter 13, if the ions in a metal were located precisely at lattice points, so that the structure would be a perfect *static* crystal, there would be no electrical resistance. However, because the ions are vibrating and/or because there are static crystalline defects such as vacancies, impurity atoms, dislocations, etc., there is resistance. We say that the electrons are scattered; they have a certain mean free path between collisions. They are scattered effectively because their wavelength is of the same size as the distance between scattering centers. We note that the drift velocity (which is superimposed on the random motion) has a magnitude of only about 1 m/sec (and usually much less); this corresponds to a current density of $10^{10}$ A/m². For a feeling for these current densities, see Example 3.10.1.

When an electrical field is applied to the bound pairs of a superconductor (which have zero velocity in the absence of the field), these all move together at *one* velocity, the drift velocity. Such Cooper pairs are therefore represented collectively by a *single* de Broglie wave. To obtain a current density of $10^{10}$ A/m², a drift velocity of about 1 m/sec is required (since the current density $J = Nev_D$, where $N$ is the number of current carriers per unit volume, $e$ is their charge, and $v_D$ is the drift velocity). This corresponds to a de Broglie wavelength of about $6 \times 10^6$ Å for the collective assembly of Cooper pairs. Since this wavelength is huge compared to the distance between scattering centers, scattering does not occur. It is the presence of this macroscopic de Broglie wave which provides the explanation for zero resistance.

## Q. 15.3.1

In a normal conductor when an electric field is applied there is (a)_____ drift velocity superimposed on (b)_____ random velocity. In a superconductor when an electric field is applied there is a small drift velocity superimposed on (c)_____. Hence Cooper pairs have (d)_____ wavelength compared to normal electrons in metals. Moreover, the wavelength of all the Cooper pairs is the (e)____. The whole assembly of Cooper pairs is called (f)_____. Because this wave has such a large wavelength relative to the spacing of scattering centers, it is (g)_____.

\* \* \*

## ANSWERS TO QUESTIONS

**15.3.1** (a) A small (1 m/sec or less), (b) a large ($10^6$ m/sec or so), (c) zero velocity, (d) a very long, (e) same, (f) a macroscopic de Broglie wave, (g) not scattered.

## 15.4 THE PENETRATION DEPTH

It was shown on theoretical grounds, by F. and H. London in 1935, that there is always some magnetic flux penetration near the surface in a bulk specimen and that $B$ varies with distance $x$ which is normal to the surface according to

$$B = B_0 e^{-x/\lambda_L}, \qquad (15.4.1)$$

where

$$\lambda_L = \left(\frac{m}{\mu_0 n_s e^2}\right)^{1/2} \qquad (15.4.2)$$

where $n_s$ is the number of electrons per unit volume in the superconducting state. The quantity $\lambda_L$ is now called the **penetration depth** or the London depth; it is of the order of 500 Å.

If instead of a bulk sample, a very thin superconductor of Type I were studied in a magnetic field, it would not exhibit the Meissner effect; i.e., flux would penetrate or $B \neq 0$. See Figure 15.4.1.

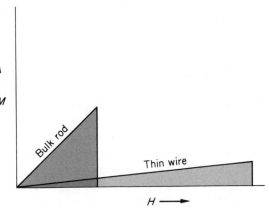

FIGURE 15.4.1. Magnetization curves of a Type I superconductor for a bulk specimen, and a wire of thickness, approximately, $\lambda$. In both cases $H$ is parallel to the rod or wire. Note that the areas are virtually the same, i.e., $-\mu_0 H_c^2/2$.

The theoretical derivation of the penetration depth and its subsequent experimental verification was one of the milestones in the understanding of superconductors. The London's derivation was based on the fact that the Gibbs free energy of the system will be least if some penetration occurs. The derivation involves the use of Maxwell's equations of electromagnetism [see P. G. de Gennes, *Superconductivity of Metals and Alloys*, W. A. Benjamin, Inc., Reading, Mass. (1966), p. 3].

If a bulk rod is placed in a magnetic field directed along the axis of the rod, there is some penetration in the radial direction. Associated with this penetration is an electrical current (circling the rod in the penetrated surface sheet). This is called a **supercurrent.** The supercurrent density can be directly computed from one of Maxwell's equations of electromagnetism.

### Q. 15.4.1

Minimum free energy in a bulk Type I superconductor in the presence of a magnetic field occurs when (a)_____ _____. This can be readily measured by studying (b)_____ superconductors. The flux penetration varies with distance $x$ normal to the surface according to (c)_____, where $\lambda_L$ is called the (d)_____. $\lambda_L$ is approximately equal to (e)_____.

\* \* \*

## ANSWERS TO QUESTIONS

**15.4.1** (a) There is some flux penetration near the surface, (b) thin film, (c) $B = B_0 e^{-x/\lambda_L}$, (d) London penetration depth, (e) 500 Å.

### 15.5 MAGNETIC FLUX QUANTIZATION

Fritz London in 1950 postulated that a supercurrent (with its associated magnetic flux field) exists only in fixed quantum states. London was the first to think of the current in superconductors as being due to macroscopic de Broglie waves. He considered an analogy with the Bohr atom of the hydrogen atom (in which the Bohr radius was taken by de Broglie to be an integral number of half-wavelengths; i.e., a stable state exists for a standing wave only). In the case of the circulating supercurrent, the quantization condition is that the flux must be quantized according to

$$\Phi = n \frac{h}{2e}, \qquad n = 1, 2, 3, \ldots . \tag{15.5.1}$$

(The flux crossing unit area is the magnetic induction $B$.) The quantum unit of flux is $2 \times 10^{-15}$ webers. Thus as flux penetration occurs (as $H$ increases), it occurs in quantum jumps. In the ordinary experiment these are too small to see. However, direct confirmation of magnetic flux quantization was made independently and almost simultaneously in 1961 by B. S. Deaver, Jr., and W. M. Fairbanks in the United States and R. Doll and M. Näbauer in Germany.

## Q. 15.5.1

When flux penetration takes place in a material in the superconducting state it actually takes place in (a)_____. The magnitude of these can be predicted by assuming that the wavelength of the circulating current, considered as a macroscopic de Broglie wave, is such that only (b)_____ are possible.

\* \* \*

## ANSWERS TO QUESTIONS

**15.5.1** (a) Quantum jumps, (b) standing waves.

### 15.6 TYPE I VERSUS TYPE II SUPERCONDUCTORS

We noted in Section 15.4 that the free energy of a system is lowered if there is some penetration. It is of interest to consider a superconductor slab as consisting of alternating layers of superconducting phase and normal phase, as in Figure 15.6.1. Let the normal phase layer be very thin (and hence

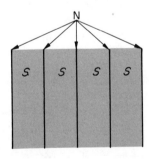

FIGURE 15.6.1. Postulated layer-like structure of superconducting phase and a thin normal phase. The mixed state.

small in volume) and let the superconducting layer have a thickness about equal to the penetration depth. The material would then be nearly completely penetrated by the magnetic field which is parallel to the length. Would having this layer like structure lower the free energy or not? The answer is: It will not if the coherence length is greater than the penetration depth (Type I behavior) but it will if the penetration depth is greater than the coherence length (Type II behavior). This latter state is called a **mixed state**.

When a superconductor is in a magnetic field there is a corresponding flux penetration. In the case of the Type I superconductor this flux penetration occurs near the external surface only, as illustrated in Figure 15.6.2. Since electrical current is carried in this region only such materials are not high current carriers.

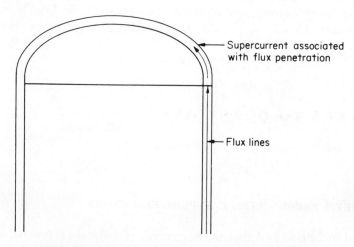

FIGURE 15.6.2. Flux penetrations in Type I superconductor.

In the case of the Type II superconductor there is flux penetration somewhat of the nature envisioned in Figure 15.6.1 but considerably different in detail. The exact picture is shown in Figure 15.6.3. This was predicted by the Russian scientist A. A. Abrikosov in 1957 based on theoretical

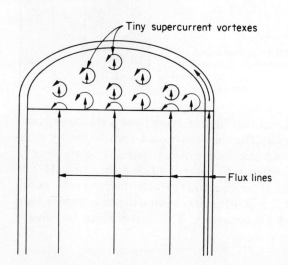

FIGURE 15.6.3. Flux penetrations in Type II superconductor.

considerations made by two other Russian theorists V. L. Ginzburg and L. D. Landau. In this picture each of the tiny vortexes has associated with it one quantum of flux. Associated with this quantum of flux which has penetrated the solid is the tiny current vortex shown. The exact nature of the vortex is shown in Figure 15.6.4. It is essentially in the normal conducting state at the center and in the superconducting state at the outside. The individual current vortex and its associated **B** field is called a **flux line**; it is also called a vortex line, a **fluxoid** or a fluxon.

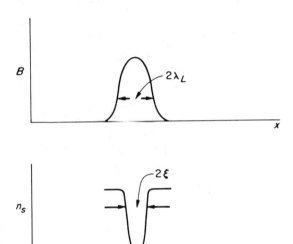

FIGURE 15.6.4. Magnetic induction and superconducting electron concentrations for a mixed state. At the center of the fluxoid the material is in the normal state; at the outside it is in the superconducting state.

In 1968 H. Träuble and U. Essmann showed that such **flux line lattices** actually exist. See Figure 15.6.5. Their method is an analog of that used in determining Bitter patterns. Tiny iron particles will be attracted to a flux line at the point where it intersects the surface. Thus the image you see in the picture is an array of iron particles.

As the field intensity is increased, additional flux lines move into the solid (presumably from the surface). Note that the fluxoid lattice is a closest packed array (which for a given number of flux lines gives the *largest* nearest-neighbor distance). As the field intensity is increased higher and higher the flux line density increases until at $H_{c_2}$, the upper critical field, the flux lines overlap and the material becomes normal.

### Q. 15.6.1

The mixed state does not occur in Type (a)―― superconductors. When a mixed state is produced it consists of (b)―――――――― fluxoid lattice. In the Type II superconductor, the flux line density (c)―――――― as the H field increases up to the field (d)――. A material will be Type I if the (e) _____ is much greater than the (f) _____ and Type II if the reverse is true.

* * *

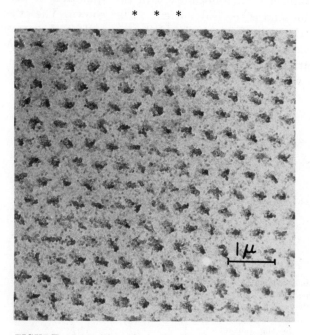

FIGURE 15.6.5. Fluxoids in Pb–6.3 at. % In. [From H. Träuble and Uwe Essmann, *Journal of Applied Physics* **39**, 4052 (1968).]

The ratio of the London penetration depth to the coherence length,

$$\kappa = \frac{\lambda_L}{\xi_0}, \qquad (15.6.1)$$

is called the **Ginsburg-Landau parameter.** It can be shown that if $\kappa > 1/\sqrt{2}$, then the superconductor is Type II, while if $\kappa < 1/\sqrt{2}$, it is Type I.

Theoretical expressions for $\kappa$ based on the quantum theory of superconductivity are given below. We shall not attempt to derive these here.

For a pure material (low residual resistivity as defined in Section 13.5)

$$\kappa = \kappa_0 = 2.7 \times 10^9 \left| \frac{dH_c}{dT} \right|_{T=T_c} \lambda_L^2(0). \qquad (15.6.2)$$

Here $H_c$ has units of amps per meter and $\lambda_L$ has units of meters.

For a material with high residual resistivity

$$\kappa = \kappa_0 + 2.4 \times 10^5 \gamma^{1/2} \rho_0. \qquad (15.6.3)$$

Here $\gamma$ is the temperature coefficient of the electronic specific heat [see (15.2.1)] and has units of joules per cubic meter-degrees Kelvin squared and $\rho_0$, the residual resistivity, has units of ohm-meters.

The derivation of the Ginsburg-Landau relation and the theoretical expressions for $\gamma$ are given in texts on superconductivity.

**EXAMPLE 15.6.1**

*Assuming* a parabolic temperature dependence for $H_c(T)$ as in Equation (3.10.1) estimate the Ginsburg-Landau parameter for lead. Use

$$\lambda_L(0) = 6.4 \times 10^{-8} \text{ m}.$$

*Answer.* We have by differentiating the expression for $H_c(T)$, substituting $T = T_c$, and taking the absolute value,

$$T_c \left| \frac{dH_c}{dT} \right|_{T=T_c} = 2H_c(0).$$

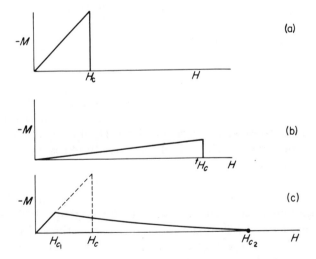

FIGURE 15.6.6. $M\text{-}H$ curves for three forms of lead. (a) Pure bulk Pb. (b) Pure thin film Pb. (c) Lead with 20 at. % In.

Since $H_c(0) = 7 \times 10^4$ A/m (from Table 3.10.1), we estimate $\kappa = 0.22$. Hence we expect pure lead to be Type I since $\kappa < 0.7$.

By adding impurities to Pb we can greatly increase its residual resistivity as was already noted in Chapter 13. Hence it is clear from (15.6.3), that $\kappa$ can be increased and in fact the lead-indium alloy with 20% indium is a Type II superconductor. We consider types of $M$-$H$ curves exhibited by lead and its alloys as shown in Figure 15.6.6. In Figure 15.6.6(a) the bulk lead exhibits the Meissner effect ($B = 0$) while the thin film of lead in Figure 15.6.6(b) does not. The addition of the alloying element to lead in Figure 15.6.6(c) increases $\kappa$ so that the alloy is Type II instead of Type I.

Note that $H_c$ for this Type II material is defined so that the area under the dashed lines equals the area under the actual $M$-$H$ curves. $H_c$ for the alloy need not be the same as for the pure lead.

## ANSWERS TO QUESTIONS

**15.6.1** (a) I, (b) a closest packed, (c) increases, (d) $H_{c_2}$, (e) coherence length, (f) penetration depth.

### 15.7 FLUXOID PINNING

The Type II superconductors have higher critical fields than Type I but the ideal Type II superconductor cannot carry large currents at large fields. This is because there is a force on the fluxoids equal to the Lorentz force,

$$\mathbf{F} = \mathbf{J} \times \mathbf{B}. \qquad (15.7.1)$$

[The student may previously have noted that the force exerted on a particle of charge $q$ moving with a velocity $\mathbf{v}$ by a magnetic induction $\mathbf{B}$ is $\mathbf{F} = q(\mathbf{v} \times \mathbf{B})$.] Thus the fluxoids in a soft Type II superconductor are readily swept out of the material if $\mathbf{J}$ and $\mathbf{B}$ are large. What is needed is a mechanism for pinning the fluxoids.

One mechanism for pinning is illustrated in Figure 15.7.1. Here the impurity precipitate and the normal region at the center of the fluxoid allow penetration. The magnetic energy of the system is lowered when the two combine. Thus it will take a force to separate them (just as it took a force to move a dislocation away from a precipitate atom). Fluxoids interact with magnetic precipitates, normal (nonmagnetic, nonsuperconductive) precipitates, voids (which are really nonmagnetic regions), dislocation arrays, dislocations, etc. All these imperfections impede the motion of

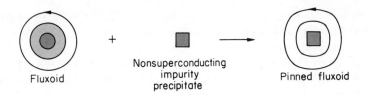

FIGURE 15.7.1. Pinning of a fluxoid.

fluxoids and hence increase the current-carrying capacity at a given field. The presence of these imperfections also produces hysteresis in the $M$-$H$ curves so these materials are called **hard superconductors,** by analogy with hard magnetic materials. They are also called **high current-carrying**

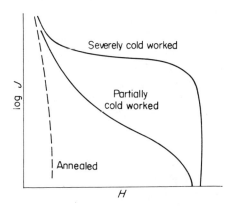

FIGURE 15.7.2. Sketch of the structure-sensitive nature of $J$ vs. $H$ for Nb–Ti alloy.

**superconductors.** Figure 15.7.2 illustrates the structure-sensitive nature of the current-carrying capacity of a superconductor.

FUNDAMENTAL THEOREM FOR HARDENING SUPERCONDUCTORS. To produce high current-carrying superconductors, disrupt the continuum. [Note that a similar theorem was applicable to making materials mechanically harder (stronger) or magnetically harder.]

In the case of the Nb–Ti alloy, which is a ductile material which can be cold-drawn into wire, cold working produces an extremely fine dislocation cell structure which is responsible for pinning fluxoids. The relative ease of manufacturing explains why Nb–60% Ti alloys are used even though $H_c$ at 4.2°K is only about $0.8 \times 10^7$ A/m compared to a value twice that high for $Nb_3Sn$. $Nb_3Sn$ is often produced by passing niobium strips through a region of tin vapor which diffuses into the heated niobium. The compound $Nb_3Sn$ is brittle.

The study of the interaction of imperfections with fluxoids is an active area of research. So is the study of the motion of fluxoids which is a dissipative process. The $J$-$H$ curves of Figure 15.7.2 refer to dc currents. When ac currents are used the fluxoids move, energy is dissipated as heat, and the material does not have high current-carrying capacity. (The heat produced is often enough to transform the material locally to the normal state; this can result in catastrophic Joule heating. Because of the low heat capacity at low temperatures and because of the small temperature interval, not much heat is required.) An examination of fluxoid lattices reveals the presence of "dislocations." One possible way by which fluxoids move is by motion of fluxoid lattice dislocations.

### Q. 15.7.1

Fluxoids in a Type II superconductor carrying a current will be under a force (a)_____. A fluxoid can be pinned by a small nonmagnetic impurity because (b)_____

_____. Almost any disruption of the superconducting crystal produces (c)_____.
Although superconductors can carry very high dc currents with *no* loss, losses do occur when (d)_____.

\* \* \*

It is hoped that future developments will include materials with higher current-carrying capacities at higher fields and at higher temperatures.

### ANSWERS TO QUESTIONS

15.7.1 (a) $\mathbf{J} \times \mathbf{B}$, (b) the magnetic energy is lowered if the normal region of the fluxoid coincides with the nonmagnetic impurity, (c) an increased hardness or increased current carrying capacity, (d) ac currents are involved.

### REFERENCES

Parks, R. D., "Quantum Effects in Superconductors," *Scientific American* (October 1965), p. 57.

Essman, U., and Träuble, H., "The Magnetic Structure of Superconductors," *Scientific American* (March 1971), p. 75. An excellent elementary discussion of fluxoids and lattices.

Livingston, J. D., and Schadler, H. W., "The Effect of Metallurgical Variables on Superconducting Properties," *Progress in Materials Science* **12**, 183 (1964).

Kittel, C., *Introduction to Solid State Physics*, John Wiley & Sons, Inc., New York (1968), Chapter 11.

Lytton, E. A., *Superconductivity*, John Wiley & Sons, Inc., New York (1969). An excellent brief introduction.

de Gennes, P. G., *Superconductivity of Metals and Alloys*, W. A. Benjamin, Inc., Reading, Mass. (1966). A good sophisticated book on Type II superconductors for the instructor.

# PROBLEMS

**15.1** Discuss five "why" questions about superconductivity.

**15.2** Create an experimental method of showing that $B = 0$ for a Type I superconductor.

**15.3** Comment on: "The superconducting state is a more ordered state than the normal state."

**15.4** A metal in the normal state exhibits two specific heat terms: One is proportional to temperature, and the other to temperature cubed. What is the origin of each term?

**15.5** What evidence is there for the statement that an energy gap exists at the Fermi surface?

**15.6** (a) Define the London penetration depth, the coherence length, and the Ginsburg-Landau parameter.
(b) What is the order of magnitude of the London penetration depth?

**15.7** Explain why a void would pin a fluxoid.

**15.8** NbTi alloys are often used for superconducting magnet materials, although their $J$-$H$ curves fall considerably below that of $Nb_3Sn$ compounds. Why?

**15.9** Invent a switch which is based on the superconducting phase transition.

# MORE INVOLVED PROBLEMS

**15.10** A question often asked is, "Is the dc-resistivity of a superconductor actually zero?" Comment on this. See, e.g., D. J. Quinn and W. B. Ittner, *Journal of Applied Physics* **33**, 748 (1962).

**15.11** Explain why the superconducting state is not just a perfect conduction state and nothing more.

**15.12** Describe the operation of the cryotron.

**15.13** Describe how $T_c$ can be measured [see L. C. Jackson, *Low Temperature Physics*, John Wiley & Sons, Inc., New York (1962)].

**15.14** Discuss the current status of superconducting magnet construction.

**15.15** Discuss the current status of the use of superconducting computer elements.

## SOPHISTICATED PROBLEMS

**15.16** Discuss the phenomenological London equation (which can be considered adjunct to Maxwell equations) and show how this leads to the penetration depth. (Use outside references.)

# EPILOGUE

> There is no law, no principle, based on past practice, which may not be overthrown in a moment by the arising of a new condition or the invention of a new material.
>
> —John Ruskin

It is a truism that: The technological development of any age is determined by the available materials and the ingenuity with which they are used. Our ability to understand, develop and properly use materials depends to a very great extent on our ability to study the structures of solids. There have been incredible advances in this area in the past two decades. Foremost among these advances were the developments of the electron microscope techniques by Peter Hirsch and others, which has played such an enormous role in the study of dislocations, point defect aggregations, precipitates, etc.; the development of the Lang X-ray technique for studying dislocations at low densities; the invention of field ion microscopy by Erwin Mueller and its development by Mueller and others so that individual atoms can be seen in high-melting metals.

The latter has been developed to a high degree by David Seidman who has been able to plot the exact positions of the many vacancies and interstitial atoms which surround a region of material in which radiation damage due to one high energy particle has occurred. Another development by Albert Crewe has made it possible to see certain individual atoms in certain molecules. We can foresee that the results of research in this area may indeed make it possible to eventually see the position of each atom in each molecule in any solid.

As we learn to understand structure and behavior, we learn to manipulate that structure to derive desired properties. The reader has already seen many examples of this. Perhaps the most exciting of these is silicon based microelectronics. However, magnetic bubbles look very promising at the present time. The author can envisage storage devices similar to magnetic bubbles in which the "storage regions" are only 25 Å in diameter. Given that they must be 100 Å apart from their centers, this would give a storage capacity of $10^{12}/cm^2$ (somewhat greater than the human brain). In fact, it may one day be possible to manipulate the nuclear spins of individual nuclei which would give an unbelievable storage density of $10^{22}/cm^3$.

It is conceivable that room temperature superconductors will be made.

It is likely that much finer grained structures will be made. It is conceivable that the size of tungsten carbide particles in sintered carbides can be decreased to 100 Å from $10^4$ Å (1 $\mu$) and that compressive stresses of the sintered carbide will increase from 70 kbars to 700 kbars. It is conceivable that methods of controlling nucleation and growth in eutectic and eutectoid structures will be developed so that grain sizes, lamellar spacing, etc., can be decreased by a factor of 100.

It is conceivable that methods for growing continuous very fine diameter whiskers at large rates will be developed. It is, for example, often said that little has come of our knowledge of dislocations but better understanding. It must be remembered that the fundamental theory of semiconductors was developed in the early 1930's and it was nearly two decades before William Shockley's great invention and another two decades before LSI was commercially available. There seems to be a gestation period after scientific developments occur. Therefore, it would not be too surprising if great advances were to be made in the mechanical properties of materials in the next decade.

It is likely that there will be developed a large number of devices and materials which are based on ideas and scientific principles which have been around for a long time. Such developments in the recent past include the laser and the magnetic bubble. The reader who wishes to pursue these matters further is encouraged to read the excellent article by J. H. Westbrook, "Materials for Tomorrow," in the book *Science and Technology in the World of the Future*, A. B. Bronwell, ed., John Wiley & Sons, New York (1970) and the references quoted therein.

Materials science and the application of materials science (molecular architecture) is in its early youth. Peter Drucker in *The Age of Discontinuity*, Harper and Row, Publishers, New York (1969), considers it to be one of four areas which will undergo revolutionary change.

# INDEX*

## A

α-amino acids, 330
α-iron, 459
Ablative compound, 161
Absorption coefficient, **618**
Absorption spectra, **182**
Acceptor, **598**
Acid, organic, 319
Acrylonitrile, 310
Activated state, **396**
Activation energy
   barrier, **396**
   for creep, **70**
   for motion, **396**
   for tracer diffusion, **402**
Active (corrosion), **503**
Addition polymerization, **303**
Adherent oxide film 415, **502**
Adipic acid, 320
Aeration cell, **497**
Aerospace, 46
Age precipitation hardened alloys, **456**
   examples, 457
Age precipitation hardening, 168
Aggregate, 347, 360
Aging, **455**
Aging heat treatment, **168**
AISI code for steel, 551
Al-Cu phase diagram, 454
Allowed bands, **581**
Alloy, **353**
Alnico V,
   microstructure of, 647
   properties of, 135
Altimeter, 390
Aluminum-copper alloys,
   Al-Cu phase diagram, 454
   crystal structure, 240, 241
   GP zone formation, 542

Aluminum-copper alloys (*Contd*)
   heat treatment, 168, 454
   precipitate strengthening, 541
   substitutional solutions, 454
Amines, **319**
Amounts of phases, 439
Ampere, 99
Amplitude, **78**
Anelastic behavior, 80
Angular momentum quantum number, **195**
Anions, **481**
Anisotropic, **41**
Anisotropy,
   of hexagonal crystals, 266
   relation to symmetry, 265
Anneal(ed), **65**
Anode, **481**
Anodizing, **502**
Antiferromagnetism, 636
Arrhenius equation, **398**
Artificial heart, 51
Asbestos, 259
Asphalt pavement, 560
ASTM grain size, **349**
   data, 350
Atactic, **311**
Atom vibrations, 392
Atomic:
   coordination, 251
   percent, 441
   radius, **253**
   radius from lattice parameter, data, 253
   scale, **341**
Austenite, **458**
Austenitic steel, **466**
Austenitizing, **168, 465**
Avalanche effect, **606**
Average molecular weight, **307**

---

*The boldface page numbers refer to the page on which the entry is defined and set in boldface type.

# B

Back voltage, **492**
Bainite, **461**
Bakelite, **325**
Band structure:
  in divalent metals, 592
  in insulators, 590
  in monovalent metals, 587
  in semiconductors, 590
Bar, **35**
Barium titanate, 345
Barometric formula, **390**
Basal plane, **245**
Base (npn junction), **609**
Basis, **237**
Batteries, 49
  references, 505
Bauschinger effect, **86**
Beta-alumina, 480
Bimetallic strip, 153
Binary phase diagram, **432**
Binary system, **432**
Biomaterials, 50
  references, 54
Bitter pattern, **342**
Bloch wall, **643**
Block polymer, **318–319**
Body-centered cubic, **238**
Bohr magnetron, **223**
Bohr postulate, **183**
Bohr radius, **185**
Boltzmann distribution, **391**
Bond:
  angles, 210
  dissociation energies, data, 212
  lengths, data, 212
Bond straightening, 316
Bonding energy:
  alkali halides, data, 207
  metals, data, 214
  secondary bonded materials, 216
Bonds:
  covalent, 208
  glassy, 284
  ionic, 203
  metallic, 213
  secondary, 214
Born repulsive potential, **204**
Boron fiber composite, 357
Boron filaments, 357, 545
Boron filament composite micrograph, 29
Boundary, grain, 279
Boyle's law, 370
Bragg:
  equation, **290**

Bragg (*Contd*)
  law, **290**
Branched polymers, **304**
Brass, 268
Bravais lattices, **236**
Breakdown, **606**
Breeder reactor, 46
Bridgman method, **337**
Brinell hardness number, 64
Brittle fracture, impact, 67
Brittle materials, **59**
Bubble raft model of crystal, 280
Bubbles, **281**
Bulk modulus, **89**
  relation to Young's modulus, 90
Burgers vector, **277**

# C

Cable construction, 115
Capacitance, 109
Capacitor, 109
Carbide, cemented, 546
Carbide precipitation, 507
Carbon-diamond phase diagram, 431
Carbon solubility in iron, 459
Carburization depth, **408**
Carriers, 593
Case:
  carburized, 408, 551
  hardened, **551**
  nitrided, 527
Cast iron:
  gray, 467, 527
  nodular, 468, 528
  white, 467
Cast irons, **467**
Casting, 350
Cathode, **481**
Cations, **481**
Caustic embrittlement, **500**
Cell:
  aeration, 497
  concentration, 483
  differential pressure, 483
  differential temperature, 483
  dissimilar electrode, 483
  oxygen, 497
  unit, 236
Cellulose, 329
Cement, Portland, 360
Cemented carbides, 546
Cementite, **460**
Ceramic materials, **259**
  references, 295
Cesium chloride crystal structure, **255**

Index    681

Chain stiffening, 314
Chain structures, 304, 306, 313
Charge:
  flux, **100**
  mobilities in semiconductors, data, 595
  mobility, **568**
Charpy test, 67
  V-notch specimen, 68
Chemical properties, 169
Chloroprene, 318
*Cis* configuration, **317**
Closest-packed:
  directions, **247**
  layer, **247**
  structure, 248
Closure domains, **641**
Coatings:
  abrasion resistant, 362
  corrosion resistant, 361
  oxidation resistant, 361
Coercive:
  field, **118**
  force, **126**
Coherence length, **660, 663**
Coherent precipitate, **455**
Coil, inductance of, 122
Cold worked, 65
Collective de Broglie wave, 663
Collector, **609**
Color, 230
Columnar grains, **352**
Comet airplane failure, 72
Component, 431, 436
Composites, **355**
  boron fiber, 357
  coatings, 361
  concrete, 360
  GRP, 355
  honeycomb, 359
  laminates, 359
  microelectronic circuit, 19, 363, 415
  plywood, 363
  prepregs, 359
  prestressed concrete, 361
  references, 365
  sintered tungsten carbide, 362
  strengthening, 528
  synthetic hard superconductor, 363
Composition, **436**
Compressive strength, 521
Concentration cell, 483
Concentration polarization, **491**
Concept of local equilibrium, **470**
Concrete, 360
  references, 366
Condensation polymerization, **320**

Conduction band, **582**
Conduction, electrical, by
  electrons, 586
  holes, 586
  ions, 412
Conductivity, electrical, **100**
  of ionic materials, 411
  of metals, 104, 587, 592
    data, 101
  of semiconductors, 107, 590, 593, 596
    data, 101
  Ohm's law, **100**
  Nernst equation, **412**
Conductivity, thermal, 154
  coefficient, **154**
  data, 155, 157, 158
  Fourier heat conduction law, **154**
Congruent melting point, **444**
Conservative:
  properties, examples, 172
  system, **78**
Constants, (*see frontispiece*)
Contact potential, **602**
Controlled eutectics, 451
Cooling curve, **475**
Cooper pairs, **660**
Coordination:
  number, **251, 257**
  polyhedron, **251**
Copolymer, **318**
Copper:
  crystal structure, 240, 241
  electrical resistivity, 101, 587
  electron mean free path, 589
  Fermi energy, 576
  melting temperature, 164
  oxidation, 170
  Poisson's ratio, 61
  shear modulus, 43
  solution with nickel, 268, 451
  thermal conductivity, 155, 157
  with beryllium, 457
  Young's modulus, 42
Copper-beryllium alloys, 457
Coring, **452**
Corrosion:
  effect of liquid velocity, 492
  fatigue, **499**
  protection, 501
  references, 505
Corrosion, types of:
  caustic embrittlement, 500
  decarburization, 499
  dezincification, 498
  fatigue, 499
  intergranular, 498
  oxygen starvation, 497

Corrosion, types of (*Contd*)
  season cracking, 500
  stress corrosion cracking, 499
Corundum, 612
Coulomb, 95
Coulomb attraction, 96
Covalent bond, **210**
C-Ni phase diagram, 433
Creep, **69**
  activation energy, **70**
  steady state, **69**
  theory of high temperature, 558
Cristobalite, 283
Critical resolved shear stress, **525**
Crosslinking, **315**
Cross slip, **532**
Cryogenic temperatures, **102**
Crystal:
  covalent, 212, 240
  homopolar, 212, 240
  ionic, 254
  metallic, 239, 240
  molecular, 216, 379
  poly-, 347
  single, 337
Crystal cell dimensions, data, 241
Crystal data, references, 294
Crystal growth:
  Bridgman method, 337
  from solution, 337
  pulling from melt, 338
  references, 365
Crystal structure, **237**
Crystal structures:
  body centered cubic, 238
  cesium chloride, 255
  diamond cubic, 240
  face centered cubic, 239, 250
  hexagonal closest packed, 241, 249
  simple cubic, 245
  sodium chloride, 206, 254
  wurtzite, 256
  zinc blende, 255
Crystal systems, 235
Crystallization:
  in castings, 350
  in polymers, strain induced, 316, 317
Crystallographic:
  direction, **242**
  plane, **243**
Cu-Au phase diagram, 441
$Cu_3Au$ ordered solution, 269
Cube-textured material, **649**
Cubic crystal:
  body-centered, 238
  diamond, 240
  face-centered, 239

Cubic crystal (*Contd*)
  simple, 245
Cubic lattice:
  body-centered, 236
  face-centered, 236
  simple, 236
Cu-Ni phase, 451
Curie:
  law, **628**
  temperature, **128**
Curie-Weiss relation, **629**
Curve:
  S-N, 71
  T-T-T, 549
Cyclic:
  loading, 71
  stress fatigue, **71**
Cyclotron frequency, **618**

**D**

δ-iron, 459
Damage, radiation, 47
Damped oscillations, 79
Darcy's equation, 77, **414**
Dark-resistance material, **617**
Data:
  angle of loss
    mechanical, 81
    electrical, 113
  ASTM grain size, 350
  atomic radius from lattice parameter, 253
  B-H data for silicon steel, 147
  bond energies, 212
  bonding energy
    ionic bonds, 207
    metals, 214
    organic compounds, 212
    secondary bonded materials, 216
  bond lengths, 212
  cell dimensions of some elements, 241
  charge mobilities in semiconductors, 595
  Curie temperature of ferromagnetic elements, 635
  diamagnetic susceptibilities, 623
  dielectric constants, 110
  dielectric loss angle, 113
  dielectric strength, 114
  diffusion:
    interstitial, 406
    self- in metals, 410
    in polymers, 414
  elastic constants
    Poisson ratio, 61

Data (Contd)
    shear modulus, 43
    Young's modulus, 42
    electric dipole moments of molecules, 631
    electric loss angle, 113
    electrical resistivity, 101
        of metals versus temperature, 102
        of semiconductors versus temperature, 107
    electron concentration, 576
    energies of vacancy formation, 387
    energies of vacancy motion, 398
    energy gaps in semiconductors, 594
    Fermi energy, 576
    Fermi temperature, 576
    Fermi velocity, 576
    ferroelectric crystal data, 118
    ferromagnetic Curie points, 128
    ferromagnetic saturation induction, 128
    ferromagnetic, soft:
        hysteresis loss, 130
        initial permeability, 130
        saturation induction, 130
    galvanic series of metals in salt water, 489
    grain size, ASTM, 350
    heat of fusion, 161
    heat of vaporization, 161
    hydrogen overvoltage, 493
    ionic radii, 257
    ionization energies of solutes in semiconductors, 598
    ionization potentials, 201
    lattice parameter of martensite versus carbon content, 462
    lattice parameters of some elements, 241
    loss angles:
        electrical, 113
        mechanical, 81
    magnetization saturation of ferromagnetic elements, 635
    maximum energy product, 135
    mean free path of gas molecules, 377
    melting temperatures, 164
        interstitial compounds, 260
        polymers, 308
    minimum radius ratios, 258
    molecular diameters, 379
    molecular speeds, 372
    Poisson ratio, 61
    polymer production, 305
    Q-factor, 84
    relative permeability, 133
    shear modulus, 43

Data (Contd)
    specific stiffness, 73
    specific tensile strength, 73
    standard oxidation potentials, 486
    strengths of filaments, wires and whiskers, 545
    superconducting critical field, 137
    superconducting critical temperature, 137, 140
    tensile strength:
        of filaments, wires and whiskers, 545
        of sapphire whiskers, size effect, 358
    thermal conductivity, 155
        of copper, 157
        of quartz, 158
    thermal emf's of junctions, 166
    thermal expansion, linear, 152
    vibrational frequencies of atoms, 394
    viscosity coefficients, 75
    void radius ratios, 253
    Young's modulus, 42
de Broglie relation, **186**
Debye, **220, 631**
Debye-Scherrer method, 291
Decarburized, **499**
Deformation:
    elastic, 35
    elastoplastic, 85
    plastic, 35
    viscous, 74
Degree of crystallinity, **288**
Degree of polymerization, **307**
Dendrites, **352**
Density of crystals, from X-ray diffraction data, 242
Density of states, **634**
Desalinization, 56
Devitrification, **288, 519**
Dezincification, **498**
Diamagnetic, **124**
Diamagnetism:
    perfect, 659
    theory of, 621
Diamond:
    band gap energy, 594
    charge mobilities, 595
    crystal structure, 240, 241
    electrical resistivity, 101
    metastability, 170
    origin of electrical resistance, 590
    production, 433
    synthesis, 433
    Young's modulus, 42
Diamond cubic structure, **240**
Diamond structure, **212**
Dielectric:
    constants, **110**

Dielectric (Contd)
  constants, data, 110
  constants, theory, 631
  loss angle, **112**
  loss angle data, 113
  loss factor, **112**
  strength, 114
  strength data, **114**
  susceptibility, **117**
  susceptibility owing to induced dipoles, 631
  susceptibility owing to permanent electric dipoles, 631
  theory, 631
Dielectric:
  ideal, 112
  nonideal, 112
Dielectrics, **109**
Differential:
  concentration cells, **483**
  oxygen cell, 497
  pressure cell, **483**
  temperature cell, **483**
Diffraction:
  angle, **290**
  by X-rays, 288
  references, 294
Diffusion:
  applications of, 418
  controlled processes, 418
  distance, **403**
  gases, 405
  in ionic crystals, 411
  in microelectronic circuit manufacture, 415
  in polymers, 413
    data, 414
  interstitial solid, 406
    data, 406
  length, **601**
  references, 424
  self- in metals, data, 410
  solids, 404
  steady-state, 400
  time-dependent, 401
  tracer, 402
  vacancies in metals, 408
Diffusionless transformation, **463**
Dipole moment, **215**
Dipoles:
  dislocations, 533
  electric field of permanent, 222
  induced electric, 221
  magnetic, 222
  permanent electric, 215
  relation of electric dipoles and polarization, 220

Dipoles (Contd)
  relation of magnetic dipoles and magnetization, 223
  torque due to electric fields, 220
Dislocation, **271**
  annihilation, 536
  barriers, 528
  climb, **409**
  cross slip, 532
  density, 276, 534
  dipole, **533**
  edge, 273
  energy per unit length, 535
  etch pits, 277
  Frank-Read source, 534
  glide, 274
  impurity atmospheres, 530
  in grain boundaries, 279
  in subcells, 537
  jogged, 409
  mixed, 276
  motion, 274, 276
  multiplication, 277, 532, 533, 534
  partial, 464
  Peierls-Nabarro stress, 526
  pileup, 522
  pinning, 529
  plastic deformation, 276
  plastic strain, 531
  rearrangement, 536, 554
  references, 561, 562
  screw, 275
Dispersion forces, **216**
Dispersion hardened alloys, 557
  SAP, 557
  TD nickel, 558
Dissimilar electrode cells, **479, 483**
Dissipative process, **81**
  examples, 173
Distinct configurations, **383**
Distribution coefficient, **469**
DNA, 332
Domain boundaries, **343**
  walls, **343, 346, 643**
Domains, **633**
  basis for occurrence of, 639
  boundary pinning of, 648
  closure, 641
  magnetic bubble, 651
  micrographs of, 26
  observing, 644
  single particle, 646
Donors, **598**
Doping, **597**
Double chain structure, **300**
Drift velocity, **568**
Drude's model, **567**

Ductile to brittle transition temperature, **68**
Ductility, **63**
  measures of, 63
  pressure induced, 63
Dulong and Petit, **160**, 395
Dye molecules, color, 230

# E

Easy direction, **642**
Edge dislocation, **273**
  moiré pattern of, 273
Effective mass, **586**
Efficiency, thermal, 69
Eigenvalue problems, 188
Einstein frequency, **395**
Einstein model of specific heats of solids, 395
Einstein relation, **182**
Elastic:
  anisotropy, **41**
  constant measurements, 120
  deformation, **35**
  energy density, 37
Elastic constants:
  bulk modulus, **89**
  Poisson's ratio, **60**
    data, 61
  shear modulus, **40**
    data, 43
  use in seismology, 90
  Young's modulus, **36**
    data, 42
Elastomers, 315
Elastoplastic materials, **85**
Electric current density, **100**
Electric dipole, 215
Electric dipole moments of molecules, data, 631
Electric intensity, **95**
Electrical conductivity, **100**
Electrical materials, references, 142
Electrical resistivity, **100**
  data, 101
  of metals versus temperature, 106
  of semiconductors versus temperature, 107
  origin, 588
Electrochemical cell, **479**
Electrode, **480**
  potential, **485**
  reaction, **485**
Electrodeposition, references, 506
Electrolytic cells, 483

Electromachining, 479
Electron:
  concentration, data, 576
  configuration, **200**
  configuration of light elements, 200
  delocalization, 209, 214
  gas, 213
  in a periodic potential, 580
  injection, **608**
  pair bond, **208**
  spin, existence of, 224
Electronic specific heat, 577
  of superconductors, 661
Electrons in solids, references, 615
Electroplating, 479
Elongated single domains, **646**
  permanent magnet, **646**
Emission spectra, **182**
Emitter, **608**
Energy:
  activation, 396
  bands, 580
  gaps, **581**
  gaps in semiconductors, data, 594
  gaps in superconductors, 660
  Gibbs free, 386
  internal, 380
  levels, 185, 573, 580, 613
  of formation of a vacancy, **269**, **386**
  of formation of vacancies, data, **387**
  of motion of a vacancy, 397
  of motion of vacancies, data, 398
  references, 53
Enthalpy, **159**
Entropy, **384**
  of mixing, **385**
Equation of state, **374**
Equiaxed, **349**
Equilibrium, **170**
Equilibrium behavior, **379**
Error function, **407**
Esaki diode, **618**
ESD magnet, 646
Etch pit techniques, **277**
Etchants, **342**
Etching, 342
Ethane molecule, 211
Ethylene, 310
Eutectic:
  alloy, **443**
  composition, 443
  controlled, 451
  lamellar, 450
  line, 443
  point, **443**
  rod-like, 450

Eutectic (*Contd*)
  structure, 447
  temperature, 443
  transformation, 443
  uniaxially solidified, 448
Eutectoid, 446
  structure, 460
Exchange interaction, 634
Excluded volume, 375
Expansion, thermal
  linear, 152
  volume, 152
Extension ratio, 33
External surface, 278
Extrinsic, 411
  semiconductor, 107, 596

## F

Face-centered cubic, 239
Failure:
  dielectric, 114
  mechanical, 67, 72
Farad, 99
Faraday equivalent, 483
Fatigue:
  cyclic stress, 71
  Comet disasters, 72
  limit, 72
Fe-C phase diagram, 459
Fe-Fe$_3$C phase diagram, 459
Fe$_3$C, 459
Fermi-Dirac distribution, 577
Fermi energy, 213, 577
  at absolute zero of temperature, 575
  data, 576
  of n-type semiconductor, 600
  of p-type semiconductor, 600
Fermi surface, 575
Fermi temperature, 575
  data, 576
Fermi velocity, 576
  data, 576
Ferrimagnetic materials, 637
Ferrimagnetism, 223, 636
Ferrite, 458
Ferrites, 636
Ferroelectric, 117
  Curie point, 118
  data, 118
  domains, 346
Ferromagnetic, 124
  Curie point, data, 128
  domain, 343
  elements, data, 635

Ferromagnetic (*Contd*)
  saturation induction, data, 128
  transition temperature, 128
Ferromagnetism, 223
  occurrence, 635
  origin, 633
  references, 654
Fibers, glass, 521, 545
Fick's first equation, 400
  law, 400
Field ion:
  micrograph, 23
  microscopy, 22
Filaments, 545
  boron, 357, 528, 545
  graphite, 528, 545
First Brillouin zone, 582
First-order transitions, 659
Flame hardening, 551
Flow:
  plastic, 59
  viscous, 74
Fluidity coefficient, 75
  dependence on temperature, 76
Fluids, Newtonian, 75
Flux, 76
Flux line, 669
Flux line lattice, 669
Flux penetration, 665
Flux quantization, 666
Fluxoid, 669
  lattice, 669
  lattice dislocation, 674
  pinning, 672
Folded-chain structure, 327
Forbidden energy bands, 581
Forces:
  electric, 95
  interatomic, 203
  magnetic, 96
  repulsive, 204
  van der Waals, 216
Formaldehyde, 324
Forsterite, 300
Forward bias, 603
Fourier heat conduction law, 154
Fractional decrease in:
  amplitude per cycle, 81
  stored energy per cycle, 81
Fracture of:
  brittle materials, 518
  ductile materials, 522
Framework structure, 300
Frank-Read source, 534
Free electrons, 213
Free electron gas model, 567
Free energy, Gibbs, 386

Frenkel defect, **270**
Frenkel's model, **517**
Frequency, 182
Fringed micelle structure, **326**
Fuel cells, 49, **495**
 references, 54, 505
Fused silica, 286

# G

GaAs:
 band gap, 594
 charge mobility, 595
 crystal structure, 597
Galloping Gertie, 83
Galvanic series, **490**
 of metals in salt water, 489
Galvanizing, **503**
Gap energy, 581, 593
Gases:
 diffusion, 405
 ideal, 369
 mixing, 382
 viscosity, 376
Gel, **360**
Gibbs free energy, **386**
Ginsburg-Landau parameter, **670**
Glass, 284, 285, **286**
 inorganic, 286
 polymeric, 316
 references, 295
Glass-reinforced plastics, (GRP), **355**
Glass transition temperature, **285**
Glaze, 287
Glide, **523**
 planes, **524**
Glide packets, **523**
GP zone, 542
Graft polymer, **319**
Grain, **347**
 boundary, 279
 boundary strengthening, 538
 columnar, 352
 equiaxed, 348, 349
 growth, **555**
 preferred orientation of, 350
 size, 360
 size, ASTM, data, 350
Grain-oriented steel, 649
Graphite, 266
Gray cast iron, **467**
Griffith theory, **519**
Ground state, 183
Grown from solutions, 337
Growth spiral, 275

Guinier-Preston zone, 542
Gutta percha, **317**

# H

Habit plane, **464**
Hair, 331
Half-cell, **485**
 potential, **485**
Hall coefficient, **103**
 anomaly, **570**
 of divalent metals, 592
 units, 104
Hall effect, **103**
Hard direction, **642**
Hard magnetic material, **134**
Hard rubber, 315
Hard superconductors, **673**
Hardenability, **550**
Hardened materials:
 magnetic, 645
 mechanical, 527
 superconductive, 672
Hardening:
 age, 541
 dispersion, 557
 flame, 551
 induction, 551
 martensite, 547
 precipitation, 541
 second phase, 541
 solute, 529
 strain, 530
Hardness, **64**
Hardness:
 Brinell, 64
 maximum, 548
 Rockwell, 64
 Vickers, 64
Harmonic motion, **78**
Harmonic oscillator:
 classical mechanical, 392
 quantum mechanical, 393
Heat capacity:
 at constant pressure, **160**
 at constant volume, **159**
Heat flow:
 steady-state, **155**
 time-dependent, **154**
Heat of:
 fusion (melting), **161**
  data, 161
 sublimation, **161**
 vaporization, **161**
  data, 161

Heat treatment, 168
  of steel, 168
Helical chain, **313**
Henry, 99
Heterogeneous nucleation, **421**
Hexamethylenediamine, 320
High current-carrying superconductors, 673
Hindered rotation, **313**
Hole, **586**
Home permanent, 332
Homogeneous nucleation, **421**
Homogenization, 452
Homopolar, 203
Honeycomb, **359**
Hooke's law, **36**
Housing, 51
Hund's rule, **635**
Hybridized orbitals, **210**
Hydrogen atom:
  Bohr radius, 185
  energy levels, 194
  quantum numbers, 195
Hydrogen bond, **217**
Hydrogen bonding, **217**
  in nylon, 321
Hydrogen-like atom, **196**
  energy levels, 196
  wave functions, 197
Hydrogen molecule, 209
Hysteresis curves, **126**
  examples, 173
Hysteresis loss, **129**

# I

Ice, 218
Ideal gas, **369**
  law, **373**
  temperature scale, **373**
Impact toughness, **67**
Imperfections, **267**
  references, 294
Impressed potential, **503**
Impurity atmospheres, **530**
Indices, Miller, 243
Indifferent point, **440**
Induced currents, 132
Induction effect, **216**
Induction hardening, 551
Induction, remanent, **126**
Induction, saturation, **126**
Injection efficiency, **610**
Insulator, 109
Insulators, electrical, **102**

Interatomic distances, 212, 253, 254
Interfacial angle, 233
Interfacial energy, 299, 347, 422, 447
Intergranular corrosion, **498**
Internal damping, 80
Internal energy, **159, 380**
Internal field, **632**
Internal friction, **80**
Internal resistance, **495**
Internal stresses due to quenching, 465
Interplanar spacings, 245, 300, 301
Interstitial compounds, **260**
  examples of, 260
Interstitial impurity, 254
Interstitial solid solution, **268**
Intrinsic, **411**
Intrinsic semiconductor, **107, 593**
Invariant points, **446**
Inversion center, **263**
Inversion operation, **263**
Inversion point, **446**
Ionic, **204**
  bond, **204**
  conductivity, 412
  diffusion, 411
  radii, **206,** 256
  radii, data, 257
Ionization energies for solutes in semi-
    conductors:
  data, 598
  theory, 597
Ionization potential, **183**
  data, 201
Iron:
  carbide, 459
  crystal structure, 239, 241, 254, 459
  ferromagnetic domains, 26, 342
  ferromagnetic properties, 128, 130, 133, 135, 635
  in cast iron, 467
  in steel, 460
  polycrystalline, 26
  solution with carbon, 268, 459
Island structure, **300**
Isotactic, **311**
Isotope effect, **660**
Isotropic, elastically, **41**

# J

Jogs, **409**
Jominy test, **550**
Joule, 99
Junction transistor, **608**

*Index* **689**

## K

Kelvin equation, **420**
Kinetic behavior, **380**
Kinetics, **170**

## L

Lamellar structure, **353**, **447**
Laminates, **359**
Langevin function, **627**
Large-angle grain boundaries, **279**
Large-scale integration, **608**
Laser, **612**
Laser, ruby, 613
Lattice, **235**
    coordinates, **239**
    parameters, **236**
        data, 241
    vector, **236**
Laue method, **291**
Level, energy, 185, 613
Lever rule, **439**
LiBr-LiCl phase diagram, 440
Light waves, 182
Limit properties, examples, 173
Line defects, **267**
Line tension, **541**
Linear behavior, **36**
Linear conservative properties, 172
Linear dissipative properties, 173
Linear polymers, **304**
Linear relationship, **36**
Linear thermal expansion coefficient, **152**
Liquidus, **436**
Local equilibrium, 428, 429
Lodestone, $Fe_3O_4$, 224, 637
Lodex, 646
Long-range interaction, 208
Loop, hysteresis, 129, 173
Lorentz force equation, **96**
Loss angle, **80**
Loss angle, relation to:
    fractional decrease in amplitude per cycle, 81
    fractional decrease in stored energy per cycle, 81
    Q-factor, 83
    relaxation time, 60
    resonance amplification, 82
    resonance peak width, 83
Losses:
    dielectrics, 112
    magnetic hysteresis, 129
    mechanical, 80

Losses (*Contd*)
    ohmic, 101
Lucite, 316

## M

Machining, role of plastic deformation, 63
Macroscopic, 30
Macrostructure, **341**
Madelung constant, 207, 228, 229
Magnesium oxide:
    crystal structure, 206
    dislocation in, 543
    melting temperature, 164
    polycrystalline, 27
    refractory, 164
    with nickel oxide, 435
Magnetic bubbles, **651**
    motion of, 653
    reference, 654
Magnetic dipole moment, **222**
Magnetic induction, **96**
    generated by magnetic dipole, 629
Magnetic materials, references, 143
Magnetic oxides, 133, 633, 650
Magnetic quantum number, **195**
Magnetic saturation, 127, 224, 628
Magnetic susceptibilities:
    diamagnetic, 623
        data, 623
    paramagnetic, 628
Magnetic susceptibility, **124**
Magnetism:
    antiferro-, 636
    dia-, 621
    ferri-, 636
    ferro-, 632
    para-, 624
Magnetization, **127**, **223**
Magnetocrystalline anisotropy, **642**
Magnetocrystalline energy, **642**
Magnetostatic energy, **639**
Magnetostriction, **133**, **640**
Magnets:
    hard, origin, 645
    soft, origin, 648
Majority carriers, **600**
Martensite, **462**
    finishes temperature, **463**
    lattice parameter, data, 462
    starts temperature, **462**
    tempered, 465
    transformation, **464**
        examples, 465

Masking, 418
Materials balance, 439, 470
Materials science, **4**
  outline of, 52
Matrix, 545
Maximum energy product, **134**
  data, 135
Maximum hardness, 548
Mean free path, **376**
  gas molecules, data, 377
Mechanical properties:
  data references, 86
  references, 86
Meissner effect, **138**
Melting temperature, **163**
  data, 164
  fraction of, **163**
  of interstitial compounds, 260
  of polymers, 307
Membrane, semipermeable, 309
Mer, **303**
Metallic bonding, 213
Metallography, **342**
  references, 365
Metals, **101**
Metastable, **347**
  structure, **171**
Metastable structures:
  age precipitation hardened solids, 456
  cold worked solids, 65, 554
  copper, in air, 170
  cored structures, 452
  diamonds, 170
  eutectic structure, 447
  eutectoid structure, 460
  grain boundaries, 347, 555
  iron carbide, $Fe_3C$, 458
  martensite, 461
  p-n junction, 617
  pearlite, 460
  supercooled solids, 454
Methane molecule, 211
MgO-NiO phase diagram, 435
Mica, 110, 114, 259
Microscopic, 30
Microduplex structure, 28
Microelectronics, 17, 45
Microelectronics technology, reference, 419
Micrograph, **342**
Microscopy references, 365
Microstructure, **341**
Microstructure, effects on:
  current carrying capacity of superconductors, 672
  hardness of magnetic materials, 645
  mechanical strength, 527

Microstructures:
  Alnico-V, 647
  dislocation tangles, 537
  domains in $BaTiO_3$, 346
  domains in iron, 26, 342
  filamentary composite, 29
  fluxoids in superconductors, 670
  foam, 29
  lamellar structures, 28
  magnetic bubbles, 651
  martensite, 464
  microduplex structure, 28
  polycrystalline iron, 26
  polycrystalline MgO, 27
  precipitates, 457
  spherulites, 27
  voids, 344
Miller indices, **243**
Miller-Bravais indices, **246**
Minimum radius ratio, **257**
  data, 258
Minority carriers, **600**
  lifetime, 600
Mixed state, **667**
Mobility, 568
Modulus:
  origin of, 513
  shear, 40
  Young's, 36
Molality, **483**
Molecular:
  architect, **45**
  beam experiments, 372
  diameter, data, 379
  solids, **214**
  speeds, data, 372
  velocity distributions, 372, 427
  weight, 307
Mooney's equation, **561**
Mott-Nabarro force, **541**
Multiphase materials, 353
Multiple slip, **531**

# N

$NaF-MgF_2$ phase diagram, 445
Natural frequency, **78**
Natural rubber, **318**
$Nb_3Sn$, 139, 140, 673
Necking, **62**
Neoprene, **318**
Nernst relation, **412**
Network-forming ions, **287**
Network-modifying ions, **287**
Network polymers, 325

Newton, 99
Newtonian fluid, **75**
Newton's equation of viscosity, **377**
Niobium, 137
Nitriding, 527
Noble coatings, 501
Nodular cast iron, **468**
Nominal strain, **34**
Noncentrosymmetric crystals, **263**
Noncrystalline materials, 354
Nonequilibrium structures, (*see* metastable structures)
"Nonmagnetic"-magnetic phase diagram, 128
Nonstoichiometric compounds, **271**
Normal stress, **34, 38**
Normal-superconductive phase diagram, 137
Notch sensitivity, 67
n-p-n junction transistor, 608
Nucleation, **420**
  references, 424
Nylon 66, **320**

## O

Octahedral
  coordination, 251
  planes, **245**
  void, **251**
Ohm, 99
Ohm's law, **100**
One-dimensional particle in a box, **191**
1-4 addition, **315**
Order of the reflection, **290**
Ordered solid solutions, **269**
Organic acids, **319**
Orientation effect, **215**
Orientation:
  electric dipole, 631
  grain, 649
  magnetic dipole, 624
  preferred, 350
Orlon, 311
Orthoferrites, 650
Osmotic pressure, **309**
Overaging, **456,** 556
Overvoltage, **494**
Overvoltage, hydrogen, data, 493
Oxidation, 415
  potentials, 486
  rates, 416, 427
  reaction, 481
Oxygen concentration cell, **496**
Oxygen starvation, **497**

## P

Packing in crystals, 247
  references, 293
Packing imperfections, **267**
Parabolic diffusion equation, **401**
Parabolic growth law, **416**
Paramagnetic, **124**
Parameter, lattice, 236
  data, 241
Particle in a box:
  one-dimensional, 191
  two-dimensional, 191
  three-dimensional, 192
Particle in a box problem, **192**
Particle in a box with infinite walls, **571**
Particle strengthening, 541
Passivation, 502
Passive, **502**
Pauli exclusion principle, **199**
Pauling's rules, **259**
Pb-Sn phase diagram, 442
Peak width, **83**
Pearlite, **460**
  formation, 460
  strength, 544
Peierls-Nabarro stress, **526**
Peltier coefficients, **167**
Peltier effect, **167**
Penetration depth, **665**
Peptide linkage, **330**
Percent:
  atomic, 441
  weight, 441
Percent reduction in area, 63
Peritectic, **446**
Peritectoid, **446**
Permanent dipole, **215**
Permanent magnets, references, 654
Permeability:
  coefficient for fluid flow, 77
  coefficient, magnetic, **122**
  flow in polymers, 413
    data, 414
  initial, **126**
  maximum, **126**
Permittivity of vacuum, **109**
Phase, **268, 353**
Phase compositions, 437
Phase diagram, **431**
Phase diagrams:
  Al-Cu, 454
  carbon-diamond, 431
  C-Ni, 433
  Cu-Au, 441
  Cu-Ni, 451

Phase diagrams (Contd)
  Fe-C, 459
  Fe-Fe$_3$C, 459
  LiBr-LiCl, 440
  MgO-NiO, 435
  NaF-MgF$_2$, 445
  "nonmagnetic"-magnetic, 128
  normal-superconductive, 137
  Pb-Sn, 442
  references, 473
Phase relationships, 438
Phase transformation, 419
Phase transformations, 474, 424
Phenol, 324
Phenol-formaldehyde plastic, **325**
Phenolic, **325**
Photons, **182**
Piezoelectric effect, **119, 264**
  inverse, **119**
Piezoelectric frequency, reference, 120
Piezoelectric motor, 121
Pileup, **522**
Pinning force, **529**
Pitting, 497
Planar density, **247**
Planck's constant, **182**
Plane defects, **267**
Plane lattice, **235**
Planes:
  crystal, 243
  slip, 523
Plastic deformation, **35, 59**
Plastic flow, 523
Plastic instability, **62**
Plexiglass, 316
Point defects, **267**
Poise, **75**
Poiseuille's equation, 74
  derivation of, 93
Poisson ratio, **60**
Polar groups, 318
Polarization, **117, 220**
  induced, 221, 631
  orientation, 220, 631
  permanent, 118
Pole mechanism, **464**
Polyacrylonitrile, 311
Polychloroprene, 318
Polycrystalline, **342**
Polycrystalline aggregate, 347
Polyethylene, **303**
Polygonization, **554**
Polyisoprene, 314
  stereoisomerism in, 317
Polymer molecules:
  cellulose, 329
  neoprene, 318

Polymer molecules (Contd)
  nylon, 320
  phenolics, 325
  polyethylene, 303
  polyisoprene, 315
  protein, 330
  silicones, 322
  Teflon, 314
  vinyl compounds, 310
Polymer structures:
  folded chain crystals, 326
  fringed micelle, 327
  spherulite, 327
Polymerization:
  addition, 303
  condensation, 320
  degree of, 307
Polymers:
  addition, 303
  amorphous, 306
  branched, 304
  condensation, 319
  copolymers, 318
  crystalline, 326
  linear, 304
  network, 324
  references, 332
  rubber-like, 315
  thermoplastic, 326
  thermosetting, 326
  vinyl compounds, 310
Polymorphism, **254**
Polypeptide chain, **330**
Polyphase materials, 353
Polystyrene, 311
Polytetrafluoroethylene, **314**
Polyurethane foam, 29
Polyvinylacetate, 311
Population inversion, **612**
Porcelain, 502
Porosity effect on transparency, 343
Portland cement, 360
Potential:
  electrode, 485
  impressed, 503
  ionization, 201
  oxidation, 485
Potential gradient, 97
Powder metallurgy, 279
Powder technique, **291**
p-n junction, **602**
  breakdown, 606
  rectifier, 605
  solar cell, 606
  thermoelectric effects, 607
  Zener diodes, 606
Precipitation hardening, 542

Pre-exponential diffusion factor, **402**
Preferred orientation, **350**
Prepregs, **359**
Pressure:
  effect on ductility, 63
  effect on viscosity coefficient, 89
Pressure gradient, **77**
Primary bonds, 203
Primitive cell, **235**
Principal quantum number, **195**
Probability density, **190**
Property:
  material, 32
  object, 32
Properties:
  chemical, 169, 479
  electrical, 94, 566
  magnetic, 122, 620
  mechanical, 33, 58, 510
  optical, 611
  superconductive, 656
  thermal, 151
Protection:
  abrasion, 362
  corrosion, 361, 501
  oxidation, 361
Protein, 330
PTFE, **314**
Pulling from the melt, **338**
Pyroceram, application of, 151
Pyroelectric effect, **165**

## Q

Q-factor, **83**
Quality factor, **83**
Quantitative microscopy, **363**
Quantized electron gas, 571
Quantum number, **183**
Quartz, 259, 265
  applications, 119
  thermal conductivity, 156, 158
Quenched, **168**
Quenching, **454**

## R

Radial probability density, **196**
Radiation damage, 47, 270
Radii:
  atomic, 253
  ionic, 257
Radius ratios for voids, 253

Random:
  chain configuration, **306**
  network, **284**
  substitutional solid solution, **268**
  walk, 403
    references, 424
Real strain, **60**
  at fracture, 63
Real stress, **60**
Recombination, **600**
Recovery processes, **556**
Recrystallization, **555**
Rectified, **605**
Reduction in area, 63
Reduction reaction, **481**
Reflection microscope, **342**
Reflection planes, **263**
Refractory materials, **163**
Reinforced:
  concrete, 361
  metals, 357, 545
  plastics, 355, 545
Relative permeability, **122**
  data, 133
Relaxation time, **80, 567**
  relation to loss angle, 80
Remanent polarization, **118**
Repulsive potential energy, 204
Residual resistivity, **588**
Resins:
  thermoplastic, 326
  thermosetting, 326
Resistance, electrical, 99
Resistivity ratio, **106**
Resolution, size, 31
Resonance amplification factor, **82**
Resonance frequency, **82**
Reverse bias, **604**
Rockwell hardness, 64, 550, 551, 552
Root mean squared diffusion distance, **403**
Root mean squared velocity, **371**
Rotation axis, **262**
Rotoinversion operation, **263**
Rubber:
  hard, 315
  soft, 315
  synthetic natural, 315
Rubber-like materials, 315
Rust inhibitors, 503

## S

Sacrificial anodes, **503**
Sacrificial coating, **503**

SAP, 557
Saturation current density, 604
Saturation magnetization, 127
Saturation polarization, 117
Schmid's law, 525
Schottky pair, 270
Schrödinger's equation, 190
Screw dislocation, 275
Season cracking, 500
Secondary bonds, 215
Second-order transition, 659
Second phase strengthening, 540
Seebeck effect, 166
Segregation:
  at grain boundaries, 281
  coefficient, 469
  during solidification, 470
  in binary alloys, 468
Selection rules, 291, 301
Self-diffusion, 402
Self-inductance, 122
Self-interstitial, 270
Semiconductors, 102
Semipermeable membrane, 309
Shear:
  modulus, 40
  strain, 39
  strength, Frenkel's model, 517
  stress, 38
  yield stress, 41
Sheet structure, 300
Short-range interaction, 207
Side chains, 304
Silica glass, 284
Silicates, 259
  cristobalite, 283
  double chain structure, 300
  framework structure, 300
  island structure, 300
  sheet structure, 300
  silica glass, 284
  single chain structure, 300
  tridymite, 283
Silicon:
  band gap energy, 594
  band structure, 590
  charge mobilities, 595
  crystal structure, 240, 241
  diffusion of phosphorus, 417
  electrical resistivity, 101
  microcircuit technology, 5, 415
  oxidation, 415
  self-diffusion, 410
Silicone plastics, 322
Silicone rubber, 322
Silk, 331
Simple cubic crystal, 245

Simple harmonic motion, 78
Single chain silicate, 300
Sinter, 279
Sintered aluminum powder, 557
Sintered carbides, 546
$SiO_2$, 283
$SiO_4^{4-}$ tetrahedron, 283
Slip, 523
  direction, 523
  planes, 523
  system, 523
  systems, in different materials, 524
Small-angle grain boundaries, 279
Small-angle tilt boundary, 279
S-N curve, 71
Snowflakes, 262
Soda glass, 287
Sodium chloride, 206
  crystal structure, 254
  structure, 206
Soft magnetic material, 130
  data, 130
Solder, 354
Solenoid, magnetic induction in, 123
Solenoidal coil, inductance of, 122
Solid solution:
  interstitial, 268
  ordered, 268
  random, 268
  substitutional, 268
  supersaturated, 454
Solid state devices, references, 615
Solid state physics, references, 615
Solidus, 436
Solubility in polymers, 318
Solubility versus temperature, 387
Solution hardening, 529
Solution heat treatment, 168
Solution treating, 168, 454
Solutionizing, 454
Sommerfeld model of electrons in metals, 577
$sp^2$ orbitals, 265
$sp^3$ orbitals, 210
Space defects, 267
Space lattice, 235
Spacings, interplanar, 245, 300, 301
Spatial imperfections, 281
Specific heat, 160
  of solids, Einstein theory, 395
  paradox, 570
Specific stiffness, 72
Specific tensile strength, 73
Spectroscopy notation, 195
Spherulites, 327
Spontaneous emission, 611
Spontaneous magnetization, 632

Squareness ratio, **145**
Stacking sequence, 248
Stainless steel, 466
  austenitic, 466
  corrosion properties, 489, 502
  intergranular corrosion, 507
  low temperature ductility, 69
Stable structure, **171**
Standard oxidation electrode potentials, **485**
  data, 486
Standard oxidation potential, data, 486
Standard reference half-cell, **487**
State function, **374**
States of matter, 31
Steady-state diffusion, **400**
Steel, **460**
  austenitic, 466
  bainite formation, 461
  Fe-Fe$_3$C phase diagrams, 459
  heat treatment, 168, 465
  martensite formation, 461
  pearlite formation, 460
  Poisson's ratio, 61
  shear modulus, 43
  tensile yield stresses, 44
  Young's modulus, 42
Stereoisomers, **311**
Steric hindrance, **313**
Stern-Gerlach experiment, 224
Stiffness constant:
  cantilever beam, 32
  coil spring, 41
  rod in tension, 37
Stiffness to weight ratio, 72
Stiffness, specific, **72**
Stimulated emission, **611**
Stirling's approximation, 384
Strain:
  energy, elastic, **37**
  extension ratio, **33**
  hardening, 60, **531**
  hardening exponent, **62**
  hardening, origin, **535**
  nominal, **34**
  rate, 70, 558
  real, **60**
  shear, **39**
Strength:
  creep, 558
    data, references, 562
  dielectric, 114
  fracture, 511, 518
  high-temperature, 553
  specific tensile, 73
  tensile, 62, 511
  yield, 515

Strength (*Contd*)
  yield, data, 44
Strength of filaments, wires and whiskers, data, 545
Strength versus microstructure, 527
Strengthening concepts, 527
Strengthening mechanisms:
  at high temperatures, 553
  composite, 528, 545
  dispersed particles, 557
  eliminate imperfections, 527
  grain boundary, 538
  lamellar structures, 544
  martensite formation, 547
  of piano wire (patented steel wire), 545, 565
  of polymers, 559
  of viscous matrices, 560
  precipitates, 540
  references, 562
  second phase, 540
  solute, 529
  strain hardening, 535
Stress:
  critical resolved shear, **525**
  normal, **34**
  real, **60**
  shear, **38**
  shear yield, **41**
  tensile, **34**
  tensile fracture, **41**
  tensile yield, 35
  tensile yield, data, 44
  thermal, 151
  units, 35
Stress concentration, **65**, 521
Stress corrosion cracking, **499**
Structure, 22
Structure insensitive property, **43**
  examples, 174
Structure sensitive property, **43**
  examples, 174
Styrene, 310
Subcells, **537**
Substitutional solid solution, **268**
Superconducting critical field, 137
  data, 137
Superconducting critical temperature, **137**
  data, 137, 140
Superconductivity, origin
  Type I, 667
  Type II, 667
Superconductivity, references, 143, 674
Superconductor, **102**
  high field, 139
  synthetic hard, 363
  Type I, 138

696  Index

Superconductor (Contd)
  Type II, 139
Supercurrent, **665**
Supersaturated solid solution, 454
Supersaturation ratio, **421**
Surface area, 299
Surface coatings, 361, 501, 502, 503
Surface energy, **278,** 299, 420
Symmetry, 261
  element, **262**
  operations, **261**
  references, 293
Syndiotactic, **311**
Synergism, **355**


# T

TD nickel, **558**
T-2 tanker, 67
Teflon, **314**
Temperature:
  change, 151, 159, 165
  field, 162
  gradient, 154
  scale, 373
Temperature, effects destroyed by, 163
Temperature effects on:
  creep rate, 69
  engine efficiency, 69
  impact toughness, 68
  strain hardened materials, 65
  viscosity coefficients, 75, 76
  yield strength, 64
Temperature, exponential effects, 163
Tempered, **168**
  glass, 527
  martensite, 465
Tempering, **465**
Tensile strength, **62**
Tensile stress, **34**
Tetrafluorethylene, 314
Tetragonal distortion, **530**
Tetrahedral bonds, **210**
Tetrahedral coordination, 251
Tetrahedral silica units:
  amorphous, 285
  chain, 300
  double chain, 300
  framework, 283, 284, 300
  island, 300
  sheet, 300
Tetrahedral void, **251**
Thermal arrest, **475**
Thermal conductivity coefficient, **154**
  data, 155, 157, 158

Thermal diffusion time, **154**
Thermal diffusivity, **154**
Thermal efficiency, 69
Thermal emf's, data, 166
Thermal expansion coefficient:
  data, 152
  linear, 152
  volume, 152
Thermal properties, references, 174
Thermal shock, **151**
Thermal stresses, 151
Thermistor, **107**
Thermocouple, **165**
  data, 166
  materials, 166
Thermodynamics, references, 424
Thermoelastic effect, **165**
Thermoelectric generator, 167
Thermoelectric power, **166**
Thermoelectric refrigerator, 167
Thermometry, 373
Thermoplastic, **326**
Thermosetting, **326**
Thin magnetic films, **136**
Tie line, **438**
Time, relaxation, 80
Time-temperature-transformation curve, **549**
Titinate, 118, 119, 345
Toughness, **64**
  impact, 67
  of infinite rods, 64
*Trans* configuration, **317**
Transducers:
  magnetostrictive, 134
  piezoelectric, 119
Transformation kinetics:
  condensation, 419
  eutectic, 447
  martensitic, 461
  nucleation, 419
  pearlitic, 460
  precipitation, 453
  solidification of alloy, 451
Transformations, 424
  eutectic, 443
  eutectoid, 446
  ferromagnetic, 128
  indifferent, 440
  inversion, 446
  melting, 446
  peritectic, 446
  peritectoid, 446
  superconductive, 137
Transformer sheet, 649
Transistor, 608

*Index* 697

Transmission electron:
  micrograph, 24
  microscopy, 31, 277
Transition, ductile to brittle, **68**
Transition temperature, ductile to brittle, **68**
Tridymite, 283
TTT curve, **549**
Tunnel diode, **618**

**U**

Ultimate shear strength, theory, 515
Ultimate tensile strength, theory, 511
Uniaxially solidified eutectics, 450
Unit cell, **235**
Universal gas constant, **373**

**V**

Vacancies, **269**
Vacancy:
  energy of formation, 386
    data, 387
  energy of motion, 397
    data, 398
  volume of formation, 386
Valence band, **582**
Valence electrons, **213**
van der Waals bonds, **217**
van der Waals equation of state, 374
Velocity, drift, 568
Velocity of ideal gas molecules, 374
Vibration fatigue, 71
Vibrational frequencies of atoms, data, 394
Vibrations in solids, 395
Vinyl chloride, 310, 311
Vinyl compounds, **310**
Viscoelastic behavior, 80
Viscosity coefficient, **75**
  data, 75
  dependence on pressure, 89
  dependence on temperature, 76
  derivation, for gases, 377
  units, 75
Viscous media, **74**
Void per atom ratio, **252**
Void radius ratio, **252**
Volt, 99
Volume fraction rule, **545**

Volume of formation of a vacancy, **386**
Volume thermal expansion coefficient, **152**
Vulcanization, **315**

**W**

Water of hydration, 360
Water molecule, 210
Waterproofing of cloth, 323
Wave function, **190**
Wave vector, **579**
Weber, 99
Westro, 647
Whiskers, 339, 358, 545
  references, 366
  size effect on strength, 358
White cast iron, **467**
Wiedemann-Franz relationship, 175, 618
Wires, 545
Work function, **378**
Work hardening, **60**
Wurtzite, **256**
Wustite, 271

**X**

Xerography, **617**
X-ray diffraction by crystals, 288
X-ray radiation, origin of, 226
X-ray small angle scattering, 456

**Y**

Yield point, **59**
Yield stress:
  in shear, **41**
  in tension, **35**
Young's modulus, **36**

**Z**

Zener diode, 606
Zeolite, 259
Zinc blende structure, **256**
Zone leveling, **473**
Zone refining, **472**
  references, 474

38
W59

314H